Functional Analysis
and Control Theory

Mathematics and its Applications *(East European Series)*

Managing Editor:

M. HAZEWINKEL

Centre for Mathematics and Computer Science, Amsterdam, The Netherlands

Editorial Board:

A. BIAŁYNICKI-BIRULA, *Institute of Mathematics, Warsaw University, Poland*
H. KURKE, *Humboldt University, Berlin, D.D.R.*
J. KURZWEIL, *Mathematics Institute, Academy of Sciences, Prague, Czechoslovakia*
L. LEINDLER, *Bolyai Institute, Szeged, Hungary*
L. LOVÁSZ, *Bolyai Institute, Szeged, Hungary*
D. S. MITRINOVIĆ, *University of Belgrade, Yugoslavia*
S. ROLEWICZ, *Institute of Mathematics, Polish Academy of Sciences, Warsaw, Poland*
Bl. H. SENDOV, *Bulgarian Academy of Sciences, Sofia, Bulgaria*
I. T. TODOROV, *Bulgarian Academy of Sciences, Sofia, Bulgaria*
H. TRIEBEL, *University of Jena, D.D.R.*

Stefan Rolewicz

Institute of Mathematics, Polish Academy of Sciences, Warsaw, Poland

Functional Analysis and Control Theory

Linear Systems

D. Reidel Publishing Company

A MEMBER OF THE KLUWER ACADEMIC PUBLISHERS GROUP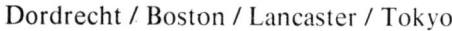

Dordrecht / Boston / Lancaster / Tokyo

PWN-Polish Scientific Publishers

Warszawa

ERRATA

Page, line	For:	Read:
X^{18}	central theory	control theory
$261_{2,3}$	(5.8.21)	(5.8.20)
481_3	every fixed a	every fixed a (u)
481_2	function of u	function of u (a)
517_5	Γ-compact set	Γ-compact set 149
517_4	Γ-topology	Γ-topology 149

S. Rolewicz, *Functional Analysis and Control Theory. Linear Systems.*

Chapters I-VII translated by Ewa Bednarczuk

English Edition published by Polish Scientific Publishers, Warszawa, Poland, in co-publication with D. Reidel Publishing Company, P. O. Box 17, 3300 AA Dordrecht, Holland.

Distributors for Albania, Bulgaria, Cuba, Czechoslovakia, German Democratic Republic, Hungary, Korean People's Democratic Republic, Mongolia, People's Republic of China, Poland, Rumania, the U.S.S.R., Vietnam, and Yugoslavia
ARS POLONA
Krakowskie Przedmieście 7, 00-068 Warszawa 1, Poland

Distributors for the U.S.A. and Canada
Kluwer Academic Publishers,
101 Philip Drive, Norwell, MA 02061, U.S.A.

Distributors for all remaining countries
Kluwer Academic Publishers Group,
P. O. Box 322, 3300 AH Dordrecht, Holland

All Rights Reserved.
Copyright © 1987 by PWN—Polish Scientific Publishers–Warszawa. No part of the material protected by this copyright notice may be reproduced or utilized in any form or by any means, electronic or mechanical including photocopying, recording or by any information storage and retrieval system, without written permission from the copyright owner.

Printed in Poland.

Table of Contents

Editor's Preface ix

Introduction xiii

CHAPTER 1. *Metric spaces* 1
  1.1. Definition and examples of metric spaces 1
  1.2. Convergence and related notions 9
  1.3. Continuous mappings 13
  1.4. Semimetric spaces 16
  1.5. Completeness of metric spaces 17
  1.6. Contractive mapping principle 23
  1.7. Sets of first and second category 27
  1.8. Spaces of modulus integrable and square integrable functions 29
  1.9. Basic facts of measure theory and the Lebesgue integral ... 32
  1.10. Separable spaces 45
  1.11. Sequentially compact spaces and compact spaces 49

CHAPTER 2. *Linear spaces and normed spaces* 55
  2.1. Basic facts about linear spaces 55
  2.2. Linear metric spaces and normed spaces 62
  2.3. Linear functionals 77
  2.4. Finite-dimensional spaces 85
  2.5. Extensions of functionals 90
  2.6. General forms of continuous linear functionals in certain spaces 100

CHAPTER 3. *Continuous linear operators in Banach spaces* 114
  3.1. The Banach–Steinhaus theorem 114

- 3.2. The Banach theorem about the continuity of the inverse operator 118
- 3.3. Closed operators 122
- 3.4. Conjugate operators 133

CHAPTER 4. *Weak topologies* 138
- 4.1. Why we introduce topological spaces? Basic topological notions 138
- 4.2. Sequentially compact spaces and compact spaces 143
- 4.3. Linear topological spaces 146
- 4.4. Weak topologies 149
- 4.5. Reflexive spaces and weak compactness 156
- 4.6. Extreme points 159

CHAPTER 5. *Optimization and observability of linear systems* . . 165
- 5.1. Linear systems 165
- 5.2. Optimization of linear systems with a fixed input operator 179
- 5.3. Sufficient conditions for the existence of an optimal input 201
- 5.4. Minimum-time problem 208
- 5.5. Reduction of the minimum-time problem to the minimum-norm problem 225
- 5.6. Observability of linear systems 231
- 5.7. Minimum-time problems of observability theory 249
- 5.8. Noncontinuous linear systems. Estimability, observability, controllability 253

CHAPTER 6. *Systems governed by ordinary differential equations* 262
- 6.1. Problems of minimization of convex functionals related to systems governed by ordinary differential equations 262
- 6.2. Controllability of finite-dimensional systems 267
- 6.3. Minimum-norm problems for sup type norms 269
- 6.4. Criteria for the uniqueness of optimal control 274
- 6.5. Bang-bang principle 277
- 6.6. Measurable multifunctions and their applications . . . 285
- 6.7. Observability of systems described by ordinary differential equations 300

6.8. Optimal observation of stationary systems 316
6.9. Functionals observed by less than n measurements . . 318
6.10. Optimal integration and differentiation formulae 322

CHAPTER 7. *Systems with distributed parameters* 328
7.1. Bases in Banach spaces 329
7.2. Eigenvalues and eigenvectors 340
7.3. Temperature distribution in a rod with zero boundary conditions 345
7.4. Nonhomogeneous heat diffusion equation with zero boundary conditions 361
7.5. Homogeneous heat diffusion equation with nonhomogeneous boundary conditions 366
7.6. Rod heating control 376
7.7. Observability of temperature distribution in a rod . . . 384
7.8. Other problems related to heat diffusion in one-dimensional objects 394
7.9. Rod vibration control 403

CHAPTER 8. *Differential and integral equations in Banach spaces* 418
8.1. Derivatives and integrals of functions with values in Banach spaces 418
8.2. Volterra integral equations and differential equations with continuous right-hand sides 421
8.3. Autonomous differential equation with a discontinuous right-hand side and semigroups of operators 428
8.4. Properties of the infinitesimal generator 440
8.5. Infinitesimal generator of bounded semigroups 445
8.6. Nonautonomous differential equations in Banach spaces 449
8.7. Controllability and approximative controllability of linear systems described by differential equations in Banach spaces 452
8.8. Long-term controllability and long-term approximative controllability 463
8.9. Linear systems with analytic semigroups 467

8.10. Optimization problems for systems described by linear differential equations in Banach spaces 472
8.11. Duality between observability and controllability of systems governed by autonomous linear differential equations in Banach spaces 476
8.12. Stability of solutions of linear differential equations in Banach spaces 481

APPENDIX. *Necessary and sufficient condition for weak duality*   495

Bibliography 505

Author index 513

Subject index 515

List of symbols 523

Editor's Preface

Approach your problems from the right end and begin with the answers. Then, one day, perhaps you will find the final question.

'The Hermit Clad in Crane Feathers' in R. Van Gulik's *The Chinese Maze Murders*.

It isn't that they can't see the solution. It is that they can't see the problem.

G.K. Chesterton, *The Scandal of Father Brown* 'The point of a Pin'.

Growing specialization and diversification have brought a host of monographs and textbooks on increasingly specialized topics. However, the "tree" of knowledge of mathematics and related fields does not grow only by putting forth new branches. It also happens, quite often in fact, that branches which were thought to be completely disparate are suddenly seen to be related.

Further, the kind and level of sophistication of mathematics applied in various sciences has changed drastically in recent years: measure theory is used (non-trivially) in regional and theoretical economics; algebraic geometry interacts with physics; the Minkowsky lemma, coding theory and the structure of water meet one another in packing and covering theory; quantum fields, crystal defects and mathematical programming profit from homotopy theory; Lie algebras are relevant to filtering; and prediction and electrical engineering can use Stein spaces. And in addition to this there are such new emerging subdisciplines as "experimental mathematics", "CFD", "completely integrable systems", "chaos, synergetics and large-scale order", which are almost impossible to fit into the existing classification schemes. They draw upon widely different sections of mathematics. This programme, Mathematics and Its Applications, is devoted to new emerging (sub)disciplines and to such (new) interrelations as exempla gratia:

— a central concept which plays an important role in several different mathematical and/or scientific specialized areas;
— new applications of the results and ideas from one area of scientific endeavour into another;
— influences which the results, problems and concepts of one field of enquiry have and have had on the development of another.

The Mathematics and Its Applications programme tries to make available a careful selection of books which fit the philosophy outlined above. With such books, which are stimulating rather than definitive, intriguing rather than encyclopaedic, we hope to contribute something towards better communication among the practitioners in diversified fields.

Because of the wealth of scholarly research being undertaken in the Soviet Union, Eastern Europe, and Japan, it was decided to devote special attention to the work emanating from these particular regions. Thus it was decided to start three regional series under the umbrella of the main MIA programme.

System and central theory, a relative newcomer in the spectrum of mathematical specialisms, is also one of the fastest growing and most immediately applicable ones. Without a doubt, finite dimensional linear systems have at the moment the greatest applicability as models of all kinds of economic, biological, chemical, and engineering processes. Equally clear is that it is sorely needed to extend the class of potential models which can be handled "completely". One way to go is towards infinite dimensional linear systems, another is towards finite dimensional nonlinear ones (there is overlap of course). Both extensions tend to have something mysterious generating some unease in the minds of engineers and others who could use these more general models. The present book is about infinite dimensional linear systems (on Banach spaces rather than Hilbert spaces), certainly a topic that among practitioners of control and system theory is usually approached with some wariness. So much can go wrong in the infinite dimensional case! There are of course a number of infinite dimensional specialists in the control and system community and there do exist a number of books on the topic written by them. As far as I know, this is the first book on the topic written by a scientist whose background is in the first place in functional analysis rather than control. This gives a unique flavour to the book and serves to make it even more evident just how powerful a tool functional analysis can be.

The German translation of this book was highly successful and I expect this improved and updated English language version to be even more so.

The unreasonable effectiveness of mathematics in science ...

Eugene Wigner

Well, if you know of a better 'ole, go to it.

Bruce Bairnsfather

What is now proved was once only imagined.

William Blake

As long as algebra and geometry proceeded along separate paths, their advance was slow and their applications limited.

But when these sciences joined company, they drew from each other fresh vitality and thenceforward marched on at a rapid pace towards perfection.

Joseph Louis Lagrange

Michiel Hazewinkel

Bussum, July 1986

Introduction

It was in the early sixties that the language and methods of Functional Analysis began to penetrate more and more deeply into both control and optimization theory.

Several books have presented the various aspects of the applications of functional analysis to control and optimization theory (see, for instance, Kulikowski, 1965, 1970; Porter, 1966; Lions, 1968; Hermes–La Salle, 1969; Balakrishnan, 1976; Curtain–Pritchard, 1977, and others).

In this book the general theory of linear systems based on the theory of linear operators in Banach spaces is developed in a consistent way.

The book is addressed not only to mathematicians but also to engineers, economists and other specialists interested in optimization and control theory.

In view of the rather restricted knowledge of Functional Analysis among non-mathematicians, the book is written in such a way that no background in Functional Analysis is required to read it. The first four chapters of the book can be treated as a short course of Functional Analysis. However, it should be pointed out that this part of the book contains solely the material which is necessary for the understanding of the rest. Thus some important topics from Functional Analysis (for instance the theory of operators in Hilbert spaces) are not included.

The first chapter of the book contains also fundamental notions and theorems (some of them without proof) in Topology and Measure Theory.

I should also mention that, although the contents of the first four chapters can be found in any handbook on Functional Analysis, what distinguishes this book from any other are the Examples and the adopted method of presentation. The book gives engineering and economical interpretations of the fundamental notions of Functional Analysis. Also, I try to convince the reader that the notions and theorems included in the book have been necessary.

The most essential part of the book is Chapter 5, where the theory of general linear systems is developed. The theory seems rather abstract. However, it permits answering some fundamental questions, for instance, what is the meaning of the Pontryagin maximum principle, and to characterize systems satisfying this principle. The chapter contains important theorems concerning the existence of universal time for time-depending systems. Optimal observability is defined in a general way and the duality between optimal observability and optimal controllability is shown.

Chapter 6 contains applications of the theory developed in Chapter 5 to systems described by ordinary differential equations. It includes not only a new exposition of the classical results but also several new results concerning optimal observability and numerical differentiations and integrations.

Chapter 7 contains applications of the theory to systems described by partial differential equations.

In Chapter 8 the fundamental facts about differential equations in Banach spaces are given. Problems of stability, optimal controllability and optimal observation for systems described by such equations are considered.

An appendix containing necessary and sufficient conditions for weak duality obtained by the Dolecki–Kurcyusz method of multifunctions is also included in the book.

I wish to express my gratitude to Professor Kazimierz Malanowski for his valuable suggestions regarding the first Polish edition of this book. My gratitude extends also to Professor Diethard Pallaschke, the translator of the book into German. Discussions with him have permitted essential improvements in the German edition. I also thank Mrs. Ewa Bednarczuk, the translator of the present English edition, and to Doctor Phan Quoc Khanh for their valuable remarks concerning this edition.

I am also greatly indebted to the staff members of the Mathematical Department of the Monash University (Clayton, Victoria, Australia) who participated in my seminars on the subject of the book. Since at that time I was preparing this edition, their suggestions were extremely helpful and had a considerable influence on the final form of the book.

Warszawa, August 1985 **Stefan Rolewicz**

Chapter 1

Metric spaces

1.1. DEFINITION AND EXAMPLES OF METRIC SPACES

Let X be a set. We say that X is a *metric space* if there is a nonnegative function ϱ of two arguments, defined on X, $\varrho(x, y) \geqslant 0$, $x, y \in X$, and called a *metric*, which satisfies the following conditions:
  (1) $\varrho(x, y) = 0$ if and only if $x = y$,
  (2) $\varrho(x, y) = \varrho(y, x)$,
  (3) $\varrho(x, y) \leqslant \varrho(x, z) + \varrho(z, y)$ (*triangle inequality*).

A metric space X with the metric ϱ will be denoted by (X, ϱ). If it is not misleading, we shall denote it briefly by X.

Given two points x, y, $\varrho(x, y)$ is called the *distance* from the point x to the point y. By (2), this distance is equal to the distance from the point y to the point x. The set of points for which the distance from a given point x_0 is less than a certain positive number r, $K_r(x_0) = \{x \in X: \varrho(x, x_0) < r\}$, is called the *ball* with radius r and centre x_0. The set of points for which the distance from x_0 is not greater than r, $\bar{K}_r(x_0) = \{x \in X: \varrho(x, x_0) \leqslant r\}$, is called the *closed ball* with radius r and centre x_0. The set of points for which the distance from x_0 is equal to r, $S_r(x_0) = \{x \in X: \varrho(x, x_0) = r\}$, is called the *sphere* with radius r and centre x_0.

Now we give examples of metric spaces.

EXAMPLE 1.1.1. The *n-dimensional real Euclidean space* R^n with the distance between points $x = (x_1, \ldots, x_n)$ and $y = (y_1, \ldots, y_n)$ defined as

$$\varrho(x, y) = \sqrt{(x_1-y_1)^2 + \ldots + (x_n-y_n)^2}$$

is a metric space. In the case $n = 3$ we obtain the classical 3-dimensional Euclidean space.

Observe that in this case a ball is a classical ball without the boundary. A closed ball is a classical ball with the boundary. A sphere is the surface of a ball, i.e., it is a sphere in the classical sense.

EXAMPLE 1.1.2. The *Euclidean plane* with distance defined as in Example 1.1.1. Observe that in this case a ball is a disc without the boundary. A closed ball is a disc. A sphere is simply a circle.

Example 1.1.2 motivates the following notion. Let (X, ϱ) be a metric space. Let X_0 be a subset of X. The metric $\varrho(x, y)$ restricted to X_0 is a metric on X_0. We shall denote this restriction by ϱ_0. The metric space (X_0, ϱ_0) is called a *subspace* of the metric space (X, ϱ).

The following examples show that the notion of distance induced by conditions (1), (2), (3) is essentially more general than the notion of the Euclidean distance.

EXAMPLE 1.1.3. Let X be the Earth's surface. Let $\varrho(x, y)$ be the distance between two points $x, y \in X$ as the crow flies. It is easy to verify that $\varrho(x, y)$ is a metric. It is not the Euclidean metric. Observe that in this metric there exist points 20 000 km away from one another while in the Euclidean metric the distance between points on the Earth's surface cannot be greater than the length of the Earth's diameter, namely 12 740 km.

EXAMPLE 1.1.4. Let X denote the set of principal towns in Poland and let $\varrho(x, y)$ denote the distance along the shortest railway line between towns x and y. It is easy to verify that $\varrho(x, y)$ is also a metric.

EXAMPLE 1.1.5. Let X be as in the previous example and let $\varrho(x, y)$ denote the road distance between towns x and y. It is easy to see that $\varrho(x, y)$ is also a metric.

EXAMPLE 1.1.6. Let X be the set of streets built along vertical and horizontal lines only (see Fig. 1.1.1). Let the distance between two points be the shortest street route. It is easy to verify that this distance

1.1. Definition and examples of metric spaces

satisfies all the conditions of a metric. Fig. 1.1.1 presents a ball with radius r and centre x_0.

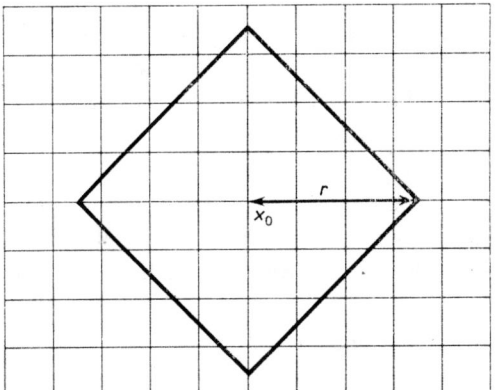

FIG. 1.1.1. The domain marked off with thick line is the ball $K_r(x_0)$ described in Example 1.1.6.

Distance, as it is defined above, is not always distance in the colloquial sense. Cost, operation time, etc. can also be interpreted as distance.

EXAMPLE 1.1.7. Let X be the set of all railway stations in Poland. Let $\varrho(x, y)$ denote the price of a ticket for travelling between railway stations x and y. It is easy to verify that $\varrho(x, y)$ is a metric.

EXAMPLE 1.1.8. Let X be the set of stops on a particular tram line. Let $\varrho(x, y)$ denote the price of a ticket for a single trip between stops x and y. In Poland it is always the same price, i.e., $\varrho(x, y) = 3$ zloties for $x \neq y$ and $\varrho(x, x) = 0$.

Example 1.1.8 motivates the following notion. Let (X, ϱ) be a metric space with the metric $\varrho(x, y)$ defined as follows:

$$\varrho(x, x) = 0, \quad \varrho(x, y) = c \text{ for } x \neq y,$$

where c is a given positive constant. A metric space with such a metric is called a *discrete space* and the metric ϱ is called a *discrete metric*.

Another example of a discrete space is the following.

EXAMPLE 1.1.9. Let X denote a factory hall and x the position of a machine. Assume that the machine must be placed on a solid concrete base. Suppose furthermore that the cost of the removal of the machine is negligible in relation to the cost of a new base. Then the cost of moving the machine from point x to point y, $\varrho(x, y)$, is equal to the cost of demolishing the old base and building a new one and it is independent of x, y. Thus, $\varrho(x, y)$ is a discrete metric.

EXAMPLE 1.1.10. Let X denote an urban area. Let $\varrho(x, y)$ denote the time which is necessary to reach point y from point x by a certain kind of city transport including the time to reach the nearest stop. As can easily be verified, $\varrho(x, y)$ is a metric. Note that this is a kind of distance which we often refer to in practice.

EXAMPLE 1.1.11. Let X be an area within the reach of a movable gantry supporting a travelling crane (Fig. 1.1.2). Let the gantry move with speed a and let the crane move on the gantry with speed b; $\varrho(x, y)$ denotes the time necessary for shifting the crane from position x to position y. It is easy to verify that $\varrho(x, y)$ is a metric.

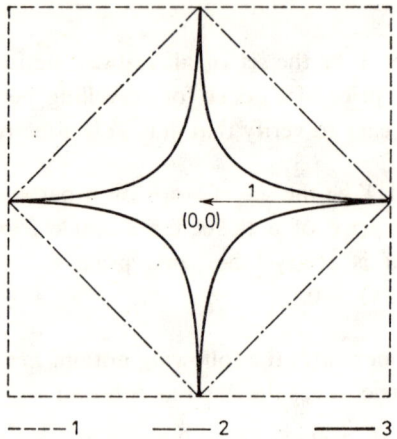

FIG. 1.1.2. Line 1 marks off the domain which is the unit ball $K_{(0,0)}(1)$ described in Example 1.1.11. Line 2 marks off the domain which is the unit ball $K_{(0,0)}(1)$ described in Example 1.1.12. Line 3 marks off the domain which is the unit ball $K_{(0,0)}(1)$ described in Example 1.1.13.

Introducing the Cartesian coordinates with axes parallel to the directions in which the gantry and the crane move, we get the formula for the metric

$$\varrho(x, y) = \max\left(\frac{|x_1-y_1|}{a}, \frac{|x_2-y_2|}{b}\right),$$

where $x = (x_1, x_2)$, $y = (y_1, y_2)$.

EXAMPLE 1.1.12. Let X be as in Example 1.1.11. We assume additionally that at each moment either the gantry or the crane is moving but not both. Then the shifting time $\varrho(x, y)$ is also a metric, but it is given by a different formula:

$$\varrho(x, y) = \frac{|x_1-y_1|}{a} + \frac{|x_2-y_2|}{b}.$$

EXAMPLE 1.1.13. Let X be as in Example 1.1.11. As in Example 1.1.12, we assume that the gantry and the crane cannot move simultaneously. We assume also that their movements are so small that instead of a speed limit an acceleration limit should be taken into account. Let a and b be maximal accelerations for the gantry and the crane, respectively. Then the formula for the shifting time is the following:

$$\varrho(x, y) = 2\left(\sqrt{\frac{|x_1-y_1|}{a}} + \sqrt{\frac{|x_2-y_2|}{b}}\right).$$

Figure 1.1.2 represents the balls which correspond to the metrics appearing in Examples 1.1.11, 1.1.12, 1.1.13 for $a = b = 1$.

EXAMPLE 1.1.14. Let X be the set of all continuous real-valued functions defined on a closed interval $[a, b]$. Let

$$\varrho(x, y) = \sup_{a \leqslant t \leqslant b} |x(t)-y(t)|. \tag{1.1.1}$$

It is easy to verify that $\varrho(x, y)$ satisfies conditions (1) and (2). We verify condition (3):

$$\begin{aligned}\varrho(x, y) &= \sup_{a \leqslant t \leqslant b} |x(t)-y(t)| \\ &\leqslant \sup_{a \leqslant t \leqslant b} |x(t)-z(t)| + \sup_{a \leqslant t \leqslant b} |z(t)-y(t)| \\ &= \varrho(x, z) + \varrho(z, y).\end{aligned}$$

We denote by $C[a, b]$ the space X with the metric $\varrho(x, y)$.

The space $C[a, b]$ often appears in applications. For instance, consider a certain physical process over a time period $[a, b]$ described by a continuous function $x(t)$. We assume that $x_0(t)$ is a pattern process and the admissible deviation during the process is r. Obviously, the size of the admissible deviation depends on the possibility of equipment destruction or (particularly in chemical processes) of product destruction or at least quality decline. Thus admissible processes $y(t)$ are those for which $\varrho(x_0, y) \leq r$.

EXAMPLE 1.1.15. If a process can be described by a system of n continuous parameters $x_1(t), \ldots, x_n(t)$ and the admissible deviation of the parameter $x_i(t)$ from the pattern process $x_i^0(t)$ is equal to r_i, then in the space of all continuous n-dimensional functions we introduce the metric

$$\varrho(x, y) = \sup_{a \leq t \leq b} \max_{1 \leq i \leq n} \frac{|x_i(t) - y_i(t)|}{r_i}.$$

It can easily be proved that $\varrho(x, y)$ is a metric. Of course, those and only those processes are admissible for which $\varrho(x, x_0) \leq 1$.

In many processes constraints imposed on different parameters are not independent. For example, in some processes deviations of temperature are admissible when not accompanied by deviations of pressure and, vice versa, deviations of pressure are admissible when not accompanied by deviations of temperature, but simultaneous deviations of pressure and temperature destroy the final product. This leads us to the following notion. Suppose that, at each moment t, $x(t)$ is an element of the n-dimensional space R^n and a metric $\tilde{\varrho}(x, y)$ is defined in R^n such that the admissible deviation at the moment t from a pattern process $x_0(t)$ cannot exceed r, i.e., admissible are those $x(t)$ for which $\tilde{\varrho}(x_0(t), x(t)) \leq r$. Then, in the space of all continuous n-dimensional functions $C_n[a, b]$ of the argument t, $a \leq t \leq b$, we introduce a metric

$$\tilde{\varrho}(x, y) = \sup_{a \leq t \leq b} \varrho(x(t), y(t)). \tag{1.1.2}$$

As in Example 1.1.14, it can be easily verified that $\tilde{\varrho}(x, y)$ satisfies conditions (1)–(3).

The space $C_n[a, b]$ with the metric $\tilde{\varrho}$ will be denoted by $C_{n,\tilde{\varrho}}[a, b]$.

EXAMPLE 1.1.16. In many processes parameters describing states of a system do not change continuously but piecewise continuously (we say that a function f is *piecewise continuous* on a closed interval $[a, b]$ if there are points $a = a_0 < a_1 < a_2 < ... < a_n = b$ such that f is continuous on each interval (a_{i-1}, a_i), $i = 1, 2, ..., n$).

In the space of piecewise continuous functions (n-dimensional piecewise continuous vector functions) the formula (1.1.1) (respectively (1.1.2)) defines a metric. This metric space will be denoted by $C'[a, b]$ (respectively $C'_{n,\tilde{\varrho}}[a, b]$).

EXAMPLE 1.1.17. In Examples 1.1.14–1.1.16 we considered the behaviour of an object in a certain limited time interval $[a, b]$. However, in many cases we are interested in satisfying the technical requirements of an object as long as it "lives" and its life-time is not defined a priori (convincing examples of such objects are hydro-electric plants). Then it is reasonable to take $b = +\infty$ and consider the spaces of bounded continuous functions $C[a, +\infty)$, $C_{n,\tilde{\varrho}}[a, +\infty)$, $C'_{n,\tilde{\varrho}}[a, +\infty)$. It is easy to verify that in each of these cases formulae of types (1.1.1) and (1.1.2) define a metric.

EXAMPLE 1.1.18. Consider a certain process in a time interval which cannot be estimated a priori. As we have mentioned in the previous example, it is advisable to consider this process in a time interval $[a, +\infty)$. However, in many cases observations are not continuous but interesting quantities are measured at certain moments. For example, the water level in a river or in a water reservoir is measured every six hours or every two hours or every hour, depending on the situation. Meteorological stations collect data on the air pressure and temperature at prescribed times. Therefore, it is reasonable to assume that a process is described by a sequence of numbers $x = (x_1, ..., x_n, ...)$. A deviation of x from a pattern process y can be measured by

$$\varrho(x, y) = \sup_n |x_n - y_n|. \tag{1.1.3}$$

It can easily be verified that formula (1.1.3) defines a metric on the space of all bounded sequences. The space of all bounded sequences with the metric $\varrho(x, y)$ will be called the *space m*.

The subspace of the space m consisting of all convergent sequences will be denoted by c.

The subspace of the space m consisting of all sequences convergent to 0 will be denoted by c_0.

When dealing with industrial objects one must take into account the fact that after a certain amount of time a given object stops its activity because of wear or of getting out of date. Hence, for such an object we can assume $\lim\limits_{n \to \infty} x_n = 0$.

EXAMPLE 1.1.19. This example concerns problems of investment efficiency. If we investigate a short term trade deal, we apply the following formula:

$$W = \frac{\text{profit}}{\text{invested capital}},$$

where W denotes the efficiency of the deal. Of course, by profit we understand net profit after cost subtraction.

However, for long term investments this formula is not adequate. A given amount of money will have a smaller value in two years than it has today. The same concerns incomes. Therefore, to compute amounts of money, either expense or incomes, it is expedient to apply the following discount formula:

$$X_n(1+r)^{-n+1} = x_n, \qquad (1.1.4)$$

where X_n denotes the amount of money spent in the n-th year in comparable prices and x_n denotes the amount of money which should now be deposited in a bank at the rate of interest r to reach in the n-th year the amount X_n. Of course, in practice we do not deposit money in a bank in advance and, moreover, r is actually assumed to be the coefficient of economy development instead of the current interest rate.

The total investment cost is

$$K = \sum_{n=1}^{\infty} k_n,$$

where k_n are the discounted costs (i.e. the costs obtained from formula (1.1.4)).

The total investment profit is given by a similar formula

$$Z = \sum_{n=1}^{\infty} z_n,$$

where z_n denotes discounted profit. Efficiency is expressed by

$$W = \frac{Z}{K}.$$

If formula (1.1.4) was not applied but the profits (or expense) were simply summed up, then the corresponding series would be divergent. If the horizon were limited to a certain number of years (eight, for example), then investments with long investment cycles, e.g., raw material investments or hydroelectric plants, would be automatically eliminated.

As a measure of deviation of the profits obtained from those which were planned we take

$$\varrho(x, y) = \sum_{n=1}^{\infty} |x_n - y_n|. \tag{1.1.5}$$

The total profit is equal to $\varrho(x, 0)$. Finally, we have the following model: X is the space of all series $x = (x_1, ..., x_n, ...)$ such that $\varrho(x, 0) < +\infty$. This space is called the *space l*.

Let a system of metric spaces $(X_1, \varrho_1), ..., (X_n, \varrho_n)$ be given. By the *Cartesian product* (briefly *product*) of spaces $X_1, ..., X_n$ we understand a metric space X whose elements are of the form $x = (x_1, ..., x_n)$, where $x_i \in X_i$. In the space X we define a metric by the formula

$$\varrho(x, y) = \max_{1 \leq i \leq n} \varrho_i(x_i, y_i).$$

It is easy to verify that this is a metric. The space X will be denoted by $X_1 \times X_2 \times ... \times X_n$.

1.2. CONVERGENCE AND RELATED NOTIONS

Let (X, ϱ) be a metric space. We say that a sequence $\{x_n\}$ of elements of the space X *converges* to an element x_0 if $\lim_{n \to \infty} \varrho(x_n, x_0) = 0$. We shall denote it by $\lim_{n \to \infty} x_n = x_0$ or $x_n \xrightarrow{\varrho} x_0$. In the sequel we shall omit the letter ϱ whenever there is no doubt with respect to which metric the sequence $\{x_n\}$ is convergent. The element x_0 is called the *limit* of the sequence $\{x_n\}$. Notice that if $\{x_n\}$ converges to x_0, then each subsequence $\{x_{n_k}\}$ of $\{x_n\}$ converges to x_0.

We say that *almost all elements of a sequence $\{x_n\}$ belong to a certain set A* if there are at most a finite number of elements of this sequence not belonging to A. In other words, there exists a number N such that $x_n \in A$ for $n \geqslant N$. Using this terminology, we may define convergence as follows: a sequence $\{x_n\}$ is convergent to x_0 if, for every positive ε, almost all elements of the sequence belong to $K_\varepsilon(x_0)$. It is easy to observe that this statement is only another formulation of the definition given before.

A set $A \subset X$ is said to be *closed* if it contains limits of all sequences $\{x_n\}$ such that $x_n \in A$. A set $B \subset X$ is said to be *open* if for every $x_0 \in B$ there exists an $\varepsilon > 0$ such that $K_\varepsilon(x_0) \subset B$.

THEOREM 1.2.1. *Let A be a closed set. Then the $CA = X \setminus A$ is open.*

Proof. Suppose that CA is not open. Hence, there exists an $x_0 \in CA$ such that, for every positive ε, $K_\varepsilon(x_0) \cap A \neq \emptyset$. Let $x_n \in K_{1/n}(x_0) \cap A$. According to the definition, $x_n \to x_0$. But A is closed; thus, $x_0 \in A$ and we get a contradiction. ∎

THEOREM 1.2.2. *Let B be an open set. Then the set $CB = X \setminus B$ is closed.*

Proof. Suppose that CB is not closed. It means that there exists a sequence of elements $x_n \in CB$ which converges to an element $x_0 \in CCB = X \setminus (X \setminus B) = B$. But B is open by the assumption; thus, there exists a positive ε such that $K_\varepsilon(x_0) \subset B$. Since $x_n \to x_0$, almost all elements of the sequence $\{x_n\}$ belong to $K_\varepsilon(x_0) \subset B$. This contradicts the assumption that $x_n \in CB$. ∎

THEOREM 1.2.3. *Let a family $\{F_\alpha\}$, $\alpha \in \mathfrak{A}$, of closed sets be given. The common part (intersection) of this family, $F = \bigcap\limits_{\alpha \in \mathfrak{A}} F_\alpha$, is closed.*

Proof. Let $\{x_n\}$ be a sequence of elements of F, convergent to a certain x_0. Of course, $x_n \in F_\alpha$ for every α. Since the sets F_α are closed, $x_0 \in F_\alpha$ for every α. Hence $x_0 \in F$. ∎

THEOREM 1.2.4. *Let $\{G_\alpha\}$, $\alpha \in \mathfrak{A}'$, be a family of open sets. The union G of this family, $G = \bigcup_{\alpha \in \mathfrak{A}'} G_\alpha$, is open.*

Proof. Let $x_0 \in G$. By the definition of G, there exists a G_α such that $x_0 \in G_\alpha$. Since G_α is open, there exists a positive ε such that $K_\varepsilon(x_0) \subset G_\alpha \subset G$. Hence G is open. ∎

The following examples show that the union of a family of closed sets need not be closed and that the intersection of a family of open sets need not be open.

EXAMPLE 1.2.1. Let X be the closed interval $[0, 1]$ with the metric $\varrho(x, y) = |x-y|$. Let F_n be a one-point set consisting of $1/n$, $F_n = \{1/n\}$. $F = \bigcup_n F_n = \{1, 1/2, 1/3, ...\}$. Observe that this sequence is convergent to 0 but 0 does not belong to F.

EXAMPLE 1.2.2. Let X be defined as in Example 1.2.1 and let $G_n = \{x: |x-1/2| < 1/n\}$. The sets G_n are open. The common part $G = \bigcap_n G_n$ is a one-point set consisting of $1/2$. Thus G is not open.

However, the following theorems can be proved.

THEOREM 1.2.5. *The union F of a finite number of closed sets, $F = F_1 \cup ... \cup F_m$, is closed.*

Proof. Let $\{x_n\}$ be a sequence of elements of F, converging to some x_0. Since F is the union of a finite number of sets, there exist a set F_r and a subsequence $\{x_{n_k}\}$ such that $x_{n_k} \in F_r$. Of course, the sequence $\{x_{n_k}\}$ is also convergent to x_0. The set F_r is closed, so, $x_0 \in F_r \subset F$. ∎

THEOREM 1.2.6. *The common part G of a finite number of open sets, $G = G_1 \cap G_2 \cap ... \cap G_m$, is open.*

Proof. Let $x_0 \in G$. It means that $x_0 \in G_i$, $i = 1, \ldots, m$. Since the sets G_i are open, there exist positive numbers $\varepsilon_1, \ldots, \varepsilon_m$ such that $K_{\varepsilon_i}(x_0) \subset G_i$, $i = 1, \ldots, m$. Let $\varepsilon = \min(\varepsilon_1, \ldots, \varepsilon_m)$. ε is positive and $K_\varepsilon(x_0) \subset G_i$ for every i; so, $K_\varepsilon(x_0) \subset G$. ∎

Let A be an arbitrary set contained in X. The *closure* $\overline{A}$ of the set A is the smallest closed set containing A. By Theorem 1.2.3, $\overline{A}$ may be defined as the common part of all closed sets containing A.

It follows from the definition that for a closed A we have $\overline{A} = A$. If $\overline{\overline{A}}$ denotes the closure of $\overline{A}$, then for every set A we have $\overline{\overline{A}} = \overline{A}$.

THEOREM 1.2.7. $\overline{A} = \{x \in X: \text{there exists a sequence } \{x_n\},\ x_n \to x,\ x_n \in A,\ n = 1, 2, \ldots\}$.

Proof. Denote by $\tilde{A}$ the set appearing on the right-hand side of the above equality; it is obvious that $A \subset \tilde{A} \subset \overline{A}$. We shall show that $\tilde{A}$ is closed. Indeed, let us take a sequence $\{x_n\}$, $x_n \in \tilde{A}$, which is convergent to a certain x_0. Let n be fixed. Since $x_n \in \tilde{A}$, there exists an $x'_n \in A$ such that $\varrho(x_n, x'_n) \leq 1/n$. Since $0 \leq \varrho(x_n, x_0) \leq \varrho(x'_n, x_n) + \varrho(x_n, x_0) \leq 1/n + \varrho(x_n, x_0)$ and $x_n \to x_0$, $x'_n \to x_0$, we have $x_0 \in \tilde{A}$. It follows from the closedness of $\tilde{A}$ that $\overline{A} \subset \tilde{A}$ and finally $\overline{A} = \tilde{A}$. ∎

The largest open set contained in A is called the *interior* of A, Int A. By Theorem 1.2.4, we can say that Int A is the union of all open sets contained in A. Every point of the interior of a set is called an *interior point* of that set. It is easy to observe that a point x_0 is an interior point of a set A if and only if there exists an $\varepsilon > 0$ such that $K_\varepsilon(x_0) \subset A$.

The following formulae are immediate consequences of Theorems 1.2.1 and 1.2.2:

$$\overline{A} = X \setminus \text{Int}(X \setminus A), \quad \text{Int} A = X \setminus \overline{(X \setminus A)}.$$

The *boundary* of a set A, Fr A, is the set

$$\text{Fr} A = \overline{A} \setminus \text{Int} A = \overline{A} \cap \overline{(X \setminus A)}.$$

We say that a set A is *dense* in the space X if $\overline{A} = X$. We say that a subset A of a set B is *dense* in B if $B \subset \overline{A}$.

1.3. Continuous mappings

Let (X, ϱ_X), (Y, ϱ_Y) be two metric spaces. Let a function (mapping) $f(x)$ map the space X into the space Y.

THEOREM 1.3.1. *The following four statements are equivalent*:

(a) *For every $x \in X$ and every positive ε, there exists a $\delta > 0$ such that $f(K_\delta(x)) \subset K_\varepsilon(f(x))$.*

(b) *If $\lim_{n\to\infty} \varrho_X(x_n, x) = 0$, then $\lim_{n\to\infty} \varrho_Y(f(x_n), f(x)) = 0$ for every $x \in X$ and every sequence $\{x_n\} \subset X$.*

(c) *The inverse image $f^{-1}(F) = \{x \in X: f(x) \in F\}$ of each closed set F is closed.*

(d) *The inverse image $f^{-1}(G) = \{x \in X: f(x) \in G\}$ of each open set G is open.*

Proof. (a) → (b). Let $\lim_{n\to\infty} \varrho_X(x_n, x) = 0$ and let ε be an arbitrary positive number. By (a), there exists a $\delta > 0$ such that $f(K_\delta(x)) \subset K_\varepsilon(f(x))$. Since $x_n \stackrel{\varrho_X}{\to} x$, almost all elements of the sequence $\{f(x_n)\}$ belong to $K_\varepsilon(f(x))$. The arbitrariness of ε implies (b).

(b) → (c). Let F be a closed subset contained in Y. Let $\{x_n\}$ be a sequence of elements of $f^{-1}(F)$ which converges to a certain x_0. In virtue of (b), $f(x_n) \stackrel{\varrho_Y}{\to} f(x_0)$ and $f(x_0) \in F$, since F is closed. Hence, $x_0 \in f^{-1}(F)$.

(c) → (d). Let G be an arbitrary open set contained in Y. By Theorem 1.2.2, the set $Y \setminus G$ is closed, hence by (c) and Theorem 1.2.1 the set $f^{-1}(G) = f^{-1}(Y \setminus (Y \setminus G)) = f^{-1}(Y) \setminus f^{-1}(Y \setminus G)$ is open.

(d) → (a). In virtue of (d), the set $f^{-1}(K_\varepsilon(f(x)))$ is open for every positive ε. Hence, there exists a $\delta > 0$ such that
$$K_\delta(x) \subset f^{-1}(K_\varepsilon(f(x))),$$
i.e., $f(K_\delta(x)) \subset K_\varepsilon(f(x))$. ∎

A mapping f which satisfies one of the conditions (a)–(d) is said to be *continuous*. If f is a one-to-one mapping and if both f and f^{-1} are continuous, then the mapping f is said to be a *homeomorphism*. Metric spaces (X, ϱ_X) and (Y, ϱ_Y) are said to be *homeomorphic* if there exists a homeomorphism which maps (X, ϱ_X) onto (Y, ϱ_Y).

Suppose we are given a metric space X with two metrics, $\varrho(x, y)$ and $\varrho'(x, y)$. Let us denote by n the natural injection, $n(x) = x$, of the space (X, ϱ) into the space (X, ϱ'). If n is continuous, then the metric $\varrho(x, y)$ is said to be *stronger* than the metric $\varrho'(x, y)$ and the metric $\varrho'(x, y)$ is said to be *weaker* than the metric $\varrho(x, y)$. If n is a homeomorphism, then the metrics $\varrho(x, y)$ and $\varrho'(x, y)$ are said to be *equivalent*.

EXAMPLE 1.3.1. If X is a finite set, then each metric defined on X is equivalent to the discrete metric

$$d(x, y) = \begin{cases} 0 & \text{for } x = y, \\ 1 & \text{for } x \neq y. \end{cases}$$

Indeed, let ϱ be an arbitrary metric. Take a sequence $\{x_n\}$ such that $x_n \xrightarrow{\varrho} x_0$. Since inf $\{\varrho(x, y): x \neq y\} > 0$, almost all x_n are equal to x_0. Hence, $\{x_n\}$ tends to x_0 in each metric.

EXAMPLE 1.3.2. Let X be the Euclidean plane and let $\varrho(x, y)$ denote the length of the closed interval with the end-points x, y (compare Example 1.1.2). Let $d(x, y)$ denote the discrete metric defined on X,

$$d(x, y) = \begin{cases} 0 & \text{for } x = y, \\ 1 & \text{for } x \neq y. \end{cases}$$

It can easily be verified that those metrics are not equivalent and the discrete metric $d(x, y)$ is stronger than the Euclidean metric $\varrho(x, y)$.

EXAMPLE 1.3.3. Let R^n be an n-dimensional space. Consider three metrics on R^n:

$$\varrho_{\sup}(x, y) = \sup_{1 \leq i \leq n} |x_i - y_i|,$$

$$\varrho_2(x, y) = \left(\sum_{i=1}^n |x_i - y_i|^2 \right)^{1/2},$$

$$\varrho_1(x, y) = \sum_{i=1}^n |x_i - y_i|.$$

It is easy to verify that

$$\varrho_{\sup}(x, y) \leq \varrho_2(x, y) \leq \varrho_1(x, y) \leq n\varrho_{\sup}(x, y). \tag{1.3.1}$$

Thus, the metrics $\varrho_{\sup}$, ϱ_1, ϱ_2 are equivalent.

EXAMPLE 1.3.4. Let $(X_1, \varrho^1), \ldots, (X_n, \varrho^n)$ be metric spaces. In the Cartesian product $X_1 \times X_2 \times \ldots \times X_n$ the metrics

$$\varrho_{\sup}(x, y) = \sup_{1 \leq i \leq n} \varrho^i(x_i, y_i),$$

$$\varrho_2(x, y) = \left(\sum_{i=1}^{n} [\varrho^i(x_i, y_i)]^2\right)^{1/2},$$

$$\varrho_1(x, y) = \sum_{i=1}^{n} \varrho^i(x_i, y_i)$$

satisfy inequality (1.3.1) and thus these metrics are equivalent.

Let (X, ϱ_X) and (Y, ϱ_Y) be metric spaces. We say that a mapping T of the space X into Y is an *isometry* if

$$\varrho_Y(Tx, Ty) = \varrho_X(x, y).$$

Clearly, each isometry is a one-to-one continuous mapping. Moreover, T^{-1} is also an isometry.

Two metric spaces X and Y are called *isometric* if there exists an isometry T which maps X onto Y.

Let (X, ϱ_X) and (Y, ϱ_Y) be metric spaces. By $C(X, Y)$ we denote the set of all continuous functions f which map X into Y and satisfy the condition

$$\sup_{x \in X} \varrho_Y(f(x), 0) < +\infty. \tag{1.3.2}$$

It can easily be shown that the set $C(X, Y)$ is a metric space with the metric $\varrho(f, g)$ given by the formula

$$\varrho(f, g) = \sup_{x \in X} \varrho_Y(f(x), g(x)) \tag{1.3.3}$$

Condition (1.3.2) ensures that $\varrho(f, g)$ is finite. The space $C(X, Y)$, where Y is the set of all real (or complex) numbers, will be denoted briefly by $C(X)$ ($C_c(X)$). If X is a closed interval $[a, b]$, then the space $C([a, b])$ will be denoted by $C[a, b]$ ($C_c[a, b]$ in the complex case).

1.4. Semimetric spaces

In practice there exist situations where a certain function of two variables $\varrho(x, y)$ defined on a set X satisfies conditions (1) and (3) of the definition of a metric but does not satisfy condition (2).

EXAMPLE 1.4.1. Let X be a network of tourist routes in the Tatra mountains. Let $\varrho(x, y)$ denote the time it takes to pass from the point x to the point y. It can be shown that conditions (1) and (3) of a metric are satisfied but condition (2) is not satisfied since the way up takes longer than the way down.

EXAMPLE 1.4.2. Let X denote a water (river) transport system. The cost of transport $\varrho(x,y)$ of a cargo unit from the point x to the point y satisfies the conditions (1), (3) of a metric. Condition (2) is not satisfied in general, since downstream transport is cheaper than upstream transport.

The above examples suggest the following definition. Suppose we are given a set X and a real-valued nonnegative function, defined on X, of two arguments $\varrho(x, y)$ which satisfies conditions (1) and (3) of the definition of a metric and the condition

(2') $\lim_{n \to \infty} \varrho(x_n, x) = 0$ if and only if $\lim_{n \to \infty} \varrho(x, x_n) = 0$.

We call the function $\varrho(x, y)$ a *semimetric*. A set X with a semimetric defined on it is called a *semimetric space*. A semimetric space X with the semimetric $\varrho(x, y)$ will be denoted by (X, ϱ) or briefly X.

Just as for metrics, we say that two semimetrics ϱ and ϱ' are *equivalent* whenever $\lim_{n \to \infty} \varrho(x_n, x) = 0$ if and only if $\lim_{n \to \infty} \varrho'(x_n, x) = 0$.

THEOREM 1.4.1. *Let (X, ϱ) be a semimetric space. Then there exists a metric $\varrho'(x, y)$ on X which is equivalent (in the sense of semimetrics) to $\varrho(x, y)$.*

Proof. Let $\varrho'(x, y) = \varrho(x, y) + \varrho(y, x)$. Obviously, $\varrho'(x, y)$ satisfies conditions (1) and (2) of a metric. We show that $\varrho'(x, y)$ satisfies also condition (3). Namely,

$$\varrho'(x, y) = \varrho(x, y) + \varrho(y, x)$$
$$\leqslant \varrho(x, z) + \varrho(z, y) + \varrho(y, z) + \varrho(z, x)$$
$$= \varrho'(x, z) + \varrho'(z, y).$$

It is a straightforward consequence of the definition that

$$0 \leqslant \varrho(x, y) \leqslant \varrho'(x, y).$$

Hence, $\lim_{n \to \infty} \varrho'(x_n, x) = 0$ entails $\lim_{n \to \infty} \varrho(x_n, x) = 0$. The converse implication is an immediate consequence of condition (2'). ∎

In view of Theorem 1.4.1, all the definitions and all the theorems of Section 1.2 and Section 1.3 can be restated by substituting the term "semimetric" for the term "metric" and after this substitution all the theorems proved in those sections remain valid.

Theorem 1.4.1 asserts that with respect to convergence, continuity etc. every semimetric space can be considered a metric space. Thus semimetric spaces are no longer interesting from the mathematical point of view. However, in many applications it is necessary to consider semimetric spaces.

1.5. COMPLETENESS OF METRIC SPACES

Let (X, ϱ) be a metric space. We say that a sequence $\{x_n\}$ of elements of the space X is a *fundamental sequence* or a *Cauchy sequence* if, for every $\varepsilon > 0$, there exists an integer N such that for $n, m > N$

$$\varrho(x_n, x_m) < \varepsilon.$$

The space (X, ϱ) is said to be *complete* if for every fundamental sequence $\{x_n\}$ of elements of X there exists an element x_0 such that $x_n \to x_0$. Obviously, if two metric spaces are isometric and one of them is complete, then the second one is also complete.

THEOREM 1.5.1. *Let (X, ϱ) be a complete metric space and let X_0 be a closed subset of X. Then the subspace (X_0, ϱ) is complete.*

Proof. Let $\{x_n\}$ be an arbitrary fundamental sequence of elements of X_0. Since X_0 is a subspace of X, the sequence $\{x_n\}$ is also a fundamental sequence in the space X. In view of the completeness of the space X, the sequence $\{x_n\}$ converges to a certain $x_0 \in X$. But X_0 is closed. Hence $x_0 \in X_0$. ∎

EXAMPLE 1.5.1. Every metric space (X, ϱ) consisting of a finite number of elements is complete. Indeed, since X consists of a finite number of elements, $\inf\{(x, y): x, y \in X, x \neq y\} > 0$. Hence a sequence $\{x_n\}$ is fundamental if and only if almost all its elements are identical and equal to a certain x_0. Then obviously $x_n \to x_0$.

EXAMPLE 1.5.2. Every discrete space is complete. This is true by the same arguments as in Example 1.5.1.

EXAMPLE 1.5.3. Consider the set X of all rational numbers with the metric $\varrho(x, y) = |x-y|$. It is easy to verify that (X, ϱ) is a metric space. (X, ϱ) is not complete. Let x_n denote the decimal approximation of $\sqrt{2}$ with the first n digits of the decimal representation of $\sqrt{2}$. Obviously, the sequence $\{x_n\}$ is fundamental since $\varrho(x_n, x_m) < 10^{-\min(n, m)}$. On the other hand, the sequence $\{x_n\}$ is not convergent since there is no rational number whose square is equal to 2.

THEOREM 1.5.2 (Cantor). *Let (X, ϱ) be a metric space. There exists a complete metric space $(\hat{X}, \hat{\varrho})$ such that X is a subspace of $\hat{X}$, the set X is dense in $\hat{X}$ and $\hat{\varrho}(x, y) = \varrho(x, y)$ for $x, y \in X$.*

Proof. Denote by $\tilde{X}$ the set of all fundamental sequences $\{x_n\}$, $x_n \in X$, $n = 1, 2, \ldots$ Let us partition $\tilde{X}$ into separate sets X_α by assigning sequences $\{x_n\}$, $\{y_n\}$ to the same X_α if and only if

$$\lim_{n \to \infty} \varrho(x_n, y_n) = 0.$$

The sets X_α are called *equivalence classes*. An equivalence class containing a sequence $\{x_n\}$ will be denoted by $[\{x_n\}]$. An arbitrary sequence $\{x_n\}$ belonging to an equivalence class X_α will be called a *representative*

1.5. Completeness of metric spaces

of that class. Denote by $\hat{X}$ the set of all equivalence classes with the metric

$$\hat{\varrho}(\tilde{x}, \tilde{y}) = \lim_{n \to \infty} \varrho(x_n, y_n)$$

where $\{x_n\}$ and $\{y_n\}$ are arbitrary representatives of the equivalence classes $\tilde{x}$ and $\tilde{y}$, respectively. Observe that $\hat{\varrho}(\tilde{x}, \tilde{y})$ does not depend upon the choice of representatives $\{x_n\}$ and $\{y_n\}$.

We first prove that $\hat{\varrho}(\tilde{x}, \tilde{y})$ is a metric. The function $\hat{\varrho}(\tilde{x}, \tilde{y})$ is non-negative. If $\tilde{x} = \tilde{y}$, then, by taking the same representatives in $\tilde{x}$ and $\tilde{y}$, we obtain $\hat{\varrho}(\tilde{x}, \tilde{y}) = 0$. On the other hand, if $\hat{\varrho}(\tilde{x}, \tilde{y}) = 0$, then representatives of $\tilde{x}$ and $\tilde{y}$ belong to the same equivalence class. Thus $\tilde{x} = \tilde{y}$. Condition (2) of the definition of a metric is satisfied since

$$\hat{\varrho}(\tilde{x}, \tilde{y}) = \lim_{n \to \infty} \varrho(x_n, y_n) = \lim_{n \to \infty} \varrho(y_n, x_n) = \hat{\varrho}(\tilde{y}, \tilde{x})$$

where, as before, $\{x_n\}$ is a representative of the class $\tilde{x}$ and $\{y_n\}$ is a representative of the class $\tilde{y}$. To verify the condition (3) of a metric let $\{z_n\}$ be a representative of a class $\tilde{z}$. We obtain

$$\hat{\varrho}(\tilde{x}, \tilde{y}) = \lim_{n \to \infty} \varrho(x_n, y_n)$$
$$\leq \lim_{n \to \infty} \varrho(x_n, z_n) + \lim_{n \to \infty} \varrho(z_n, y_n) = \hat{\varrho}(\tilde{x}, \tilde{z}) + \hat{\varrho}(\tilde{z}, \tilde{y}).$$

Hence, $\hat{\varrho}(\tilde{x}, \tilde{y})$ is a metric. An element $x \in X$ is identified with the equivalence class generated by the sequence $\{x, \ldots, x, \ldots\}$ and is denoted by $[x]$. Clearly, $\hat{\varrho}([x], [y]) = \varrho(x, y)$.

The space $(\hat{X}, \hat{\varrho})$ is complete. To see this, let $\{x^m\}$ be a Cauchy sequence and let $\{x_n^m\}$ be a representative of x^m. The sequence $\{x_n^m\}$ is a Cauchy sequence with respect to n for $m = 1, 2, \ldots$ Hence, for each m there is an integer $N(m)$ such that

$$\varrho(x_n^m, x_{n_1}^m) < 1/2^m$$

for all $n, n_1 > N(m)$. Let $x^0 = \{x_{N(m)}^m\}$. Now we show that the sequence $\{x_{N(m)}^m\}$ is a Cauchy sequence in X. Let ε be an arbitrary positive number. Since $\{x^m\}$ is a Cauchy sequence in $\hat{X}$, there exists an index m_0 such that for $m_1, m_2 > m_0$

$$\hat{\varrho}(x^{m_1}, x^{m_2}) < \varepsilon/3.$$

Without loss of generality we may assume $1/2^{m_0} < \varepsilon/3$. By the definition of $\hat{\varrho}$, there is an integer M such that

$$\varrho(x_n^{m_1}, x_n^{m_2}) < \varepsilon/3 \quad \text{for } n > M.$$

Denoting by n an arbitrary integer greater than $\max(M, N(m_1), N(m_2))$ we have

$$\varrho(x_{N(m_1)}^{m_1}, x_{N(m_2)}^{m_2}) \leqslant \varrho(x_{N(m_1)}^{m_1}, x_n^{m_1}) + \varrho(x_n^{m_1}, x_n^{m_2}) + \varrho(x_n^{m_2}, x_{N(m_2)}^{m_2})$$
$$< \varepsilon/3 + \varepsilon/3 + \varepsilon/3 = \varepsilon.$$

Therefore $\{x_{N(m)}^m\}$ is a Cauchy sequence and, by the definition of $\hat{\varrho}$, we have

$$\lim_{m \to \infty} \hat{\varrho}(x^m, x^0) = \lim_{m \to \infty} \lim_{n \to \infty} \varrho(x_n^m, x_{N(n)}^n).$$

The definition of $N(n)$ entails

$$\varrho(x_n^m, x_{N(n)}^n) < 1/2^n$$

for $n > m$. Therefore

$$\hat{\varrho}(x^m, x^0) < 1/2^n$$

and

$$\lim_{m \to \infty} \hat{\varrho}(x^m, x^0) = 0.$$

Hence, $(\hat{X}, \hat{\varrho})$ is complete. To finish the proof we should show that the set X' of all equivalence classes of the form $[x]$, $x \in X$, is dense in $\hat{X}$. Let $\tilde{x} \in \hat{X}$ and let $\{x_n\}$ be a representative of $\tilde{x}$. Let ε be an arbitrary positive number. Since the sequence $\{x_n\}$ satisfies the Cauchy condition, there exists an index N such that $\varrho(x_n, x_N) < \varepsilon$ for $n > N$. Evidently $\hat{\varrho}(\tilde{x}, [x_N]) < \varepsilon$. Thus, the set X' is dense in $\hat{X}$, which completes the proof. ∎

The space $\hat{X}$ is called the *completion* of the space X.

Now we give examples of complete metric spaces.

EXAMPLE 1.5.4. The completion of the set of all rational numbers is called the *set of real numbers*. This definition is slightly different from the usual definition exploiting sections (cf. Kuratowski, 1961). Here we show the equivalence of these definitions. Let r be a real number given by a section, i.e., by a partition of rational numbers into disjoint sets A and B such that $x < y$ for all $x \in A$ and $y \in B$. Let $\{x_n\}$ be a sequence of elements of the set A such that for every n there exists a $y \in B$ satisfying $|x_n - y| < 1/n$. Assign to a number r the class x represented by $\{x_n\}$. By associating in the same way a real number r_1 with the equivalence class $\tilde{y}$ represented by the sequence $\{y_n\}$ we obtain

$$|r - r_1| = \varrho(\tilde{x}, \tilde{y}).$$

To complete the proof it is enough to show that to every equivalence class x represented by a fundamental sequence $\{x_n\}$ corresponds a certain section. Let A be the set of rational numbers a such that $a \leq x_n$ for almost all n and let B be the set of rational numbers b such that $b > x_n$ for almost all n. Since the sequence $\{x_n\}$ is fundamental, $A \cup B$ is the set of all rational numbers.

EXAMPLE 1.5.5. The n-dimensional real space R^n, i.e., the space of all n-tuples $(x_1, \ldots, x_n)$ of real numbers with a metric given by the formula

$$\varrho((x_1, \ldots, x_n), (y_1, \ldots, y_n)) = \max_{1 \leq i \leq n} |x_i - y_i|$$

is a complete space. The verification that ϱ is a metric is simple. Let $\{(x_1^m, \ldots, x_n^m)\}$ be a fundamental sequence. It follows immediately from the definition of a metric that for every i, $i = 1, 2, \ldots, n$, the sequence of real numbers $\{x_i^m\}$ is fundamental. Since the space of all real numbers is complete, $x_i^m \to x_i$ for $i = 1, \ldots, n$. Hence,

$$\lim_{m \to \infty} \varrho((x_1^m, \ldots, x_n^m), (x_1, \ldots, x_n)) = \lim_{m \to \infty} \max_{1 \leq i \leq n} |x_i^m - x_i|$$
$$= \max_{1 \leq i \leq n} \lim_{m \to \infty} |x_i^m - x_i| = 0,$$

which implies the completeness of the space R^n.

EXAMPLE 1.5.6. The set of all complex numbers C with a metric given by the formula

$$\varrho(z, z') = |z - z'| = \sqrt{(a - a')^2 + (b - b')^2}$$

where $z = a + bi$, $z' = a' + b'i$, is a complete metric space. The proof that $\varrho(z, z')$ is a metric can be found in every textbook on elementary algebra. Consider another metric $\varrho'(z, z')$ in C, given by the formula

$$\varrho'(z, z') = \max(|a - a'|, |b - b'|).$$

Observe that $\varrho'(z, z') \leq \varrho(z, z') \leq \sqrt{2}\varrho'(z, z')$. This implies that any sequence is convergent in the metric ϱ if and only if it is convergent in the metric ϱ' and any sequence is fundamental in the metric ϱ if and only if it is fundamental in the metric ϱ'. Since C with the metric ϱ' is isometric to R^2, and R^2 is complete, C is complete in the metric ϱ' and hence in the metric ϱ.

EXAMPLE 1.5.7. The n-dimensional complex space C^n of ordered n-tuples $(z_1, \ldots, z_n)$ of complex numbers with the metric
$$\varrho((z_1, \ldots, z_n), (z'_1, \ldots, z'_n)) = \max_{1 \leq i \leq n} |z_i - z'_i|$$
is complete. It is easy to verify that ϱ is a metric. Just as in the previous example, we introduce the metric
$$\varrho'((z_1, \ldots, z_n), (z'_1, \ldots, z'_n)) = \max_{1 \leq i \leq n} \max(|a_i - a'_i|, |b_i - b'_i|)$$
and, by similar arguments, any sequence is fundamental in the metric ϱ if and only if it is fundamental in the metric ϱ' and any sequence converges in the metric ϱ if and only if it converges in the metric ϱ'. Moreover, the space C^n with the metric ϱ' is isomorphic to R^{2n}. This implies the completeness of C^n.

EXAMPLE 1.5.8. The space of all continuous functions $C[a, b]$ is complete. Let $\{x_n(t)\}$ be a fundamental sequence of elements of $C[a, b]$. Hence for every t the sequence $\{x_n(t)\}$ is a fundamental sequence of real numbers. By the completeness of reals, the sequence $\{x_n(t)\}$ converges to a certain number $x(t)$. We shall prove that $x(t)$ is a continuous function of t. Let ε be an arbitrary positive number. Since the sequence $\{x_n(t)\}$ is fundamental, there exists an index N such that for all $n > N$
$$|x_n(t) - x_N(t)| < \varepsilon/3.$$
Passing to the limit with n tending to $+\infty$, we get
$$|x(t) - x_N(t)| \leq \varepsilon/3$$
for all t. Let δ be a positive number such that $|x_N(t) - x_N(t')| < \varepsilon/3$, provided $|t - t'| < \delta$. Then
$$|x(t) - x(t')| \leq |x(t) - x_N(t)| + |x_N(t) - x_N(t')| + |x_N(t') - x(t')|$$
$$\leq \varepsilon/3 + \varepsilon/3 + \varepsilon/3 = \varepsilon.$$

Thus, $x(t)$ is uniformly continuous. Moreover, inequality (1.5.1) implies $\varrho(x, x_N) \leqslant \varepsilon/3$. Since ε was taken arbitrarily, $\{x_n\}$ converges to x.

THEOREM 1.5.3. *The product of n complete metric spaces, $X_1 \times X_2 \times \ldots \times X_n$, is complete.*

Proof. The proof is similar to that in Example 1.5.5. ∎

1.6. CONTRACTIVE MAPPING PRINCIPLE

As an application of the concept of completeness we give the contractive mapping principle.

THEOREM 1.6.1 (Banach, 1922). *Let (X, ϱ) be a complete metric space. Let T be a mapping of the space X into itself such that*

$$\varrho(T(x), T(y)) \leqslant a\varrho(x, y) \tag{1.6.1}$$

where a is a certain number, $0 < a < 1$. Then there exists a unique element $x \in X$ such that $T(x) = x$.

Proof. Let x_0 be an arbitrary element of X. Write $x_n = T^n(x_0)$. We shall prove that the sequence $\{x_n\}$ is fundamental. For $m > n$ we have

$$\begin{aligned} \varrho(x_n, x_m) &= \varrho(T^n(x_0), T^m(x_0)) \\ &\leqslant \varrho(T^n(x_0), T^{n+1}(x_0)) + \ldots + \varrho(T^{m-1}(x_0), T^m(x_0)) \\ &= (a^n + \ldots + a^{m-1})\varrho(T(x_0), x_0) \\ &\leqslant (a^n/(1-a))\varrho(T(x_0), x_0). \end{aligned} \tag{1.6.2}$$

Let ε be an arbitrary positive number. By letting

$$N = \frac{1}{\log a} \log \frac{(1-a)\varepsilon}{\varrho(T(x_0), x_0)},$$

it is easy to verify, in view of (1.6.2), that $\varrho(x_n, x_m) < \varepsilon$ for $n, m > N$. Thus, the sequence $\{x_n\}$ is fundamental. By the completeness of X, the sequence $\{x_n\}$ converges to a certain element x, i.e., $\lim_{n \to \infty} T^n(x_0) = x$. By (1.6.1), the mapping T is continuous and hence

$$T(x) = T(\lim_{n \to \infty} T^n(x_0)) = \lim_{n \to \infty} T^{n+1}(x_0) = \lim_{n \to \infty} T^n(x_0) = x.$$

We prove the uniqueness of x by contradiction. In fact, suppose that there are two elements x, y such that $T(x) = x$, $T(y) = y$. Hence $\varrho(x, y) = \varrho(T(x), T(y)) = a\varrho(x, y)$. Since $0 < a < 1$, this is possible only when $\varrho(x, y) = 0$, i.e., when $x = y$. ∎

Elements x such that $T(x) = x$ are called *fixed points* (or *invariants*) *of T*. A mapping T satisfying (1.6.1) is called a *contractive mapping*. Therefore, Theorem 1.6.1 can be formulated as follows:

A contractive mapping of a complete metric space has a unique fixed point.

THEOREM 1.6.2. *Let* (X, ϱ) *be a complete metric space. Let* T_0, T_1 *be contractive mappings*

$$\varrho(T_0(x), T_0(y)) \leqslant a\varrho(x, y),$$

$$\varrho(T_1(x), T_1(y)) \leqslant a\varrho(x, y),$$

$0 < a < 1$, *satisfying the condition*

$$\varrho(T_0(x), T_1(x)) \leqslant b \tag{1.6.3}$$

for all $x \in X$ *and a certain* b. *Let* x_0, x_1 *denote the fixed points of* T_0, T_1 *respectively. Then*

$$\varrho(x_0, x_1) \leqslant b/(1-a). \tag{1.6.4}$$

Proof. By hypothesis,

$$\begin{aligned}\varrho(x_0, x_1) &= \varrho(T_0(x_0), T_1(x_1)) \\ &\leqslant \varrho(T_0(x_0), T_1(x_0)) + \varrho(T_1(x_0), T_1(x_1)) \\ &\leqslant b + a\varrho(x_0, x_1),\end{aligned}$$

which trivially implies (1.6.4). ∎

Now we shall apply the contractive mapping principle to the theory of the Volterra integral and differential operators.

EXAMPLE 1.6.1. Let R^n be the n-dimensional real Euclidean space with the standard metric ϱ (see Example 1.1.1). Let $[a, b]$ be a closed interval. Let $K(t, s, x_1, \ldots, x_n)$ be a continuous function defined on the Cartesian product $[a, b]^2 \times R^n$ which takes on values in the n-dimensional space R^n.

1.6. Contractive mapping principle

Assume that the function $K(t, s, x_1, \ldots, x_n)$ satisfies the Lipschitz condition with respect to $x = (x_1, \ldots, x_n)$, uniformly with respect to t, i.e., there exists an $L(s) > 0$ such that

$$\varrho\big(K(t, s, x_1, \ldots, x_n), K(t, s, y_1, \ldots, y_n)\big) \leqslant L(s)\varrho(x, y).$$

Moreover, assume that the function $L(s)$ is integrable. Consider now the *Bielecki space* (see Bielecki 1956, 1956b), i.e., the space of all real-valued n-dimensional continuous functions defined on the interval $[a, b]$ with the following metric, called the *Bielecki metric*:

$$\varrho_{B,p}(x, y) = \sup_{a \leqslant t \leqslant b} \varrho\big(x(t), y(t)\big) \cdot e^{-p \int_a^t L(s)ds}.$$

Clearly, the Bielecki metric is equivalent to the standard metric

$$\tilde{\varrho}(x, y) = \sup_{a \leqslant t \leqslant b} \varrho\big(x(t), y(t)\big).$$

Thus the Bielecki space can be regarded as the product of n spaces $C[a, b]$. These spaces are complete (Example 1.5.8) and therefore, by Theorem 1.5.3, the Bielecki space is also complete.

Consider now the Volterra integral operator

$$(Kx)|_t = \int_a^t K\big(t, s, x_1(s), \ldots, x_n(s)\big)ds.$$

Under our hypothesis the operator K maps the Bielecki space into itself. We shall prove that for sufficiently large p the operator K is a contractive mapping (Bielecki 1956, 1956b). We have

$$\varrho_{B,p}(Kx, Ky) = \sup_{a \leqslant t \leqslant b} e^{-p \int_a^t L(s)ds} \int_a^t [K(t, s, x_1(s), \ldots, x_n(s)) -$$
$$- K(t, s, y_1(s), \ldots, y_n(s))]ds$$
$$\leqslant \sup_{a \leqslant t \leqslant b} e^{-p \int_a^b L(s)ds} \int_a^t L(s) \cdot \varrho\big(x(s), y(s)\big)ds$$
$$\leqslant \sup_{a \leqslant t \leqslant b} e^{-p \int_a^t L(s)ds} \int_a^t L(s) e^{-p \int_a^t L(s)ds} e^{p \int_a^t L(s)ds} \times$$
$$\times \varrho\big(x(s), y(s)\big)ds$$

$$\leq \sup_{a\leq t\leq b} e^{-p\int_a^t L(s)ds} \varrho_{B,p}(x(\cdot), y(\cdot)) \int_a^t L(s) e^{p\int_a^s L(u)du} ds$$

$$= \varrho_{B,p}(x, y)(1/p) \qquad (1.6.5)$$

since

$$\int_a^t L(s) e^{p\int_a^s L(u)du} ds = (1/p) e^{p\int_a^u L(s)ds}\Big|_a^t = (1/p) e^{p\int_a^t L(s)ds} - 1/p$$

$$\leq (1/p) e^{p\int_a^t L(s)ds}. \qquad (1.6.6)$$

Therefore for $p > 1$ the operator K is a contractive mapping in the Bielecki space. This implies that the equation

$$x(t) = \int_a^t K(t, s, x(s)) ds \qquad (1.6.7)$$

has a unique solution. Note moreover that, for any n-dimensional continuous function f, the operator

$$K_f(x) = K(x) + f$$

is also a contractive mapping and has a unique fixed point x_f. In view of Theorem 1.6.2, x_f depends continuously on f and moreover we can prove the Lipschitz continuity of x_f with respect to f, i.e., the existence of an $M > 0$ such that for arbitrary f, g

$$\varrho_0(x_f, x_g) \leq M\varrho_0(f, g)$$

for any metric ϱ_0 of the type

$$\varrho_0(x(\cdot), y(\cdot)) = \sup_{a\leq t\leq b} \varrho_1(x(t), y(t)),$$

where ϱ_1 denotes the Euclidean metric in R^n.

The results obtained in Example 1.6.1 can be applied to differential equations, as is shown by the following example.

EXAMPLE 1.6.2. Consider the differential equation

$$\frac{dx}{dt} = f(t, x), \qquad (1.6.8)$$

where x is an n-dimensional vector function defined on some interval $[a, b]$ with an initial condition

$$x(a) = x_0. \qquad (1.6.9)$$

Assume that f is a continuous n-dimensional vector function of $n+1$ variables satisfying the Lipschitz condition with respect to x, i.e., there exists an integrable function $L(t)$ such that

$$\varrho(f(t, x), f(t, y)) \leqslant L(t)\varrho(x, y). \qquad (1.6.10)$$

The system of equations (1.6.8) with the initial condition (1.6.9) is equivalent to the Volterra integral equation

$$x(t) = x_0 + \int_0^t f(t, x(t)) dt. \qquad (1.6.11)$$

Equation (1.6.11) satisfies all the requirements from Example 1.6.1. Therefore, equation (1.6.11) has a unique solution which depends continuously on the initial value x_0.

1.7. SETS OF THE FIRST AND SECOND CATEGORY

Given a metric space (X, ϱ), we say that a set $A \subset X$ is *nowhere dense* if A is not dense in any open set. If A is nowhere dense, then each open set $U \subset X$ contains a ball K which is disjoint with A. It can easily be shown that the union of a finite number of nowhere dense sets is also a nowhere dense set. A set $B \subset X$ is said to be *of the first category* if B can be expressed as the union of a countable number (sequence) of nowhere dense sets, $B = \bigcup_{n=1}^{\infty} B_n$. It is an immediate consequence of the definition that the union of a countable number (sequence) of sets of the first category is also of the first category. Sets which are not of the first category are said to be *of the second category*.

THEOREM 1.7.1. *If a set $A \subset X$ is not nowhere dense, then there exists an open set U such that A is dense in U.*

Proof. If A is not nowhere dense, then there exists an open set U such that every open ball contained in U has a nonempty intersection with A. Thus, by definition, A is dense in U. ∎

COROLLARY 1.7.2. *Every closed set of the second category contains a ball.*

Proof. A set A of the second category is not nowhere dense and therefore it is dense in a certain open set U. Since A is closed, it contains the whole set U. Furthermore, U contains a certain ball K because U is open. Finally $K \subset A$. ∎

THEOREM 1.7.3 (Baire, 1899). *Let (X, ϱ) be a complete metric space. Then the whole space X is of the second category in itself.*

The proof is based on the following lemma.

LEMMA 1.7.4. *Let (X, ϱ) be a complete metric space. Let $\{K_{r_n}(x_n)\}$ be a sequence of balls satisfying*

$$\lim_{n \to \infty} r_n = 0, \tag{1.7.1}$$

$$\overline{K_{r_n}(x_n)} \subset \overline{K_{r_{n-1}}(x_{n-1})}, \quad n = 1, 2, \ldots \tag{1.7.2}$$

Then

$$\bigcap_{n=1}^{\infty} \overline{K_{r_n}(x_n)} \neq \varnothing. \tag{1.7.3}$$

Proof. Let $\{y_n\}$ be an arbitrary sequence such that $y_n \in \overline{K_{r_n}(x_n)}$. By (1.7.2), $y_m \in \overline{K_{r_n}(x_n)}$ for all $m > n$. It follows from (1.7.1) that the sequence $\{y_n\}$ is fundamental. Since X is complete, the sequence $\{y_n\}$ converges to a certain $y \in X$. Since $y_m \in \overline{K_{r_n}(x_n)}$ for almost all m, $y \in \overline{K_{r_n}(x_n)}$ for $n = 1, 2, \ldots$ This means that $y \in \bigcap_{n=1}^{\infty} \overline{K_{r_n}(x_n)}$. ∎

Proof of Theorem 1.7.3. Suppose that X is of the first category, i.e., X can be represented as the union of nowhere dense sets A_n. Without loss of generality we assume that $A_n \subset A_{n+1}$. Since A_1 is nowhere dense, there exists a ball $K_{r_1}(x_1)$ whose closure is disjoint with A_1. Since A_2

is nowhere dense, there exists a ball $K_{r_2}(x_2)$, $\overline{K_{r_2}(x_2)} \subset \overline{K_{r_1}(x_1)}$ which is disjoint with A_2; in the same way we can choose a ball $K_{r_3}(x_3)$, $\overline{K_{r_3}(x_3)} \subset \overline{K_{r_2}(x_2)}$ which is disjoint with A_3. Generally, by induction, we can choose a ball $K_{r_n}(x_n)$, $\overline{K_{r_n}(x_n)} \subset \overline{K_{r_{n-1}}(x_{n-1})}$, which is disjoint with A_n. By Lemma 1.7.4, the intersection of all these balls is nonempty and, by construction, it is disjoint with the union of the sets A_n. Therefore, the union of A_n cannot be equal to the whole space X. ∎

1.8. SPACES OF MODULUS INTEGRABLE AND SQUARE INTEGRABLE FUNCTIONS

In this section we give other examples of metric spaces which are important for applications. We start with two engineering applications which make use of the concept of a metric space.

EXAMPLE 1.8.1. Consider a fuel container with a moving cover. Assume that the fuel flows in by the input valve with a time dependent velocity $x(t)$ and flows out by the output valve with a time dependent velocity $y(t)$.

The distance travelled by the cover in a time interval $[a, b]$ is expressed by the formula

$$\varrho_1(x, y) = k \int_a^b |x(t) - y(t)| \, dt$$

(the coefficient k depends on the diameter of the container). It can easily be shown that $\varrho_1(x, y)$ is a metric on the space of all continuous functions $C[a, b]$.

EXAMPLE 1.8.2. Let R be a resistance which separates two time-depending potentials $x(t)$ and $y(t)$. The heat production caused by the electric current in a time interval $[a, b]$ can be expressed by the formula

$$\tilde{\varrho}_2(x, y) = \frac{Q}{R} \int_a^b (x(t) - y(t))^2 dt,$$

where Q is the heat transformation coefficient. We show that the function

$\varrho_2(x, y) = \sqrt{\tilde{\varrho}_2(x, y)}$, is a metric in the space of all continuous functions $C[a, b]$. The validity of conditions (1) and (2) of the definition of a metric is immediate. A slightly more complicated proof is that of the triangle inequality. To prove it we first show the *Schwarz inequality*

$$\left|\int_a^b x(t)y(t)\,dt\right| \leq \left(\int_a^b x^2(t)\,dt\right)^{1/2}\left(\int_a^b y^2(t)\,dt\right)^{1/2}. \tag{1.8.1}$$

Observe that the function

$$F(s) = \int_a^b (x(t)+sy(t))^2\,dt$$

$$= \int_a^b x^2(t)\,dt + 2s\int_a^b x(t)y(t)\,dt + s^2\int_a^b y^2(t)\,dt$$

is a nonnegative quadratic function of s. Hence,

$$0 \geq \Delta = b^2 - 4ac = 4\left(\int_a^b x(t)y(t)\,dt\right)^2 - 4\int_a^b x^2(t)\,dt\int_a^b y^2(t)\,dt, \tag{1.8.2}$$

which immediately gives (1.8.1).

The Schwarz inequality implies the *Cauchy inequality*

$$\left(\int_a^b (x(t)+y(t))^2\,dt\right)^{1/2} \leq \left(\int_a^b x^2(t)\,dt\right)^{1/2} + \left(\int_a^b y^2(t)\,dt\right)^{1/2}. \tag{1.8.3}$$

Since both sides of inequality (1.8.3) are nonnegative, it holds if and only if the square of the expression on the left is not greater than the square of the expression on the right, i.e.,

$$\int_a^b x^2(t)\,dt + 2\int_a^b x(t)y(t)\,dt + \int_a^b y^2(t)\,dt$$

$$\leq \int_a^b x^2(t)\,dt + 2\left(\int_a^b x^2(t)\,dt\right)^{1/2}\left(\int_a^b y^2(t)\,dt\right)^{1/2} + \int_a^b y^2(t)\,dt.$$

1.8. Spaces of modulus integrable and square integrable functions 31

After elementary arithmetic transformations we get

$$\int_a^b x(t)y(t)\,dt \leqslant \left(\int_a^b x^2(t)\,dt\right)^{1/2} \left(\int_a^b y^2(t)\,dt\right)^{1/2},$$

which is exactly inequality (1.8.1).

Substituting in inequality (1.8.3) the expression $x(t)-z(t)$ for $x(t)$ and $z(t)-y(t)$ for $y(t)$, we immediately obtain the triangle inequality for the metric $\varrho_2(x,y)$.

Hence, the expressions

$$\varrho_1(x,y) = \int_a^b |x(t)-y(t)|\,dt$$

and

$$\varrho_2(x,y) = \left(\int_a^b |x(t)-y(t)|^2\,dt\right)^{1/2}$$

define metrics in the space $C[a,b]$. Unfortunately, the space $C[a,b]$ is not complete in any of these metrics. To see this consider the sequence of functions

$$x_n(t) = \begin{cases} -1 & \text{for } a \leqslant t \leqslant \dfrac{a+b}{2} - \dfrac{1}{n}, \\ n\left(t - \dfrac{a+b}{2}\right) & \text{for } \dfrac{a+b}{2} - \dfrac{1}{n} < t \leqslant \dfrac{a+b}{2} + \dfrac{1}{n}, \\ +1 & \text{for } \dfrac{a+b}{2} + \dfrac{1}{n} < t \leqslant b. \end{cases}$$

It can easily be verified that the sequence $\{x_n(t)\}$ is fundamental. It is convergent in the metrics ϱ_1 and ϱ_2 to the function

$$x(t) = \begin{cases} -1 & \text{for } a \leqslant t < \dfrac{a+b}{2}, \\ 0 & \text{for } t = \dfrac{a+b}{2}, \\ +1 & \text{for } \dfrac{a+b}{2} < t \leqslant b. \end{cases}$$

The function $x(t)$ is not continuous and hence the space $C[a, b]$ is not complete either in ϱ_1 or in ϱ_2.

The completion of the space $C[a, b]$ in the metric ϱ_1 is called the *space of modulus integrable functions*. Elements of this space are called *modulus integrable functions*. The space of all modulus integrable functions on the interval $[a, b]$ will be denoted by $L^1[a, b]$. The completion of the space $C[a, b]$ in the metric ϱ_2 will be called the *space of square integrable functions on the interval* $[a, b]$. Elements of this space will be called *square integrable functions*. The space of all square integrable functions on the interval $[a, b]$ will be denoted by $L^2[a, b]$.

The definitions of the spaces $L^1[a, b]$ and $L^2[a, b]$ can be extended to the case $a = -\infty$ or $b = +\infty$. Namely, we take the same definitions of ϱ_1 and ϱ_2 considering the fact that in this case the integrals are improper. The spaces $L^1[a, b]$, $L^2[a, b]$ are then obtained by taking the completion of the space $C_f[a, b]$ of all continuous functions $x(t)$ with finite supports (i.e. $\sup \{|t|: x(t) \neq 0\} < +\infty$) in the corresponding metrics ϱ_1 and ϱ_2.

1.9. Basic Facts of Measure Theory and the Lebesgue Integral

In the previous section we introduced modulus integrable functions and square integrable functions. However, the definitions given there are far from natural. It would be of interest to define modulus integrable functions (square integrable functions) as those for which the module (square of the module) is integrable.

Unfortunately, the Riemann integral is not an adequate tool for this purpose. The first problem is that in the class $\mathscr{R}_1$ ($\mathscr{R}_2$) of modulus (square) Riemann integrable functions the metric ϱ_1 (ϱ_2) does not necessarily distinguish certain functions.

This disadvantage can easily be eliminated by the following frequently used procedure. Namely, we identify functions $x(t)$ and $y(t)$ if $\varrho_1(x, y) = 0$ ($\varrho_2(x, y) = 0$). In other words, instead of considering $\mathscr{R}_1$ ($\mathscr{R}_2$) we consider $\tilde{\mathscr{R}}_1$ ($\tilde{\mathscr{R}}_2$) whose elements are sets of functions $\tilde{x}(t)$ such that functions $x_1(t)$ and $x_2(t)$ belong to the same $\tilde{x}(t)$ if and only if $\varrho_1(x_1, x_2)$

1.9. Measure theory and the Lebesgue integral

$= 0$ ($\varrho_2(x_1, x_2) = 0$). A metric $\tilde\varrho_1$ ($\tilde\varrho_2$) in the space $\tilde{\mathscr{R}}_1$ ($\tilde{\mathscr{R}}_2$) is given by the formula

$$\tilde\varrho_1(\tilde x, \tilde y) = \varrho_1(x, y), \quad (\tilde\varrho_2(\tilde x, \tilde y) = \varrho_2(x, y)),$$

where x, y are arbitrary elements such that $x \in \tilde x$, $y \in \tilde y$. It can be shown that $\tilde\varrho_1$ and $\tilde\varrho_2$ are both well-defined, i.e., $\tilde\varrho_1$ and $\tilde\varrho_2$ do not depend upon the choice of the elements $x \in \tilde x$, $y \in \tilde y$, which are called *representatives* of $\tilde x$ and $\tilde y$. Hence the function $\tilde\varrho_1$ ($\tilde\varrho_2$) is a metric on the set of all equivalence classes $\tilde{\mathscr{R}}_1$ ($\tilde{\mathscr{R}}_2$). Unfortunately, the spaces thus obtained are not complete. This can be seen by taking the fundamental sequence of continuous functions

$$x_n(t) = \begin{cases} \sqrt[4]{n} & \text{for } a \leqslant t < a + \dfrac{1}{n}, \\ \sqrt[4]{\dfrac{1}{|t-a|}} & \text{for } a + \dfrac{1}{n} \leqslant t \leqslant b, \end{cases}$$

which does not converge to a function integrable in the sense of Riemann (since unbounded functions are not Riemann integrable).

If, in addition to Riemann integrable functions, we consider also functions which have improper integrals, then in such an enlarged space the sequence $\{x_n\}$ converges to the function

$$x(t) = \frac{1}{\sqrt[4]{|t-a|}}$$

in the metrics ϱ_1 and ϱ_2. However, this enlarged space is not complete, either. To see this, consider the following example.

Order all the rational numbers from the interval $[a, b]$ into a sequence $\{w_n\}$. Let

$$y_n(t) = \sqrt[4]{\frac{1}{|t-w_n|}}$$

and

$$x_n(t) = \sum_{i=0}^{n} \frac{1}{i!} y_i(t).$$

Since the convergence of the sequence $\left\{\dfrac{1}{n!}\right\}$ is very rapid, the sequence $\{x_n\}$ is fundamental both in the metric ϱ_1 and in the metric ϱ_2. Since its elements are increasing functions, it converges or at some points diverges to $+\infty$. Let $x(t)$ denote the limit of the sequence $\{x_n\}$. In a neighborhood of each point t_0 the function $x(t)$ is unbounded. Hence, an improper Riemann integral of the function $x(t)$ does not exist.

The above examples motivate a more general notion of integrability than the integrability in the sense of Riemann. In a book devoted to functional analysis one cannot give a complete description of measure theory and integral theory. Therefore we present without proof only those elements of those theories which will be useful in the sequel.

We start with some general notions, which allow us to obtain the Lebesgue measure as a particular measure.

Let Ω be a given set. A family Σ of subsets of the set Ω is called a σ-*field* (a *countably additive field*) if $\Omega \setminus A \in \Sigma$ for every $A \in \Sigma$ and $\bigcup_{n=1}^{\infty} A_n \in \Sigma$ for every sequence of sets $\{A_n\}$, $A_1, \ldots, A_n, \ldots \in \Sigma$. A nonnegative function μ defined on elements of the family Σ is called a *measure* if it is a countably additive function, i.e.,

$$\mu\left(\bigcup_{n=1}^{\infty} A_n\right) = \sum_{n=1}^{\infty} \mu(A_n)$$

for each sequence $\{A_n\}$ of pairwise disjoint sets $A_1, \ldots, A_n, \ldots \in \Sigma$. We admit $+\infty$ as a value of the function μ.

A function $x(t)$ defined on Ω which takes on values in the field of real (or complex) numbers is said to be *measurable* if, for every open set U of real (or complex) numbers, the inverse image $x^{-1}(U) = \{t\colon x(t) \in U\}$ belongs to Σ. The sum as well as the product of any two measurable functions is measurable.

We say that a sequence $\{x_n(t)\}$ of measurable functions *converges almost everywhere* to a measurable function $x(t)$ if it converges to that function everywhere except at most a certain set A of measure zero.

Let $x(t)$ be a real-valued function defined on Ω which is measurable and nonnegative. Let $0 = a_1 < a_2 < \ldots < a_n$ be an arbitrary system

1.9. Measure theory and the Lebesgue integral

of reals. The number

$$S = \sup\left[\sum_{i=1}^{n-1} a_i \mu(\{t: a_i \leqslant x(t) \leqslant a_{i+1}\}) + a_n \mu(\{t: x(t) \geqslant a_n\})\right]$$

we shall call the *integral* of the function $x(t)$ and we shall denote it by

$$S = \int_\Omega x(t)\,d\mu.$$

Let $y(t)$ be any real-valued measurable function. By $y_+(t) = \frac{1}{2}(|y(t)|+y(t))$ and $y_-(t) = \frac{1}{2}(|y(t)|-y(t))$ we denote the positive part and the negative part of $y(t)$, respectively. If both functions are *integrable* (i.e., both integrals are finite) we define the *integral* of the function $y(t)$ by the formula

$$\int_\Omega y(t)\,d\mu \stackrel{df}{=} \int_\Omega y_+(t)\,d\mu - \int_\Omega y_-(t)\,d\mu.$$

Note that the above definition of the integral implies that a function $x(t)$ is integrable if and only if the function $|x(t)| = x_+(t)+x_-(t)$ is integrable.

A complex-valued function $z(t) = x(t)+iy(t)$ is said to be *integrable* if the real part $x(t)$ and the imaginary part $y(t)$ are both integrable. By the integral

$$\int_\Omega z(t)\,d\mu$$

we mean the number

$$\int_\Omega x(t)\,d\mu + i\int_\Omega y(t)\,d\mu.$$

The integral defined above is a linear mapping which maps the set of all integrable functions into the set of real (or complex) numbers, i.e.,

$$\int_\Omega (ax(t)+by(t))\,d\mu = a\int_\Omega x(t)\,d\mu + b\int_\Omega y(t)\,d\mu,$$

where a, b are numbers.

Let E be a subset of Ω and an element of Σ. By the integral over the set E we mean

$$\int_E x(t)\,d\mu \stackrel{df}{=} \int_\Omega \chi_E(t)x(t)\,d\mu,$$

where χ_E is the *characteristic function* of the set E, i.e.,

$$\chi_E(t) = \begin{cases} 1 & \text{for } t \in E, \\ 0 & \text{for } t \notin E. \end{cases}$$

It is an immediate consequence of the definition that, for disjoint sets E_1 and E_2, we have

$$\int_{E_1 \cup E_2} x(t)\,d\mu = \int_{E_2} x(t)\,d\mu + \int_{E_1} x(t)\,d\mu.$$

Given two real-valued integrable functions $x(t)$, $y(t)$ such that $x(t) \leq y(t)$, we get

$$\int_\Omega x(t)\,d\mu \leq \int_\Omega y(t)\,d\mu.$$

The Riemann integral also has the above properties. We omit the proof of the following important theorem.

THEOREM 1.9.1 (Lebesgue). *Let $\{x_n(t)\}$ be a sequence of measurable functions which converges almost everywhere to a function $x(t)$. Moreover, $|x_n(t)| \leq y(t)$ where $y(t)$ is an integrable function. Then the function $x(t)$ is integrable and*

$$\lim_{n \to \infty} \int_\Omega x_n(t)\,d\mu = \int_\Omega x(t)\,d\mu.$$

The following corollary is a consequence of Theorem 1.9.1.

COROLLARY 1.9.2. *Consider a set Ω with a measure μ defined on it. Assume that the measure of the whole Ω is finite. Let $\{x_n(t)\}$ be a uniformly convergent sequence of integrable functions. Then the limit of this sequence, $x(t)$, is an integrable function.*

1.9. Measure theory and the Lebesgue integral

Proof. By hypothesis, the sequence $\{x_n(t)\}$ converges almost everywhere to the function $x(t)$ and there exists an N such that for $n > N$

$$|x_N(t) - x_n(t)| \leqslant 1$$

for all t. This implies that $|x_n(t)| \leqslant |x_N(t)| + 1$. Since the measure of the whole Ω is finite, the function $|x_N(t)| + 1$ is integrable. Hence, by Theorem 1.9.1, the function $x(t)$ is integrable. ∎

By $\tilde{L}^1(\Omega, \Sigma, \mu)$ ($\tilde{L}^2(\Omega, \Sigma, \mu)$) we denote the set of all modulus integrable (square integrable) functions.

Let us partition the set $\tilde{L}^1(\Omega, \Sigma, \mu)$ ($\tilde{L}^2(\Omega, \Sigma, \mu)$) into disjoint sets in such a way that two functions, $\tilde{x}_1(t)$ and $\tilde{x}_2(t)$, belong to the same set x if and only if they differ only on a set of measure zero. The sets x are called *equivalence classes* and each function $\tilde{x}(t)$ such that $\tilde{x}(t) \in x$ is called a *representative* of the class x. The spaces of equivalence classes will be denoted by $L^1(\Omega, \Sigma, \mu)$ ($L^2(\Omega, \Sigma, \mu)$). Let x and y be elements of the space $L^1(\Omega, \Sigma, \mu)$ ($L^2(\Omega, \Sigma, \mu)$). The distance between x and y is given by the formula

$$\varrho_1(x, y) = \int_\Omega |\tilde{x}(t) - \tilde{y}(t)| \, d\mu$$

$$\left(\varrho_2(x, y) = \left(\int_\Omega |\tilde{x}(t) - \tilde{y}(t)|^2 \, d\mu \right)^{1/2} \right),$$

where $\tilde{x}(t)$, $\tilde{y}(t)$ are arbitrary representatives of the classes x and y.

Since it is not misleading, we regard elements of the spaces $L^1(\Omega, \Sigma, \mu)$ and $L^2(\Omega, \Sigma, \mu)$ as functions and by the integral $\int_\Omega x(t) \, d\mu$ we mean the integral of an arbitrary representative of the class $x(t)$. Observe that ϱ_1 (ϱ_2) is a metric on $L^1(\Omega, \Sigma, \mu)$ ($L^2(\Omega, \Sigma, \mu)$). The proof of this fact for the space $L^1(\Omega, \Sigma, \mu)$ is immediate. For the space $L^2(\Omega, \Sigma, \mu)$ the proof is similar to that given in Example 1.8.2 and relies on the Schwarz inequality and the Cauchy inequality.

Now we prove

THEOREM 1.9.3. *If the measure $\mu(\Omega)$ of the whole set Ω is finite, then the spaces $L^p(\Omega, \Sigma, \mu)$ for $p = 1$ or $p = 2$ are complete.*

Proof. Let $\{x_n\}$ be a fundamental sequence of elements of the space $L^p(\Omega, \Sigma, \mu)$ for $p = 1$ or $p = 2$. Thus, it contains a subsequence $\{x_{n_k}\}$ such that

$$\varrho_p(x_{n_k}, x_{n_{k+1}}) < \frac{1}{2^{2k+1}},$$

i.e.,

$$\int_\Omega |x_{n_k}(t) - x_{n_{k+1}}(t)|^p d\mu \leq \frac{1}{2^{kp+p}}. \tag{1.9.1}$$

By formula (1.9.1), $\mu(A_k) \leq 1/2^k$, where

$$A_k = \left\{ t : |x_{n_{k+1}}(t) - x_{n_k}(t)| > \frac{1}{2^k} \right\}.$$

Let

$$B_k = \bigcap_{i=k}^{\infty} (\Omega \setminus A_i) = \Omega \setminus \left(\bigcup_{i=k}^{\infty} A_i \right).$$

Then we obtain

$$\mu(B_k) \geq \mu(\Omega) - \sum_{i=k}^{\infty} \mu(A_i) \geq \mu(\Omega) - \frac{1}{2^{k-1}}$$

and

$$|x_{n_{m+1}}(t) - x_{n_m}(t)| \leq \frac{1}{2^m}$$

for all $m > k$ and for $t \in B_k$. Hence the sequence $\{x_{n_m}\}$ converges uniformly on the set B_k.

Obviously, $B_k \subset B_{k+1}$ and $\mu\left(\bigcup_{k=1}^{\infty} B_k \right) = \mu(\Omega)$. This implies that the sequence of numbers $\{x_{n_m}(t)\}$ is fundamental for almost all t. Thus, the sequence $\{x_{n_m}(t)\}$ converges almost everywhere to a certain $x(t)$, which is a measurable function.

The sequence $\{x_{n_m}\}$ is fundamental as a subsequence of the sequence $\{x_n\}$. For any $\varepsilon > 0$, there exists an index $N(\varepsilon)$ such that

$$\int_\Omega |x_{n_m}(t) - x_{n_{m'}}(t)| d\mu < \varepsilon$$

for $m, m' > N(\varepsilon)$. Since the integrand is nonnegative,

$$\int_{B_k} |x_{n_m}(t) - x_{n_{m'}}(t)| \, d\mu < \varepsilon$$

for every k. Passing to the limit with m' tending to $+\infty$, we get

$$\int_{B_k} |x_{n_m}(t) - x(t)| \, d\mu \leq \varepsilon$$

for every k. Furthermore,

$$\int_{\cup B_k} |x_{n_m}(t) - x(t)| \, d\mu \leq \varepsilon$$

since the family B_k forms an ascending sequence of sets. Thus,

$$\int_{\Omega} |x_{n_m}(t) - x(t)| \, d\mu \leq \varepsilon. \tag{1.9.2}$$

This implies that $x_{n_m} - x \in L^p(\Omega, \Sigma, \mu)$ and by the Cauchy inequality an element x,

$$x = (x - x_{n_m}) + x_{n_m},$$

belongs to $L^p(\Omega, \Sigma, \mu)$. Since ε was arbitrary in formula (1.9.2), the sequence $\{x_{n_m}\}$ converges to x. To complete the proof we have to show that $x_n \to x$. Select any $\varepsilon > 0$. Since the sequence $\{x_n\}$ is fundamental and $x_{n_m} \to x$, there exists an index $n(\varepsilon)$ such that $\varrho_p(x_n, x_{n'}) < \varepsilon$ for $n, n' > n(\varepsilon)$ and $\varrho_p(x_{n_m}, x) < \varepsilon$ for $n_m > n(\varepsilon)$. Putting $n' = n_m$, we get

$$\varrho_p(x_n, x) \leq \varrho_p(x_n, x_{n_m}) + \varrho_p(x_{n_m}, x) < 2\varepsilon.$$

Thus, $x_n \to x$. ∎

A measure μ defined on a σ field Σ of subsets of a set Ω is called *σ-finite* if Ω can be represented as the union of sets Ω_n, $n = 1, 2, \ldots$,

$$\Omega = \bigcup_{n=1}^{\infty} \Omega_n,$$

such that all Ω_n belong to Σ and $\mu(\Omega_n) < +\infty$ for $n = 1, 2, \ldots$

We say that a set $A \in \Sigma$ is of *σ-finite measure* if the measure μ restricted to the subsets of A is σ-finite.

THEOREM 1.9.4. *Let μ be a σ-finite measure. The space $L^p(\Omega, \Sigma, \mu)$ for $p = 1$ or $p = 2$ is complete.*

Proof. Let $\{x_n\}$ be a fundamental sequence in the space $L^p(\Omega, \Sigma, \mu)$ ($p = 1$ or 2). By assumption, $\Omega = \bigcup_{k=1}^{\infty} \Omega_k$, where the measure of the sets Ω_k, $k = 1, 2, \ldots$, is finite. Let $\Sigma_k = \{A\colon A \in \Sigma, A \subset \Omega_k\}$ $= \{A\colon A = \Omega_k \cap B, B \in \Sigma\}$ and let μ_k be the restriction of the measure μ to the family Σ_k. In view of Theorem 1.9.3, the spaces $L^p(\Omega_k, \Sigma_k, \mu_k)$ are complete. Hence, the sequence of functions $\{x_n \chi_{\Omega_k}\}$ converges to a certain function x^k which is defined on Ω_k. Clearly, on the intersection $\Omega_k \cap \Omega_i$ the functions x^k and x^i may differ only on a set of measure zero, and up to a certain set of measure zero the function x^k can be regarded as the restriction to Ω_k of a certain function $x(t)$ defined on the whole set Ω, $\Omega = \bigcup_{k=1}^{\infty} \Omega_k$.

Choose any $\varepsilon > 0$. Since the sequence $\{x_n\}$ is fundamental in the space $L^p(\Omega, \Sigma, \mu)$ ($p = 1$ or 2), there exists an $n(\varepsilon)$ such that

$$\int_{\Omega} |x_n(t) - x_m(t)|^p d\mu < \varepsilon$$

for $n, m > n(\varepsilon)$. Hence

$$\int_{\Omega_1 \cup \ldots \cup \Omega_k} |x_n(t) - x_m(t)|^p d\mu < \varepsilon.$$

Passing to the limit with m tending to $+\infty$, we get

$$\int_{\Omega_1 \cup \ldots \cup \Omega_k} |x_n(t) - x(t)|^p d\mu \leqslant \varepsilon.$$

The above inequality holds for arbitrary k. So

$$\int_{\Omega} |x_n(t) - x(t)|^p d\mu \leqslant \varepsilon.$$

Just as in Theorem 1.9.3, using the Cauchy inequality, we can show that $x \in L^p(\Omega, \Sigma, \mu)$ and the sequence $\{x_n\}$ converges to x in the metric of the space $L^p(\Omega, \Sigma, \mu)$. ∎

1.9. Measure theory and the Lebesgue integral

THEOREM 1.9.5. *The space $L^p(\Omega, \Sigma, \mu)$, $p = 1$ or 2, is complete for an arbitrary measure μ.*

Proof. Let $\{x_n\}$ be a fundamental sequence in the space $L^p(\Omega, \Sigma, \mu)$. Since each function $x_n(t)$ belongs to $L^p(\Omega, \Sigma, \mu)$, the set $A_{n,k} = \{t: |x_n(t)| > 1/k\}$ is of finite measure for every k, and consequently the support A_n of the function x_n, $A_n = \{t: x_n(t) \neq 0\} = \bigcup_{k=1}^{\infty} A_{n,k}$, is a set of σ-finite measure. Thus, the set Ω_0, $\Omega_0 = \bigcup_{n=1}^{\infty} A_n$, is a set of σ-finite measure. Let x_n^0 denote the restriction of the function x_n to the set Ω_0. By Theorem 1.9.3, the sequence $\{x_n^0\}$ converges to a certain function $x^0(t)$ defined on Ω_0. Letting

$$x(t) = \begin{cases} x^0(t) & \text{for } t \in \Omega_0, \\ 0 & \text{for } t \notin \Omega_0, \end{cases}$$

we obtain $x_n \to x$ since $x_n(t) = 0$ for $n = 1, 2, \ldots$ and $t \notin \Omega_0$. ∎

For different sets Ω and measures μ we obtain different spaces.

EXAMPLE 1.9.1. Let Ω be the set of all natural numbers and let Σ be the family of all subsets of Ω. Let

$\mu(E) = $ the cardinality of the set E.

Every function defined on Ω can be treated as a sequence $\{x(n)\} = \{x_n\}$. Every function of this form is measurable and

$$\int_\Omega |x(t)|^p d\mu = \sum_{n=1}^{\infty} |x_n|^p.$$

In this case the space $L^p(\Omega, \Sigma, \mu)$ is denoted briefly by l^p.

Now the question arises whether a family Σ and a measure μ can be defined on the interval $[a, b]$ in such a way that the space $L^p(\Omega, \Sigma, \mu)$ could be identified with the space $L^p[a, b]$ introduced in the previous section.

To get a positive answer to this question we must construct a measure μ satisfying the following conditions:

(I) for continuous functions the integral with respect to the measure μ coincides with the Riemann integral,

(II) the space of all continuous functions is dense in the space $L^p(\Omega, \Sigma, \mu)$ ($p = 1$ or 2).

If, for a given measure, conditions (I) and (II) are satisfied, the space $L^p(\Omega, \Sigma, \mu)$ is the completion of the space of all continuous functions in the corresponding metric and hence, by definition, $L^p(\Omega, \Sigma, \mu)$ is equal to the space $L^p[a, b]$.

The construction of a σ-field Σ and a measure μ satisfying the conditions (I) and (II) is based on the concept of outer measure and the Carathéodory theorem.

Let a set Ω be given. A nonnegative function μ_e (which can take the value $+\infty$), defined on the family of all subsets of Ω is called an *outer measure* if μ_e satisfies the following conditions:

(a) $\mu_e(\emptyset) = 0$,

(b) if $A \subset \bigcup_{n=1}^{\infty} A_n$, then $\mu_e(A) \leq \sum_{n=1}^{\infty} \mu_e(A_n)$.

THEOREM 1.9.6 (Carathéodory). *Let μ_e be an outer measure on Ω and let Σ be a family of those subsets A of Ω which satisfy the equality*

$$\mu_e(Z) = \mu_e(A \cap Z) + \mu_e(Z \setminus A)$$

for each $Z \subset \Omega$. Then Σ is a σ-field. The outer measure μ_e considered on Σ is a measure. Sets of outer measure zero belong to Σ and are of measure zero.

We omit the proof of this theorem; it can be found e.g. in Sikorski (1969).

Now we define the outer Lebesgue measure on the interval $[a, b]$ by the formula

$$\mu_e(E) = \inf \left\{ \sum_{n=1}^{\infty} |I_n| : I_n \text{ denotes an open interval } (a_n, b_n), |I_n| = |b_n - a_n|, E \subset \bigcup_{n=1}^{\infty} I_n \right\}.$$

The measure induced by the outer Lebesgue measure μ_e is called

1.9. Measure theory and the Lebesgue integral

the *Lebesgue measure* and is denoted by μ. On the basis of this definition it can be verified that

$$\mu((c,d)) = \mu([c,d)) = \mu((c,d]) = \mu([c,d]) = d-c.$$

Thus, all intervals, open, closed, or half-open, belong to the σ-field Σ, called the *σ-field of measurable sets*.

Let $x(t)$ be a nonnegative continuous function such that the inverse image of an arbitrary point is a finite set. Let $0 = a_1 < a_2 < \ldots < a_n$ be a finite sequence of numbers. The sets $\{t: a_i \leqslant x(t) < a_{i+1}\}$ are the unions of finite numbers of intervals. Thus

$$\sum_{i=1}^{n-1} a_i \mu(\{t: a_i \leqslant x(t) < a_{i+1}\}) + a_n \mu(\{t: a_n \leqslant x(t)\})$$

is a certain lower Darboux sum. This implies that the integral with respect to the measure μ is not greater than the Riemann integral. By increasing n, we condense subdivisions $0 = a_1 < a_2 < \ldots < a_n$ so as to obtain a certain normal sequence of subdivisions of the interval $[a,b]$. For such a sequence the lower Darboux sum tends to the Riemann integral. This entails equality between the Riemann integral and the Lebesgue integral for nonnegative functions with finite inverse images. Such functions approximate uniformly arbitrary nonnegative continuous functions. The theorems on passing to limits for uniformly convergent sequences hold for both types of integrals. Hence, both types of integrals coincide for nonnegative continuous functions. Moreover, the integral of the difference of any two functions is equal to the difference of the integrals of those functions for both types of integrals. Thus, both types of integrals coincide for arbitrary continuous function. This proves (I).

We forgo here the proof of (II), which relies on some properties of measurable functions which are of no use in the sequel.

So far we have considered only the spaces $L^p(\Omega, \Sigma, \mu)$ for $p = 1$ or $p = 2$. In most textbooks on functional analysis the spaces $L^p(\Omega, \Sigma, \mu)$ are also considered for $1 < p < +\infty$ with the metrics

$$\varrho_p(x,y) = \left(\int_\Omega |x(t)-y(t)|^p d\mu\right)^{1/p}.$$

The proof that $\varrho_p(x, y)$ is a metric requires stronger inequalities than the Schwarz inequality and the Cauchy inequality, namely, the Hölder inequality and the Minkowski inequality. The *Hölder inequality* is of the form

$$\int_\Omega |x(t)y(t)|\,d\mu \leq \left(\int_\Omega |x(t)|^p d\mu\right)^{1/p} \left(\int_\Omega |y(t)|^q\,d\mu\right)^{1/q},$$

where $1 < p, q < +\infty$, and both are related by the equality

$$\frac{1}{p} + \frac{1}{q} = 1,$$

and the *Minkowski inequality* is of the form

$$\left(\int_\Omega |x(t)+y(t)|^p d\mu\right)^{1/p} \leq \left(\int_\Omega |x(t)|^p d\mu\right)^{1/p} + \left(\int_\Omega |y(t)|^q d\mu\right)^{1/q},$$

where p and q satisfy the same relation as before. The proofs of these inequalities can be found in every course of functional analysis, for example in Dunford and Schwartz (1958). We omit those proofs because, in practice, we usually encounter the spaces $L^p(\Omega, \Sigma, \mu)$ for $p = 1$ and $p = 2$.

It is worth mentioning that the proofs of completeness of the spaces $L^p(\Omega, \Sigma, \mu)$ for $1 < p < +\infty$ are similar to those for $p = 1$ and $p = 2$.

One more space is related to the concept of a measure. It is the space of essentially bounded functions.

EXAMPLE 1.9.2. Suppose we are given a set Ω and a σ-field Σ of subsets of Ω and a measure μ defined on Σ. Let $\tilde{M}(\Omega, \Sigma, \mu)$ denote the set of all measurable functions $x(t)$ such that for each of them there exists a set E_x of measure zero, $\mu(E_x) = 0$, such that the function $x(t)$ is bounded on the set $\Omega \setminus E_x$. Just as in the space $L^p(\Omega, \Sigma, \mu)$, we identify functions which differ only on a set of measure zero. The space of all equivalence classes is denoted by $M(\Omega, \Sigma, \mu)$. In this space a metric is given by the formula

$$\varrho(x, y) = \text{ess sup}|x(t)-y(t)| = \inf_{E, \mu(E)=0} \sup_{t \in \Omega \setminus E} |x(t)-y(t)|.$$

As can easily be seen the function $\varrho(x, y)$ satisfies all the conditions of the definition of a metric. The space $M(\Omega, \Sigma, \mu)$ is complete in this metric. To see this, let $\{x_n\}$ be a fundamental sequence of elements of the space $M(\Omega, \Sigma, \mu)$. Denote by $A_{n,m}$ a set of measure zero such that

$$\varrho(x_n, x_m) = \sup_{t \in \Omega \setminus A_{n,m}} |x_n(t) - x_m(t)|.$$

Let $A = \bigcup_{n=1}^{\infty} \bigcup_{m=1}^{\infty} A_{n,m}$. By the countable additivity of μ, the set A is of measure zero. On the set $\Omega \setminus A$ the sequence of functions $\{x_n(t)\}$ is fundamental in uniform convergence. Denote by $x^0(t)$ its limit function and by $x(t)$ an arbitrary extension of $x^0(t)$ to the whole set Ω. Taking $E_x = A$, we obtain $x \in M(\Omega, \Sigma, \mu)$. Since

$$0 \leqslant \lim_{n \to \infty} \varrho(x, x_n) \leqslant \lim_{n \to \infty} \sup_{t \in \Omega \setminus A} |x(t) - x_n(t)| = 0,$$

we obtain $x_n \to x$.

Let Ω be the set of natural numbers, Σ the family of all subsets of Ω and $\mu(A)$ the cardinality of A. Then the space $M(\Omega, \Sigma, \mu)$ is the space of all bounded sequences. It will be denoted briefly by m (compare Example 1.1.18).

Let $\Omega = [a, b]$, Σ the σ-field of Lebesgue measurable subsets of Ω, and μ the Lebesgue measure. In this case the space $M(\Omega, \Sigma, \mu)$ will be denoted briefly by $M[a, b]$.

1.10. SEPARABLE SPACES

Let (X, ϱ) be a metric space. The space (X, ϱ) is called *separable* if there exists a sequence $\{x_n\}$ of elements of X which is dense in X. In other words, for every $x \in X$ and every $\varepsilon > 0$ there exists an index n_0 such that $\varrho(x, x_{n_0}) < \varepsilon$.

THEOREM 1.10.1. *A subspace of a separable space is separable.*

Proof. Let (X, ϱ) be a separable space and let Y be a subspace of X. Denote by $x_{n,m}$ an arbitrary point of Y such that $\varrho(x_n, x_{n,m}) < 1/m$ if such a point exists. Consider the set $\{x_{n,m}\}$. Clearly, the point $x_{n,m}$ need not exist for each pair (n, m). The set $\{x_{n,m}\}$ forms a sequence which is dense in Y. Let ε be an arbitrary positive number. Take m such that

$2/m < \varepsilon$. Since the sequence $\{x_n\}$ is dense in X, for every $y \in Y \subset X$ there exists an x_{n_0} such that $\varrho(y, x_{n_0}) < 1/m$. Then
$$\varrho(y, x_{n_0, m}) \leqslant \varrho(y, x_{n_0}) + \varrho(x_{n_0}, x_{n_0, m}) < 2/m < \varepsilon,$$
which completes the proof. ∎

THEOREM 1.10.2. *Let (X, ϱ) be a metric space. If X is a separable space, then the completion $\tilde{X}$ of the space X is also separable.*

Proof. Let $\{x_n\}$ be a dense sequence of elements of X. Select any $x \in \tilde{X}$ and $\varepsilon > 0$. Since X is dense in $\tilde{X}$, there exists an $x_0 \in X$ such that $\varrho(x, x_0) < \varepsilon/2$. Moreover, since X is separable, there exists an index n_0 such that $\varrho(x_0, x_{n_0}) < \varepsilon/2$. Hence, $\varrho(x, x_{n_0}) \leqslant \varrho(x, x_0) + \varrho(x_0, x_{n_0}) < \varepsilon$. ∎

For a given metric space (X, ϱ), a set Ω contained in X is called an *ε-net* if for every $x \in X$ there is a $y \in \Omega$ such that $\varrho(x, y) < \varepsilon$.

THEOREM 1.10.3. *A metric space (X, ϱ) is separable if and only if for every $\varepsilon > 0$ there exists a countable ε-net Ω_0.*

Proof. Necessity. Assume that X is separable. By definition, X contains a countable dense subset Ω_0. Clearly, Ω_0 forms an ε-net for every $\varepsilon > 0$.

Sufficiency. Let Ω_ε be a countable ε-net. Then $\Omega_0 = \bigcup_{m=1}^{\infty} \Omega_{1/m}$ is countable and dense in X. ∎

COROLLARY 1.10.4. *Let (X, ϱ) be a metric space. If there exist an uncountable set Ω and a number $\varepsilon > 0$ such that $\varrho(x, y) \geqslant \varepsilon$ for $x, y \in \Omega$ and $x \neq y$, then the space X is not separable.*

Proof. Let $\Omega_{\varepsilon/2}$ be an arbitrary $\varepsilon/2$-net. Assign to each element $x \in \Omega_{\varepsilon/2}$, an element $y \in \Omega$ such that $\varrho(x, y) < \varepsilon/2$, if one exists. This correspondence is defined on part of $\Omega_{\varepsilon/2}$ only. Moreover, by the definition of Ω, an element $y \in \Omega$ is uniquely determined (though the cor-

1.10. Separable spaces

respondence is not necessarily one-to-one). This implies that the cardinality of $\Omega_{\varepsilon/2}$ is greater than the cardinality of Ω. Hence, $\Omega_{\varepsilon/2}$ is an uncountable set and, in view of Theorem 1.10.3, the space X is not separable. ∎

Metric spaces which are not separable are called *nonseparable spaces*. Now we give examples of separable and nonseparable spaces.

EXAMPLE 1.10.1. The space $C[a, b]$ is separable. Let $a = w_0 < w_1 < \ldots < w_{n-1} < w_n = b$. By $L(w_0, \ldots, w_n; r_0, \ldots, r_n)$ we denote a piecewise-linear curve which takes on the values r_i at w_i, i.e., L is a function $f(x)$ such that
(1) $f(w_i) = r_i$, $i = 1, 2, \ldots, n$,
(2) $f(x)$ is a linear function on each interval $[w_{i-1}, w_i]$.

The set of all functions $L(w_0, \ldots, w_n; r_0, \ldots, r_n)$ with $w_1, \ldots, w_{n-1}$ and $r_0, \ldots, r_n$ rational is dense in $C[a, b]$ and forms a sequence. Hence, the space $C[a, b]$ is separable.

EXAMPLE 1.10.2. The space $L^p[a, b]$ ($p = 1$ or $p = 2$) is separable. Divide the interval $[a, b]$, $a = w_0 < w_1 < \ldots < w_n = b$, and define the function $S(w_0, \ldots, w_n; r_1, \ldots, r_n; x)$ by the formula

$$S(w_0, w_1, \ldots, w_n; r_1, \ldots, r_n; x)$$
$$= \begin{cases} r_i & \text{for } w_{i-1} < x \leqslant w_i \quad (i = 1, \ldots, n), \\ r_1 & \text{for } x = w_0. \end{cases}$$

The set of all functions S for which $w_1, \ldots, w_{n-1}, r_1, \ldots, r_n$ are rational, is dense in $L^p[a, b]$ and countable. This implies the separability of the space $L^p[a, b]$.

EXAMPLE 1.10.3. The space l^p is separable. Let $\mathfrak{U}$ denote the set of elements $\{r_1, \ldots, r_n, 0, 0, \ldots\}$, where $r_1, \ldots, r_n$ are rational. The set $\mathfrak{U}$ is countable and dense in the space l^p.

EXAMPLE 1.10.4. The space $M[a, b]$ is nonseparable. Consider the family of functions of the form $x_c = \chi_{[a,c]}$, where c runs over the set $(a, b]$. The set of all such functions is uncountable and $\varrho(x_c, x_{c'}) = 1$ for $c \neq c'$. By Corollary 1.10.4, the space $M[a, b]$ is nonseparable.

EXAMPLE 1.10.5. The space m is nonseparable. Let A be an arbitrary subset of natural numbers. Denote by $x^A = \{x_n^A\}$ a sequence such that

$$x_n^A = \begin{cases} 1 & \text{for } n \in A, \\ 0 & \text{for } n \notin A. \end{cases}$$

The set of elements of the form x^A is uncountable. Moreover, $\varrho(x^A, x^{A'}) = 1$ for $A \neq A'$. Hence, by Corollary 1.10.4, the space m is nonseparable.

Let a metric space (X, ϱ) be given. A family of open sets H_t, $t \in \Omega$, is said to be a *covering* if $X \subset \bigcup_{t \in \Omega} H_t$.

THEOREM 1.10.5 (Lindelöf). *Let H_t, $t \in \Omega$, be a covering of a separable metric space (X, ϱ). Then there exists a sequence of indices $\{t_p\}$ such that $X \subset \bigcup_{p=1}^{\infty} H_{t_p}$.*

Proof. By assumption, X contains a dense sequence $\{x_n\}$. A family of balls $K_r(x_n)$, where r runs over all rational numbers, is countable. Hence, the sets belonging to this family form a sequence $\{G_m\}$. Let $\{G_{m_p}\}$ be a subsequence of the sequence $\{G_m\}$ such that for every G_{m_p}, there exists a t_p such that $G_{m_p} \subset H_{t_p}$. We shall show that $X \subset \bigcup_{p=1}^{\infty} H_{t_p}$. Let x be an arbitrary element of X. Since by assumption a family H_t, $t \in \Omega$, forms a covering, there is a t_0 such that $x \in H_{t_0}$. Since H_{t_0} is open, there is an r such that $K_r(x) \subset H_{t_0}$. Since $\{x_n\}$ is dense in X, there exists an x_{n_0} such that $\varrho(x, x_{n_0}) < r/2$. Moreover,

$$x \in K_{r/2}(x_{n_0}) \subset K_r(x) \subset H_{t_0}, \tag{1.10.1}$$

and $K_{r/2}(x_{n_0})$ belongs to the family G_m, provided r is rational. Then, by inclusion (1.10.1), the ball $K_{r/2}(x_{n_0})$ also belongs to the family G_{m_p}. Hence $x \in H_{t_p}$, but clearly t_p is not necessarily equal to t_0. ∎

Theorem 1.10.5 can also be formulated as follows:

If (X, ϱ) is a separable metric space, then every covering of X contains a countable covering.

1.11. SEQUENTIALLY COMPACT SPACES AND COMPACT SPACES

A metric space (X, ϱ) is called *sequentially compact* if every sequence $\{x_n\}$ of elements of X contains a subsequence $\{x_{n_k}\}$ which converges to a certain element $x_0 \in X$. A subset $A \subset X$ is said to be *sequentially compact* if (A, ϱ) is a sequentially compact metric space.

THEOREM 1.11.1. *Every closed subset of a sequentially compact space is sequentially compact.*

Proof. Let $\{x_n\}$ be an arbitrary sequence of elements of a closed set A. Since $A \subset X$ and X is sequentially compact, there exists a subsequence $\{x_{n_k}\}$ which converges to a certain $x_0 \in X$. Since $x_{n_k} \in A$, the closedness of A implies that $x_0 \in A$. ∎

THEOREM 1.11.2. *Every sequentially compact space is complete.*

Proof. Let $\{x_n\}$ be a fundamental sequence of elements of a sequentially compact space X. By definition, there exists a subsequence $\{x_{n_k}\}$ of a sequence $\{x_n\}$ which converges to a certain x_0, i.e., for every $\varepsilon > 0$ there exists a K such that $\varrho(x_{n_k}, x_0) < \varepsilon$ for $k > K$. On the other hand, since $\{x_n\}$ is a fundamental sequence, there exists an N such that $\varrho(x_n, x_m) < \varepsilon$ for $n, m > N$. Taking $m = n_k$ for $k > K$, we obtain

$$\varrho(x_n, x_0) \leqslant \varrho(x_n, x_{n_k}) + \varrho(x_{n_k}, x_0) < 2\varepsilon,$$

which completes the proof. ∎

THEOREM 1.11.3. *Every sequentially compact space X is separable.*

Proof. Let ε be an arbitrary positive number. Let $A_\varepsilon = \{p_1^\varepsilon, ..., p_n^\varepsilon\}$ be a set containing all points with the following property: for every pair of points $p_i^\varepsilon, p_j^\varepsilon, i, j = 1, ..., n, p_i^\varepsilon \neq p_j^\varepsilon$, we have $\varrho(p_i^\varepsilon, p_j^\varepsilon) \geqslant \varepsilon$. By the sequential compactness of X, the set A_ε is finite. Since A_ε contains all the points satisfying the above condition, A_ε is an ε-net. Suppose that A_ε is not an ε-net. This means that there exists an x such that $\varrho(x, p_i^\varepsilon) > \varepsilon$, $i = 1, 2, ..., n$. Putting $x = p_{n+1}^\varepsilon$, we find that the set A_ε contains more than n elements such that the distance between any two of them is not less than ε. This contradicts the definition of the set A_ε. The set $A_0 = \bigcup_{n=1}^{\infty} A_{1/n}$ is countable and dense in X. ∎

THEOREM 1.11.4 (Cantor). *Let (X, ϱ) be a sequentially compact space. Let $\{F_n\}$ be a descending sequence of closed subsets of X, $F_n \supset F_{n+1}$. The intersection $\bigcap_{n=1}^{\infty} F_n$ is nonempty.*

Proof. Let $x_n \in F_n$. Since X is sequentially compact, there exists a subsequence $\{x_{n_k}\}$ of the sequence $\{x_n\}$ which converges to a certain $x_0 \in X$. Observe that $x_{n_k} \in F_m$ for $n_k > m$ and $m = 1, 2, \ldots$ Hence, by the closedness of F_m, $x_0 \in F_m$ for $m = 1, 2, \ldots$ and $x_0 \in \bigcap_{m=1}^{\infty} F_m$. ∎

THEOREM 1.11.5 (Borel). *Let (X, ϱ) be a sequentially compact space and let $\{G_n\}$ be a sequence of open sets such that $X \subset \bigcup_{n=1}^{\infty} G_n$. There exists a finite number of sets $G_{n_1}, G_{n_2}, \ldots, G_{n_k}$ such that*
$$X \subset \bigcup_{j=1}^{k} G_{n_j}.$$

Proof. Write $F_n = X \setminus (G_1 \cup \ldots \cup G_n)$. The sets F_n are closed and form a descending sequence of sets. In view of Theorem 1.11.4, if all F_n were nonempty, there would exist an x_0 belonging to the intersection of the sets F_n. This would contradict the assumption that the union of G_i coincides with the whole set X. Hence, for a certain m', the set $F_{m'}$ must be empty, which implies that
$$X = G_1 \cup \ldots \cup G_{m'}. \blacksquare$$

THEOREM 1.11.6 (Borel, Lebesgue). *Let (X, ϱ) be a sequentially compact space and let $\{G_t\}$, $t \in T$, be a family of open sets such that $X \subset \bigcup_{t \in T} G_t$. Then there exist a finite number of indices $t_1, \ldots, t_k$ such that $X \subset \bigcup_{i=1}^{k} G_{t_i}$.*

1.11. Sequentially compact spaces and compact spaces

Proof. In virtue of the Lindelöf theorem (Theorem 1.10.5) and by the separability of the space (X, ϱ) (Theorem 1.11.3), a sequence of sets $\{G_{t_n}\}$ can be found such that $X \subset \bigcup_{n=1}^{\infty} G_{t_n}$. Now the conclusion of the theorem is an immediate consequence of the Borel theorem (Theorem 1.11.5). ∎

A fully equivalent formulation of Theorem 1.11.6 is the following

THEOREM 1.11.6'. *Every covering of a sequentially compact space (X, ϱ) contains a finite covering.*

Theorem 1.11.6 implies

THEOREM 1.11.7. *If X is a sequentially compact space, then for every positive ε there exists a finite ε-net in X.*

Proof. Consider the covering of X consisting of the balls $K_\varepsilon(x)$ with centres at points $x \in X$ and radius ε. By Theorem 1.11.6, there exists a finite covering contained in this covering. The centres of the balls form a finite ε-net. ∎

Clearly, there are sets which are not sequentially compact, but have, for every positive ε, a finite ε-net, e.g., an open segment (a, b).

THEOREM 1.11.8. *If (X, ϱ) is a complete metric space and, for every $\varepsilon > 0$, there exists a finite ε-net in X, then X is a sequentially compact space.*

Proof. Let $\{x_n\}$ be an arbitrary infinite sequence. Since there exists a finite covering consisting of balls of radius $1/2$, the sequence $\{x_n\}$ contains an infinite subsequence $\{x_n^1\}$ with the property that $\varrho(x_i^1, x_j^1) < 1$ for $i, j = 1, 2, \ldots$ Since there exists also a finite covering consisting of balls of radius $1/4$, the sequence $\{x_n^1\}$ contains an infinite subsequence $\{x_n^2\}$ such that $\varrho(x_i^2, x_j^2) < 1/2$ for $i, j = 1, 2, \ldots$ etc. Therefore, we can construct by induction a family of sequences $\{x_n^k\}$ with the property that $\{x_n^k\}$ is an infinite subsequence of the sequence $\{x_n^{k-1}\}$ and $\varrho(x_i^k, x_j^k) < 1/2^k$.

Consider now the sequence $\{x_n^n\}$. It is a fundamental sequence since for all k large enough it is a subsequence of the sequence $\{x_n^k\}$. By the completeness of the space X, the sequence $\{x_n^n\}$ converges to a certain $x_0 \in X$. Hence, according to the definition, X is a sequentially compact space. ∎

The next theorem is converse to Theorem 1.11.6'.

THEOREM 1.11.9. *Let (X, ϱ) be a metric space and let every covering of X contain a finite covering. Then the space X is sequentially compact.*

Proof. First, we prove that if X satisfies the assumptions of the theorem then X is complete. Suppose that X is not complete. Let $\hat{X}$ be its completion and let x_0 be an arbitrary point of $\hat{X} \setminus X$.

Let

$$R_n = \left\{x \in X: \frac{1}{n+1} < \varrho(x, x_0) < \frac{1}{n-1}\right\}, \quad n = 1, 2, \ldots$$

with the convention that $1/0 = +\infty$.

It is easy to verify that the sets R_n form a covering of the space X and this covering does not contain any finite covering.

Thus, we have proved that the space X satisfying the assumption of this theorem is complete. Let ε be an arbitrary positive number. By the assumption, the covering of X consisting of the balls with radii ε and centres at points $x \in X$ contains a finite covering. This means that there exists a finite ε-net in X. Hence, by Theorem 1.11.8, the space X is sequentially compact. ∎

A metric space is said to be *compact* if its every covering contains a finite covering.

As a consequence of Theorem 1.11.6' and 1.11.9 we have

THEOREM 1.11.10. *A metric space is compact if and only if it is sequentially compact.*

1.11. Sequentially compact spaces and compact spaces

In subsequent theorems we consider other properties of compact (sequentially compact) spaces. Since the two definitions coincide for metric spaces (see Theorem 1.11.10), further results will be formulated only for compact spaces.

THEOREM 1.11.11 (Riesz). *Let (X, ϱ) be a compact space. If $\{F_t\}$, $t \in T$, is a family of closed subsets of X with the property that its every finite subfamily has a nonempty intersection, $F_{i_1} \cap \ldots \cap F_{i_n} = \emptyset$, then the intersection of the whole family is nonempty, $\bigcap_{t \in T} F_t \neq \emptyset$.*

Proof. Write $G_t = X \setminus F_t$. Assume that the intersection of the family $\{F_t\}$, $\bigcap_{t \in T} F_t$, is empty. This means that the family $\{G_t\}$ forms a covering of the space X. Hence, by definition, there exists a finite covering contained in $\{G_t\}$, $X = G_{t_1} \cup \ldots \cup G_{t_n}$. This implies that $F_{t_1} \cap \ldots \cap F_{t_n}$ is empty, which contradicts the assumption. ∎

THEOREM 1.11.12. *Let (X, ϱ) be a compact metric space and let f be a continuous mapping of the space X into a metric space Y. The image of the space X, $f(X)$, is a compact set.*

Proof. Let $\{y_n\}$ be an arbitrary sequence of elements of the set $f(X)$, i.e., for every y_n there exist an $x_n \in X$ such that $f(x_n) = y_n$. Since X is compact, by Theorem 1.11.9, there exists a subsequence $\{x_{n_k}\}$ which converges to a certain $x_0 \in X$. By the continuity of f,

$$\lim_{k \to \infty} y_{n_k} = \lim_{k \to \infty} f(x_{n_k}) = f(x_0) \in f(X). \blacksquare$$

COROLLARY 1.11.13. *Let (X, ϱ) be a compact metric space and let f be a continuous mapping of the space X into a metric space Y. Then f maps closed sets into closed sets.*

Proof. Let F be a closed subset of X. By Theorem 1.11.1, the set F is compact. Hence, by Theorem 1.11.12, the image of F, $f(F)$, is also compact, and thus closed. ∎

COROLLARY 1.11.14. *Let X be a compact space. Let f be a one-to-one continuous mapping of X onto a metric space Y. Then the inverse mapping f^{-1} is also continuous.*

Proof. Write $g = f^{-1}$; thus $g^{-1} = f$ maps closed sets into closed sets. Hence, by Theorem 1.3.1, the mapping $g = f^{-1}$ is continuous. ∎

THEOREM 1.11.15. *Let X be a compact space and let f be a real-valued function defined on X. Then f attains on X its supremum (infimum).*

Proof. We prove that any compact set A of real numbers is bounded. If it were not so, there would exist a sequence of elements of A which diverges to $+\infty$ or $-\infty$. It is easy to observe that no subsequence of such a sequence can be convergent, since its every subsequence tends to $+\infty$ (or $-\infty$ respectively). Hence, A is bounded. As a compact set, A is closed and contains its lower and upper bounds. For $f(X) = A$ the proof results trivially from Theorem 1.11.12. ∎

Chapter 2

Linear spaces and normed spaces

2.1. BASIC FACTS ABOUT LINEAR SPACES

In many engineering applications we are faced with problems of vector addition. If, for instance, we say that a load is to be moved three metres forward, two metres to the left, one metre back and three metres to the right, then we have a typical application of vector addition. As another example we can consider interference of waves emitted from two different sources.

Let $Q_i(t, x)$, $i = 1, 2$, denote the deviation from the equilibrium at a moment t at a point x caused by a wave emitted from the i-th source ($i = 1, 2$). Then the total deviation caused by the waves emitted from both sources can be expressed by the formula

$$Q(t, x) = Q_1(t, x) + Q_2(t, x).$$

Observe that in the first example we deal with addition of two-dimensional vectors, an operation we are familiar with, while in the second example the vectors are, in general, not finite-dimensional.

In the above examples multiplication of vectors by scalars may also appear. For instance, if in the first case we decide that a load should be shifted in the same direction but twice as far, or if in the second case we conclude that the increase of the power of a wave source by 50% causes a 50% increase of deviation, then, in fact, we multiply vectors by the corresponding scalars.

These and other examples motivate the usefulness of the notion of a linear space.

By a *linear space* we mean a set X with two operations defined on X; the operation of addition of elements of X, $x+y$, and the operation of multiplication of elements of X by numbers (real or complex), tx, which satisfy the following axioms:

(1) The operation of addition is associative, i.e., for all $x, y, z \in X$

$$(x+y)+z = x+(y+z).$$

This element will be denoted by $x+y+z$.

(2) For every $x, y \in X$ there exists a $z \in X$ such that $x+z = y$; the element z will be called the *difference* of elements y and x and will be denoted by $y-x$.

(3) The operation of addition is commutative, i.e., $x+y = y+x$.

(4) $1 \cdot x = x$.

(5) The operation of multiplication by numbers is associative, i.e., for an arbitrary element $x \in X$ and arbitrary numbers a, b,

$$a(bx) = (ab)x.$$

(6) The operation of addition is distributive with respect to multiplication by numbers, i.e., for arbitrary $x, y \in X$ and an arbitrary number t

$$t(x+y) = tx+ty.$$

(7) The operation of addition of numbers is distributive with respect to multiplication by elements of X, i.e., for arbitrary numbers a, b, and an arbitrary $x \in X$,

$$(a+b)x = ax+bx.$$

Elements of linear spaces shall be sometimes called *points*.

If A, B are two subsets of a linear space X, then by $A+B$ $(A-B)$ we shall denote the sets

$$A \pm B = \{z \in X \colon z = x \pm y,\ x \in A, y \in B\}.$$

The set $\{x\} \pm A$ will be denoted by $x \pm A$. By tA we shall denote the set

$$tA = \{z \in X \colon z = tx,\ x \in A,\ t \text{ is a fixed number}\}.$$

If we define multiplication only by real numbers, then the resulting linear space is said to be *over the field of real numbers*. If we define multiplication by complex numbers, then the resulting linear space is said to be *over the field of complex numbers*. For the sake of brevity we shall say: *linear real space* or *linear complex space*.

2.1. Basic facts about linear spaces

Since in most applications we deal with real spaces, in the sequel we shall assume that a linear space means a real linear space if it is not stated otherwise.

Let x be an arbitrary element of X. By 0_x we denote $x-x$. Observe that, for every $y \in X, y+0_x = y$. In fact, by definition, $x+0_x = x$, and by adding to both sides $(y-x)$ we get $(y-x)+x+0_x = (y-x)+x$ and by the commutativity of addition,

$$y+0_x = y \text{ for all } y \in X. \tag{2.1.1}$$

The elements 0_x satisfying (2.1.1) are called *zeros* (or *neutral elements*). We shall show that in every linear space X (real or complex) there exists a unique zero. Assume that there exist two zeros, 0_1 and 0_2; then $0_1 = 0_1 + 0_2 = 0_2$. Traditionally, we denote this element by 0. The element $0-x$ we denote by $-x$. Observe that the uniqueness of 0 implies the uniqueness of $0-x$.

Let X be a linear space (real or complex). Let $x_1, \ldots, x_n$ be elements of X and let $t_1, \ldots, t_n$ be a system of numbers (real or complex). The element $t_1 x_1 + \ldots + t_n x_n$ will be called a *linear combination*. We say that elements $x_1, \ldots, x_n$ are *linearly dependent* if there exist numbers $t_1, \ldots, t_n$ such that $|t_1| + \ldots + |t_n| > 0$ and $t_1 x_1 + \ldots + t_n x_n = 0$. Elements $x_1, \ldots, x_n$ are *linearly independent* if $x_1, \ldots, x_n$ are not linearly dependent, in other words, a linear combination of $x_1, \ldots, x_n$ is equal to zero only if all t_i are equal to zero.

A linear space X (real or complex) is called an *n-dimensional space* (n finite) if a linearly independent system of elements of X of greatest cardinality contains exactly n elements. If the cardinality of linearly independent systems of elements of X is not bounded from above, then we say that the space X is *infinite-dimensional*.

The number n is called the *dimension* of an n-dimensional space and we write $\dim X = n$.

If a space X is infinite-dimensional, we say that its dimension is *infinite* and we write $\dim X = +\infty$.

It is easy to prove by induction that every infinite-dimensional space contains a sequence $\{x_n\}$ of elements which are linearly independent.

If a set is simultaneously regarded as a real space and a complex space, then the dimension of the real space is twice as great as the dimension of the complex space. This follows from the fact that the set of complex numbers regarded as a real linear space (with usual addition and multiplication by reals) is of dimension 2.

Let X be a linear space, real or complex. A set (system) $\mathfrak{A}$ of elements of X is called *linearly independent* if each finite set of its elements is linearly independent. A set (system) $\mathfrak{B}$ of elements of a space X is called an *algebraic basis* (shortly: *basis*) if each element of X can be represented uniquely as a finite linear combination of elements $x_1, \ldots, x_n$ belonging to $\mathfrak{B}$.

The uniqueness of such a representation immediately implies that the set $\mathfrak{B}$ contains linearly independent elements.

THEOREM 2.1.1. *Let X be an n-dimensional linear space (real or complex). Every system of n linearly independent elements $x_1, \ldots, x_n$ forms a basis.*

Proof. Assume that y cannot be represented as a linear combination of $x_1, \ldots, x_n$. Then a linear combination $t_1 x_1 + \ldots + t_n x_n + ty$ is equal to zero only for $t = 0$. This implies, by the linear independence of $x_1, \ldots, x_n$, that $t_1 = t_2 = \ldots = t_n = 0$. Hence, the elements $x_1, \ldots, x_n, y$ constitute a linearly independent system of cardinality $n+1$ which contradicts the assumption that X is n-dimensional.

The uniqueness of the representation of each element of X in the form of a linear combination is an immediate consequence of the linear independence of elements $x_1, \ldots, x_n$. If $y = t_1 x_1 + \ldots + t_n x_n = t'_1 x_1 + \ldots + t'_n x_n$, then $(t_1 - t'_1) x_1 + \ldots + (t_n - t'_n) x_n = 0$ and all $(t_i - t'_i)$ are simultaneously equal to zero. Thus, $t_i = t'_i$. ∎

Let X be an n-dimensional linear space (real or complex) and let $\{e_1, \ldots, e_n\}$ be a system of n linearly independent elements of X. By Theorem 2.1.1, elements $\{e_1, \ldots, e_n\}$ form a basis in X. Hence, each element $x \in X$ can be represented in the form $x = t_1 e_1 + \ldots + t_n e_n$. Therefore, there is a one-to-one correspondence between elements of X and ordered n-tuples of numbers $(t_1, \ldots, t_n)$. Thus, from the algebraical point of view, there is an equivalence between an n-dimensional space and systems of n numbers $(t_1, \ldots, t_n)$.

Let X be a linear space (real or complex) and let Y be a subset of X such that for $x, y \in Y$ we have $x+y \in Y$ and $tx \in Y$ for each number t.

2.1. Basic facts about linear spaces

Then Y is called a *linear subset* (*subspace*) of X. Obviously, Y is a linear space and $\dim Y \leqslant \dim X$.

If A is a subset of X, then a subspace *spanned by A* or *a space spanned by A* is the smallest subspace Y containing A. We shall denote it by $\lin A$, $Y = \lin A$.

THEOREM 2.1.2.
$$\lin A = \{t_1 x_1 + \ldots + t_n x_n \colon x_1, \ldots, x_n \in A, \ t_1, \ldots, t_n \text{ — numbers}\}.$$

Proof. Since $\lin A$ is a linear space containing A, $\lin A$ contains all linear combinations of elements of A. Since the set of all linear combinations of elements of A is a linear space, it contains, by the definition, $\lin A$. ∎

COROLLARY 2.1.3. *Let X be a linear space (real or complex). Let A be a set of linearly independent elements. Then A forms a basis in $\lin A$.*

Proof. By Theorem 2.1.2, each element of A can be represented as a linear combination of elements of A. The uniqueness of the representation follows from the linear independence of elements of A. ∎

THEOREM 2.1.4. *Let X be a linear space (real or complex). Let $\{e_1, \ldots, e_n\}$ be a system of n linearly independent elements. Then the space $\lin \{e_1, \ldots, e_n\}$ is of dimension n.*

Proof. Since $\{e_1, \ldots, e_n\}$ are linearly independent, the dimension of the space $\lin \{e_1, \ldots, e_n\}$ is not less than n. On the other hand, let $\{f_1, \ldots, f_m\}$ be a set of m elements of $\lin \{e_1, \ldots, e_n\}$. We shall prove that if $m > n$, then the elements $f_1, \ldots, f_m$ are linearly dependent. Write $f_i = t_1^i e_1 + \ldots + t_n^i e_n$ ($i = 1, \ldots, m$). Since the system of equations $a_1 t_j^1 + \ldots + a_m t_j^m = 0$ ($j = 1, \ldots, m$) has more variables than equations, there exists a solution $a'_1, \ldots, a'_m$ such that not all a'_i are equal to zero. This implies that the linear combination $a'_1 f_1 + \ldots + a'_m f_m$ is equal to zero with some nonzero a'_i and hence $f_1, \ldots, f_m$ are linearly dependent. ∎

The following example illustrates the importance of the notion of a subspace in engineering applications.

EXAMPLE 2.1.1. Consider a network of n resistors. Assume that at k points of this network we can attach sources of direct current. What are the distributions of currents resulting from the varying voltages and changing orientations of the sources? Denote by I_i the intensity of the current passing through the resistor R_i. The set of all possible current intensities in the network forms an n-dimensional space. Let $I^j = (I_1^j, ..., I_n^j)$, $j = 1, 2, ..., k$, denote a distribution determined by switching off all current sources except the j-th, where the voltage of the j-th source is equal to 1 V.

It is easy to verify that by varying the voltages of current sources and changing their orientations we can get those and only those distributions which belong to the space lin $\{I^1, ..., I^k\}$. The dimension of this space is not greater than k, and it is equal to k if and only if elements $I^1, ..., I^k$ are linearly independent.

Let X be a linear space. Let $x_1, ..., x_n \in X$. We say that $a_1 x_1 + ... + a_n x_n$ is a *convex combination of elements* $x_1, ..., x_n$ if $a_1, ..., a_n \geqslant 0$ and $a_1 + ... + a_n = 1$. A set $A \subset X$ will be called *convex* if for any elements $x_1, ..., x_n \in A$ their convex combinations belong to A.

PROPOSITION 2.1.5. *A set A is convex if and only if for any elements $x_1, x_2 \in A$ any convex combination $a_1 x_1 + a_2 x_2$ of those elements belongs to A.*

Proof. If A is convex, then each convex combination of two elements of A also belongs to A. Now suppose that each convex combination of two elements of A belongs to A. We shall show by induction that a convex combination of arbitrary n elements of A also belongs to A. For $n = 2$ it is our assumption. Suppose that it is true for an index n. Take now an arbitrary convex combination of $n+1$ elements

$$a_1 x_1 + ... + a_{n+1} x_{n+1} = (1 - a_{n+1}) \left(\frac{a_1}{1 - a_{n+1}} x_1 + ... + \frac{a_n}{1 - a_{n+1}} x_n \right) +$$

$+ a_{n+1} x_{n+1}$. The element $\frac{a_1}{1-a_{n+1}} x_1 + ... + \frac{a_n}{1-a_{n+1}} x_n$ belongs to A, by the inductive assumption. Then $a_1 x_1 + ... + a_{n+1} x_{n+1} \in A$ and the set A is convex. ∎

2.1. Basic facts about linear spaces

Let A be an arbitrary set. By the *convex hull*, conv A, of the set A we mean the smallest convex set containing A.

PROPOSITION 2.1.6.
$$\mathrm{conv}\, A = \{a_1 x_1 + \ldots + a_n x_n \colon x_1, \ldots, x_n \in A,\ a_1, \ldots, a_n \geqslant 0,$$
$$a_1 + \ldots + a_n = 1\}.$$

Proof. Let us denote the set in the right-hand side by B. We show that B is convex. Let $y_1, y_2 \in B$. Thus, $y_1 = \alpha_1 x_1 + \ldots + \alpha_n x_n$, $y_2 = \alpha_{n+1} x_{n+1} + \ldots + \alpha_{n+m} x_{n+m}$, $\alpha_1, \ldots, \alpha_{n+m} \geqslant 0$, $\alpha_1 + \ldots + \alpha_n = 1$, $\alpha_{n+1} + \ldots + \alpha_{n+m} = 1$, $x_1, \ldots, x_{n+m} \in A$. Let b_1, b_2 be arbitrary numbers such that $b_1, b_2 \geqslant 0$, $b_1 + b_2 = 1$. Then
$$b_1 y_1 + b_2 y_2 = b_1 \alpha_1 x_1 + \ldots + b_1 \alpha_n x_n + b_2 \alpha_{n+1} x_{n+1} + \ldots +$$
$$+ b_2 \alpha_{n+m} x_{n+m} \in B$$
because $b_1 \alpha_1 + \ldots + b_1 \alpha_n + b_2 \alpha_{n+1} + \ldots + b_2 \alpha_{n+m} = b_1 + b_2 = 1$. Thus B is a convex set.

By the definition of convexity, each convex set containing A contains also all convex combinations of its elements and therefore it contains B. Thus, B is the smallest convex set containing A. ∎

Let Y be a linear subset (subspace) of a linear space X (real or complex). By *cosets* (or *equivalence classes*) we mean subsets of X defined as follows: two elements x and y belong to the same coset if and only if their difference belongs to Y. The coset containing x is denoted by $[x]$. In the set of all cosets we introduce the operations $[x] + [y] = [x+y]$ and $t[x] = [tx]$. It is easy to verify that those operations have all the properties required for operations defined in linear spaces. The zero element is the coset Y. The set of all cosets with operations defined as above is called a *quotient space*. The quotient space is denoted by X/Y.

Let $X_1, \ldots, X_n$ be a system of linear spaces, all of them defined either over the field of real numbers or over the field of complex numbers. By the Cartesian product of the spaces $X_1, \ldots, X_n$ we mean a space X whose elements are systems of elements $x = (x_1, \ldots, x_n)$, where $x_i \in X_i$.

Addition in X is defined as
$$(x_1, \ldots, x_n) + (y_1, \ldots, y_n) = (x_1 + y_1, \ldots, x_n + y_n),$$

and multiplication by scalars as

$$t(x_1, \ldots, x_n) = (tx_1, \ldots, tx_n).$$

It is easy to verify that the space X with operations defined as above is a linear space over the field of real or complex numbers.

The space X, i.e., the Cartesian product of spaces $X_1, \ldots, X_n$, is denoted by

$$X = X_1 \times \ldots \times X_n.$$

If all spaces X_i ($i = 1, 2, \ldots, n$) are finite-dimensional, then the following formula is valid

$$\dim(X_1 \times \ldots \times X_n) = \dim X_1 + \ldots + \dim X_n. \tag{2.1.2}$$

If at least one space X_i is infinite-dimensional, then the product $X_1 \times \ldots \times X_n$ is infinite-dimensional and also in this case formula (2.1.2) is valid.

2.2. LINEAR METRIC SPACES AND NORMED SPACES

Let (X, ϱ) be a metric space. Suppose that X is also a linear space. We say that (X, ϱ) is a *linear metric space* if the operations of addition and multiplication by a scalar are continuous, i.e., if $x_n \to x$, $y_n \to y$ and $t_n \to t$, then $x_n + y_n \to x+y$, $t_n x_n \to tx$.

EXAMPLE 2.2.1. The classical Euclidean space with addition defined as the addition of vectors and multiplication by scalars defined as the multiplication of vectors by numbers is a linear metric space.

EXAMPLE 2.2.2. Given a set of resistors R_i ($i = 1, \ldots, n$). We denote by I_i the intensity of the current passing through the resistor R_i. As a measure of the distance between two different distributions of current $I = (I_1, \ldots, I_n)$, $I' = (I'_1, \ldots, I'_n)$ we take the number $\varrho(I, I') = \sum_{i=1}^{n} |I_i - I'_i|$.
It is easy to verify that in this way we obtain a linear metric space.

EXAMPLE 2.2.3. Let X be a set of positions which can be reached by a gantry. As the metric $\varrho(x, y)$ we take the time necessary to shift a load

2.2. Linear metric spaces and normed spaces

from position x to position y (see Examples 1.1.11 and 1.1.12). It is easy to verify that X is a linear metric space.

EXAMPLE 2.2.4. The space $C[a, b]$ (Example 1.1.14) is a linear metric space with addition defined as the standard addition of functions and multiplication by scalars defined as the multiplication of functions by numbers.

EXAMPLE 2.2.5. If in Example 2.2.3 we take into account also the time necessary to lift up a load, we obtain a linear and metric space which is not a linear metric space, since the operation of multiplication by a scalar is not continuous. Indeed, $\varrho(tx, x) \geqslant a > 0$ for $t \neq 1$, where a denotes the time of lifting up a load. Hence, $t_n x$ does not tend to x even if t_n tends to 1.

Let X be a linear and metric space over the field of real or complex numbers with a metric $\varrho(x, y)$. The metric $\varrho(x, y)$ is called *invariant* if for all x, y, z

$$\varrho(x+z, y+z) = \varrho(x, y).$$

In Examples 2.2.1–2.2.4 we dealt with invariant metrics. Clearly, in engineering applications linear metric spaces with non-invariant metrics may also appear.

EXAMPLE 2.2.6. Let X be a certain district. In X we can define addition and multiplication by a scalar as the addition of vectors and the multiplication of vectors by numbers. Let $\varrho(x, y)$ denote the time of passing from x to y. X with the metric ϱ is a linear metric space but $\varrho(x, y)$ is not invariant since it depends not only on the Euclidean distance between points but also on the available roads and their quality.

Let (X, ϱ) be a linear metric space over the field of real or complex numbers with the invariant metric $\varrho(x, y)$. The function $||x|| = \varrho(x, 0)$ is called an *F-norm*. The properties of a metric imply the following properties of *F*-norms:
1. $||x|| = 0$ if and only if $x = 0$.
2. $||x|| = ||-x||$.
3. $||x+y|| \leqslant ||x|| + ||y||$.

Continuity of multiplication by scalars implies the following property:

4. if $|t_n-t| \to 0$ and $||x_n-x|| \to 0$, then $||t_n x_n - tx|| \to 0$.

Now let X be a linear space with a function $||x||$ defined on X, which satisfies conditions 1–3. Then the expression $\varrho(x, y) = ||x-y||$ defines an invariant metric. Observe that condition 3 entails continuity of addition. In fact, let $x_n \to x$, $y_n \to y$; then

$$0 \leqslant ||(x_n+y_n)-(x+y)|| \leqslant ||x_n-x||+||y_n-y|| \to 0,$$

thus, $x_n+y_n \to x+y$.

Conditions 1–3 do not ensure continuity of multiplication by scalars. For instance, the F-norm

$$||x|| = \begin{cases} 0 & \text{for } x = 0, \\ 1 & \text{for } x \neq 0, \end{cases}$$

defines the metric $\varrho(x, y) = ||x-y||$, i.e., the already mentioned discrete metric. Obviously, if we assume also that condition 4 holds, then we automatically obtain continuity of multiplication.

Instead of saying that a sequence $\{x_n\}$ is convergent with respect to an invariant metric generated by an F-norm $||\ ||$ we say that it is *convergent with respect to an F-norm* $||\ ||$. Similarly, instead of saying that invariant metrics defined by two F-norms $||\ ||_1$ and $||\ ||_2$ are equivalent we say shortly that *F-norms $||\ ||_1$ and $||\ ||_2$ are equivalent*. If an invariant metric induced by an F-norm $||\ ||_1$ is stronger (weaker) than an invariant metric induced by an F-norm $||\ ||_2$, then we say that the *F-norm $||\ ||_1$ is stronger (weaker) than the F-norm $||\ ||_2$*.

A space X with an F-norm $||\ ||$ defined on X is called an F^*-*space* and we shall denote it by $(X, ||\ ||)$, or briefly X.

A subspace of a linear metric space is a linear metric space.

Let a system of linear metric spaces $X_1, \ldots, X_n$ over the field of real or complex numbers be given. Let $\varrho_i(x, y)$ be a metric in the space X_i, $i = 1, \ldots, n$. As we have already shown, the Cartesian product $X_1 \times \ldots \times X_n$ is a linear space (see Section 2.2) and a metric space (see Section 1.1).

To show that $X_1 \times \ldots \times X_n$ is a linear metric space it is necessary to prove that the operations of addition and multiplication by scalars are continuous.

Assume that a sequence $\{x^m\} = \{(x_1^m, \ldots, x_n^m)\}$ is convergent to an

element $x^0 = (x_1^0, \ldots, x_n^0)$ and a sequence $\{y^m\} = \{(y_1^m, \ldots, y_n^m)\}$ is convergent to an element $y^0 = (y_1^0, \ldots, y_n^0)$. Then, by the definition of a metric in the product space (see Section 1.1),
$$\varrho(x^m+y^m, x^0+y^0) = \max_{1 \leqslant i \leqslant n} \varrho^i(x_i^m+y_i^m, x_i^0+y_i^0).$$

By continuity of addition in each space X_i, the right-hand side of the above equality tends to zero. Hence, the left-hand side also tends to zero, which entails continuity of addition. In a similar way we can also prove continuity of multiplication by scalars.

Let X be a (real or complex) linear space with an F-norm, $||x||$, defined on X, which satisfies conditions 1–3.

We say that an F-norm $||x||$ is *homogeneous* if

4′. $||tx|| = |t|\,||x||$ for all numbers t (real or complex).

Condition 4′ immediately implies condition 4. Indeed, let $t_n \to t$ and $x_n \to x$. Then
$$||t_n x_n - tx|| = ||t_n(x_n-x)+(t_n-t)x||$$
$$\leqslant |t_n|\,||x_n-x||+|t_n-t| \cdot ||x|| \to 0,$$
since the sequence $\{t_n\}$ is bounded.

F-norms which are homogeneous are shortly called *norms*. The metrics given in Examples 2.2.1, 2.2.2 and 2.2.4 define homogeneous F-norms. The same situation is in Example 2.2.3, provided that the metric is given as in Example 2.1.11.

The F-norm generated by the metric considered in Example 2.1.12 is not homogeneous.

An F^*-space with a homogeneous F-norm (i.e. with a norm) is called a *normed space* or a B^*-*space*.

Now we give an important example of a normed space.

EXAMPLE 2.2.7. Let X be a linear space over the field of real (or complex) numbers. By an *inner product* we mean a function of two arguments (x, y), $x, y \in X$, having the following properties:
(1) $(x, x) \geqslant 0$ and $(x, x) = 0$ only if $x = 0$,
(2) $(ax, y) = a(x, y)$,
(3) $(x_1+x_2, y) = (x_1, y)+(x_2, y)$,
(4) $(x, y) = \overline{(y, x)}$, where $\bar{z} = a-ib$ for $z = a+ib$.

Clearly, if X is a linear space over the field of real numbers, then (4) takes the form: $(x, y) = (y, x)$.

Properties (1)–(4) immediately imply

(5) $(x, ay) = \bar{a}(x, y)$,

(6) $(x, y_1+y_2) = (x, y_1)+(x, y_2)$.

Two elements $x, y \in X$ will be called *orthogonal* if $(x, y) = 0$. Let X_0, X_1 be subspaces of X. We say that X_0 *is orthogonal to* X_1 if $(x, y) = 0$ for all $x \in X_0$, $y \in X_1$. If $(y, x) = 0$ for all $x \in X_0$, we say that y *is orthogonal to the subspace* X_0. The subspace X_1 is called an *orthogonal complement of* X_0, if X_0, X_1 are orthogonal and $X_0+X_1 = X$.

In the space X we define a norm by the formula

$$\|x\| = \sqrt{(x, x)}. \qquad (2.2.1)$$

It follows from (1) that $\|x\| = 0$ if and only if $x = 0$. (2) and (5) imply the homogeneity of the norm $\|x\|$.

The triangle inequality can also be proved. Namely, by (1) we obtain for arbitrary $x, y \in X$ and an arbitrary real number t

$$0 \leqslant (x+ty, x+ty) = (x, x)+2t[(y, x)+(x, y)]+t^2(y, y)$$
$$= (x, x)+2t\operatorname{Re}(x, y)+t^2(y, y). \qquad (2.2.2)$$

Hence,

$$\frac{\Delta}{4} = \operatorname{Re}(x, y)^2 - (x, x)(y, y) \geqslant 0. \qquad (2.2.3)$$

If

$$a = \frac{(x, y)}{|(x, y)|}, \qquad (2.2.4)$$

then

$$(x, ay) = \bar{a}(x, y) = \frac{\overline{(x, y)}(x, y)}{|(x, y)|} = |(x, y)|. \qquad (2.2.5)$$

On the other hand, $|a| = 1$ and $(ay, ay) = (y, y)$. Substituting in (2.2.3) ay for y we obtain the *Cauchy inequality*

$$|(x, y)|^2 \leqslant (x, x)(y, y). \qquad (2.2.6)$$

2.2. Linear metric spaces and normed spaces

From inequality (2.2.6) we immediately obtain the inequality
$$\begin{aligned}(x+y, x+y) &= (x, x) + 2\mathrm{Re}(x, y) + (y, y) \\ &\leqslant (x, x) + 2\sqrt{(x, x)(y, y)} + (y, y) \\ &\leqslant (\sqrt{(x, x)} + \sqrt{(y, y)})^2.\end{aligned} \qquad (2.2.7)$$

By taking the square roots of both sides of inequality (2.2.7) we obtain the triangle inequality
$$\sqrt{(x+y, x+y)} \leqslant \sqrt{(x, x)} + \sqrt{(y, y)}. \qquad (2.2.8)$$

Traditionally, the above inequality is called the *Schwarz inequality*.

By the above argument, the space X with the norm given by formula (2.2.1) is a normed space. We shall call it a *pre-Hilbert space*.

For instance, the space $L^2(\Omega, \Sigma, \mu)$ with arbitrary Ω, Σ, μ is a pre-Hilbert space. The inner product in this space can be defined as
$$(x, y) = \int_\Omega x(t)\overline{y(t)}\,d\mu.$$

A subspace of a normed space is clearly a normed space. It is easy to verify that the Cartesian product of n normed spaces (real or complex) is also a normed space.

F^*-spaces which are complete are called *F-spaces*. Complete normed spaces are called *Banach spaces* or shortly *B-spaces*. Complete pre-Hilbert spaces are called *Hilbert spaces*.

A closed linear subspace of an F-space (a Banach space, a Hilbert space) is an F-space (a Banach space, a Hilbert space) (see Theorem 1.5.1).

The Cartesian product of n F-spaces (Banach spaces), real or complex, is an F-space (Banach space), real or complex (see Theorem 1.5.3).

Suppose we are given a system of n real or complex pre-Hilbert spaces $H_1, \ldots, H_n$. Denote by $(x, y)_i$ the inner product in the space H_i, $i = 1, 2, \ldots, n$. Now take the Cartesian product of the spaces H_i, $H_1 \times \ldots \times H_n$, and define an inner product by the formula
$$(x, y) = \sum_{i=1}^n (x_i, y_i)_i,$$
where $x = (x_1, \ldots, x_n)$ and $y = (y_1, \ldots, y_n)$ are elements of the Carte-

sian product space. It is easy to verify that (x, y) defines an inner product in $H_1 \times \ldots \times H_n$. Hence, the Cartesian product $H_1 \times \ldots \times H_n$ with the inner product (x, y) is a pre-Hilbert space.

By Theorem 1.5.3, just as for F^*-spaces and normed spaces, we infer that the Cartesian product $H_1 \times \ldots \times H_n$ of Hilbert spaces H_i is a Hilbert space.

THEOREM 2.2.1. *Let $(X, \|\ \|)$ be a normed space (real or complex). The unit ball $K = \{x\colon \|x\| < 1\}$ is a set that is open, convex and symmetric (i.e., if $x \in K$, then $-x \in K$).*

Proof. The openness of the ball K follows from the fact that any ball in a metric space is open (cf. Section 1.2). Condition (2) of a norm entails symmetry of the ball. Condition (3) (the triangle inequality) implies the convexity of K. In fact, $x, y \in K$, $\|x\| < 1$, $\|y\| < 1$, whence, for every t, $0 \leqslant t \leqslant 1$,
$$\|tx+(1-t)y\| \leqslant |t|\,\|x\|+|1-t|\,\|y\| \leqslant 1$$
and
$$tx+(1-t)y \in K. \blacksquare$$

In real normed spaces a partially converse result holds.

THEOREM 2.2.2. *Let X be a real normed space and let U be an open convex and symmetric set. There exists a function $\|x\|'$ defined on X, which satisfies conditions 2, 3, 4' of a norm and is such that*
$$U = \{x\colon \|x\|' < 1\}.$$

Proof. Let
$$\|x\|' = \inf\left\{t > 0\colon \frac{x}{t} \in U\right\}.$$

Condition 2 is satisfied, by the symmetry of U. We prove now that condition 4' (homogeneity) is also satisfied:
$$\|ax\|' = \inf\left\{t > 0\colon \frac{ax}{t} \in U\right\} = |a|\inf\left\{\frac{t}{|a|} > 0\colon \frac{|a|x}{t} \in U\right\}$$
$$= |a|\inf\left\{s > 0\colon \frac{x}{s} \in U\right\} = |a|\,\|x\|'.$$

It remains to verify condition 3 (the triangle inequality). Let ε be an arbitrary positive number less than 1. By the definition of $\|x\|'$,
$$(1-\varepsilon)\frac{x}{\|x\|'}, (1-\varepsilon)\frac{y}{\|y\|'} \in U$$
for arbitrary x, y such that $\|x\|' \neq 0$, $\|y\|' \neq 0$.
Since U is convex, we have
$$(1-\varepsilon)\frac{x}{\|x\|'} \cdot \frac{\|x\|'}{\|x\|'+\|y\|'} + (1-\varepsilon)\frac{y}{\|y\|'} \cdot \frac{\|y\|'}{\|x\|'+\|y\|'}$$
$$= (1-\varepsilon)\frac{x+y}{\|x\|'+\|y\|'} \in U.$$
It follows from the definition of $\|x\|'$ that
$$(1-\varepsilon)\frac{\|x+y\|'}{\|x\|'+\|y\|'} \leq 1,$$
and hence,
$$\|x+y\|' \leq \frac{1}{1-\varepsilon}(\|x\|'+\|y\|').$$
Since ε was chosen arbitrarily, the above inequality implies the triangle inequality. ∎

A function satisfying conditions 2, 3, 4' is called a *pseudonorm*. Thus, we have proved that each open, convex and symmetric set defines a certain pseudonorm. Clearly, a pseudonorm need not be a norm. It is easy to verify that a pseudonorm defined by a set U is a norm if and only if U does not contain lines (i.e., if for every $x \neq 0$ there exists an $r > 0$ such that $rx \notin U$).

We recall that two norms defined on a linear space X (real or complex) are *equivalent* if the metrics defined by those norms are equivalent. Norms are not always equivalent.

EXAMPLE 2.2.8. Let X be the space of all continuous functions $C[a, b]$ (see Example 1.1.14). Suppose we are given two norms: the first norm defined by the metric introduced in the Example 1.1.14,
$$\|x\| = \sup_{a \leq t \leq b} |x(t)|$$

and the second norm given by the formula

$$\|x\|_1 = \int_a^b |x(t)|\,dt.$$

These norms are not equivalent since, for instance, the sequence of functions

$$x_n = \left(\frac{t}{\max(|a|,|b|)}\right)^n$$

converges to 0 in the norm $\|x\|_1$ and is not convergent in the norm $\|x\|$.

THEOREM 2.2.3. *Let X be a linear space (real or complex). Two norms $\|x\|$ and $\|x\|'$ defined on X are equivalent if and only if there exist constants $A, B > 0$ such that*

$$A \leqslant \frac{\|x\|'}{\|x\|} \leqslant B. \tag{2.2.9}$$

Proof. If (2.2.9) holds and $\|x_n\| \to 0$, then $\|x_n\|' \leqslant B\|x_n\|$ and $\|x_n\|' \to 0$. Applying the second inequality, we deduce in the same way that if $\|x_n\|' \to 0$ then $\|x_n\| \to 0$.

Assume that the second inequality of (2.2.9) does not hold. Then there exists a sequence $\{x_n\}$ such that $\|x_n\|' \geqslant n^2 \|x_n\|$.

Let

$$y_n = \frac{1}{n} \cdot \frac{x_n}{\|x_n\|}.$$

Clearly, $\|y_n\| \to 0$ but $\|y_n\|'$ does not tend to zero. Similarly, we get a contradiction if the first inequality does not hold. ∎

COROLLARY 2.2.4. *Let X be a linear space (real or complex). Two norms $\|x\|$ and $\|x\|'$ are equivalent if and only if there exist positive constants A and B such that*

$$\{x \in X : \|x\| < 1\} \subset B\{x \in X : \|x\|' < 1\},$$
$$\{x \in X : \|x\|' < 1\} \subset A^{-1}\{x \in X : \|x\| < 1\}.$$

As an immediate consequence of Corollary 2.2.4 we obtain

2.2. Linear metric spaces and normed spaces

THEOREM 2.2.5. *Let $(X, \|\ \|)$ be a real normed space. Let U be an open, convex and symmetric set in X. A pseudonorm $\|x\|'$ defined by U is equivalent to the norm $\|x\|$ if and only if*

$$\sup_{x \in U} \|x\| < +\infty.$$

Let $(X, \|\ \|)$ be a normed space, real or complex. A set $A \subset X$ is said to be a *bounded set* if $\sup_{x \in A} \|x\| < +\infty$. The disadvantage of the above definition is that it is not applicable in linear metric spaces which are not normed. Hence, the following definition is frequently used. A set A contained in a linear metric space X (real or complex) is *bounded* if, for every sequence of numbers $\{t_n\}$ tending to 0 and every sequence $\{x_n\}$ of elements of the set A, the sequence $\{t_n x_n\}$ tends to 0.

Observe that if X is a normed space, the two definitions are equivalent.

If A is bounded, A does not contain any line. Indeed, if we had $tx \in A$ for a certain $x \neq 0$ and all real t, then $x_n = nx \in A$ but the sequence $\left\{\frac{1}{n} x_n\right\} = \{x\}$ would not tend to zero whereas the sequence $\left\{\frac{1}{n}\right\}$ tends to 0.

THEOREM 2.2.6 (Kolmogorov, 1934). *Let (X, ϱ) be a real linear metric space. Assume that there exists an open, convex and bounded set U contained in X. Then there exists a norm $\|x\|$ on X such that the metric defined by $\|x\|$ is equivalent to the metric $\varrho(x, y)$.*

Proof. Let $V = U - x_0$, where x_0 is an arbitrary point from the interior of U and let $W = V \cap (-V)$. The set W is open, convex, bounded and symmetric.

Let

$$\|x\| = \inf\left\{t > 0: \frac{x}{t} \in W\right\}.$$

As in Theorem 2.2.2, it can be verified that $\|x\|$ is a pseudonorm. Since W is bounded, it does not contain any line and thus $\|x\|$ is a norm. It remains to prove that this norm induces a metric which is equivalent to the metric $\varrho(x, y)$, i.e., $\|x_n\| \to 0$ if and only if $\varrho(x_n, 0) \to 0$.

First, we shall show that $\varrho(x_n, 0) \to 0$ implies $\|x_n\| \to 0$. Indeed, let ε be an arbitrary positive number. Since W is open and 0 is an interior point of W, there exists an index N such that $x_n \in \varepsilon W$ for $n > N$. This implies that $\|x_n\| < \varepsilon$.

On the other hand, for every sequence $\{x_n\}$ of elements of the space X, $\dfrac{x_n}{2\|x_n\|}$ belongs to the bounded set W. Hence, if $\|x_n\| \to 0$, $2\|x_n\| \cdot \dfrac{x_n}{2\|x_n\|} = x_n$ tends to 0 in the metric $\varrho(x, y)$. ∎

THEOREM 2.2.7 (Kolmogorov, 1934). *Let (X, ϱ) be a complex linear metric space. Assume that there exists an open convex bounded set U contained in X. Then there exists a norm $\|x\|$ on X such that the metric defined by $\|x\|$ is equivalent to the metric $\varrho(x, y)$.*

Proof. Clearly, the space (X, ϱ) can be considered a real linear metric space. Hence, by Theorem 2.2.6, there exists a real norm $\|x\|_0$ in X (i.e., a norm which is homogeneous with respect to multiplication by real numbers) such that the metric generated by $\|x\|_0$ is equivalent to the metric $\varrho(x, y)$.

Let
$$\|x\| = \sup_{|a|=1} \|ax\|_0.$$

It is easy to verify that $\|x\|$ is a norm, provided $\|x\|_0$ is finite. To show that $\|x\|$ exists and is equivalent to $\|x\|_0$ it is enough to show that

$$M = \sup_{x \in X} \frac{\sup_{|a|=1} \|ax\|_0}{\|x\|_0} < +\infty. \tag{2.2.10}$$

(2.2.10) implies immediately that $\|x\| \leqslant M \cdot \|x\|_0$ and hence $\|x\|$ exists. Moreover, $\|x\|_0 \leqslant \|x\|$. Thus, the norms $\|x\|$ and $\|x\|_0$ are equivalent.

Now we prove (2.2.10). Assume that (2.2.10) does not hold. Thus, there exist a sequence of elements $\{x_n\}$ and a sequence of numbers $\{a_n\}$, $|a_n| = 1$, such that

$$\|a_n x_n\|_0 \geqslant n \cdot \|x_n\|_0.$$

2.2. Linear metric spaces and normed spaces

Clearly, the sequence $\{x_n/\|x_n\|_0\}$ is bounded and the sequence of numbers $\{a_n/n\}$ tends to zero. But, on the other hand, norms of elements of the sequence $\left\{\dfrac{a_n}{n} \cdot \dfrac{x_n}{\|x_n\|_0}\right\}$ are not less than 1 and hence this sequence does not tend to zero. We obtain a contradiction, which proves (2.2.10). ∎

In this section we have so far considered normed spaces and we have shown that in the case of linear real normed spaces there is a one-to-one correspondence between all norms equivalent to a given norm and open symmetric convex bounded sets.

It turns out that similar considerations are applicable to semimetric spaces.

Suppose we are given a linear space X (real or complex) with a semimetric $\varrho(x, y)$ (see Section 1.4). We say that (X, ϱ) is a *linear semimetric space* if the operations of addition and multiplication by scalars are continuous with respect to the semimetric $\varrho(x, y)$. We say that a semimetric $\varrho(x, y)$ is *invariant* if $\varrho(x+z, y+z) = \varrho(x, y)$ for all x, y, z. If $\varrho(x, y)$ is an invariant semimetric, then $\varrho(x, 0)$ is called a *semi-F-norm* and is denoted by $\|x\|$. If a semi-F-norm $\|x\|$ is positively homogeneous, i.e., $\|tx\| = t\|x\|$ for $t > 0$, then $\|x\|$ is called a *seminorm*.

A seminorm obviously satisfies the following conditions:

1° $\|x\| = 0$ if and only if $x = 0$,
3° $\|x+y\| \leqslant \|x\| + \|y\|$,
4° $\|tx\| = t \cdot \|x\|$ for $t > 0$.

(In the above enumeration number 2° is omitted in order to point out the analogy with the conditions of a norm.)

A function $\|x\|$ defined on X which satisfies conditions 3° and 4° is called a *pseudoseminorm*.

Since for every semimetric there exists an equivalent metric (Theorem 1.4.1), all the results concerning norms, normed spaces, equivalency of norms, etc. can be applied to seminorms and seminormed spaces, where by *seminormed* spaces we mean linear semimetric spaces with semimetrics generated by seminorms.

Now we prove

THEOREM 2.2.8. *Let X be a real semimetric space with a metric generated by a seminorm $||x||$. Then there exists a positive constant A such that*
$$||-x|| \leqslant A \cdot ||x||.$$

Proof. Assume that there is no such a constant. Then there exists a sequence of elements $\{x_n\}$ such that
$$||-x_n|| \geqslant n^2 \cdot ||x_n||.$$
Let
$$y_n = \frac{x_n}{n \cdot ||x_n||}.$$
Then the sequence $\{y_n\}$ converges to 0 but $\left\|-\frac{1}{n}y_n\right\| \geqslant n$. This contradicts continuity of multiplication by scalars. ∎

THEOREM 2.2.9. *Let $(X, ||\ ||)$ be a seminormed space. Then the set*
$$K = \{x\colon ||x|| < 1\}$$
is an open and convex set which contains 0.

The proof is exactly the same as the proof of Theorem 2.2.1.

THEOREM 2.2.10. *Let X be a real linear metric space and let U be an open set containing 0. Then there exists a continuous pseudoseminorm $||x||$ such that*
$$U = \{x\colon ||x|| < 1\}.$$

The proof is the same as the proof of Theorem 2.2.2.

If a given set U does not contain any half-line (i.e., for every $x \neq 0$ there exists a positive t such that $tx \notin U$), then the pseudoseminorm defined in Theorem 2.2.10 is a seminorm. Moreover, if U is bounded, then a semimetric generated by $||x||$ is equivalent to the original metric.

Let $(X, ||\ ||)$ be a linear metric space. If Y is a subspace of X, then Y is a linear metric space with the metric $\varrho(x, y)$. Assume that Y is closed and consider the quotient space X/Y. The question arises whether we can define such a metric on the quotient space that:

2.2. Linear metric spaces and normed spaces

(1) the quotient space with that metric is a linear metric space,
(2) if $x_n \to x$, then $[x_n] \to [x]$.

Assume that $\varrho(x, y)$ is invariant and define a metric $\tilde{\varrho}(\tilde{x}, \tilde{y})$ on the quotient space by the formula

$$\tilde{\varrho}(\tilde{x}, \tilde{y}) = \inf\{\varrho(x, y): x \in \tilde{x}, y \in \tilde{y}\}. \tag{2.2.11}$$

We shall verify that $\tilde{\varrho}(\tilde{x}, \tilde{y})$ is a metric. Indeed, if $\tilde{x} = \tilde{y}$, then taking $x = y \in \tilde{x} = \tilde{y}$ we obtain $\tilde{\varrho}(\tilde{x}, \tilde{y}) \leq \varrho(x, y) = 0$. Suppose that $\tilde{\varrho}(\tilde{x}, \tilde{y}) = 0$. This means that there exist sequences of elements $\{x_n\}$, $\{y_n\}$, $x_n \in \tilde{x}$, $y_n \in \tilde{y}$ such that $\lim_{n \to \infty} \varrho(x_n, y_n) = 0$. By the invariance of the metric, we have $\lim_{n \to \infty} \varrho(x_n - y_n, 0) = 0$. Observe that $x_n - y_n \in \tilde{x} - \tilde{y}$. Moreover, by the closedness of Y and the continuity of addition, all equivalence classes are closed. Hence, $\tilde{x} - \tilde{y}$ contains 0, which entails $\tilde{x} - \tilde{y} = 0$. Thus $\tilde{x} = \tilde{y}$.

Invariance of the metric $\varrho(x, y)$ implies the invariance of the metric $\tilde{\varrho}(\tilde{x}, \tilde{y})$. In this case, the triangle inequality of the metric is equivalent to the inequality

$$\tilde{\varrho}(0, \tilde{x} + \tilde{y}) \leq \tilde{\varrho}(0, \tilde{x}) + \tilde{\varrho}(0, \tilde{y}).$$

Let $x_n \in \tilde{x}$ and $y_n \in \tilde{y}$ be sequences such that $\lim_{n \to \infty} \varrho(0, x_n) = \tilde{\varrho}(0, \tilde{x})$ and $\lim_{n \to \infty} \varrho(0, y_n) = \tilde{\varrho}(0, \tilde{y})$. Then

$$\tilde{\varrho}(0, \tilde{x} + \tilde{y}) \leq \inf \varrho(0, x_n + y_n) \leq \inf[\varrho(0, x_n) + \varrho(0, y_n)]$$
$$= \tilde{\varrho}(0, \tilde{x}) + \tilde{\varrho}(0, \tilde{y}).$$

In the same way continuity of multiplication by scalars can be verified. Indeed, if $\tilde{x}_n \to 0$, then there exists a sequence $x_n \in \tilde{x}_n$ such that $x_n \to 0$. Let $\{t_n\}$ be a sequence of scalars tending to 0. Then $t_n x_n \to 0$. Hence, the sequence $\tilde{\varrho}(0, [t_n x_n])$ tends to zero, since $\tilde{\varrho}(0, [t_n x_n]) \leq \varrho(0, t_n x_n)$.

Thus, we have proved that the quotient space with the invariant metric $\tilde{\varrho}(\tilde{x}, \tilde{y})$ is a linear metric space. The metric $\tilde{\varrho}(\tilde{x}, \tilde{y})$ has property (2) since $\tilde{\varrho}([x], [y]) \leq \varrho(x, y)$.

If Y is not a closed set, then $\tilde{\varrho}([x], [y])$ does not satisfy the first condition of a metric. For this reason we shall mainly consider closed sets Y.

We should point out here the terminological ambiguities related to different uses of the term "subspace". Namely, in functional analysis this term is used exclusively for linear subsets which are closed. This does not correspond to the definition of subspaces introduced in Section 1.1 for linear subsets. However, following tradition, we shall hold to this ambiguity. Since in the sequel we shall consider only linear metric spaces and their generalizations, namely linear topological spaces, we shall use the term subspace for a closed linear set.

If a metric $\varrho(x, y)$ is not invariant, formula (2.2.11) need not give a metric as is shown by the following example.

EXAMPLE 2.2.7. Let $X = R^2$ be a two-dimensional real space (a plane) with the metric

$$\varrho((x_1, x_2), (y_1, y_2)) = |\tan^{-1}x_1 - \tan^{-1}y_1| + |\tan^{-1}x_2 - \tan^{-1}y_2|.$$

Let Y be a subspace defined as follows: $Y = \{(x_1, x_2): x_1 = x_2\}$. The cosets are of the form $Y_r = \{(x_1, x_2): x_1 = x_2 + r\}$.

Formula (2.2.11) does not define a metric since $\tilde{\varrho}(Y_r, Y_{r_0}) = \inf_t (|\tan^{-1} t - \tan^{-1} t| + |\tan^{-1}(t+r) - \tan^{-1}(t+r_0)|) = 0$.

The difficulties involved in defining a metric on quotient spaces emphasize the importance of the Kakutani theorem, cited below without proof.

THEOREM 2.2.11 (Kakutani, 1936). *Let (X, ϱ) be a linear metric space (real or complex). Then there exists an invariant metric $\varrho'(x, y)$ in X which is equivalent to $\varrho(x, y)$.*

The proof of this theorem can be found e.g. in Rolewicz (1985b).

Let X be a normed space, real or complex, and let Y be a closed subspace of X. Consider the quotient space X/Y. Define a norm on X/Y by the formula

$$||\tilde{x}||' = \inf\{||x||: x \in \tilde{x}\} \qquad (2.2.12)$$

Since there is a one-to-one correspondence between invariant metrics and F-norms, in virtue of the previous arguments $||x||'$ is an F-norm. Observe that the homogeneity of $||x||$ implies the homogeneity of $||\tilde{x}||'$. Hence $||\tilde{x}||'$ is a norm.

Let X be a real seminormed space. In a similar way as above we can verify that formula (2.2.12) defines a seminorm on a quotient space.

Let $(X_1, \varrho^1), \ldots, (X_n, \varrho^n)$ be linear metric spaces. We assume that the metrics $\varrho^1, \ldots, \varrho^n$ are invariant. Thus, we can define on the Cartesian product $X_1 \times \ldots \times X_n$ the following metrics (see Example 1.3.4)

$$\varrho_{\sup}(x, y) = \sup_{1 \leq i \leq n} \varrho^i(x_i, y_i), \tag{2.2.13}$$

$$\varrho_1(x, y) = \sum_{i=1}^{n} \varrho^i(x_i, y_i), \tag{2.2.14}$$

$$\varrho_2(x, y) = \left(\sum_{i=1}^{n} (\varrho^i(x_i, y_i))^2\right)^{1/2}, \tag{2.2.15}$$

where $x = (x_1, \ldots, x_n)$, $y = (y_1, \ldots, y_n)$. Since ϱ^i, $i = 1, \ldots, n$, are invariant, the metrics $\varrho_{\sup}, \varrho_1, \varrho_2$ are also invariant. The corresponding F-norms are

$$||x||_{\sup} = \sup_{1 \leq i \leq n} ||x_i||_i, \tag{2.2.16}$$

$$||x||_{l_1} = \sum_{i=1}^{n} ||x_i||_i, \tag{2.2.17}$$

$$||x||_{l_2} = \left(\sum_{i=1}^{n} ||x_i||_i^2\right)^{1/2}, \tag{2.2.18}$$

where $||x||_i$ denotes the norm on X_i corresponding to the metric ϱ^i, $i = 1, \ldots, n$.

If the norms $||\ ||_i$, $i = 1, \ldots, n$, are homogeneous, then the norms $||x||_{\sup}, ||x||_{l_1}, ||x||_{l_2}$ are also homogeneous. Since

$$||x||_{\sup} \leq ||x||_{l_2} \leq ||x||_{l_1} \leq n \cdot ||x||_{\sup},$$

the norms $||x||_{\sup}, ||x||_{l_1}, ||x||_{l_2}$ are equivalent.

2.3. LINEAR FUNCTIONALS

Let X and Y be linear spaces (real or complex). A mapping f of the space X into the space Y such that

$$f(x+y) = f(x) + f(y) \text{ for } x, y \in X,$$
$$f(tx) = tf(x) \text{ for } x \in X \text{ and all numbers } t,$$

is called a *linear operator*. If Y is a scalar field, f is called a *linear functional*.

Let X be a linear space (real or complex). A subset Y of the space X is called a *linear manifold* if $ax+by \in Y$ for $x, y \in Y$ and all numbers a, b such that $a+b = 1$.

Let A be an arbitrary set contained in X. By the *affine hull* of the set A, aff(A), we mean the smallest linear manifold containing A. In the same way as in Theorem 2.1.2 we can show that

$$\text{aff}(A) = \{t_1 x_1 + \ldots + t_n x_n \colon x_1, \ldots, x_n \in A, t_1 + \ldots + t_n = 1\}.$$

THEOREM 2.3.1. *Let X be a linear space (real or complex) and let Y be a linear manifold in X. Let x_0 be an arbitrary element of Y. Then the set $Y_0 = Y - x_0$ forms a linear subspace which is independent of the choice of x_0.*

Proof. Let $y_1, y_2 \in Y_0$. Then $y_1 = y_1' - x_0$, $y_2 = y_2' - x_0$ for some $y_1', y_2' \in Y$. Let a, b be arbitrary numbers. Consider

$$ay_1 + by_2 + x_0 = ay_1' + by_2' + (1-a-b)x_0.$$

Since $a+b+(1-a-b) = 1$, the above element belongs to Y and hence $ay_1 + by_2 \in Y_0$. This means that Y_0 is a subspace. Let x_1 be an arbitrary element of Y. Then $x_1 - x_0 \in Y_0$ and $Y - x_0 = Y - x_0 + (x_0 - x_1) = Y - x_1$ since Y_0 is a subspace. ∎

We say that a linear manifold Y *is of dimension n* if the dimension of Y_0 is equal to n. We say that a linear manifold Y *has codimension n* (or *is of codimension n*) if the quotient space X/Y_0 is of dimension n, where, as above, Y_0 is the subspace $Y - x_0$. Manifolds of codimension 1 are called *hyperplanes*.

THEOREM 2.3.2. *Let X be a linear space (real or complex). Let f be a linear functional defined on X. Then the set*

$$H_f = \{x \in X \colon f(x) = 1\}$$

is a hyperplane which does not contain the element 0. Conversely, if H is a hyperplane which does not contain 0, then there exists a linear functional f such that $H = \{x \in X \colon f(x) = 1\}$.

2.3. Linear functionals

Proof. Let $x, y \in H_f$ and let a and b be numbers such that $a+b = 1$. Then $f(ax+by) = af(x)+bf(y) = a+b = 1$, i.e., $ax+by \in H_f$. The set $H_f^0 = H_f - x_0$ is a subspace for $x_0 \in H_f$ and $H_f^0 = \{x \in X: f(x) = 0\}$. Consider now the quotient space X/H_f^0. On each element of X/H_f^0, i.e., on each coset, the functional f is constant. Moreover, if the functional f takes the same value on two elements of X, then those elements belong to the same coset. Hence, a linear one-to-one correspondence can be established between numbers and cosets. This proves that the quotient space is one-dimensional. Moreover, $f(0) = 0$ and $0 \notin H_f$.

Suppose now that a hyperplane H is given, i.e., H is a linear manifold such that the quotient space X/H^0 is of dimension one, H^0 being a subspace generated by H. That is, each element of the quotient space can be represented as te, where e is an arbitrary nonzero element of the quotient space. By assumption, $H \neq [0]$ and H can be represented by the element e. By defining a functional f on X as $f(x) = t$ for $x \in te$ it is easy to verify that f is a linear functional and $H_f = H$. ∎

Let X be a linear metric space (real or complex). A linear functional (operator) f defined on the space X is called a *continuous linear functional* (*operator*) if it is a continuous mapping.

THEOREM 2.3.3. *If a linear functional (operator) f is continuous at a certain point $x_0 \in X$, then it is continuous on the whole space.*

Proof. Assume that a sequence $\{x_n\}$ is convergent to $x \in X$. Then by continuity of addition, the sequence $\{x_n'\}$, $x_n' = x_n + (x_0 - x)$, converges to x_0 and, by assumption, $\{f(x_n')\}$ converges to $f(x_0)$. By the additivity of f

$$\lim_{n \to \infty} f(x_n) + f(x_0) - f(x) = f(x_0),$$

and finally

$$\lim_{n \to \infty} f(x_n) = f(x). \blacksquare$$

THEOREM 2.3.4. *Let X be a linear metric space (real or complex). A linear functional f defined on X is continuous if and only if the linear manifold $H_f = \{x \in X: f(x) = 1\}$ is closed.*

Proof. Assume that f is continuous and take a sequence $\{x_n\}$ of elements of H_f which converges to $x \in X$. Then $f(x_n) = 1$, $n = 1, 2, ...$, and, by the continuity of f, $f(x) = 1$. Hence $x \in H_f$.

Suppose now that f is not continuous. This implies that f is not continuous at 0. Hence, there exists a sequence of points $\{x_n\}$ convergent to zero and a positive number ε such that $|f(x_n)| > \varepsilon$. The sequence $\{1/f(x_n)\}$ is bounded and, by the Bolzano–Weierstrass theorem, it contains a convergent subsequence, say $\{1/f(x_{n_m})\}$. Continuity of multiplication by numbers implies that the sequence $\{y_m\} = \{x_{n_m}/f(x_{n_m})\}$ converges to 0. Moreover, $f(y_m) = 1$, $y_m \in H_f$ but $0 \notin H_f$. ∎

It may happen that in a linear metric space there are no continuous linear functionals which are not identically equal to 0. However, as we shall see later on, in normed spaces continuous linear functionals different from 0 always exist.

Let X be a linear space (real or complex). The set of all linear functionals defined on X is called a *dual space* and denoted by X'. In this set operations of addition and multiplication by numbers are defined as follows: the sum of elements $f, g \in X$, $(f+g)$, is a linear functional defined by the formula

$$(f+g)(x) = f(x)+g(x),$$

the product of a functional f by a number t is a linear functional defined by the formula

$$(tf)(x) = tf(x).$$

It is evident that the set X' with the operations defined above forms a linear space (real or complex).

Let X be a linear metric space (real or complex). By X^* we denote the subset of the dual space consisting of all continuous linear functionals. Since the sum of two continuous mappings is continuous and the product of a continuous mapping by a number is a continuous mapping, the set X^* is a linear space. This space is called the *conjugate space* of X.

Now the question arises whether a metric $\varrho(f, g)$ can be introduced on X^* in such a way that:

1° X^* is a metric space with the metric $\varrho(f, g)$,

2.3. Linear functionals

2° a mapping f is continuous with respect to both variables, i.e., if $f_n \to f$ and $x_n \to x$, then $f_n(x_n) \to f(x)$.

It is not always possible to define such a metric. However, in the case which interest us most, namely in the case of normed spaces, a metric having properties 1° and 2° can be defined.

Let $(X, \|\ \|)$ be a normed space (real or complex). By the *norm of a functional* f defined on X we mean the number

$$\|f\| = \sup_{\|x\| \leq 1} |f(x)|. \tag{2.3.1}$$

Observe that for every continuous linear functional f the number given by formula (2.3.1) is finite. Suppose, on the contrary, that this number is not finite for some f. Then there exists a sequence of elements $\{x_n\}$, $\|x_n\| \leq 1$, such that $f(x_n) \geq n^2$. Since the sequence $\{x_n\}$ is bounded, the sequence $\{x_n/n\}$ tends to zero. On the other hand, the sequence $\{f(x_n/n)\}$ diverges to infinity, which contradicts the continuity of f.

Now we show that $\|f\|$ is a norm. Indeed, for the functional identically equal to 0, $\|0\| = \sup_{\|x\| \leq 1} 0 = 0$, and conversely, if $\|f\| = 0$, then for $x \neq 0$

$$|f(x/\|x\|)| \leq \|f\| = 0.$$

Hence, $|f(x)| = 0$ and consequently $f = 0$.

The triangle inequality holds since

$$\|f+g\| = \sup_{\|x\| \leq 1} |f(x)+g(x)| \leq \sup_{\|x\| \leq 1} |f(x)| + \sup_{\|x\| \leq 1} |g(x)|$$
$$= \|f\| + \|g\|,$$

and $\|f\|$ is homogeneous:

$$\|tf\| = \sup_{\|x\| \leq 1} |tf(x)| = |t| \cdot \sup_{\|x\| \leq 1} |f(x)| = |t| \cdot \|f\|.$$

Hence, $\|f\|$ is a norm. The definition implies that $|f(x)| \leq \|f\| \cdot \|x\|$. This inequality is obvious for $x = 0$. Let $x \neq 0$; then $|f(x/\|x\|)| \leq \|f\|$ and $|f(x)| \leq \|f\| \|x\|$.

This inequality implies immediately that if $x_n \to x$ and $f_n \to f$ then $f_n(x_n) \to f(x)$. We have

$$|f_n(x_n) - f(x)| \leq \|f_n\| \cdot \|x_n - x\| + \|f_n - f\| \cdot \|x\|.$$

Since the sequence $\{f_n\}$ converges, the sequence of norms of f_n is bounded. This implies that the right-hand side of the above inequality tends to zero.

If $(X, \|\ \|)$ is a real normed space, then

$$\|f\| = \sup_{\|x\|\leq 1} f(x). \tag{2.3.2}$$

Indeed, by the definition of a norm there exists a sequence $\{x_n\}$, $\|x_n\| \leq 1$, such that $\lim_{n\to\infty} |f(x_n)| = \|f\|$.

Let

$$y_n = \begin{cases} x_n & \text{if } f(x_n) \geq 0, \\ -x_n & \text{if } f(x_n) < 0. \end{cases}$$

Then $\|y_n\| \leq 1$, $f(y_n) = |f(x_n)|$ and $\lim_{n\to\infty} f(y_n) = \|f\|$, which implies formula (2.3.2).

For complex normed spaces formula (2.3.2) is of the form

$$\|f\| = \sup_{\|x\|\leq 1} \operatorname{Re} f(x), \tag{2.3.2'}$$

where $\operatorname{Re} a$ denotes the real part of a complex number a. The proof of formula (2.3.2') is similar to that of formula (2.3.2). Namely, we choose a sequence $\{x_n\}$, $\|x_n\| \leq 1$, such that $\lim_{n\to\infty} |f(x_n)| = \|f\|$. Putting

$$y_n = \frac{|f(x_n)|}{f(x_n)} x_n,$$

we obtain $\|y_n\| = 1$ and $\operatorname{Re} f(y_n) = f(y_n) = |f(x_n)|$. This and the inequality $\operatorname{Re} f(x) \leq |f(x)|$ immediately imply formula (2.3.2').

Observe that in formulae (2.3.1), (2.3.2), (2.3.2') the supremum over the closed unit ball can be replaced by the supremum over the open unit ball (which follows immediately from the continuity of functionals) or by the supremum over the sphere $S = \{x: \|x\| = 1\}$ (which follows from the homogeneity of functionals).

Now we turn to investigations of seminormed spaces, i.e., real spaces $(X, \|\ \|)$ with seminorms $\|\ \|$. It turns out that formula (2.3.2) defines a seminorm in a conjugate space. Clearly, for such a seminorm the formula $f(x) \leq \|f\| \cdot \|x\|$ holds.

2.3. Linear functionals

Let $(X, \|\ \|)$ be a real normed (or seminormed) space. The ball
$$K = \{x \in X: \|x\| < 1\}$$
defined by the norm (or seminorm) $\|\ \|$ is a set that is open, bounded, convex and symmetric (containing 0) and there is a one-to-one correspondence between $\|\ \|$ and such sets.

Suppose we are given a functional with the norm equal to 1 in X (i.e., $\|f\| = 1$). As we have shown, the functional f defines the closed hyperplane $H_f = \{x \in X: f(x) = 1\}$. Now the geometrical question arises: how are the ball K and the hyperplane H_f situated with respect to one another?

First, observe that if $x \in K$ then $\|x\| < 1$ and by the definition of $\|f\|$, the inequality $f(x) \leq \|f\| \cdot \|x\| < 1$ holds. This means that the ball K is disjoint with the hyperplane H_f. On the other hand, there exist sequences $\{x_n\}$, $\{y_n\}$, $x_n \in K$, $y_n \in H_f$, such that $\lim_{n \to \infty} \|x_n - y_n\| = 0$. Indeed, by the definition of $\|f\|$, there exists a sequence of elements $x_n \in K$ such that $\lim_{n \to \infty} f(x_n) = \|f\| = 1$.

Letting
$$y_n = \frac{1}{f(x_n)} \cdot x_n,$$
we have $y_n \in H_f$ and
$$\lim_{n \to \infty} \|x_n - y_n\| = \lim_{n \to \infty} \left(1 - \frac{1}{f(x_n)}\right) \cdot \|x_n\| = 0.$$

We say that a closed hyperplane H is a *supporting hyperplane of a convex open set K* if H has no points in common with K and there exist two sequences $\{x_n\}$, $\{y_n\}$, $x_n \in K$, $y_n \in H$, such that $\lim_{n \to \infty} \|x_n - y_n\| = 0$.

Thus, we have shown that, for a functional f with $\|f\| = 1$, H_f is a supporting hyperplane.

Assume now that H is a supporting hyperplane for an open unit ball K. By Theorem 2.3.2 and Theorem 2.3.4, there exists a continuous linear functional f such that $H = H_f = \{x \in X: f(x) = 1\}$. We shall prove that the norm of this functional is equal to 1. Let $\|x\| < 1$. Thus,

$x \in K$ and $f(x) < 1$ since K is disjoint with H. On the other hand, there exist two sequences $\{x_n\}$, $\{y_n\}$, $x_n \in K$, $y_n \in H$ such that $\lim_{n \to \infty} ||x_n - y_n|| = 0$. Since $y_n \in H$, we have $f(y_n) = 1$. Hence, by the continuity of f, $\lim_{n \to \infty} f(x_n - y_n) = \lim_{n \to \infty} f(x_n) - 1 = 0$ and, consequently,

$$||f|| = \sup_{||x|| < 1} f(x) = 1.$$

As we see, there is a one-to-one correspondence between closed hyperplanes supporting a unit ball and functionals with the norm equal to 1.

In a similar way we can prove that, in general, the hyperplane H_f is a supporting hyperplane of the ball K_r,

$$K_r = \{x : ||x|| < r\}, \quad \text{where } r = \frac{1}{||f||}.$$

Indeed, by the definition of the norm we have

$$f(x) \leq ||f|| \, ||x|| < 1$$

for $x \in K_r$, which implies that K_r and H_f are disjoint. On the other hand, choose a sequence $\{x_n\}$ such that $||x_n|| < 1$, $f(x_n) \to ||f||$. Let

$$y_n = \frac{x_n}{||f||}, \quad z_n = \frac{1}{f(x_n)} x_n.$$

Obviously, $y_n \in K_r$, $z_n \in H_f$ and $\lim_{n \to \infty} ||y_n - z_n|| = 0$.

We say that a closed hyperplane H is *tangent to an open convex set* K if $H \cap K = \emptyset$ and there exists a sequence $\{x_n\}$, $x_n \in K$, which converges to an element $x \in H$. Clearly, each tangent hyperplane is also a supporting hyperplane. However, there may exist supporting hyperplanes which are not tangent. To prove this, observe first that in the case where K is the unit ball the definition of a tangent hyperplane can be formulated in the following way. A closed hyperplane H is tangent to the unit ball K if and only if there exists an element $x \in H$ with the norm equal to 1.

EXAMPLE 2.3.1. Let $X = L^1[0, 1]$ and let a functional f be of the form

$$f(x) = \int_0^1 t x(t) \, dt.$$

We have
$$f(x) \leq \int_0^1 |tx(t)|\,dt < \int_0^1 |x(t)|\,dt = ||x||. \tag{2.3.3}$$
This means that $||f|| \leq 1$. On the other hand, letting
$$x_n = n \cdot \chi_{(1-1/n,\,1)}$$
where χ_A denotes the characteristic function of the set A, we obtain
$$||x_n|| = \int_0^1 x_n(t)\,dt = n \cdot \frac{1}{n} = 1$$
and
$$f(x_n) = \int_0^1 tx_n(t)\,dt = n \cdot \int_{1-\frac{1}{n}}^1 t\,dt > n\left(1 - \frac{1}{n}\right) \cdot \frac{1}{n} = \left(1 - \frac{1}{n}\right).$$
Thus,
$$||f|| = \sup_{||x|| \leq 1} |f(x)| \geq 1$$
and finally $||f|| = 1$.

Since inequality (2.3.3) is strict, there is no element x with the norm equal to 1 such that $f(x) = ||f|| = 1$.

2.4. FINITE-DIMENSIONAL SPACES

Let X be an n-dimensional linear metric space, real or complex, with a basis $\{e_1, \ldots, e_n\}$.

Suppose we are given a sequence of n-tuples of numbers
$$t^m = (t_1^m, \ldots, t_n^m)$$
such that
$$\lim_{m \to \infty} t_i^m = t_i, \quad i = 1, \ldots, n. \tag{2.4.1}$$

Continuity of addition and scalar multiplication imply that the sequence $\{x^m\}$,
$$x^m = t_1^m e_1 + \ldots + t_n^m e_n, \tag{2.4.2}$$

converges to the element

$$x = t_1 e_1 + \ldots + t_n e_n. \tag{2.4.3}$$

THEOREM 2.4.1. *If a sequence of elements of form* (2.4.2) *converges to an element x given by formula* (2.4.3), *then* (2.4.1) *holds*.

Proof. By continuity of addition it is enough to prove the theorem for $x = 0$. Suppose that all sequences $\{t_i^m\}$, $i = 1, \ldots, n$, are bounded but not all converge to zero. Hence, by the Bolzano–Weierstrass theorem, we can choose a sequence of indices $\{m_k\}$ such that all sequences $\{t_i^{m_k}\}$ converge to some t_i and not all t_i are equal to zero. Then, by continuity of addition and scalar multiplication we obtain

$$t_1 e_1 + \ldots + t_n e_n = 0,$$

which contradicts the linear independence of elements $e_1, \ldots, e_n$.

Suppose now that the sequences $\{t_i^m\}$ are unbounded for all i. Then we can choose a sequence of indices $\{m_k\}$ and an index i_0 such that

$$\lim_{k \to \infty} |t_{i_0}^{m_k}| = +\infty, \tag{2.4.4}$$

$$|t_{i_0}^{m_k}| \geq |t_i^{m_k}|, \quad i = 1, \ldots, n. \tag{2.4.5}$$

Write

$$b_k = (t_{i_0}^{m_k})^{-1}.$$

The sequence $\{b_k\}$ tends to zero and hence $b_k x^{m_k} \to 0$. By (2.4.5), the sequences $\{b_k t_i^{m_k}\}$ are bounded for $i = 1, \ldots, n$ and $b_k t_{i_0}^{m_k} = 0$, $k = 1, 2, \ldots$ In this way we reduce this case to the previous one, namely to the case of bounded sequences $\{t_i^m\}$. ∎

As a consequence of Theorem 2.4.1 we obtain

THEOREM 2.4.2. *Every finite-dimensional space is complete*.

Proof. Let $\{e_1, \ldots, e_n\}$ be a basis in X and let $\{x^m\}$, $x^m = t_1^m e_1 + \ldots + t_n^m e_n$, be a fundamental sequence. This implies that the sequences $\{t_i^m\}$ are fundamental and hence each $\{t_i^m\}$ converges to a certain

t_i, $i = 1, \ldots, n$. By continuity of addition and scalar multiplication, the sequence $\{x^m\}$ converges to an element $x = t_1 e_1 + \ldots + t_n e_n$. ∎

THEOREM 2.4.3. *Let X be a finite-dimensional linear metric space. Then there exists a neighbourhood of 0 in X with the compact closure.*

Proof. Let $\{e_1, \ldots, e_n\}$ be a basis in X. The set
$$U = \{x \colon x = t_1 e_1 + \ldots + t_n e_n, \quad |t_i| < 1, i = 1, \ldots, n\}$$
is a neighbourhood of zero. By the Bolzano–Weierstrass theorem, the closure of the set U is compact. ∎

Let X be an n-dimensional space, real or complex, with a basis $\{e_1, \ldots, e_n\}$. Let f be a linear functional defined on X. Then f can be expressed in the form
$$f(x) = a_1 t_1 + \ldots + a_n t_n, \tag{2.4.6}$$
where $x = t_1 e_1 + \ldots + t_n e_n$ and $a_i = f(e_i)$, $i = 1, \ldots, n$. On the other hand, each formula of form (2.4.6) defines a linear functional. Let f_i be linear functionals such that
$$f_i(t_1 e_1 + \ldots + t_n e_n) = t_i.$$

By (2.4.6), every linear functional defined on X can be expressed in the form
$$f = a_1 f_1 + \ldots + a_n f_n, \tag{2.4.7}$$
where $a_i = f(e_i)$ and the representation (2.4.7) is unique. Thus, the dual space X' is also n-dimensional.

Let X be an n-dimensional linear metric space (real or complex) and let f be a linear functional defined on X. Let $\{x^m\}$, $x^m = t_1^m e_1 + \ldots + t_n^m e_n$, be a sequence of elements of X convergent to $x = t_1 e_1 + \ldots + t_n e_n$. Then, by Theorem 2.4.1, $\lim_{m \to \infty} t_i^m = t_i$, $i = 1, \ldots, n$, and, by (2.4.6), $f(x^m) = a_1 t_1^m + \ldots + a_n t_n^m$. Therefore we have
$$f(x^m) \to f(x) = a_1 t_1 + \ldots + a_n t_n.$$

Thus, we have shown

THEOREM 2.4.4. *In a finite-dimensional linear metric space (real or complex) every linear functional is continuous.*

In other words, in finite-dimensional spaces $X^* = X'$. Using the axiom of choice we can show that $X^* \neq X'$ in any infinite-dimensional space.

THEOREM 2.4.5. *Let $(X, ||\ ||)$ be a finite-dimensional normed space (real or complex) and let K be an open, convex and bounded set contained in X. Every hyperplane H supporting K is tangent to K.*

Proof. By definition, $H \cap K = \emptyset$ and there exist sequences of elements $\{x_n\}$, $\{y_n\}$, $x_n \in K$, $y_n \in H$ such that $\lim_{n\to\infty} ||x_n - y_n|| = 0$. By Theorem 2.4.3, there exists a neighbourhood U of zero whose closure is compact. Since K is bounded, there exists a number b such that $K \subset bU$. Clearly, the set $b\overline{U}$ is compact. Hence, the set $\overline{K}$ is compact and the sequence $\{x_n\}$ contains a subsequence $\{x_{n_k}\}$ which is convergent to a certain $x \in X$. Hence,

$$\lim_{k\to\infty} ||y_{n_k} - x|| \leq \lim_{k\to\infty} ||y_{n_k} - x_{n_k}|| + \lim_{k\to\infty} ||x_{n_k} - x|| = 0,$$

which proves that $y_{n_k} \to x$. By the closedness of H we obtain $x \in H$. ∎

COROLLARY 2.4.6. *If $(X, ||\ ||)$ is a finite-dimensional normed or seminormed space (real or complex), then for every linear functional f there exists an element x with the norm equal to 1 such that $f(x) = ||f||$.*

Let $(X, ||\ ||)$ be a real normed (or seminormed) space and let K be the unit ball, $K = \{x: ||x|| < 1\}$. Consider now the dual space $X' = X^*$. The question arises whether we can determine the unit ball in the conjugate space. Clearly, if the norm $||x||$ is defined analytically by a certain formula, we can evaluate the respective norm of functionals in the conjugate space and determine the unit ball in that space. However, in many engineering applications balls are determined by experiments as certain convex sets. The problem is in determining the respective balls in the conjugate space by means of geometrical constructions and not analytically.

2.4. Finite-dimensional spaces

The spaces X and X' are n-dimensional, and these spaces generate in the space R^n norms $||x||$, $||x||'$ and balls $K = \{x: ||x|| < 1\}$, $K' = \{x: ||x||' < 1\}$. Suppose that we know the ball K and our aim is to determine the ball K' geometrically.

The space R^n is equipped with the inner product $(x, y) = x_1 y_1 + \ldots + x_n y_n$. The construction is as follows. Take a hyperplane H which is tangent to K and project 0 orthogonally on H. All such projections form a figure K_0. Transform K_0 by inversion, i.e., by assigning to an arbitrary point $x \in K_0$ a point $x' = ax$, where a is the inverse of the square of the Euclidean distance between x and 0.

The reader who is not familiar with n-dimensional spaces may argue that the above reasoning is more analytic than geometrical (especially for $n > 3$). However, for $n = 2, 3$ we obtain nice geometrical constructions.

Now we prove that x' obtained in the above way can be interpreted as a linear functional with the norm equal to 1.

Let H be a supporting hyperplane of K. Then there exists a linear functional f, $f(x) = a_1 t_1 + \ldots + a_n t_n$, such that $H = \{x: f(x) = 1\}$.

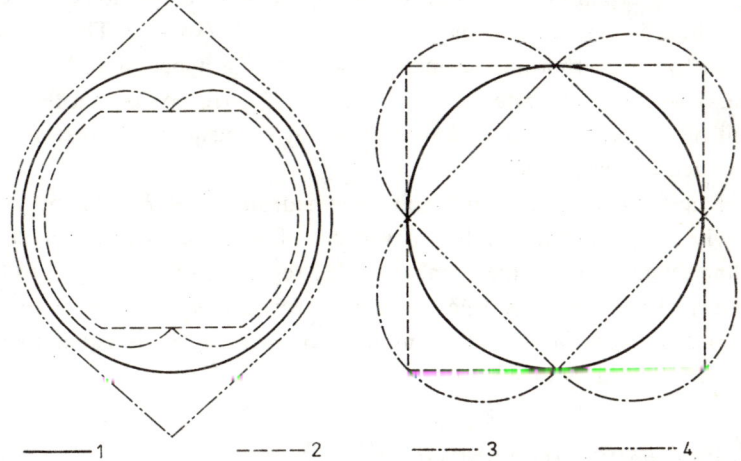

FIGS. 2.4.1 and 2.4.2. Line 1—the boundary of the unit ball of the Euclidean space. Line 2—the boundary of the ball K. Line 3—the figure consisting of orthogonal projections of the centre on the supporting hyperplanes. Line 4—the boundary of the ball K' in the conjugate space.

Moreover, we can identify f with an n-tuple of numbers $(a_1, \ldots, a_n)$. This vector is orthogonal to the hyperplane H and hence it has the same direction as $x \in K_0$ but its length is unknown. Put $f(x) = ax$. But $x \in H$ and consequently $f(x) = 1$, whence $a(x, x) = 1$.

Figures 2.4.1–2.4.2 illustrate the application of the above construction to the construction of unit balls in conjugate spaces (we omit the pictures of those solids which are obtained by rotating the figures shown in Figs. 2.4.1–2.4.2 around the vertical axis for such pictures would hardly be readable in one-colour print). These figures deserve a short comment. Recall that inversions map circles on circles, spheres on spheres and a line (plane) can be regarded as a particular case of a circle (sphere) with an infinite radius. Moreover, it is clear that circles (spheres) passing through 0 are mapped on lines (planes). The following fact from elementary geometry is useful in our considerations. Let a line segment be given. The geometrical locus of points which are right-angle vertices of a triangle with the given line segment as the base is a circle (sphere) having that line segment as its diameter.

If the ball K has an acute vertex, then there are more than one supporting hyperplanes passing through that vertex. Projections of zero onto those hyperplanes form a section of a circle (sphere). This section is transformed into a line segment (a section of a hyperplane) by inversion. Therefore, an acute vertex of the ball K corresponds to a flat segment of the ball K'. And vice-versa, a flat segment of K corresponds to an acute vertex of K'.

Henceforth, if K is a polygon (polyhedron), then K' is a polygon (polyhedron) and the number of vertices of K' is equal to the number of line segments of K, the number of line segments of K' is equal to the number of vertices of K (the number of vertices of K' is equal to the number of faces of K and the number of faces of K' is equal to the number of vertices of K).

2.5. Extensions of functionals

Let $(X, \| \ \|)$ be a normed (seminormed) real space. Let X_0 be a subspace of X and let f be a continuous linear functional defined on X_0. The question is whether it is possible to extend f onto the whole space X,

2.5. Extensions of functionals

i.e., whether there exists a continuous linear functional F defined on the whole space X such that $F(x) = f(x)$ for $x \in X_0$. The functional F is called an *extension* of f. If moreover $||F|| = ||f||$, we say that the extension F is *norm-preserving*.

THEOREM 2.5.1 (Hahn, 1927; Banach, 1929). *Let $(X, ||\ ||)$ be a real normed (seminormed) space. Let X_0 be a subspace of the space X of codimension 1. Then every continuous linear functional f defined on X_0 can be extended to a continuous linear functional F defined on X and such that $||F|| = ||f||$.*

Proof. Let y be an arbitrary element of X which does not belong to X_0. Since X_0 is of codimension 1, every element $x \in X$ can be uniquely represented in the form

$$x = ty+x', \tag{2.5.1}$$

where $x' \in X_0$. Let x', x'' be arbitrary elements of the space X_0. By the definition of norm,

$$f(x')+f(x'') = f(x'+x'') \leq ||f||\, ||(y+x')+(-y+x'')|| $$
$$\leq ||f||\,(||y+x'||+||-y+x''||).$$

Hence,

$$f(x'')-||f||\,||-y+x''|| \leq -f(x')+||f||\,||y+x'||.$$

Since the above inequality holds for arbitrary $x', x'' \in X_0$, we have

$$A = \sup_{x'' \in X_0} [f(x'')-||f||\,||-y+x''||]$$
$$\leq \inf_{x' \in X_0} [-f(x')+||f||\,||y+x'||] = B.$$

Let c be an arbitrary number such that $A \leq c \leq B$. Define a functional F on X by the formula

$$F(x) = tc+f(x'),$$

where, according to formula (2.5.1), x is of the form $x = ty+x'$, $x' \in X_0$. Take $t > 0$. Then

$$F(x) = tc+f(x') \leq tB+f(x')$$
$$\leq t\left(-f\left(\frac{x'}{t}\right)+||f||\cdot\left\|y+\frac{x'}{t}\right\|\right)+f(x')$$
$$= -f(x')+||f||\cdot||ty+x'||+f(x') = ||f||\cdot||x||.$$

For $t < 0$ the argumentation is the same with the only difference that the inequality $A \leqslant c$ is used instead of $c \leqslant B$. ∎

Observe that if Theorem 2.5.1 holds for all continuous linear functionals with the norm equal to 1 then it holds for all continuous linear functionals. For continuous linear functionals f with the norm equal to 1 the proof of Theorem 2.5.1 has a nice geometrical interpretation. Namely, let $X_1 = \{x \in X_0 \colon f(x) = 0\}$; X_1 is a subspace of X_0 and hence a subspace of X. Consider the quotient space X/X_1. Since X_1 is a subspace of X_0 of codimension 1 and X_0 is a subspace of X of codimension 1, the quotient space X/X_1 is two-dimensional. Since $(X, \|\ \|)$ is a normed (seminormed) space, the norm (seminorm) $\|\ \|$ induces a norm (seminorm) $\|\ \|'$ on X/X_1. Take the unit ball K in X/X_1. The ball K is an open convex symmetric set containing 0. The space X_0 generates a line in the space X/X_1. The set of all $x \in X_0$ such that $f(x) = 1$ determines a point lying on the intersection of this line with the sphere $S = \{y \in X/X_1 \colon \|y\|' = 1\}$.

Theorem 2.5.1 can be interpreted as follows: suppose we are given a two-dimensional open convex set and a point belonging to its boundary. Then there exists a line passing through the given point and not touching the set. The proof of Theorem 2.5.1 is actually an analytical proof of this fact.

As an immediate consequence of Theorem 2.5.1 we have the following theorem.

THEOREM 2.5.2 (Hahn, 1927; Banach, 1929). *Let $(X, \|\ \|)$ be a real separable normed (seminormed) space. Let X_0 be a subspace of the space X and let f be a continuous linear functional defined on X_0. Then there exists an extension of the functional f to a continuous linear functional F defined on X and such that $\|f\| = \|F\|$.*

Proof. Let $\{x_n\}$ be a dense sequence in X and let $X_n = \text{lin}\,\{X_0 \cup \{x_1, \ldots, x_n\}\}$. Clearly, either $X_n = X_{n+1}$ or X_n is a subspace of X_{n+1} of codimension 1. Hence, by induction with respect to n, using Theorem 2.5.1, we can extend the functionals f to a continuous linear functional

F defined on the set $\tilde{X} = \bigcup_{n=1}^{\infty} X_n$, which is dense in X. Moreover, $\|F\| = \|f\|$.

Let $y \in X$ and let $\{y_n\}$ be a sequence of elements of $\tilde{X}$ convergent to y. Since the norm of F is bounded, the sequence $\{F(y_n)\}$ is fundamental and $\{F(y_n)\}$ converges to a certain number a. Putting $F(y) = a$, we define a functional on the whole space X. It is easy to verify that F is a continuous linear functional with the norm equal to $\|f\|$. ∎

If $(X, \|\ \|)$ is a real nonseparable normed (seminormed) space, the proof of an analogous result to that of Theorem 2.5.2 is a little more complicated. All known proofs are based on an axiom of the set theory called the *axiom of choice* or on any other equivalent axiom. P. Cohen has shown the independence of this axiom from the other axioms of the set theory. In what follows we shall assume the axiom of choice. For the proof of Theorem 2.5.3, which is an analogue of Theorem 2.5.2 in nonseparable spaces, we shall use an axiom which is equivalent to the axiom of choice, namely the *Kuratowski–Zorn lemma*.

Let a set X be given. We say that an *order relation* $\prec$ is defined on X if the relation $\prec$ has the following properties: $x \prec x$, $x \prec y$ and $y \prec x$ imply $x = y$, $x \prec y$ and $y \prec z$ imply $x \prec z$. We say that a subset Y of the set X is *totally ordered* if, for any pair of elements $x, y \in Y$, either $x \prec y$ or $y \prec x$.

Let A be a subset of X. An element $x_A \in A$ is called a *maximal element* of the set A if $x_A \prec x$ for an $x \in A$ implies $x = x_A$. Clearly, the set A may have many maximal elements.

KURATOWSKI–ZORN LEMMA. *Let X be a set with an order relation $\prec$ defined on it. Assume that for each totally ordered subset A of the set X there exists an element a such that $x \prec a$ for all $x \in A$. Then X has a maximal element.*

With this axiom (observe that, in spite of its name, we regard the Kuratowski–Zorn lemma as an axiom) we can now prove

THEOREM 2.5.3 (Hahn, 1927; Banach, 1929). *Let $(X, \|\ \|)$ be a real normed (seminormed) space. Let X_0 be a subspace of X and let f be a continuous* *linear functional defined on X_0. Then there exists a continuous linear functional F which is a norm-preserving extension of f.*

Proof. Let $\mathfrak{A}$ be the set of all possible norm-preserving extensions of f. Thus, the set $\mathfrak{A}$ consists of all continuous linear functionals F such that F is defined on a certain subspace X_F containing X_0 and moreover $||f|| = ||F||$. We can order the set $\mathfrak{A}$ by the following relation: $F \prec G$ if and only if $X_F \subset X_G$. Consider now an arbitrary totally ordered subset $\mathfrak{B}$ of $\mathfrak{A}$. For any $F, G \in \mathfrak{B}$ either F is an extension of G or G is an extension of F.

Let $F_\mathfrak{B}$ be a functional defined on $X_{F_\mathfrak{B}} = \bigcup_{F \in \mathfrak{B}} X_F$ by the formula

$$F_\mathfrak{B}(x) = F(x) \quad \text{for} \quad x \in X_F.$$

Since $\mathfrak{B}$ is totally ordered, the functional $F_\mathfrak{B}$ is well defined and $F \prec F_\mathfrak{B}$ for each $F \in \mathfrak{B}$. Hence, by the Kuratowski–Zorn lemma there exists a maximal element F_0 in $\mathfrak{A}$. We shall prove that $X_{F_0} = X$. Suppose that $X_{F_0} \neq X$ and let y be an arbitrary element of X not belonging to X_{F_0}. Let $X_1 = \text{lin}\{X_{F_0} \cup \{y\}\}$. Clearly, X_{F_0} is a subspace of the space X_1 and X_{F_0} is of codimension 1. Hence, by Theorem 2.5.1, the functional F_0 can be extended to a continuous linear functional F_1 defined on the space X_1 and such that $||F_1|| = ||f||$. But this contradicts the fact that F_0 is a maximal element. ∎

THEOREM 2.5.4 (Bohnenblust, Sobczyk, 1939). *Let $(X, ||\;||)$ be a complex normed space. Let X_0 be a subspace of X and let f be a continuous linear functional defined on X_0. Then there exists a continuous linear functional F defined on X which is a norm-preserving extension of f.*

Proof. Clearly, the space X can be regarded as a real linear space. Then the functional $\text{Re} f$ (the real part of f) is a real-valued continuous linear functional defined on X_0. By Theorem 2.5.3 (or Theorem 2.5.2 for separable spaces), there exists a continuous linear real-valued functional $\tilde{F}$ defined on the whole space X which is such an extension of the functional $\text{Re} f$ that $||\tilde{F}|| = ||\text{Re} f||$. Let $F(x) = \tilde{F}(x) - i\tilde{F}(ix)$. The functional F is continuous, additive (i.e. $F(x+y) = F(x) + F(y)$) and

2.5. Extensions of functionals

homogeneous over reals. We shall prove that F is also homogeneous over complex numbers. Indeed,

$$\begin{aligned}F[(a+ib)x] &= \tilde{F}[(a+ib)x] - i\tilde{F}[i(a+ib)x] \\ &= a\tilde{F}(x) + b\tilde{F}(ix) - ia\tilde{F}(ix) + ib\tilde{F}(x) \\ &= (a+ib)\tilde{F}(x) - (a+ib)i\tilde{F}(ix) = (a+ib)F(x).\end{aligned}$$

Now we shall prove that F is an extension of f. Observe first that the homogeneity of f entails the equality $f(ix) = if(x)$ and hence

$$\operatorname{Im} f(x) = -\operatorname{Re} f(ix).$$

Thus,

$$\begin{aligned}F(x) &= \tilde{F}(x) - i\tilde{F}(ix) = \operatorname{Re} f(x) - i\operatorname{Re} f(ix) = \operatorname{Re} f(x) + i\operatorname{Im} f(x) \\ &= f(x)\end{aligned}$$

for $x \in X_0$ and F is an extension of f. This implies that $||F|| \geqslant ||f||$. To complete the proof it is enough to prove the opposite inequality. We have

$$\begin{aligned}||F|| &= \sup_{||x|| \leqslant 1} |F(x)| = \sup_{||x|| \leqslant 1} \sup_{|a|=1} |F(ax)| \\ &= \sup_{||x|| \leqslant 1} \sup_{|a|=1} \operatorname{Re} F(ax) = \sup_{||x|| \leqslant 1} \operatorname{Re} F(x) = ||\operatorname{Re} F||.\end{aligned}$$

Since $\operatorname{Re} F(x) = \tilde{F}(x)$, we obtain $||F|| = ||\tilde{F}|| = ||\operatorname{Re} f|| \leqslant ||f||$. ∎

Theorem 2.5.3 and Theorem 2.5.4 have numerous important applications.

THEOREM 2.5.5. *Let $(X, || \ ||)$ be a (real or complex) normed (or real seminormed) space. Then for every element $x_0 \in X$ there exists a continuous linear functional f with the norm equal to 1 such that $f(x_0) = x_0$.*

Proof. Let $X_0 = \operatorname{lin}\{x_0\}$. Define a functional $\tilde{f}$ on the space X_0 by the formula $\tilde{f}(x) = a||x_0||$, where $x = ax_0$. By taking an extension of $\tilde{f}$, defined on the whole space X, we find the functional having the required properties. ∎

COROLLARY 2.5.6. *Let $(X, || \ ||)$ be a (real or complex) normed (or real seminormed) space. Then*

$$||x|| = \sup_{||f|| \leqslant 1} |f(x)| \quad (||x|| = \sup_{||f|| \leqslant 1} f(x), \text{ respectively}).$$

Let $(X, \|\ \|)$ be a (real or complex) normed space (or a real seminormed space). Let X^* denote the conjugate space of the space X. As we have proved (see Section 2.3), the space X^* is a (real or complex) normed space (or a real seminormed space). Let $X^{**} = (X^*)^*$ denote the conjugate space of X^*. This space is called the *second conjugate space*.

Let x_0 be a fixed element of X. We shall prove that the formula
$$F(f) = f(x_0)$$
defines a continuous linear functional on X^*. The linearity of this functional follows immediately from the definition of addition of functionals and from the definition of multiplication of functionals by scalars. The continuity is a consequence of the inequality
$$F(f) = |f(x_0)| \leq \|f\| \|x_0\| \tag{2.5.2}$$
(or $\quad F(f) = f(x_0) \leq \|f\| \|x_0\|$).

The correspondence between x_0 and F induces the *canonical injection* n of X into X^{**} given by the formula $n(x_0) = F$. By (2.5.2), $\|n(x_0)\| \leq \|x_0\|$. Moreover, by the definition of the canonical injection and by Corollary 2.5.6, we have
$$\|F\| = \sup_{\|f\| \leq 1} |F(f)| = \sup_{\|f\| \leq 1} |f(x_0)| = \|x_0\|$$
(or $\quad \|F\| = \sup_{\|f\| \leq 1} F(f) = \sup_{\|f\| \leq 1} f(x_0) = \|x_0\|$).

Hence, the canonical injection is an isometry.

The separation theorems proved by Mazur (1933) are another important consequence of Theorem 2.5.3.

Let $(X, \|\ \|)$ be a normed (or seminormed) space. We say that two convex sets A and B can be *separated*, if there exist a continuous linear functional f and a constant c such that
$$f(x) \begin{cases} \leq c & \text{for } x \in A, \\ \geq c & \text{for } x \in B. \end{cases} \tag{2.5.3}$$

We say that the sets A and B can be *strongly separated* if there exist a continuous linear functional f, a constant c and a positive constant ε such that
$$f(x) \begin{cases} \leq c & \text{for } x \in A, \\ \geq c+\varepsilon & \text{for } x \in B. \end{cases} \tag{2.5.4}$$

2.5. Extensions of functionals

THEOREM 2.5.7. *Let $A \subset X$ be a convex set with a nonempty interior, Int $A \neq \emptyset$. Let x be an arbitrary point not belonging to Int A. Then the sets $\{x\}$ and A can be separated.*

Proof. Let x_0 be an arbitrary point of Int A. Then the set $A - x_0$ contains 0 in its interior. Hence, the set $A - x_0$ generates a seminorm $\|x\|$. Observe that $x - x_0$ does not belong to the set $A - x_0$. By Theorem 2.5.5, there exists a continuous linear functional f with the seminorm equal to 1 such that $f(x - x_0) = \|x - x_0\| \geq 1$. On the other hand,

$$f(y - x_0) \leq 1 \quad \text{for} \quad y \in A,$$

because the seminorm $\|f\|$ is equal to 1. This implies

$$f(y) \leq c = 1 + f(x_0)$$

for $y \in A$ and $f(x) \geq c$. ∎

COROLLARY 2.5.8. *Let A be a convex set with a nonempty interior, Int $A \neq \emptyset$. Assume that $0 \notin $ Int A. Then there exists a continuous linear functional f such that $f(y) \geq 0$ for $y \in A$.*

Proof. By Theorem 2.5.7, zero and the set A can be separated by a certain continuous linear functional $\tilde{f}$, $\tilde{f}(0) = 0$. Hence, $\tilde{f}(y)$ is either nonnegative for all $y \in A$ or nonpositive for all $y \in A$. In the first case we take $f = \tilde{f}$ and in the second case we take $f = -\tilde{f}$. ∎

THEOREM 2.5.9. *Let X be a real normed space. Let A and B be convex sets such that Int $A \neq \emptyset$ and Int $A \cap B = \emptyset$. Then the sets A and B can be separated.*

Proof. The set Int $A - B$ does not contain 0 because the sets Int A and B are disjoint. Moreover, the set Int $A - B$ has a nonempty interior. Hence, by Corollary 2.5.8, there exists a continuous linear functional f such that $f(x - y) \geq 0$ for $x \in $ Int A and $y \in B$. Thus, by the continuity of f, $f(x) \leq c$ for $x \in A$, and $f(y) \geq c$ for $y \in B$, where c is an arbitrary number such that

$$\sup_{y \in B} f(y) \leq c \leq \inf_{x \in A} f(x). \quad \blacksquare$$

The assumption that at least one of the sets A, B has a nonempty interior is essential and one can easily verify this by constructing approprate examples.

THEOREM 2.5.10. *Let $(X, \|\ \|)$ be a real normed space. Let A and B be disjoint closed convex sets in X. Assume that A is compact. Then there exists a continuous linear functional f which strongly separates the sets A and B.*

Proof. Since A is compact, B is closed and both sets are disjoint, we have

$$\inf_{\substack{x \in A \\ y \in B}} \|x-y\| = r > 0. \tag{2.5.5}$$

Let $K_r = \{x \in X: \|x\| < r\}$. By (2.5.5), the open set $A + K_r$ is disjoint with B. Hence, by Theorem 2.5.4, there exists a continuous linear functional f which separates the sets $A + K_r$ and B, i.e., there exists a c such that

$$f(x) \begin{cases} \leqslant c & \text{for } x \in A+K_r, \\ \geqslant c & \text{for } x \in B. \end{cases} \tag{2.5.6}$$

Observe that (2.5.6) implies $f(x) \leqslant c - \|f\| \cdot r$ for $x \in A$. This means that f strongly separates the sets A and B. ∎

COROLLARY 2.5.11. *Let X be a real normed space. Let B be a closed convex set and let x_0 be an arbitrary element which does not belong to B. Then there exist a continuous linear functional f, constants c and ε, $\varepsilon > 0$, such that*

$$f(x_0) \leqslant c,$$
$$f(x) \geqslant c+\varepsilon \quad \text{for} \quad x \in B.$$

Theorems and Corollaries 2.5.7–2.5.11 can be also formulated in complex spaces. Namely, we consider a given complex space as a real space. The assumptions of the above theorems and corollaries imply the existence of a functional $\tilde{f}$, continuous and linear (with respect to muliplication by real numbers), which separates (or strongly separates) corresponding sets. Let $f(x) = \tilde{f}(x) - i\tilde{f}(ix)$.

As follows from the proof of Theorem 2.5.4, the functional f is linear with respect to multiplication by complex numbers and

$$\tilde{f}(x) = \mathrm{Re}\, f(x).$$

Hence, Theorems and Corollaries 2.5.7–2.5.11 can be formulated in complex spaces by replacing a continuous linear functional f by its real part $\mathrm{Re}\, f$.

Corollary 2.5.11 implies

COROLLARY 2.5.12. *Let X be a real (or complex) normed space. Suppose that Y is a subspace of X (i.e., Y is a closed linear subset of X). Then there exists a continuous linear functional f, $f \neq 0$, defined on X, which is equal to zero on Y.*

Proof. Let x_0 be an arbitrary element which does not belong to Y. By Corollary 2.5.11, there exist a continuous linear functional f and a constant c such that

$$f(x) \leqslant c \quad (\mathrm{Re}\, f(x) \leqslant c) \quad \text{for} \quad x \in Y$$

and

$$f(x_0) > c \quad (\mathrm{Re}\, f(x_0) > c).$$

Since Y is a linear subspace, the first inequality implies that $f(x)$ is equal to zero on Y. Indeed, if for some $y \in Y$, $f(y) \neq 0$ then, taking

$$t = (c+1)\frac{1}{f(y)}$$

we obtain $f(ty) = c+1 > c$, which contradicts the second inequality. The second inequality implies that f is not identically equal to zero. ∎

As a consequence of Corollary 2.5.12 we get

THEOREM 2.5.13. *Let $(X, \|\ \|)$ be a normed space. If the conjugate space X^* is separable, then X is separable.*

Proof. Suppose that X is a nonseparable space. We shall prove that under this hypothesis the space X^* is also nonseparable. Let F denote a system of functionals such that $f_0 = 0 \in F$ and

$$F = \{f_\alpha \colon \|f_\alpha - f_\beta\| > 1 \text{ for } \alpha \neq \beta\}.$$

If we order all such systems by inclusion, then every totally ordered family of such systems has a maximal element. By the Kuratowski–Zorn Lemma there exists a maximal element F_0 in the family of all such systems. We shall prove that F_0 contains an uncountable number of elements. Suppose that F_0 is countable and $F_0 = \{f_n\}, n = 0, 1, \ldots, f_0 = 0$. Clearly, $\|f_n\| > 1$ for $n = 1, 2, \ldots$ By the definition of a norm, there exists an element x_n with the norm equal to 1 such that $f_n(x_n) > 1$. Let $Y = \overline{\operatorname{lin}\{x_n\}}$. The space Y is separable and $X \neq Y$ because X is nonseparable. By Corollary 2.5.12, there exists a nonzero functional f such that $f(y) = 0$ for $y \in Y$. Without loss of generality we can assume $\|f\| > 1$. Hence, $\|f - f_0\| = \|f\| > 1$ and moreover, for $n \neq 0$,

$$\|f - f_n\| = \sup_{\|x\| \leq 1} |f(x) - f_n(x)| \geq |f(x_n) - f_n(x_n)| > 1.$$

Thus, F_0 cannot be a maximal element, which contradicts the assumption. Finally, by Corollary 1.10.4, the space X^* is nonseparable. ∎

The result converse to that of Theorem 2.5.13 is not valid, as will be shown in the next section.

2.6. General forms of continuous linear functionals in certain spaces

Finite-dimensional spaces. By the results of Section 2.4, every continuous linear functional f defined on an n-dimensional linear metric space X can be expressed by the formula

$$f(x) = a_1 x_1 + \ldots + a_n x_n, \tag{2.6.1}$$

where $x = x_1 e_1 + \ldots + x_n e_n$ is an element of X and $(a_1, \ldots, a_n)$ is an n-tuple of numbers. Conversely, each formula of the form (2.6.1) defines a continuous linear functional on the space X.

2.6. Forms of continuous linear functionals

The space c_0. Let f be a continuous linear functional defined on the space c_0. Every element x belonging to c_0 can be expressed as

$$x = \sum_{n=1}^{\infty} x_n e_n, \qquad (2.6.2)$$

where $e_n = (0, ..., \underset{n\text{-th position}}{0, 1, 0}, ...)$. The series (2.6.2) converges in c_0 and the representation of x in the form (2.6.2) is unique.

By formula (2.6.2) and the continuity of the functional f we get

$$f(x) = \sum_{n=1}^{\infty} a_n x_n, \qquad (2.6.3)$$

where

$$a_n = f(e_n). \qquad (2.6.4)$$

Let z_N be the following element of the space c_0 for $N = 1, 2, ...$

$$z_N = \{z_{N,n}\}, \quad \text{where} \quad z_{N,n} = \begin{cases} 0 & \text{for } n > N, \\ \operatorname{sgn} a_n & \text{for } n \leqslant N. \end{cases}$$

We recall that $\operatorname{sgn} r = 0$ for $r = 0$, $\operatorname{sgn} r = 1$ for $r > 0$ and $\operatorname{sgn} r = -1$ for $r < 0$.

Clearly, all z_N for $N = 1, 2, ...$ belong to the space c_0 and $\|z_N\| = 1$ for $N = 1, 2, ...$ This implies that

$$\sum_{n=1}^{N} |a_n| = f(z_N) \leqslant \|f\|.$$

Since N is arbitrary in the above formula, the series $\sum_{n=1}^{\infty} |a_n|$ converges and

$$\sum_{n=1}^{\infty} |a_n| \leqslant \|f\|. \qquad (2.6.5)$$

On the other hand, if the series $\sum_{n=1}^{\infty} |a_n|$ converges, formula (2.6.3) defines a continuous linear functional and

$$|f(x)| \leqslant \left(\sum_{n=1}^{\infty} |a_n| \right) \cdot \|x\|. \qquad (2.6.6)$$

Formulae (2.6.5) and (2.6.6) imply the equality

$$\|f\| = \sum_{n=1}^{\infty} |a_n|. \tag{2.6.7}$$

Hence, there exists a linear isometry between the spaces $(c_0)^*$ and l.

The space l. Let f be a continuous linear functional defined on l and let x be an arbitrary element of the space l. As before, we can express x in the form (2.6.2) with the only difference that the convergence of the series (2.6.2) is now considered in the space l. This implies formula (2.6.3).

Now the elements z_N are defined in a slightly different manner, namely, $z_N = (\operatorname{sgn} a_N) e_N$, for $N = 1, 2, \ldots$ Clearly, $\|z_N\| = 1$. Since for all N, $f(z_N) = |a_N| \leq \|f\|$, we have

$$A = \sup_n |a_n| \leq \|f\|. \tag{2.6.8}$$

On the other hand, we find by simple calculations that if a sequence $\{a_n\}$ satisfies (2.6.8), then formula (2.6.3) represents a continuous linear functional f defined on l and moreover $\|f\| \leq A$. Thus, $\|f\| = A$ and l^* and m are isometric.

The space l^2. The reasoning is similar to that of the previous cases. Namely, we express an arbitrary element of l^2 in the form (2.6.2), which entails (2.6.3). Next, we define z_N, $N = 1, 2, \ldots$, as follows:

$$z_N = \{z_{N,n}\}, \quad \text{where} \quad z_{N,n} = \begin{cases} 0 & \text{for } n > N, \\ a_n & \text{for } n \leq N. \end{cases}$$

Clearly, $z_N \in l^2$ and

$$\|z_N\| = \left(\sum_{n=1}^{N} a_n^2\right)^{1/2}. \tag{2.6.9}$$

Hence,

$$\sum_{n=1}^{N} a_n^2 = f(z_N) \leq \|f\| \cdot \|z_N\| = \|f\| \left(\sum_{n=1}^{N} a_n^2\right)^{1/2}$$

and

$$\left(\sum_{n=1}^{N} a_n^2\right)^{1/2} \leq \|f\|. \tag{2.6.10}$$

2.6. Forms of continuous linear functionals

Since (2.6.10) is valid for an arbitrary natural N, we obtain

$$\left(\sum_{n=1}^{\infty} a_n^2\right)^{1/2} \leqslant ||f||. \tag{2.6.11}$$

On the other hand, if the series appearing in formula (2.6.11) converges, then, by the Schwarz inequality (see Section 1.8), formula (2.6.3) represents a continuous linear functional f defined on the space l^2 and, moreover,

$$|f(x)| \leqslant \left(\sum_{n=1}^{\infty} a_n^2\right)^{1/2} \cdot ||x||. \tag{2.6.12}$$

This implies that

$$||f|| = \left(\sum_{n=1}^{\infty} a_n^2\right)^{1/2}, \tag{2.6.13}$$

which means that there is an isometry between the spaces $(l^2)^*$ and l^2.

By replacing the Schwarz inequality by the Hölder inequality, the above reasoning can be applied to the spaces l^p, $1 < p < +\infty$, and it can be proved that each continuous linear functional f defined on l^p is given by formula (2.6.3), where the sequence $\{a_n\}$ belongs to the space l^q, $\dfrac{1}{p} + \dfrac{1}{q} = 1$, and vice-versa, each sequence which belongs to the space l^q determines, by formula (2.6.3), a continuous linear functional defined on l^p.

The spaces $L^p[0, 1]$, $p = 1$ or $p = 2$. To derive the general form of functionals in $L^p[0, 1]$ we shall use the notion of absolutely continuous functions and the Radon–Nikodym theorem.

We say that a function $f(t)$, $0 \leqslant t \leqslant 1$, is *absolutely continuous* if for every positive ε there exists a $\delta > 0$ such that for every finite set of disjoint intervals $I_i = [a_i, b_i]$, $i = 1, 2, \ldots, n$, satisfying $\sum\limits_{i=1}^{n} |b_i - a_i| < \delta$ we have $\sum\limits_{i=1}^{n} |f(b_i) - f(a_i)| < \varepsilon$.

RADON–NIKODYM THEOREM. *If a function $f(t)$ is absolutely continuous, then its derivative $f'(t)$ exists almost everywhere (i.e., except at most* a set of measure zero) and moreover,

$$f(t) = f(0) + \int_0^t f'(t)\,dt.$$

The proof of this theorem can be found in elementary textbooks of calculus.

Now by the Radon–Nikodym theorem, we shall derive the general form of continuous linear functionals in the spaces $L^p[0, 1]$, $p = 1$ or $p = 2$.

Let f be a continuous linear functional defined on the space $L^1[0, 1]$ ($L^2[0, 1]$). Let u_t be the characteristic function of an interval $[0, t]$, $u_t = \chi_{[0, t]}$. Consider a function g of the variable t which is defined by the equality $g(t) = f(u_t)$. We shall prove that the function $g(t)$ is absolutely continuous. For an arbitrary finite set of disjoint intervals $I_i = [a_i, b_i]$, $i = 1, 2, \ldots, n$ we obtain

$$J_n = \sum_{i=1}^n |g(b_i) - g(a_i)| = \sum_{i=1}^n f\big(\operatorname{sgn} f(\chi_{I_i})\chi_{I_i}\big)$$

$$\leqslant \|f\| \cdot \Big\|\sum_{i=1}^n \operatorname{sgn} f(\chi_{I_i})\chi_{I_i}\Big\|. \qquad (2.6.14)$$

But in the space $L^1[0, 1]$

$$\Big\|\sum_{i=1}^n \operatorname{sgn} f(\chi_{I_i})\chi_{I_i}\Big\| = \sum_{i=1}^n |b_i - a_i| \qquad (2.6.15)$$

(in the space $L^2[0, 1]$

$$\Big\|\sum_{i=1}^n \operatorname{sgn} f(\chi_{I_i})\chi_{I_i}\Big\| = \Big(\sum_{i=1}^n |b_i - a_i|^2\Big)^{1/2}). \qquad (2.6.15')$$

Formulae (2.6.14) and (2.6.15) ((2.6.15′)) imply the absolute continuity of the function $g(t)$. Hence, by the Radon–Nikodym theorem, $g(t)$ is almost everywhere differentiable and

$$g(t) = g(0) + \int_0^t g'(t)\,dt = \int_0^t g'(t)\,dt$$

since $g(0) = f(0) = 0$. For the sake of brevity we introduce the notation $a(t) = g'(t)$. Hence, by the definition of $g(t)$, we have

$$f(u_t) = g(t) = \int_0^t a(t)\,dt = \int_0^1 u_t a(t)\,dt.$$

Since f is a linear functional, the above inequality implies that for every function $z(t)$ of the form

$$z(t) = \sum_{i=1}^n c_i(u_{b_i} - u_{a_i}) \tag{2.6.16}$$

we have

$$f(z) = \int_0^1 z(t)a(t)\,dt. \tag{2.6.17}$$

Let $x(t)$ be an arbitrary measurable bounded function and let $\{z_n\}$ be a sequence of functions of the form (2.6.16), which converges uniformly to the function x almost everywhere. Clearly, the sequence $\{z_n\}$ converges also in the spaces $L^p[0, 1]$, $p = 1$ or $p = 2$. Since f is a continuous functional, by the Lebesgue theorem we obtain

$$f(x) = \lim_{n \to \infty} f(z_n) = \lim_{n \to \infty} \int_0^1 z_n(t)a(t)\,dt$$
$$= \int_0^1 \lim_{n \to \infty} z_n(t)a(t)\,dt = \int_0^1 x(t)a(t)\,dt. \tag{2.6.18}$$

Let E be an arbitrary set of positive measure. Putting

$$z_E = \frac{1}{|E|}\,\text{sgn}\,a(t)\chi_E,$$

where $|E|$ denotes the Lebesgue measure of the set E, we obtain

$$\|z_E\| = 1$$

and

$$f(z_E) = \int_E |a(t)|\,dt \leqslant \|f\|.$$

Since the above formula holds for every set E, we obtain that $a(t) \in M[0, 1]$ and
$$\operatorname*{ess\,sup}_{0 \leq t \leq 1} |a(t)| \leq \|f\|. \tag{2.6.19}$$

If f is a continuous linear functional defined on the space $L^2[0, 1]$ then, by taking for all natural N
$$x_N(t) = \begin{cases} a(t) & \text{if } |a(t)| \leq N, \\ N \cdot \operatorname{sgn} a(t) & \text{if } |a(t)| \geq N, \end{cases}$$
we obtain the respective formulae
$$\int_0^1 (x_N(t))^2 dt \leq \int_0^1 x_N(t) a(t) dt = f(x_N) \leq \|f\| \cdot \|x_N\|$$
$$= \|f\| \cdot \left(\int_0^1 (x_N(t))^2 dt\right)^{1/2}$$
and
$$\left(\int_0^1 (x_N(t))^2 dt\right)^{1/2} \leq \|f\|. \tag{2.6.19'}$$

Since the latter formula holds for all natural N, we obtain
$$\left(\int_0^1 a^2(t) dt\right)^{1/2} \leq \|f\|. \tag{2.6.19''}$$

On the other hand, if the left-hand side of inequality (2.6.19) ((2.6.19')) is bounded, then formula (2.6.17) defines on the space $L^1[0, 1]$ ($L^2[0, 1]$) a continuous linear functional. This follows immediately from the inequality
$$\left|\int_0^1 x(t) a(t) dt\right| \leq \operatorname*{ess\,sup}_{0 \leq t \leq 1} |a(t)| \cdot \int_0^1 |x(t)| dt \tag{2.6.20}$$

(respectively, from the Schwarz inequality
$$\left|\int_0^1 x(t) a(t) dt\right| \leq \left(\int_0^1 a^2(t) dt\right)^{1/2} \left(\int_0^1 x^2(t) dt\right)^{1/2} \tag{2.6.20'}$$

— see Section 1.8).

2.6. Forms of continuous linear functionals

As a consequence of formula (2.6.20) ((2.6.20′)) we have

$$\|f\| = \underset{0 \leq t \leq 1}{\operatorname{ess\,sup}} |a(t)| \tag{2.6.21}$$

(respectively

$$\|f\| = \left(\int_0^1 a^2(t)\,dt\right)^{1/2}). \tag{2.6.21′}$$

Clearly, similar formulae hold for the spaces $L^1[a, b]$ and $L^2[a, b]$.

By similar arguments based on the Hölder inequality instead of the Schwarz inequality we can prove that every continuous linear functional F defined on the space $L^p[a, b]$, $1 < p < +\infty$, is of the form

$$F(x) = \int_a^b f(t)x(t)\,dt, \tag{2.6.22}$$

where

$$f(t) \in L^q[a, b], \quad \frac{1}{p} + \frac{1}{q} = 1,$$

and conversely for every function $f(t)$ which belongs to the space $L^q[a, b]$ formula (2.6.22) determines a continuous linear functional on the space $L^p[a, b]$. Moreover, the norm of the functional F is equal to $\|f\|_{L^q}$, $\|F\| = \|f\|_{L^q}$.

So far we have analysed the general forms of continuous linear functionals defined on the spaces l^p and $L^p[a, b]$. It can be seen that results similar to those established above hold for continuous linear functionals F defined on $L^p(\Omega, \Sigma, \mu)$, $1 < p < +\infty$. Namely, we have

$$F(x) = \int_\Omega f(t)x(t)\,d\mu, \tag{2.6.22′}$$

where $f(t) \in L^q(\Omega, \Sigma, \mu)$, $\dfrac{1}{p} + \dfrac{1}{q} = 1$ for $p > 1$ and $f(t) \in M(\Omega, \Sigma, \mu)$ for $p = 1$.

And conversely, formula (2.6.22′) defines a continuous linear functional F on the space $L^p(\Omega, \Sigma, \mu)$ and the norm of the functional F is equal to the norm of f in the space $L^q(\Omega, \Sigma, \mu)$ (or $M(\Omega, \Sigma, \mu)$).

The proof of this fact is similar to the proof of the corresponding fact for the Lebesgue measure on an interval. However, it requires the notion of absolute continuity of set functions and the general form of the Radon–Nikodym Theorem. A detailed proof can be found in textbooks of calculus.

Clearly, in virtue of the Hölder inequality, every function f, $f \in L^q(\Omega, \Sigma, \mu)$ ($f \in M(\Omega, \Sigma, \mu)$), defines a continuous linear functional F on the space $L^p(\Omega, \Sigma, \mu)$, $1 < p < +\infty$ ($L^1(\Omega, \Sigma, \mu)$) by formula (2.6.22′) and, moreover, the norm of the functional F is equal to the norm of f in the appropriate spaces.

The space $C[0, 1]$. We begin the analysis of continuous linear functionals defined on the space $C[0, 1]$ with two definitions. Firstly, we introduce the notion of bounded variation of a function.

Let $g(t)$ be a real-valued function defined on the closed interval $[0, 1]$. Let $\Delta = \{a_0, a_1, \ldots, a_n\}$ be a partition of the interval $[0, 1]$, i.e.,
$$0 = a_0 < a_1 < \ldots < a_n = 1.$$

The function $g(t)$ is said *to be of bounded variation* if
$$\operatorname*{Var}_0^1 g(t) = \sup_{\Delta} \sum_{k=1}^n |g(a_k) - g(a_{k-1})| < +\infty,$$
where the supremum is taken over all partitions Δ of the interval $[0, 1]$. The number $\operatorname*{Var}_0^1 g(t)$ is called the *variation* of the function $g(t)$.

Secondly, we define the Riemann–Stieltjes integral. Suppose we are given a sequence $\{\Delta^m\}$ of partitions, $\Delta^m = \{a_{m,0}, \ldots, a_{m,n_m}\}$. We say that the sequence $\{\Delta^m\}$ is a *normal sequence of partitions* if the length of the greatest interval in the n-th partition tends to zero, i.e.,
$$\lim_{m \to \infty} \sup_{1 \leq i \leq n_m} |a_{m,i} - a_{m,i-1}| = 0.$$

Let $f(t)$ be a function defined on the interval $[0, 1]$. Consider the sum
$$I_m = \sum_{k=1}^{n_m} f(t_k)(g(a_k) - g(a_{k-1})),$$
where t_k is an arbitrary point which belongs to the interval $[a_{k-1}, a_k]$.

2.6. Forms of continuous linear functionals

As in the definition of the Riemann integral, by the *Riemann–Stieltjes integral* of the function $f(t)$ with respect to the function $g(t)$ we mean the limit of the sequence $\{I_m\}$ provided that this limit does not depend on the choice of t_k. The Riemann–Stieltjes integral will be denoted by $\int_0^1 f(t)\,dg(t)$. As in the classical theory of the Riemann integral it can be shown that the definition of the Riemann–Stieltjes integral does not depend on the choice of a normal sequence of partitions and, moreover,

(1) if the function $f(t)$ is continuous and the function $g(t)$ is of bounded variation, then the integral $\int_0^1 f(t)\,dg(t)$ exists;

(2) $\int_0^1 f_1(t)\,dg(t) + \int_0^1 f_2(t)\,dg(t) = \int_0^1 (f_1+f_2)(t)\,dg(t)$, provided that all integrals exist;

(3) $\int_0^1 af(t)\,dg(t) = a \cdot \int_0^1 f(t)\,dg(t)$ for every number a;

(4) $\int_0^1 f(t)\,dg(t) \leqslant \sup_{0\leqslant t\leqslant 1} |f(t)| \cdot \operatorname*{Var}_0^1 g(t)$.

More information about the properties of functions of bounded variation can be found in textbooks of calculus.

It follows from properties (1)–(4) that every function $g(t)$ of bounded variation defines, by means of the Riemann–Stieltjes integral, a continuous linear functional $\tilde{g}$ on the space of all continuous functions $C[0, 1]$ by the formula

$$\tilde{g}(x) = \int_0^1 x(t)\,dg(t). \tag{2.6.23}$$

The norm of the functional $\tilde{g}$ is not greater than the variation of the function $g(t)$,

$$\|\tilde{g}\| \leqslant \operatorname*{Var}_0^1 g(t). \tag{2.6.24}$$

We shall show that the converse assertion is valid. Namely, every continuous linear functional $\tilde{g}(x)$ defined on the space $C[0, 1]$ generates a certain function of bounded variation $g(t)$ such that formula (2.6.23) holds and

$$\|\tilde{g}\| \geqslant \operatorname*{Var}_{0}^{1} g(t). \tag{2.6.25}$$

Let $\tilde{g}(x)$ be a continuous linear functional defined on the space $C[0,1]$. The space $C[0,1]$ can be regarded as a subspace of the space $M[0,1]$. Hence, in virtue of the Hahn–Banach theorem (Theorem 2.5.3), there exists a norm-preserving extension of the functional $\tilde{g}$ to the whole space $M[0,1]$. Since it is not misleading, the extension of the functional $\tilde{g}$ will also be denoted by $\tilde{g}$.

Let u_t be the characteristic function of the interval $[0, t)$, i.e.,

$$u_t(\tau) = \begin{cases} 1 & \text{for } 0 \leqslant \tau < t, \\ 0 & \text{for } t \leqslant \tau \leqslant 1. \end{cases}$$

Let

$$g(t) = \tilde{g}(u_t).$$

We shall show that the function $g(t)$ is of bounded variation. Let $\varDelta = \{a_0, \ldots, a_n\}$ be a partition of the interval $[0, 1]$. Then

$$\sum_{i=1}^{n} |g(a_i) - g(a_{i-1})| = \sum_{i=1}^{n} |\tilde{g}(u_{a_i}) - \tilde{g}(u_{a_{i-1}})|$$

$$= \sum_{i=1}^{n} \tilde{g}\left(\operatorname{sgn}[g(a_i) - g(a_{i-1})] \cdot (u_{a_i} - u_{a_{i-1}})\right) \leqslant \|\tilde{g}\|,$$

since the function $\sum_{i=1}^{n} \operatorname{sgn}[g(a_i) - g(a_{i-1})] \cdot (u_{a_i} - u_{a_{i-1}})$ belongs to the space $M[0,1]$ and its norm is equal to 1.

Since $\varDelta$ is an arbitrary partition, we have

$$\operatorname*{Var}_{0}^{1} g(t) \leqslant \|\tilde{g}\|. \tag{2.6.26}$$

Let $x(t)$ be an arbitrary continuous function and let

$$x_n(t) = \sum_{k=1}^{n} x\left(\frac{k}{n}\right)\left(u_{\frac{k}{n}}(t) - u_{\frac{k-1}{n}}(t)\right).$$

2.6. Forms of continuous linear functionals

The continuity of the function $x(t)$ implies that the sequence $\{x_n\}$ converges uniformly to the function $x(t)$. In other words, the sequence $\{x_n\}$ converges to x in the norm of the space $M[0, 1]$. The continuity of the functional $\tilde{g}$ immediately implies that

$$\tilde{g}(x) = \lim_{n\to\infty} \tilde{g}(x_n) = \lim_{n\to\infty} \sum_{k=1}^{n} x\left(\frac{k}{n}\right)\left[g\left(\frac{k}{n}\right) - g\left(\frac{k-1}{n}\right)\right]$$

$$= \int_0^1 x(t)\,dg(t).$$

Thus we have proved the Riesz theorem (Riesz, 1909):

Every continuous linear functional $\tilde{g}$ defined on the space of all continuous functions $C[0, 1]$ is of the form

$$\tilde{g}(x) = \int_0^1 x(t)\,dg(t),$$

where $g(t)$ is a function of bounded variation. Moreover,

$$\|\tilde{g}\| = \operatorname*{Var}_0^1 g(t).$$

Clearly, similar formulae can be established for the space $C[a, b]$ of all continuous functions on the interval $[a, b]$.

Hilbert space. Let X be a Hilbert space with an inner product (x, y) (see Example 2.2.7). Of course, each element y of the space X induces a linear functional f_y by the formula

$$f_y(x) = (x, y). \tag{2.6.27}$$

By the Cauchy inequality (2.2.6), we have

$$|f_y(x)| \leq \|x\| \cdot \|y\|. \tag{2.6.28}$$

Moreover, f_y is a continuous linear functional and

$$\|f_y\| \leq \|y\|. \tag{2.6.29}$$

On the other hand,

$$|f_y(y)| = \|y\|^2$$

and finally $\|f_y\| = \|y\|$.

Now we shall prove that each continuous linear functional defined on the space X is of the form (2.6.27).

Let f be an arbitrary continuous linear functional defined on X and let $H = \{x\colon f(x) = 0\}$. Suppose that we are able to construct an element $y \neq 0$ such that $(y, x) = 0$ for all $x \in H$. Of course, $y \notin H$ and hence the element $y_f = \dfrac{f(y)}{(y, y)} y$, does not belong to H. Moreover, each element $x \in X$ can be represented in the form $ty_f + h$, where $h \in H$ and t is a number. Thus,

$$f(x) = f(ty_f+h) = tf(y_f) = t\frac{f^2(y)}{(y,y)} = t(y_f, y_f)$$
$$= (ty_f+h, y_f) = (x, y_f).$$

Hence, the problem is to construct an element $y \neq 0$ such that $(y, x) = 0$ for all $x \in H$. Take an arbitrary $y_0 \notin H$. Let $\{x_n\}$ be a sequence of elements of H such that

$$\lim_{n\to\infty} ||y_0 - x_n|| = \inf\{||y_0 - x||\colon x \in H\}. \tag{2.6.30}$$

In the sequel we shall put $\delta = \inf\{||y_0 - x||\colon x \in H\}$.

For arbitrary x_n, x_m we have an estimation

$$\delta \leqslant \left\|y_0 - \frac{x_n+x_m}{2}\right\| \leqslant \tfrac{1}{2}||y_0 - x_n|| + \tfrac{1}{2}||y_0 - x_m||.$$

Hence, by (2.6.30)

$$\lim_{n,m\to\infty} \left\|y_0 - \frac{x_n+x_m}{2}\right\| = \delta. \tag{2.6.31}$$

Using the definition of a norm in a Hilbert space, we obtain the equality

$$||x+y||^2 + ||x-y||^2 = 2(||x||^2 + ||y||^2) \tag{2.6.32}$$

for arbitrary $x, y \in X$. Substituting $x = y_0 - x_m$, $y = y_0 - x_n$ in formula (2.6.32), we obtain

$$||x_n - x_m||^2 = 2(||y_0 - x_n||^2 + ||y_0 - x_m||^2) - 4 \cdot \left\|y_0 - \frac{x_n+x_m}{2}\right\|^2. \tag{2.6.33}$$

By (2.6.30) and (2.6.31), the right-hand side of (2.6.33) tends to zero, if n, m tend to $+\infty$. This means that the sequence $\{x_n\}$ is fundamental.

2.6. Forms of continuous linear functionals

Since X is complete, the sequence $\{x_n\}$ tends to a certain element x_0. Evidently

$$\|y_0 - x_0\| = \inf\{\|y_0 - x\|: x \in H\}. \tag{2.6.34}$$

We shall show that $y_0 - x_0$ is the required element, i.e., $(y_0 - x_0, x) = 0$ for all $x \in H$. Suppose on the contrary that $(y_0 - x_0, x_1) = \sigma \neq 0$ for a certain $x_1 \in H$. Let $\bar{x} = x_0 + \dfrac{\sigma}{(x_1, x_1)} x_1$. Then

$$\begin{aligned}
\|y_0 - \bar{x}\|^2 &= \left(y_0 - x_0 - \frac{\sigma}{(x_1, x_1)} x_1, y_0 - x_0 - \frac{\sigma}{(x_1, x_1)} x_1\right) \\
&= \|y_0 - x_0\|^2 - \frac{\sigma}{(x_1, x_1)}(y_0 - x_0, x_1) - \\
&\quad - \frac{\sigma}{(x_1, x_1)}(x_1, y_0 - x_0) + \frac{|\sigma|^2}{(x_1, x_1)} \\
&= \|y_0 - x_0\|^2 - \frac{\sigma^2}{(x_1, x_1)} \\
&< \|y_0 - x_0\|^2.
\end{aligned}$$

This contradicts (2.6.34) and hence $y_0 - x_0$ is the required element.

Chapter 3

Continuous linear operators in Banach spaces

3.1. THE BANACH–STEINHAUS THEOREM

Let $(X, \|\ \|_X)$ and $(Y, \|\ \|_Y)$ be normed spaces (real or complex). By $B(X \to Y)$ we denote the set of all continuous linear operators which map the space X into the space Y. In the set $B(X \to Y)$ we can introduce operations of addition and multiplication by scalars in the following way:

$$(T_1+T_2)(x) = T_1(x)+T_2(x),$$
$$(aT)(x) = aT(x).$$

The set $B(X \to Y)$ with the operations defined as above forms a linear space. The space $B(X \to X)$ will be denoted briefly by $B(X)$. For $T \in B(X \to Y)$ we denote

$$\|T\|_{B(X \to Y)} = \sup_{\|x\|_X \leq 1} \|T(x)\|_Y. \tag{3.1.1}$$

It follows from the continuity of T that the number $\|T\|_{B(X \to Y)}$ is finite. Indeed, if $\|T\|_{B(X \to Y)} = +\infty$, we can choose a sequence $\{x_n\}$ such that $\|x_n\|_X = 1$, $n = 1, 2, \ldots$, and $\|T(x_n)\|_Y > n^2$. Moreover, the sequence $\left\{\dfrac{x_n}{n}\right\}$ tends to zero and $\left\|T\left(\dfrac{x_n}{n}\right)\right\|_Y \geq n$. But this contradicts the continuity of the operator T. By simple calculations it can be verified that $\|T\|_{B(X \to Y)}$ satisfies all conditions of a norm. $\|T\|_{B(X \to Y)}$ will be called the *norm of the operator* T (cf. the definition of the norm of a continuous linear functional, Section 2.3). The space $B(X \to Y)$ with this norm is a normed space. In the sequel we shall denote the norms $\|\ \|_X$, $\|\ \|_Y$, $\|\ \|_{B(X \to Y)}$ by the symbol $\|\ \|$ provided that it does not lead to ambiguities.

3.1. The Banach-Steinhaus theorem

Observe that if X and Y are real seminormed spaces then formula (3.1.1) defines a seminorm in the space $B(X \to Y)$.

Let $(X, \| \ \|_X)$, $(Y, \| \ \|_Y)$, $(Z, \| \ \|_Z)$ be normed spaces. Let $A \in B(X \to Y)$ and $B \in B(Y \to Z)$. Hence $BA \in B(X \to Z)$. Moreover

$$\|BA\| = \sup_{\|x\| \leqslant 1} \|BA(x)\| \leqslant \sup_{\|y\| \leqslant \|A\|} \|B(y)\| = \|B\| \|A\|.$$

In the particular case where $X = Y = Z$, $B = A$ we obtain

$$\|A^2\| \leqslant \|A\|^2$$

and

$$\|A^k\| \leqslant \|A\|^k \quad \text{for} \quad k = 1, 2, \ldots$$

If X is a space over the field of real (complex) numbers and Y is the field of real (complex) numbers, then the space $B(X \to Y)$ is, in fact, the conjugate space of X, X^*. On X^* the norm given by formula (3.1.1) is simply the norm (or in the case of seminormed spaces—the seminorm) of functionals.

Let X and Y be Banach spaces, i.e., normed complete spaces. We shall prove that the space $B(X \to Y)$ is also complete, i.e., $B(X \to Y)$ is a Banach space.

We begin with the following theorem.

THEOREM 3.1.1 (Banach and Steinhaus, 1927). *Let X and Y be Banach spaces and let $\mathfrak{A} \subset B(X \to Y)$ be a set of continuous linear operators such that for all $x \in X$*

$$\sup \{\|A(x)\| : A \in \mathfrak{A}\} < +\infty. \tag{3.1.2}$$

Then

$$\sup \{\|A\| : A \in \mathfrak{A}\} < +\infty. \tag{3.1.3}$$

Proof. Let

$$K_m = \{x : \|A(x)\| \leqslant m \text{ for } A \in \mathfrak{A}\} = \bigcap_{A \in \mathfrak{A}} \{x : \|A(x)\| \leqslant m\}.$$

The sets K_m are closed for $m = 1, 2, \ldots$ Moreover, by (3.1.2), the union of all K_m is equal to the whole space X,

$$X = \bigcup_{m=1}^{\infty} K_m. \tag{3.1.4}$$

Since the space X is complete, in virtue of the Baire theorem (Theorem 1.7.3), the space X is of the second category. This implies that there is an index m_0 such that the set K_{m_0} is of the second category. Since the set K_{m_0} is closed, it contains a certain closed ball $K_r(x_0)$, $\overline{K_r(x_0)} \subset K_{m_0}$. Hence, if $||x|| \leqslant r$, then for each operator A which belongs to the set $\mathfrak{A}$ we have

$$||A(x+x_0)|| \leqslant m_0. \tag{3.1.5}$$

Thus,

$$||A(x)|| \leqslant m_0 + ||A(x_0)|| \leqslant 2m_0 \tag{3.1.6}$$

and

$$||A|| = \sup_{||x|| \leqslant 1} ||A(x)|| = \frac{1}{r} \sup_{||x|| \leqslant r} ||A(x)|| \leqslant \frac{2m_0}{r}. \blacksquare \tag{3.1.7}$$

COROLLARY 3.1.2 (Banach and Steinhaus, 1927). *Let X and Y be Banach spaces. Let $\{A_n\}$, $A_n \in B(X \to Y)$ for $n = 1, 2, \ldots$, be a sequence of continuous linear operators such that the sequence $\{A_n(x)\}$ converges for each $x \in X$. Then the operator A defined as $A(x) = \lim_{n \to \infty} A_n(x)$ is a continuous linear operator.*

Proof. It is an immediate consequence of the properties of the limit that A is a linear operator. Now we prove that A is continuous. To see this, observe that for every $x \in X$

$$||A(x)|| = \lim_{n \to \infty} ||A_n(x)|| \leqslant \sup_n ||A_n(x)|| \leqslant \sup_n ||A_n|| \cdot ||x||$$
$$\leqslant M \cdot ||x||, \tag{3.1.8}$$

where

$$M = \sup_n ||A_n|| \tag{3.1.9}$$

is finite by (3.1.3). $\blacksquare$

THEOREM 3.1.3. *Let X and Y be Banach spaces. Then the space $B(X \to Y)$ is a complete space (i.e., $B(X \to Y)$ is a Banach space).*

3.1. The Banach-Steinhaus theorem

Proof. Let $\{A_n\}$ be a fundamental sequence in the space $B(X \to Y)$. This means that for every $\varepsilon > 0$ there exists an index n_0 such that for $n, m \geqslant n_0$

$$\|A_n(x) - A_m(x)\| < \varepsilon \tag{3.1.10}$$

for all x, $\|x\| \leqslant 1$. In particular, for each $x \in X$ the sequence $\{A_n(x)\}$ is fundamental. Since Y is complete, the sequence $\{A_n(x)\}$ converges for every $x \in X$, $\|x\| \leqslant 1$, to a limit point $A(x)$. By Corollary 3.1.2, the operator A is continuous and linear. Passing to the limit in inequality (3.1.10) with m tending to infinity, we obtain

$$\|A_n(x) - A(x)\| \leqslant \varepsilon \tag{3.1.11}$$

for all $x \in X$, $\|x\| \leqslant 1$. By the definition of a norm, we conclude that

$$\|A_n - A\| \leqslant \varepsilon \tag{3.1.12}$$

and, since ε was taken arbitrarily, the sequence $\{A_n\}$ converges to A in the space $B(X \to Y)$. ∎

COROLLARY 3.1.4. *For each Banach space X the conjugate space X^* is complete (i.e., X^* is a Banach space).*

PROPOSITION 3.1.5. *Suppose that T is an operator which maps a Banach space X into itself. If $\|T\| < 1$ then the operator $I - T$ is invertible and*

$$(I-T)^{-1} = \sum_{i=0}^{\infty} T^i. \tag{3.1.13}$$

Proof. Let $\|T\| = a$. The sequence $\{S_n\}$, $S_n = \sum_{i=0}^{n} T^i$, is fundamental since for $m < n$

$$\|S_n - S_m\| \leqslant \sum_{i=m+1}^{n} \|T^i\| \leqslant \sum_{i=m+1}^{n} a^i \leqslant a^m \frac{1}{1-a}.$$

Hence, by Theorem 3.1.3, the series (3.1.13) converges. Moreover,

$$(I-T)S_n = I - T^{n+1}$$

and the left-hand side of the equality tends to the identity operator I with n tending to infinity. This entails (3.1.13). ∎

The series appearing in the right-hand side of formula (3.1.13) is called the *von Neumann series*. As a consequence of Proposition 3.1.5 we obtain

PROPOSITION 3.1.6. *Let X and Y be two Banach spaces. Let A be an invertible operator mapping X onto Y. Then there is a constant $a > 0$ such that, for any operator B such that $||B|| < a$, the sum $A+B$ is an invertible operator.*

Proof.
$$A+B = A(I+A^{-1}B).$$
Take $a = ||A^{-1}||^{-1}$ and put $T = A^{-1}B$. Then $||T|| < 1$ and $I+A^{-1}B$ is invertible, which completes the proof. ∎

3.2. THE BANACH THEOREM ABOUT THE CONTINUITY OF THE INVERSE OPERATOR

THEOREM 3.2.1 (Banach, 1927). *Let X and Y be Banach spaces and let T be a continuous linear operator which maps the space X onto the whole space Y. Then the image of an open subset of X is an open subset of Y.*

Proof. We prove that the closure of the image of a neighbourhood U of zero, $\overline{T(U)}$, contains a neighbourhood of zero.

The continuity of addition implies that there exists a neighbourhood of zero, say V, such that $V-V \subset U$. Since V is a neighbourhood of zero, the continuity of multiplication by numbers implies that, for every $x \in X$, there exists a natural n such that $\dfrac{x}{n} \in V$. Hence

$$X = \bigcup_{n=1}^{\infty} nV \tag{3.2.1}$$

and

$$Y = \bigcup_{n=1}^{\infty} nT(V), \tag{3.2.2}$$

where the latter equality follows from the hypothesis that T maps X onto the whole space Y.

3.2. Continuity of the inverse operator

Thus, in virtue of the Baire theorem (Theorem 1.7.3), one of the sets $nT(V)$, say $n_0 T(V)$, is of the second category. Hence, the set $\overline{n_0 T(V)}$ has a nonempty interior. Further, since the mapping $y \to n_0 y$ is a homeomorphism of the space Y onto itself, the set $\overline{T(V)}$ has a nonempty interior and consequently

$$\overline{T(U)} \supset \overline{T(V)} - \overline{T(V)} \supset \overline{T(V)} - \overline{T(V)} \supset \operatorname{Int} T(V) - \operatorname{Int} T(V). \tag{3.2.3}$$

The set $\operatorname{Int} T(V) - \operatorname{Int} T(V)$ is open and contains zero, whence it is a neighbourhood of zero.

In particular, the closure of the image of an arbitrary ball with centre at zero contains a ball with centre at zero.

Put

$$X_r = \{x \in X: \|x\| < r\}, \quad Y_r = \{y \in Y: \|y\| < r\} \tag{3.2.4}$$

for an arbitrary positive number r. Select any positive number ε_0 and positive numbers ε_i, $i = 1, 2, \ldots$ such that

$$\sum_{i=1}^{\infty} \varepsilon_i < \varepsilon_0. \tag{3.2.5}$$

As we have already shown, for every ε_i there is a positive number η_i such that

$$\overline{T(X_{\varepsilon_i})} \supset Y_{\eta_i}, \quad i = 0, 1, 2, \ldots \tag{3.2.6}$$

Let y be an arbitrary element of the set Y_{ε_0}. It follows from the inclusion (3.2.6) that there is an element $x_0 \in X_{\varepsilon_0}$ such that

$$\|y - T(x_0)\| < \eta_1. \tag{3.2.7}$$

In other words, $y - T(x_0) \in Y_{\eta_1}$. Using formula (3.2.6) again we find that there is an element $x_1 \in X_{\varepsilon_1}$ such that

$$\|y - T(x_0) - T(x_1)\| < \eta_2. \tag{3.2.8}$$

Repeating this procedure, we can define a sequence $\{x_n\}$ such that $x_n \in X_{\varepsilon_n}$ and

$$\left\| y - \sum_{i=1}^{n} T(x_i) \right\| < \eta_{n+1} \tag{3.2.9}$$

for $n = 1, 2, \ldots$

Let
$$z_m = x_0 + x_1 + \ldots + x_m. \qquad (3.2.10)$$

The sequence $\{z_m\}$ is fundamental since for $m' > m$

$$||z_m - z_{m'}|| < \varepsilon_m + \varepsilon_{m+1} + \ldots + \varepsilon_{m'}$$

and, in view of (3.2.5), the right-hand side of the inequality tends to zero with m tending to infinity.

Since the space X is complete, the sequence $\{z_m\}$ converges to a certain element $z \in X$. Moreover, by inequality (3.2.5)

$$||z|| \leq \sum_{i=0}^{\infty} \varepsilon_i < 2\varepsilon_0.$$

This expresses the fact that $T(X_{2\varepsilon_0}) \supset Y_{\eta_0}$. Since ε_0 was taken arbitrarily, we conclude that the image of an arbitrary neighbourhood V of zero, $V \subset X$, contains a certain neighbourhood W of zero, $W \subset Y$.

Now, let U be an arbitrary open set contained in X and let x be an arbitrary element of U. Since the set U is open, there is an open neighbourhood N of zero such that $x+N \subset U$. Choosing a neighbourhood M of zero, $M \subset Y$, such that $T(N) \supset M$, we get

$$T(U) \supset T(x+N) \supset T(x) + T(N) \supset T(x) + M. \qquad (3.2.11)$$

Hence, if $y \in T(U)$, then there exists a neighbourhood V of y which is contained in $T(U)$. Thus, by definition, the set $T(U)$ is open. ∎

COROLLARY 3.2.2 (Banach, 1929, 1931). *Let X and Y be Banach spaces. Suppose that T is a one-to-one continuous linear operator which maps the space X onto the space Y. Then the inverse operator T^{-1} is a continuous linear operator which maps Y onto X.*

Proof. The inverse operator T^{-1} is well-defined since T is one-to-one. T^{-1} is linear because T is linear. By Theorem 3.2.1, the image of any open set under T is open. Let V be an arbitrary open set contained in X. Then the inverse image $(T^{-1})^{-1}(V)$ is open since $(T^{-1})^{-1}(V) = T(V)$. Therefore, by definition, T^{-1} is continuous. ∎

COROLLARY 3.2.3. *Let $(X, \| \ \|)$ be a Banach space. Assume that there is another norm $\| \ \|_1$ defined on X which is weaker than $\| \ \|$ (i.e., for any sequence $\{x_n\}$, $\|x_n\| \to 0$ entails $\|x_n\|_1 \to 0$). Moreover, suppose that the space X is complete with respect to the norm $\| \ \|_1$. Then the two norms are equivalent.*

Proof. Denote by Y the space X with the norm $\| \ \|_1$. Let T be a mapping of the space X onto Y such that $T(x) = x$ for any $x \in X$. The continuity of the operator T follows from the hypothesis that the norm $\| \ \|_1$ is weaker than the norm $\| \ \|$. Hence, by Corollary 3.2.2, the operator T^{-1} is continuous. This implies, in turn, that the norm $\| \ \|$ is weaker than the norm $\| \ \|_1$. Thus, the two norms are equivalent. ∎

COROLLARY 3.2.4. *Let X be a normed space which is complete with respect to a norm $\| \ \|$ and a norm $\| \ \|_1$. Assume that the sets of all linear functionals which are continuous in their respective norms coincide. Then the norms $\| \ \|$ and $\| \ \|_1$ are equivalent.*

Proof. Let $\| \ \|_2 = \| \ \| + \| \ \|_1$. Clearly, $\| \ \|_2$ defines a norm on X which is stronger than any of the norms $\| \ \|$, $\| \ \|_1$. Denote by Z the space X with the norm $\| \ \|_2$. We claim that Z is a complete space. To see this, let us take an arbitrary fundamental sequence $\{x_n\}$ of elements of the space Z. Since the norm $\| \ \|_2$ is stronger than any of the norms $\| \ \|$, $\| \ \|_1$, the sequence $\{x_n\}$ is fundamental with respect to each of the norms $\| \ \|$, $\| \ \|_1$. Since X is complete with respect to $\| \ \|$ and $\| \ \|_1$, the sequence $\{x_n\}$ converges to a certain element x in the norm $\| \ \|$ and to a certain element x_1 in the norm $\| \ \|_1$. To show that $x = x_1$ we exploit the hypothesis that the continuity of linear functionals in the norm $\| \ \|$ implies the continuity of linear functionals in the norm $\| \ \|_1$ and conversely, the continuity of linear functionals in the norm $\| \ \|_1$ implies the continuity of linear functionals in the norm $\| \ \|$. Suppose that $x \neq x_1$. Then, by the separation theorems, there exists a continuous linear functional f such that $f(x) \neq f(x_1)$. By the assumption, f is continuous in the norms $\| \ \|$, $\| \ \|_1$. Thus, $f(x) = \lim_{n \to \infty} f(x_n) = f(x_1)$, which is a contradiction. Hence, the sequence $\{x_n\}$ converges to x in the norms $\| \|$, $\| \|_1$. This implies that $\{x_n\}$ converges to x in the norm $\| \|_2$. This means that the space Z is complete. Moreover, by Corollary 3.2.3, the norm $\|x\|_2$ is equivalent to $\|x\|$ and $\|x\|_1$. Thus, $\|x\|$ and $\|x\|_1$ are equivalent. ∎

Note that for the proof of Corollary 3.2.4 it is not essential to assume the equality of the set of all linear functionals which are continuous with respect to the first norm and the set of all linear functionals which are continuous with respect to the second norm. What is really essential is the existence of a certain family of linear functionals f_α which are continuous with respect to both norms and are such that the condition $f_\alpha(x) = 0$ for all α entails $x = 0$. Thus, we can formulate

COROLLARY 3.2.4′. *Let X be a Banach space with respect to the norms $\| \|$, $\| \|_1$. Suppose that there exists a family f_α of linear functionals continuous with respect to both norms, which is total, i.e., $f_\alpha(x) = 0$ for all α, entails $x = 0$. Then the norms $\| \|$, $\| \|_1$ are equivalent.*

3.3. CLOSED OPERATORS

Let X and Y be normed spaces over the field of real or complex numbers. Let A be a linear operator defined on a linear subset D_A of the space X, which takes values in the space Y. The set D_A is called the *domain* of the operator A, the set $R_A = A(D_A)$ is called the *range* of the operator A. By the *graph*, G_A, of the operator A we mean a linear subset of the Cartesian product $X \times Y$ defined as

$$G_A = \{(x, y) \in X \times Y : x \in D_A, y = A(x)\}. \tag{3.3.1}$$

A linear operator A is called *closed*, if its graph G_A is closed (here, the topology on $X \times Y$ is the standard product topology, see Section 2.2).

In other words, an operator A is closed if the fact that $x_n \to x$, $x_n \in D_A$ and $A(x_n) \to y$ implies that $x \in D_A$ and $y = A(x)$.

Suppose now that we are given a one-to-one linear operator A. Then the inverse operator A^{-1} is well-defined. Evidently, the domain of A^{-1}

coincides with the range of A and the range of A^{-1} coincides with the domain of A. Observe that the graph $G_{A^{-1}}$,

$$G_{A^{-1}} = \{(y, x) \in Y \times X: y \in R_A, A^{-1}(y) = x\}$$
$$= \{(y, x) \in Y \times X: x \in D_A, y = A(x)\}$$

is closed if and only if the graph of the operator A is closed. Thus we have

PROPOSITION 3.3.1. *Let X and Y be normed spaces. Let A be a closed operator which maps $D_A \subset X$ into Y. If A is a one-to-one operator, then the inverse operator A^{-1} is also closed.*

PROPOSITION 3.3.2. *Let X and Y be normed spaces. Let A be a continuous linear operator which maps its domain D_A into Y. The operator A is closed if and only if the domain D_A is closed.*

Proof. Necessity. Suppose that D_A is not closed. Let $x_0 \in \bar{D}_A \setminus D_A$ and let $x_n \to x_0$, $x_n \in D_A$. Since A is continuous, we have $A(x_n) \to A(x_0)$. But $x_0 \notin D_A$ and A is not closed.

Sufficiency. Suppose that D_A is closed. Let $\{x_n\}$ be a sequence of elements of D_A which converges to x_0. Clearly, $x_0 \in D_A$. By the continuity of A, the sequence $\{A(x_n)\}$ converges to $A(x_0)$. Hence, if $A(x_n) \to y_0$, then $A(x_0) = y_0$. ∎

PROPOSITION 3.3.3. *Let $(X, \| \|_X)$ and $(Y, \| \|_Y)$ be normed spaces. Let $\| \|'_X$ ($\| \|'_Y$) be a norm defined on X (Y) which is stronger than $\| \|_X$ ($\| \|_Y$). Let A be a linear operator which maps $D_A \subset X$ into Y. If A is closed as an operator acting between the spaces $(X, \| \|_X)$ and $(Y, \| \|_Y)$, then it is also closed as an operator acting between the spaces $(X, \| \|'_X)$ and $(Y, \| \|'_Y)$.*

Proof. The norm $\| \|_1$ induced on $X \times Y$ by the norms $\| \|'_X$ and $\| \|'_Y$ is stronger than the norm $\| \|$ induced on $X \times Y$ by the norms $\| \|_X$ and $\| \|_Y$. Since the graph of A, G_A, is closed in the norm $\| \|$, it is also closed in the norm $\| \|_1$. ∎

PROPOSITION 3.3.4. *Let X and Y be Banach spaces. Suppose that a linear operator D mapping a linear subset D_D into Y is right invertible (i.e., there is a linear operator R defined on the whole Y such that DR is the identity operator on Y). Suppose that the kernel of the operator D, $\ker D = \{x: D(x) = 0\}$, is closed. If R is continuous, then D is closed.*

Proof. Let X_1 denote the quotient space $X/\ker D$. Since $\ker D$ is closed, it follows that X_1 is a normed space and the mapping g assigning to each x the coset $[x]$ which contains x, $g(x) = [x]$, is continuous (see Section 2.2). The operators D and R induce an operator D_1 mapping $D_{D_1} = \{[x] \in X_1 : x \in D_D\}$ into Y, and an operator R_1 mapping Y into D_{D_1}, which are defined as follows:
$$D_1([x]) = D(x), \quad R_1(y) = [R(y)].$$
It is easy to verify that R_1 is continuous as the superposition of continuous operators, $R_1(x) = g(R(x))$, and that the operator D_1 is invertible and its inverse is equal to the operator R_1. By Proposition 3.3.2, the operator R_1 is closed. By Proposition 3.3.1, the operator D_1 is closed, i.e., the graph G_{D_1} is closed.

The mapping g induces a mapping $\tilde{g}$ which maps $X \times Y$ onto $X_1 \times Y$ defined as $\tilde{g}(x, y) = (g(x), y)$. The operator $\tilde{g}$ is continuous. Thus, the graph G_D of the operator D is closed as it is the inverse image of the closed set G_{D_1} under the continuous operator $\tilde{g}$, $G_D = \tilde{g}^{-1}(G_{D_1})$. This means that D is closed. ∎

By Proposition 3.3.4, we obtain an important example of a closed operator which is not continuous.

EXAMPLE 3.3.1. Let $X = Y = C[0, 1]$ (or $L^2[0, 1]$ or $L[0, 1]$). Consider the differential operator $D = \dfrac{d}{dt}$. The domain of D consists of all differentiable (or almost everywhere continuously differentiable) functions $x(t)$ such that their derivatives are continuous (belong to $L^2[0, 1]$ or $L[0, 1]$). The operator D is right invertible. The operator of integration $R(x) = \int_0^1 x(s)\,ds$ is continuous in each of the above spaces. Moreover, the kernel of D is one-dimensional, and hence closed.

Another application of Proposition 3.3.4 is provided by the following

EXAMPLE 3.3.2. Let $\Omega = [0, 1] \times [0, 1]$ be the unit square and let $X = Y = C(\Omega)$. Consider the operator

$$D(x(t, s)) = \frac{\partial^2}{\partial t \, \partial s} x(t, s).$$

We shall show that the operator D is closed. Indeed, the kernel, ker D, consists of functions of the form $x(s)+y(t)$ and is isomorphic to $C[0, 1] \times C[0, 1]$, whence it is closed. The operator D has the right inverse

$$R(x(t, s)) = \int_0^t \int_0^s x(u, v) \, dv \, du.$$

The right inverse is continuous and thus, by Proposition 3.3.4, the operator D is closed.

Another examples of right invertible operators can be found in Przeworska–Rolewicz (1987).

However, not all operators which are important for applications are closed, as is shown by the following example.

EXAMPLE 3.3.3. Let $X = L^2[0, 1]$, $Y = R$. Let f be a functional defined on the domain D_f, $D_f = C[0, 1]$, by the formula

$$f(x) = x(0). \tag{3.3.2}$$

The functional f is a linear operator. This operator is not closed. To see this, let us take

$$x_n = (1-t)^n. \tag{3.3.3}$$

The sequence $\{x_n\}$ tends to 0 in $L^2[0, 1]$, $f(x_n) = 1$ but $f(0) = 0 \neq 1$.

Observe that the closure $\bar{G}_f$ of the graph of f contains all elements of the form $(0, r)$, where r is an arbitrary real number. Indeed, replacing in (3.3.3) x_n by rx_n we obtain $(rx_n, f(rx_n)) \to (0, r)$. Since $\bar{G}_f$ is linear, we have $(x(t), r) \in \bar{G}_f$ for all continuous functions. The set of all continuous functions is dense in $L^2[0, 1]$ and hence $\bar{G}_f = L^2[0, 1] \times R$.

Consequently, the operator f does not admit a closed extension, i.e., there is no closed operator A defined on $L^2[0, 1]$ such that f is the restriction of A to the space of all continuous functions.

THEOREM 3.3.5 (Banach, 1929, 1931). *Let $(X, \|\ \|_X)$ and $(Y, \|\ \|_Y)$ be Banach spaces. Let A be a closed operator which takes values in Y such that the domain D_A of A coincides with the whole space X, $D_A = X$. Then the operator A is continuous.*

Proof. Since the spaces X and Y are complete, the product space $(X \times Y, \|\ \|_{X \times Y})$, where $\|(x, y)\|_{X \times Y} = \|x\|_X + \|y\|_Y$, is complete (Theorem 1.5.3). Thus, the graph of A, G_A, is a complete space as a closed subset of the complete space $X \times Y$. By assigning to each element $(x, A(x))$ of the graph G_A the element x we define a continuous mapping since

$$\|(x, A(x))\|_{X \times Y} = \|x\|_X + \|A(x)\|_Y \geqslant \|x\|_X.$$

By Corollary 3.2.2, the inverse mapping is also continuons, i.e., $x_n \to 0$ implies $\|x_n\|_X + \|A(x_n)\|_Y \to 0$. This implies $\|A(x_n)\|_Y \to 0$ and finally A is continuous. ∎

COROLLARY 3.3.6. *Let $(X, \|\ \|_X)$ and $(Y, \|\ \|_Y)$ be Banach spaces. Let A be a closed operator which maps the domain $D_A \subset X$ onto the whole space Y, $R_A = Y$. If A is a one-to-one mapping, then there exists a constant $K > 0$ such that*

$$K\|A(x) - A(y)\|_Y \geqslant \|x - y\|_X \quad \text{for} \quad x, y \in D_A.$$

Proof. The inverse operator A^{-1} is closed and, by Theorem 3.3.5, A^{-1} is continuous. Let $K = \|A^{-1}\|$. Thus,

$$\|x - y\|_X = \|A^{-1}A(x) - A^{-1}A(y)\|_X \leqslant K\|A(x) - A(y)\|_Y. \blacksquare$$

If A is not a one-to-one mapping, we obtain

PROPOSITION 3.3.7. *Let $(X, \|\ \|_X)$ and $(Y, \|\ \|_Y)$ be Banach spaces. Let A be a closed operator which maps its domain $D_A \subset X$ onto the whole space Y. Then there exists a constant $K > 0$ such that*

$$\{y \in Y \colon \|y\|_Y \leqslant 1\} \subset KA(\{x \in D_A \colon \|x\|_X \leqslant 1\}).$$

Proof. We introduce the norm
$$\|x\|_1 = \|x\|_X + \|A(x)\|_Y$$
on the set D_A. As in the proof of Theorem 3.3.5, we can show that $(D_A, \|\ \|_1)$ is complete. Thus, by Theorem 3.2.1, there is an $\alpha > 0$ such that
$$A(\{x \in D_A \colon \|x\|_1 \leq 1\}) \supset \alpha\{y \in Y \colon \|y\|_Y < \alpha\}.$$
Let $K = \dfrac{1}{\alpha}$. Thus,
$$\{y \in Y \colon \|y\|_Y \leq 1\} \subset KA(\{x \in D_A \colon \|x\|_1 \leq 1\})$$
$$\subset KA(\{x \in D_A \colon \|x\|_X \leq 1\}). \blacksquare$$

PROPOSITION 3.3.8. *Let $(X, \|\ \|_X)$ and $(Y, \|\ \|_Y)$ be Banach spaces. Let A and B be linear operators which map the domains D_A and D_B into Y. Suppose that A is closed, $D_A \subset D_B$ and there are constants $K > 0$, $0 < \alpha < 1$ such that*
$$\|B(x)\|_Y \leq K\|x\|_X + \alpha\|A(x)\|_Y \tag{3.3.4}$$
for all $x \in D_A$. Then the sum $A+B$ is a closed operator.

Proof. Let $\{x_n\}$ be a sequence of elements of the set D_A which converges to a certain x_0. Suppose that the sequence $\{(A+B)(x_n)\}$ converges to a certain y. Clearly, the sequences $\{x_n\}$, $\{A(x_n)+B(x_n)\}$ are fundamental. By (3.3.4), we have
$$\|A(x)\|_Y \leq \|B(x)\|_Y + \|(A+B)(x)\|_Y$$
$$\leq K\|x\|_X + \alpha\|A(x)\|_Y + \|(A+B)(x)\|_Y \tag{3.3.5}$$
and
$$\|A(x)\|_Y \leq \frac{1}{1-\alpha}\bigl(K\|x\|_X + \|(A+B)(x)\|_Y\bigr). \tag{3.3.6}$$

Formula (3.3.6) implies that the sequence $\{A(x_n)\}$ is fundamental and $A(x_n) \to A(x_0)$ for A is closed. Using (3.3.4) we get
$$\|B(x_n - x_0)\|_Y \leq K\|x - x_0\|_X + \alpha\|A(x_n) - A(x_0)\|_Y \to 0 \tag{3.3.7}$$
and $B(x_n) \to B(x_0)$. Thus, $y = \lim\limits_{n \to \infty} A(x_n) + B(x_n) = A(x_0) + B(x_0)$. $\blacksquare$

COROLLARY 3.3.9. *Let X and Y be Banach spaces. Let A be a closed operator mapping its domain $D_A \subset X$ into Y. Let B be a continuous linear operator mapping X into Y. Then the sum $A+B$ is a closed operator.*

However, the sum of closed operators need not be closed, as is shown by the following example.

EXAMPLE 3.3.4. Let $X = Y = L^2[0, 1]$. Let $A = \dfrac{d}{dt}$ be the operator of differentiation with the domain D_A equal to the space of all continuous and almost everywhere differentiable functions $x(t)$ such that $x'(t) \in L^2[0, 1]$ (see Example 3.3.1). Let $B(x) = \alpha \cdot x(0)$, $0 < \alpha < 1$. Let t_x be a point at which $|x(t)|$ reaches its infimum. Then

$$|x(t_x)| \leqslant \left(\int_0^1 |x(s)|^2 ds\right)^{1/2} = ||x||_{L^2}.$$

Moreover,

$$x(0) - x(t_x) = -\int_0^{t_x} x'(s) ds.$$

Hence,

$$|x(0)| \leqslant |x(t_x)| + \int_0^{t_x} |x'(s)| ds \leqslant ||x||_{L^2} + \int_0^1 |x'(s)| ds$$

$$\leqslant ||x||_{L^2} + \left(\int_0^1 |x'(s)|^2 ds\right)^{1/2} \leqslant ||x||_{L^2} + ||x'||_{L^2}.$$

By Proposition 3.3.8, the operator $-A+B$ is closed. The sum of A and $-A+B$ is equal to B and, by Example 3.3.3, this sum is not closed. Moreover, there is no closed extension of the operator B.

The superposition BA of a closed operator A and a continuous linear operator B need not be closed, as is shown by the following example.

EXAMPLE 3.3.5. Let $Y = C[0, 1]$, $X = L^2[0, 1]$. The natural injection of Y into X is a continuous linear operator. Thus, the inverse operator

3.3. Closed operators

A is closed. We recall that the domain of A is the set of those functions belonging to $L^2[0, 1]$ which are continuous. The range of A is the space $Y = C[0, 1]$. We can define a linear continuous functional f on Y by the formula

$$f(x) = x(0)$$

(see Example 3.3.3).

Observe that the superposition $fA(x)$ which maps X into reals is a nonclosed functional as described in Example 3.3.3.

PROPOSITION 3.3.10. *Let X, Y, Z be Banach spaces. Let A be a continuous linear operator mapping X into Y. Let B be a closed operator mapping the domain $D_B \subset Y$ into Z. Then the operator BA with the domain $D_{AB} = A^{-1}(D_B)$ is closed.*

Proof. Let $x_n \in D_{AB}$ for $n = 1, 2, \ldots$, $x_n \to x$, $BA(x_n) \to z$. Since the operator A is continuous, $A(x_n) \to A(x)$ and $BA(x_n) \to z$ implies $BA(x_n) \to BA(x)$. ∎

THEOREM 3.3.11 (Douglas, 1966). *Let X, Y, Z be Banach spaces. Let A be a continuous linear operator which maps X into Z. Let B be a closed operator which maps the domain $D_B \subset Y$ into Z. Suppose that the range of B contains the range of A, $R_A \subset R_B$, and the operator B is a one-to-one mapping. Then there is a continuous linear operator C which maps X into Y and is such that*

$$A(x) = BC(x). \tag{3.3.8}$$

Proof. Since B is a one-to-one mapping, the inverse operator B^{-1} is well defined. By Proposition 3.3.1, B^{-1} is closed. The domain of B^{-1} coincides with the range of B, $R_B \subset Z$. Since $R_A \subset R_B$, the operator $C(x) = B^{-1}A(x)$ is well defined for all $x \in X$. By Proposition 3.3.10, C is closed. Since C is defined on the whole X, it follows that C is continuous (see Proposition 3.3.5). ∎

THEOREM 3.3.12 (Douglas, 1966). *Let X, Y, Z be Banach spaces. Let A be a continuous linear operator which maps X into Z. Let B be a closed* operator mapping its domain D_B into Z. Assume that there is a continuous linear projection P of Y onto $\ker B = \{y \in D_B \colon B(y) = 0\}$, i.e., a continuous linear operator P such that $P^2 = P$, and $P(Y) = \ker B$. If $R_A \subset R_B$, then there is a continuous linear operator C which maps X into Y and is such that the range of C is contained in the domain of B, $R_C \subset D_B$ and

$$A(x) = BC(x).$$

Proof. Let $Y_1 = (I-P)(Y)$. It is easy to see that the operator B restricted to Y_1 is a one-to-one mapping. Therefore, by Proposition 3.3.11, there is a continuous linear operator C_1 mapping X into Y_1 and such that

$$A(x) = BC_1(x).$$

Clearly, C_1 can be considered a continuous linear operator C which maps X into Y and this operator C has the required properties. ∎

THEOREM 3.3.13 (Douglas, 1966). *Let* $(X, \| \ \|_X)$, $(Y, \| \ \|_Y)$, $(Z, \| \ \|_Z)$ *be Banach spaces. Let A be a closed operator which maps its domain $D_A \subset X$ into Z. Let B be a continuous linear operator which maps $D_B \subset Y$ into Z. Suppose that there exists a continuous linear projection P of the space Y onto* $\ker B = \{x \colon B(x) = 0\}$. *If the range of the operator A is contained in the range of the operator B, $R_A \subset R_B$, then there is a closed operator C such that the domain of C is equal to the domain of A, $D_C = D_A$, the range of the operator C is contained in the domain of the operator B, $R_C \subset D_B$ and*

$$A(x) = BC(x) \quad \text{for} \quad x \in D_A. \tag{3.3.9}$$

Proof. We introduce the norm

$$\|x\|_1 = \|x\|_X + \|A(x)\|_Y$$

on the set D_A. The set D_A with this norm is a complete space, since $(D_A, \| \ \|_1)$ is isometric to $(G_A, \| \ \|_{X \times Y})$ (cf. the proof of Theorem 3.3.5). Thus, we can apply Theorem 3.3.11 and find a continuous linear operator C which maps $(D_A, \| \ \|_1)$ into $(Y, \| \ \|_Y)$ and is such that (3.3.8) holds.

3.3. Closed operators

Now we shall show that the operator C regarded as an operator from the space $(X, \|\ \|_X)$ into $(Y, \|\ \|_Y)$ is closed. Take any $x_n \to x$, $x_n \in D_C = D_A$ such that $C(x_n) \to z$. By the continuity of B, $BC(x_n) \to B(z)$. But $BC(x_n) = A(x_n)$ and hence the closedness of A implies $B(z) = A(x) = BC(x)$. Since $x_n \to x$, $A(x_n) \to A(x)$ the sequence $\{x_n\}$ converges to x also in the space $(D_A, \|\ \|_1)$ and, by the continuity of C, $C(x_n) \to C(x)$. Hence $z = C(x)$. ∎

In Theorem 3.3.12 and Theorem 3.3.13 we have assumed that there is a continuous linear projection on the kernel of B. In Hilbert spaces this hypothesis is automatically satisfied. Namely, we have

PROPOSITION 3.3.14. *Let X be a Hilbert space. Let X_0 be a subspace of X. Then there is a continuous linear projection of X onto X_0 with the norm equal to 1.*

Proof. By successive repetitions of the construction given in Section 2.6 we can find for each element $y_0 \notin X_0$ an element x_0 such that $(y_0 - x_0, x) = 0$ for $x \in X_0$. Of course, the set

$$X_1 = \{x: (x, y) = 0,\ y \in X_0\}$$

is a linear space. Each element x of the space X can be uniquely represented in the form $x = x_0 + x_1$, where $x_0 \in X_0$, $x_1 \in X_1$. Thus, the mapping $P(x) = x_0$ is a linear projection of X onto X_0. Since

$$\|x\| = \sqrt{\|x_0\|^2 + \|x_1\|^2} \geqslant \|x_0\|,$$

$\|P(x)\| \leqslant \|x\|$ and the norm of P is equal to 1. ∎

The projection P constructed in Proposition 3.3.14 will be called an *orthogonal projection*.

In an arbitrary Banach space Proposition 3.3.14 might not be valid. Moreover, if a Banach space has the property that there is a continuous linear projection on each subspace, then X is a Hilbert space (Lindenstrauss–Tzafiri, 1972).

The existence of a continuous linear projection on the kernel of B is essential in the above theorems (P. Wojtaszczyk, 1984, private communication). The corresponding example is based on two classical facts (Banach, 1931), which we shall quote here without proof.

(1) For each separable Banach space X there is a continuous linear operator T_X mapping l^1 onto X.

(2) Each infinite-dimensional subspace of l^1 contains a subspace isomorphic to l^1.

Now, let $X = Z = l^2$. Let A be the identity mapping I. Let $Y = l^1$ and let $B = T_{l^2}$ be a continuous linear operator mapping l^1 onto l^2. The existence of C which maps l^2 into l^1 and satisfies the condition $BC = I$ implies that l^2 is isomorphic to a subspace of l^1, which contradicts the fact (2).

PROPOSITION 3.3.15. *Let X and Y be Banach spaces. Let A be a closed operator defined on the domain $D_A \subset X$ which maps D_A onto Y. If A is a one-to-one mapping, then there is a constant $a > 0$ such that, for each continuous linear operator B belonging to $B(X \to Y)$ and satisfying the condition $\|B\| < a$, the operator $A+B$ is a one-to-one mapping of the domain D_A onto Y and the inverse operator $(A+B)^{-1}$ is continuous.*

Proof. By Proposition 3.3.1, the operator A^{-1} is closed. Hence, by Theorem 3.3.5, A^{-1} is continuous. Let $a = \dfrac{1}{\|A^{-1}\|}$. Take any operator B, $B \in B(X \to Y)$, $\|B\| < a$. The operator $A^{-1}B$ maps X into itself and $\|A^{-1}B\| \leqslant \|A^{-1}\| \cdot \|B\| < 1$. To be precise, the operator $A^{-1}B$ maps D_A into itself because A^{-1} maps Y into D_A. Thus, by Proposition 3.1.5, the operator $I+A^{-1}B$ is an isomorphism which maps X onto itself and $(I+A^{-1}B)(D_A) \subset D_A$. Further, by Proposition 3.3.10, the operator

$$A+B = A(I+A^{-1}B)$$

is closed. Moreover, $A+B$ is a one-to-one mapping as the superposition of one-to-one mappings. Finally, its inverse

$$(A+B)^{-1} = (I+A^{-1}B)^{-1}A^{-1}$$

is continuous as the superposition of two continuous operators. ∎

3.4. CONJUGATE OPERATORS

Let X and Y be two real or complex linear spaces. Let A be a linear operator which maps X into Y and let f be a linear functional defined on Y. As can easily be verified, the functional F defined as

$$F(x) = f(A(x)) \tag{3.4.1}$$

is a linear functional defined on X. Hence, formula (3.4.1) may be viewed as the definition of an operator A' which assigns to linear functionals defined on Y linear functionals defined on X.

Denote by X' and Y' the spaces of all linear functionals defined on the spaces X and Y. The spaces X' and Y' are called the *dual spaces* of the spaces X and Y, respectively. The operator A' is called the *dual operator* of the operator A.

The spaces X' and Y', equipped with the operations of addition and multiplication by scalars defined as follows

$$(f+g)(x) = f(x)+g(x),$$
$$(af)(x) = af(x),$$

form linear spaces. It is easy to verify that the dual operator A' is a linear operator.

Let X and Y be Banach spaces over the field of real or complex numbers. Let A be a continuous linear operator which maps the space X into Y. The restriction of the operator A' to the conjugate space $Y^* \subset Y'$ will be called the *conjugate operator*. The conjugate operator will be denoted by A^*.

Note that if $f \in Y^*$ then $A^*f = A'f$ is a continuous linear functional defined on X since the superposition of two continuous functions is continuous. Hence, the range of the conjugate operator is a subset of X^*.

THEOREM 3.4.1. $\|A^*\| = \|A\|$.

Proof. By Corollary 2.5.6,

$$\|A^*\| = \sup_{\|f\| \leq 1} \|A^*f\| = \sup_{\|f\| \leq 1} \sup_{\|x\| \leq 1} |f(A(x))|$$
$$= \sup_{\|x\| \leq 1} \sup_{\|f\| \leq 1} |f(A(x))| = \sup_{\|x\| \leq 1} \|A(x)\| = \|A\|. \blacksquare$$

Observe that Theorem 3.4.1 is also valid in real seminormed spaces. As a consequence of the equality proved above we have

THEOREM 3.4.2. *Let X and Y be Banach spaces. Let A be a linear operator which maps X into Y. If the restriction of the dual operator A' to the space Y^* is a continuous mapping of Y^* into X^*, then A is continuous.*

Now we shall analyse conjugate operators of closed operators in a more detailed way.

Let X and Y be Banach spaces. Let A be a linear operator which maps the domain $D_A \subset X$ into Y. Let η be an arbitrary continuous linear functional defined on Y, $\eta \in Y^*$. Clearly, $\eta(A(x))$ need not be continuous. By D_{A^*} we shall denote the set of all $\eta \in Y^*$ such that $\eta(A(x))$ is continuous. Observe that

$$\xi(x) = \eta(A(x)) \tag{3.4.2}$$

is a continuous linear functional defined on D_A. This functional can be uniquely extended onto the closure of the domain, $\bar{D}_A$. Thus, if we want to determine the conjugate operator

$$\xi = A^*\eta$$

by formula (3.4.2) we need to assume that $\bar{D}_A = X$. Consequently, we shall henceforth assume that $\bar{D}_A = X$.

The graph of the conjugate operator

$$G_{A^*} = \{(\xi, \eta) \in X^* \times Y^*: \xi(x) = \eta(A(x)) \text{ for all } x \in D_A\}$$
$$= \bigcap_{x \in D_A} \{(\xi, \eta) \in X^* \times Y^*: \xi(x) = \eta(A(x))\}$$

is closed even if A is not closed. Hence, we obtain

PROPOSITION 3.4.3. *Let X and Y be normed spaces. Let A be a linear operator such that the domain of A, D_A, is dense in X. Then the conjugate operator $\xi = A^*\eta$, defined by formula (3.4.2), is a well-defined closed operator.*

Let X, Y and Z be Banach spaces. Let A be a linear operator which maps the domain $D_A \subset X$ into Y. Let B be a linear operator which maps the domain $D_B \subset Y$ into Z. We assume that D_A is dense in X, the domain D_B is dense in Y and the range of A is contained in the domain of B. Thus, the superposition BA is well-defined and the domain of BA is equal to the domain of A, $D_{BA} = D_A$.

We claim that

$$(BA)^* = A^*B^*. \tag{3.4.3}$$

To prove this, it is enough to observe that for arbitrary $x \in D_A = D_{BA}$ and $\zeta \in Z^*$ we have

$$\zeta(BA(x)) = (B^*\zeta)(A(x)) = (A^*(B^*\zeta))(x), \tag{3.4.4}$$

which implies (3.4.3).

Let X and Y be Banach spaces. Let A be a one-to-one closed linear operator which maps the domain D_A, $D_A \subset X$, into Y and $\bar{D}_A = X$. We assume that the range of the operator A is dense in Y, $\bar{R}_A = Y$.

Let A^* and $(A^{-1})^*$ be the conjugate operators of A and A^{-1}, respectively. The operator $A^{-1}A$ is equal to the identity operator on D_A and AA^{-1} is equal to the identity operator on R_A. By (3.4.3) we have

$$I_{X^*} = (A^{-1}A)^* = A^*(A^{-1})^*$$

and

$$I_{Y^*} = (AA^{-1})^* = (A^{-1})^*A^*.$$

Thus, $(A^{-1})^* = (A^*)^{-1}$.

THEOREM 3.4.4. *Let X and Y be Banach spaces. Let A be a closed operator defined on the domain D_A which is dense in X. The range R_{A^*} of the conjugate operator A^* coincides with the whole X^* if and only if the operator A is a one-to-one mapping and the inverse A^{-1} is continuous.*

Proof. Necessity. Suppose that $R_{A^*} = X^*$. Since the operator A^* is closed, in view of Proposition 3.3.7 there is a constant $K > 0$ such that

$$\{\xi: \|\xi\|_{X^*} \leqslant 1\} \subset KA^*(\{\eta \in D_A: \|\eta\|_{Y^*} \leqslant 1\}). \tag{3.4.5}$$

If A is non-invertible or A^{-1} is not continuous, then there is a sequence

$\{x_n\}$, $x_n \in D_A$, such that $||x_n|| \geq 1$ and $||A(x_n)|| \to 0$. But, by (3.4.5), we have

$$1 \leq ||x_n|| = \sup_{||\xi|| \leq 1} |\xi(x_n)| = \sup_{\substack{||\eta|| \leq K \\ \eta \in D_{A^*}}} |\eta(A(x_n))| \leq K \cdot ||A(x_n)|| \to 0,$$

which leads to a contradiction.

Sufficiency. Suppose that A^{-1} is continuous. Let ξ be an arbitrary continuous linear functional $\xi \in X^*$. Clearly, $\xi(A^{-1}(y))$ is continuous on the range of A, R_A. Denoting by η an arbitrary extension of this functional, we have $\eta(A(x)) = \xi(x)$ on D_A. Hence $\eta \in D_{A^*}$ and $A^*\eta = \xi$, which completes the proof. ∎

The following theorems are, in a way, dual to Theorem 3.4.4.

THEOREM 3.4.5. *Let X and Y be Banach spaces. Let A be a closed operator defined on the domain D_A which is dense in X, $\bar{D}_A = X$. If the range of the operator A coincides with the whole space Y, then the conjugate operator A^* is invertible and $(A^*)^{-1}$ is continuous.*

Proof. Let ξ be an arbitrary element of the range of A^*, $\xi \in R_{A^*}$. This means that there is an $\eta \in Y^*$ such that $\xi = A^*\eta$, i.e.,

$$\xi(x) = \eta(A(x)) \quad \text{for} \quad x \in D_A. \tag{3.4.6}$$

Putting $y = A(x)$, we infer in view of Proposition 3.3.7 that there is a constant $K > 0$ such that for all $y \in Y$

$$y = A(x) \quad \text{and} \quad ||x|| \leq K \cdot ||y||.$$

Thus,

$$||\xi|| = \sup_{||x||_X \leq 1} |\xi(x)| = \sup_{\substack{||x||_X \leq 1 \\ x \in D_A}} |\xi(x)| = \sup_{\substack{||x||_X \leq 1 \\ x \in D_A}} |\eta(A(x))|$$
$$\leq \sup_{||y||_Y \leq K} |\eta(y)| \leq K \cdot ||\eta||.$$

Hence, A^* is invertible and $(A^*)^{-1}$ is continuous. ∎

THEOREM 3.4.6 (Goldberg, 1959; Kato, 1958; Rota, 1958; see also Goldberg, 1966). *Let X and Y be Banach spaces. Let A be a closed operator such that the domain of A, D_A, is dense in X, $\bar{D}_A = X$, and the values of*

3.4. Conjugate operators

A are in the space Y. If the conjugate operator A^* has the continuous inverse, then $R_A = Y$.

The proof of Theorem 3.4.6 is based on the following lemmas.

LEMMA 3.4.7. *Let $(X, \|\ \|_X)$ and $(Y, \|\ \|_Y)$ be Banach spaces. Let A be a linear operator mapping $D_A \subset X$ into Y. If $(A^*)^{-1}$ exists and $(A^*)^{-1}$ is continuous, then the closure of the image of the unit ball contains the ball of radius $r = \dfrac{1}{\|(A^*)^{-1}\|}$,*

$$\overline{A(\{x\colon \|x\|_X \leqslant 1\})} \supset \{y\colon \|y\|_Y \leqslant r\}. \tag{3.4.7}$$

Proof. Suppose that (3.4.7) does not hold. Then there is a $y \in Y$, $\|y\|_Y < r$, such that $y \notin \overline{A(\{x\colon \|x\|_X \leqslant 1\})}$. In view of the separation theorems, there is a functional $\eta \in Y^*$ such that

$$\operatorname{Re}\eta(y) \geqslant 1,$$
$$\operatorname{Re}\eta(z) \leqslant 1 \quad \text{for} \quad z \in A(\{x\colon \|x\|_X \leqslant 1\}).$$

Thus, $|\eta(A(x))| \leqslant 1$ for all $x \in D_A$, $\|x\|_X \leqslant 1$. Hence, $\eta \in D_{A^*}$ and

$$\|\eta\|_{Y^*}\|y\|_Y \geqslant |\eta(y)| \geqslant \sup_{\|x\|_X \leqslant 1} |\eta(A(x))| = \|A^*\eta\| \geqslant r\|\eta\|$$

and we get a contradiction with the hypothesis that $\|y\| < r$. ∎

LEMMA 3.4.8. *Under the assumptions of the previous lemma the image of the unit ball contains a ball.*

Proof. The proof follows the same line as the second part of the proof of the Banach theorem (Theorem 3.2.1). ∎

The proof of Theorem 3.4.6 is an immediate consequence of Lemma 3.4.8.

Chapter 4

Weak topologies

4.1. WHY WE INTRODUCE TOPOLOGICAL SPACES? BASIC TOPOLOGICAL NOTIONS

In the context of optimization we often deal with the following problem. Let X be a Banach space and let f be a continuous linear functional defined on X. We seek an element $\bar{x}$ belonging to X, $\|\bar{x}\| = 1$, such that

$$f(\bar{x}) = \|f\|. \qquad (4.1.1)$$

An element $\bar{x}$ satisfying (4.1.1) does not always exist. We want to establish conditions ensuring the existence of $\bar{x}$.

If the closed unit ball $\{x\colon \|x\| \leqslant 1\}$ were compact, the existence of $\bar{x}$ would be an immediate consequence of the properties of continuous functions on compact sets (cf. Theorem 1.11.15).

Unfortunately, if X is an infinite-dimensional space, the closed unit ball is not compact. To see this, we can construct by induction a sequence $\{x_n\}$ such that

(1) $1/2 \leqslant \|x_n\| \leqslant 1$, $n = 1, 2, \ldots,$

(2) $\|x_i - x_j\| \geqslant 1/2$, $i \neq j$.

Let x_1 be an arbitrary element satisfying condition (1). Suppose that we have chosen elements $x_1, \ldots, x_n$ satisfying (1) and (2). Let us denote by X_n the space spanned by the elements $x_1, \ldots, x_n$, $X_n = \lin\{x_1, \ldots, x_n\}$.

Consider now the quotient space X/X_n. It is a normed space. Since X is infinite-dimensional, the quotient space X/X_n contains nonzero elements and consequently there is an element (an equivalence class) with the norm equal to $1/2$. By the definition of the norm in the quotient space, this equivalence class contains an element, say x_{n+1}, satisfying

4.1. Basic topological notions

(1). Observe that $x_1, \ldots, x_n$ belong to the equivalence class containing 0. Hence,

(2') $\|x_{n+1} - x_i\| \geq 1/2$, $i = 1, 2, \ldots, n$.

Each sequence satisfying conditions (1), (2) belongs to the closed unit ball and, by (2), such a sequence does not contain any convergent subsequence. Thus, the closed unit ball in any infinite-dimensional Banach space is not compact.

Now we can ask whether it is possible, at least in some Banach spaces, to define a metric such that

1. the closed unit ball is compact,
2. all linear continuous functionals remain continuous.

The answer is negative. Therefore, we would like to modify the notion of metric to ensure the existence of elements satisfying (4.1.1) at least in some Banach spaces. This approach leads us to the notion of weak topologies.

Before we give the definition of weak topologies, let us introduce the basic topological notions.

Let X be a given set and let θ be a family of sets, called *open sets*, which satisfy the following conditions:

(1) the empty set and the whole set X are open,
(2) the intersection of two open sets is open,
(3) the union of an arbitrary number of open sets is open.

The family θ is called a *topology*.

A set X with a topology is called a *topological space*.

Let X be a topological space and let x be an arbitrary element of X. An arbitrary open set containing x is called a *neighbourhood* of x. Since families θ could be relatively large, it is often convenient to deal with *bases of neighbourhoods*. Namely, we say that a family of neighbourhoods $B(x)$ at a point $x \in X$ is a *basis of neighbourhoods at* x if for every neighbourhood U of x there is a neighbourhood V belonging to $B(x)$ such that $V \subset U$.

Suppose we are given a family of bases of neighbourhoods $B(x)$, $x \in X$. In a certain sense this family determines a topology. Indeed, in view of (3), the unions $\bigcup U_\alpha$, where $U_\alpha \in B(x_\alpha)$, are open. And conversely, every open set U can be represented as the union of basic neighbourhoods since

$$U = \bigcup_{x \in U} V,$$

where $V \in B(x)$ is an arbitrary neighbourhood of x such that $V \subset U$.

Let X be a topological space. A set F contained in the topological space X is called *closed* if its complement $X \setminus F$ is open. The intersection of an arbitrary family of closed sets is closed. The union of two closed sets is closed. The empty set and the whole space are closed.

Let A be an arbitrary set contained in a topological space X. The closure of the set A, $\bar{A}$, is the smallest closed set containing A. Clearly

$$\bar{A} = \bigcap \{F : F \text{ closed}, A \subset F\}.$$

The *interior* of the set A, Int A, is the largest open set contained in A. Clearly,

$$\text{Int}\, A = \bigcup \{G : G \text{ open}, G \subset A\} = X \setminus \overline{(X \setminus A)}.$$

The definitions given above are not identical with those given in Section 1.2 but, in virtue of the theorems formulated in Section 1.2, they are equivalent. Thus, a metric space with a topology given by open sets as defined in Section 1.2 is a topological space.

In the sequel we shall restrict ourselves to *Hausdorff topological spaces*, i.e. topological spaces which satisfy the following condition:

for every pair of points, $x_1, x_2, x_1 \neq x_2$, there exist disjoint neighbourhoods U_1, U_2 of points x_1, x_2 such that $U_1 \cap U_2 = \emptyset$.

Observe that metric spaces regarded as topological spaces are Hausdorff spaces.

Let X be a topological space. Let Y be a subset of the set X. We can induce a topology on Y by defining open sets as follows: $U_Y = Y \cap \cap U_X$, where U_X is an arbitrary open set in X. The space Y with the induced topology is called a *subspace* of X.

Let Y be a closed subset of a topological space X. Let Z be a closed subset of Y regarded as a subspace of X. Then the set Z is closed in X. To see this, observe first that the set Z is of the form $Y \setminus G$, where G is an open set in Y. Since Y is a subspace of X, we have $G = Y \cap G_0$, where G_0 is an open set in X. Further, $Y = X \setminus G_1$, where G_1 is open

4.1. Basic topological notions

in X since Y is a closed subset of X, and finally
$$Z = X \setminus (G_0 \cup G_1)$$
is closed since the union of two open sets is open.

Let X be a topological space. A set $A \subset X$ is called *dense* in X if $A \cap U \neq \emptyset$ for every point $x \in X$ and every neighbourhood U of x. The space X is called *separable* if it contains a countable dense set.

Let X be a topological space and let A be a dense subset of X. Then the closure of A, $\bar{A}$, is equal to the whole space X.

THEOREM 4.1.1. *Let X and Y be topological spaces. Let f be a function which maps the space X into the space Y. The following conditions are equivalent*:

(a) *for every x_0 and every neighbourhood U of $f(x_0)$ there is a neighbourhood V of x_0 such that $f(V) \subset U$*;

(b) *the inverse image $f^{-1}(G) = \{x : f(x) \in G\}$ is open for every open set G*;

(c) *the inverse image $f^{-1}(F) = \{x : f(x) \in F\}$ is closed for every closep set F*.

Proof. (a) $\Rightarrow$ (b). Let G be an open set and $x \in f^{-1}(G)$. Then $f(x) \in G$. Since G is open, there is a neighbourhood U of $f(x)$ such that $U \subset G$. By (a), there exists a neighbourhood V_x of x such that $f(V_x) \subset U \subset G$. Hence, $V_x \subset f^{-1}(G)$ and consequently,
$$f^{-1}(G) = \bigcup_{x \in f^{-1}(G)} V_x$$
is open as the union of open sets.

(b) $\Rightarrow$ (a). It is enough to take $V = f^{-1}(U)$.

(b) $\equiv$ (c). Let F be a closed set (G an open set). Then the set $Y \setminus F$ is open (the set $Y \setminus G$ is closed) and, in virtue of (b), (of (c)) the set $f^{-1}(Y \setminus F)$ is open (the set $f^{-1}(Y \setminus G)$ is closed). Hence, the set
$$f^{-1}(F) = X \setminus f^{-1}(Y \setminus F)$$
is closed (the set $f^{-1}(G) = X \setminus f^{-1}(Y \setminus G)$ is open). ∎

A function f which satisfies one of the conditions (a), (b), (c) is called a *continuous function*. A function f which is a one-to-one mapping of a topological space X onto a topological space Y is called a *homeomorphism* if the functions f and f^{-1} are both continuous. We say that the spaces X and Y are *homeomorphic* if there exists a homeomorphism which maps X onto Y.

The question arises whether it is possible to define convergence of sequences in topological spaces. The answer is positive.

Let X be a topological space. We say that a sequence $\{x_n\}$ converges to an x_0, if for every neighbourhood U of x_0 there exists an index $n(U)$ such that $x_n \in U$ for $n > n(U)$.

If the space X is a Hausdorff space (i.e., if there are sufficiently many open sets in the topology of X), then the limit point x_0 is unique.

If a sequence $\{x_n\}$ converges to a certain x_0, then every its subsequence converges to the same x_0.

Let X and Y be topological spaces. A function f which maps the space X into the space Y is called *sequentially continuous* if f transforms convergent sequences into convergent sequences.

THEOREM 4.1.2. *Every continuous function f which maps a topological space X into a topological space Y is sequentially continuous.*

Proof. Let x_0 be an arbitrary element of X and let $\{x_n\}$ be a sequence which converges to x_0. Let U be an arbitrary neighbourhood of $f(x_0)$. By the continuity of f, there exists a neighbourhood V of x_0 such that $f(V) \subset U$. Since the sequence $\{x_n\}$ converges to x_0, there is an index N such that $x_n \in V$ for $n > N$. Hence, $f(x_n) \in U$, which entails the convergence of the sequence $\{f(x_n)\}$ to $f(x_0)$. ■

The converse result to that of Theorem 4.1.2 is not true, i.e., there exist sequentially continuous mappings which are not continuous. An example will be given in Section 4.4.

Let A be a closed set contained in a topological space X. Let $\{x_n\}$ be a sequence of elements of A which converges to a certain $x_0 \in X$. Then $x_0 \in A$. To see this, suppose on the contrary that $x_0 \notin A$. Since the set $X \setminus A$ is open and $x_0 \in X \setminus A$, there is a neighbourhood U of x_0 such that $U \cap A = \emptyset$. But this contradicts the fact that the sequence $\{x_n\}$ converges to x_0. Thus, for an arbitrary set A contained in X, the set $\tilde{A}$, $\tilde{A} = \{x: x = \lim_{n\to\infty} x_n, x_n \in A\}$, is contained in the closure of A, $\bar{A}$. For metric spaces we have proved a stronger result, namely $\tilde{A} = \bar{A}$ (Theorem 1.2.7). In the general case this equality does not hold. Moreover, it may happen that the set $\tilde{\tilde{A}}$, $\tilde{\tilde{A}} = \{x \in X: x = \lim_{n\to\infty} x_n, x_n \in \tilde{A}\}$, is essentially larger than $\tilde{A}$.

4.2. SEQUENTIALLY COMPACT SPACES AND COMPACT SPACES

In Section 1.11 we gave two definitions of compact spaces which are equivalent in metric spaces. Namely, a space X is said to be *sequentially compact* if every sequence of elements of X contains a convergent subsequence. Furthermore, a space X is said to be *compact* if every covering of X contains a finite covering of X.

These definitions are also valid in Hausdorff topological spaces but in those spaces they are not equivalent. Examples of sequentially compact spaces which are not compact can be found, e.g., in Engelking (1968, p. 148).

Examining the proofs of the theorems given in Section 1.11, we observe that many of those theorems remain valid in topological spaces. Here we formulate some of them in this new setting.

THEOREM 4.2.1. *A closed subset of a sequentially compact space is sequentially compact* (cf. Theorem 1.11.1).

THEOREM 4.2.2. *Let X be a sequentially compact space. Every descending sequence of closed sets has a nonempty common part* (cf. Theorem 1.11.4).

THEOREM 4.2.3. *Let X be a sequentially compact space. Every countable covering of X contains a finite covering of X* (cf. Theorem 1.11.5).

THEOREM 4.2.4. *Let X be a sequentially compact space and let f be a sequentially continuous function which maps X onto a Hausdorff space Y. Then the space Y is sequentially compact* (cf. Theorem 1.11.12).

THEOREM 4.2.5. *If f is a real-valued function defined on a sequentially compact space, then f attains its infimum and supremum on this space* (cf. Theorem 1.11.15).

THEOREM 4.2.6. *Let f be a sequentially continuous one-to-one function which maps a sequentially compact space X onto a Hausdorff space Y. Then the inverse function f^{-1} is also sequentially continuous.*

Proof. Let $\{y_n\}$ be a sequence of elements of the space Y which converges to some $y_0 \in Y$. Putting $x_0 = f^{-1}(y_0)$ and $x_n = f^{-1}(y_n)$, we have to show that the sequence $\{x_n\}$ converges to x_0. Suppose that the converse statement is true. Then there is a neighbourhood U of x_0 such that infinitely many elements of the sequence $\{x_n\}$ do not belong to U. Since the space X is sequentially compact, the sequence consisting of those elements contains a convergent subsequence, say $\{x_{n_k}\}$, which converges to some $\bar{x}$, and $\bar{x} \notin U$. By the sequential continuity of f, the sequence $\{y_{n_k}\}$, $y_{n_k} = f(x_{n_k})$, converges to $f(\bar{x})$. Since $\{y_{n_k}\}$ is a subsequence of the sequence $\{y_n\}$, we obtain $f(\bar{x}) = f(x_0)$, which contradicts the hypothesis that f is a one-to-one function. ∎

Since sequentially compact spaces and compact spaces form different classes of spaces we shall show the analogues of the above theorems in compact spaces.

THEOREM 4.2.7. *Let X be a compact space. Let $\{F_t\}$, $t \in T$, be a family of closed subsets of the space X such that every finite subfamily $F_{t_1}, \ldots, F_{t_n}$ has a nonempty common part, $F_{t_1} \cap \ldots \cap F_{t_n} \neq \emptyset$. Then the intersection of the whole family is nonempty.*

Proof. The proof is the same as the proof of Theorem 1.11.9. ∎

THEOREM 4.2.8. *If in a Hausdorff space X each family of closed sets has the property described in the statement of Theorem 4.2.7 (i.e., if every finite subfamily has a nonempty intersection then the intersection of the whole family is nonempty), then the space X is compact.*

Proof. Let $\{G_t\}$, $t \in T$, be a covering of the set X. Let $F_t = X \setminus G_t$. The sets F_t are closed and the intersection of all F_t, $t \in T$, is empty. Hence, by hypothesis, there is a finite subfamily $F_{t_1}, \ldots, F_{t_n}$, whose intersection is empty. This implies that $X \subset G_{t_1} \cup \ldots \cup G_{t_n}$. ∎

THEOREM 4.2.9. *Let X be a compact topological space and let Y be a closed subset of X. Then Y is a compact topological space.*

Proof. Let $\{F_t\}$, $t \in T$, be an arbitrary family of closed subsets of Y such that the intersection of each finite subfamily is nonempty. Observe that each set F_t is a closed subset of X since Y is closed. Hence, by Theorem 4.2.7, the intersection of the whole family is nonempty and, by Theorem 4.2.8, Y is a compact topological space. ∎

THEOREM 4.2.10. *Let X be a compact topological space. Let f be a continuous function mapping the space X into a Hausdorff space Y. Then $f(X)$ is compact.*

Proof. Let $\{G_t\}$, $t \in T$, be an arbitrary covering of $f(X)$. Since f is continuous, the sets $f^{-1}(G_t)$ are open and form a covering of the space X. Since X is compact, this covering contains a finite covering $f^{-1}(G_{t_1}), \ldots, f^{-1}(G_{t_n})$. This in turn implies that the sets $G_{t_1}, \ldots, G_{t_n}$ form a finite covering of $f(X)$. ∎

COROLLARY 4.2.11. *Let X be a compact topological space and let f be a continuous mapping of the space X into a Hausdorff space Y. Then f transforms closed sets into closed sets.*

COROLLARY 4.2.12. *Let f be a continuous one-to-one mapping of a compact topological space X into a Hausdorff space Y. Then the inverse mapping f^{-1} is continuous.*

Corollary 4.2.12 immediately implies that Theorem 1.11.15 holds for compact topological spaces.

Let X be a topological space. A subset $A \subset X$ is called a *compact set* (a *sequentially compact set*) if A is a compact subspace of X.

Let $\{X_\alpha\}$, $\alpha \in A$, be a family of Hausdorff topological spaces. By $\underset{\alpha \in A}{\times} X_\alpha$ we denote the set of all sequences of the form $\{x_\alpha\}$, $x_\alpha \in X_\alpha$. In the space $\underset{\alpha \in A}{\times} X_\alpha$ we introduce a topology in the following way. Let $x^0 = \{x_\alpha^0\}$ be a fixed element of this space and let $U_{\alpha, \beta}$ be a basis of neighbourhoods of x_α^0 in the space X_α. We define a basis of neighbourhoods at x^0 as a family of all sets V of the form:

$$V = \Big\{ \{x_\alpha\} \in \underset{\alpha \in A}{\times} X_\alpha \colon x_{\alpha_n} \in U_{\alpha_n, \beta_n} \quad \text{for some finite number} \\ \text{of indices } \alpha_1, \beta_1, \ldots \alpha_n, \beta_n \Big\}.$$

It is easy to verify that $\underset{\alpha \in A}{\times} X_\alpha$ is a Hausdorff topological space. It is called the *Tikhonov product space* of the spaces $\{X_\alpha\}$.

Now we state without proofs the following significant theorems.

THEOREM 4.2.13 (Tikhonov). *Let $\{X_\alpha\}$ be a family of compact topological spaces. Then the Tikhonov product space $\underset{\alpha \in A}{\times} X_\alpha$ is a compact space.*

One can find the proof of this theorem in topology textbooks (see e.g. Engelking, 1968, p. 112).

The above theorem justifies the conclusion that the class of compact spaces is more important than the class of sequentially compact spaces. There exist examples of two sequentially compact spaces whose product is not sequentially compact (see Engelking, 1968, p. 148).

In this section we have considered compact spaces and sequentially compact spaces separately. It can be shown that every compact space is sequentially compact. We omit the proof of this fact since it is not needed in the present book.

4.3. LINEAR TOPOLOGICAL SPACES

Let X be a linear space over the field of real (complex) numbers and, moreover, let X be a Hausdorff topological space. We say that X is a *linear topological space* if the operations of addition and multiplication by scalars are continuous. In other words,

(1) for every $x, y \in X$ and every neighbourhood U_{x+y} of $x+y$ there exist a neighbourhood U_x of x and a neighbourhood U_y of y such that
$$U_x + U_y \subset U_{x+y};$$

(2) for every $x \in X$ and every neighbourhood U_{tx} of tx, where t is an arbitrary number, there exist a neighbourhood U_x of x and a number $\varepsilon > 0$ such that $t'U_x \subset U_{tx}$ for all t' satisfying $|t-t'| < \varepsilon$.

Continuity of addition implies that for any basis U of neighbourhoods at zero the sets $\{x+U_\alpha\}$, $x \in X$, form a basis of neighbourhoods of the whole space.

Let a topological space X be given. We say that the topology of X is *invariant* if for every open set G and every $x \in X$ the set $x+G$ is open. Clearly, continuity of addition implies that the topology is invariant. The converse is not true. The invariance of a topology does not lead to continuity of addition. For invariant topologies, continuity of operations amounts to continuity of operations at zero, i.e., conditions (1), (2) can be replaced by the following conditions:

(1') for every neighbourhood U of zero there exists a neighbourhood V of zero such that
$$V+V \subset U;$$

(2') for every neighbourhood U of zero and every $\varepsilon > 0$ there exists a neighbourhood V of zero such that $tV \subset U$ for $|t| < \varepsilon$.

Let X be a linear topological space. Recall that a set $A \subset X$ is convex if for $x, y \in A$, $a, b \geq 0$, $a+b = 1$ we have $ax+by \in A$.

A linear topological space X is called *locally convex* if for every neighbourhood U of zero there exists a convex neighbourhood V of zero such that $V \subset U$.

Let X be a locally convex space over the field of real numbers. Let U be an arbitrary neighbourhood of zero and let V be a convex neighbourhood of zero such that $V \subset U$. The continuity of multiplication implies that the set $W = \bigcap_{|\varepsilon|=1} \varepsilon V$ is also a neighbourhood of zero.

The set W defines a pseudonorm $||x||$ in the space X as described in Theorem 2.2.2. Let $X_0 = \{x \in X : ||x|| = 0\}$ and $X' = X/X_0$. In virtue of the Hahn–Banach theorem, there exists a linear functional in X' which is continuous with respect to $||x||$. Since the set W is open in X, that functional generates a linear functional in X which is continuous in X. This construction is applicable to any neighbourhood U of zero.

Since X is locally convex, for arbitrary $x \neq 0$ there exists a neighbourhood U of zero which does not contain x. Hence, we obtain

THEOREM 4.3.1. *Let X be a locally convex space over the field of real numbers. Then, for every $x \neq 0$, there exists a continuous linear functional f such that $f(x) \neq 0$.*

Just as in Theorem 2.5.4, we obtain

THEOREM 4.3.2. *Let X be a locally convex space over the field of complex numbers. Let $x \neq 0$. Then there exists a continuous linear functional f such that $f(x) \neq 0$.*

Separation theorems are also valid in locally convex spaces.

THEOREM 4.3.3 (Mazur, 1936). *Let X be a locally convex space over the field of real (complex) numbers. Let A and B be convex sets. If $\operatorname{Int} A \neq \emptyset$ and $(\operatorname{Int} A) \cap B = \emptyset$, then there exists a continuous functional f which separates the sets A and B, i.e.,*

$$\sup_{x \in A} f(x) \leqslant \inf_{x \in B} f(x) \quad (\sup_{x \in A} \operatorname{Re} f(x) \leqslant \inf_{x \in B} \operatorname{Re} f(x)).$$

THEOREM 4.3.4. *Let X be a locally convex space over the field of real (complex) numbers. Let U be a closed convex set contained in X and $x_0 \notin U$. Then there exist a continuous linear functional f, a real number c and a positive number ε such that*

$$f(x) \leqslant c \quad \text{for} \quad x \in U \quad \text{and} \quad f(x_0) > c + \varepsilon$$
$$(\operatorname{Re} f(x) \leqslant c \quad \text{for} \quad x \in U \quad \text{and} \quad \operatorname{Re} f(x_0) > c + \varepsilon).$$

The proofs of the above theorems proceed in the same way as the proofs of the corresponding results in normed spaces.

4.4. WEAK TOPOLOGIES

Let X be a linear space over the field of real (complex) numbers. Let Γ be a family of linear functionals defined on X which satisfies the conditions:

(1) the family Γ is *linear*, i.e., if $f, g \in \Gamma$, then $f+g \in \Gamma$ and $af \in \Gamma$ for every scalar a;

(2) the family Γ is *total*, i.e., $f(x) = 0$ for each $f \in \Gamma$ entails $x = 0$.

We can introduce a Γ-*topology* on X by defining a basis of neighbourhoods at zero as

$$U = \{x \in X : |f_i(x)| \leqslant c_i, \ i = 1, 2, \ldots, n,$$
$$\text{where } f_1, \ldots, f_n \in \Gamma, \ c_i > 0\}.$$

This family of sets forms a basis of neighbourhoods since the intersection of two sets of the above form contains a set of the above form.

Since X is a linear space, a basis of neighbourhoods at zero fully determines a topology. It is easy to verify (since the family Γ is total) that X is a Hausdorff topological space.

In our further considerations two examples of Γ-topologies will play a particularly important role. Both of them are related to Banach spaces.

Let X be a Banach space and $\Gamma = X^*$, i.e., the family Γ is the conjugate space of X. The Γ-topology thus obtained is called the *weak topology* of the space X.

Let $X = Y^*$ be the conjugate space of a certain Banach space Y. Let $\Gamma = \{f \in X^* : f(x) = x(y)\}$. The corresponding Γ-topology is called the *weak-∗-topology*. More precisely, we should call this topology the *weak-∗-topology with respect to the space Y*, since it may happen that two nonisomorphic spaces Y and Y_1 have conjugate spaces which are isomorphic. However, if it is not misleading the first term will be used.

A sequentially compact (compact) set in a Γ-topology will briefly be called *sequentially Γ-compact* (*Γ-compact*). In particular, sequentially compact (compact) sets in a weak topology will be called *sequentially weakly compact* (*weakly compact*) sets. Sequentially compact (compact) sets in a weak-∗-topology will be called *sequentially weakly-∗-compact* (*weakly-∗- compact*) sets.

THEOREM 4.4.1. *Let X be a linear space over the field of real (complex) numbers. Let Γ be a family of linear functionals which satisfies conditions* (1), (2). *Then a linear functional f is continuous in the Γ-topology if and only if $f \in \Gamma$.*

Proof. *Sufficiency*. By the definition of neighbourhoods in the Γ-topology, if $f \in \Gamma$, then f is a linear functional continuous in the Γ-topology.

Necessity. Since f is continuous in the Γ-topology, there exists in this topology a neighbourhood U of zero such that
$$\sup_{x \in U} |f(x)| < 1.$$
But U is of the form
$$U = \{x \in X : |f_i(x)| < c_i, \ i = 1, \ldots, n\},$$
and hence U contains a subspace
$$X_0 = \{x \in X : f_i(x) = 0, \ i = 1, \ldots, n\}.$$
Since f is bounded, the homogeneity of f implies that f is identically equal to zero on the whole X_0. Consider now the quotient space X/X_0. By definition, the space X_0 is at most n-dimensional. Each of the functionals f_i, $i = 1, \ldots, n$, induces a linear functional $\tilde{f}_i$ which is defined on the space X/X_0. In particular, the functional f (which is equal to zero on X_0) induces a functional $\tilde{f}$ on the quotient space. The particular form of the space X_0 enables us to represent $\tilde{f}$ as a linear combination of functionals $\tilde{f}_i$. This means that the functional f can be expressed as a linear combination of linear functionals f_i. Hence, by linearity of Γ, $f \in \Gamma$. ∎

THEOREM 4.4.2. *Let X and Y be linear spaces over the field of real (complex) numbers. Let Γ_X- and Γ_Y-topologies be defined on X and Y, respectively. Let A be a linear operator which maps X into Y. The operator A is a continuous operator which maps the space X with the Γ_X-topology into the space Y with the Γ_Y-topology if and only if for every $f \in \Gamma_Y$ the functional $fA(x)$ is linear and belongs to Γ_X (i.e., A' maps Γ_Y into Γ_X).*

Proof. *Necessity*. Let $f \in \Gamma_Y$. Then, by the continuity of the operator A, the functional $fA(x)$ is a linear functional defined on X which is continuous in the Γ'_X-topology. Hence, by Theorem 4.4.1, the functional $fA(x)$ belongs to Γ_X.

Sufficiency. Suppose that U is an arbitrary neighbourhood of zero in Y in the Γ_Y-topology, i.e., U is of the form

$$U = \{y \in Y: |f_i(y)| < c_i, \ i = 1, \ldots, n, \ f_i \in \Gamma_Y\}.$$

The inverse image of U,

$$A^{-1}(U) = \{x \in X: |f_i A(x)| < c_i, \ i = 1, \ldots, n\}$$

is a neighbourhood of zero in X in the Γ_X'-topology. ∎

COROLLARY 4.4.3. *Let X and Y be Banach spaces. A linear continuous operator A which maps X into Y is weakly continuous, i.e., A is continuous as a mapping of the space X with the weak topology into the space Y with the weak topology.*

Proof. If f is a continuous linear functional defined on Y, then $fA(x)$ is a continuous linear functional defined on X. It means that the operator A' maps Y^* into X^*. Hence, by the definition of weak continuity and by Theorem 4.4.2, the operator A is weakly continuous. ∎

COROLLARY 4.4.4. *Let X be a Banach space and the conjugate of a given space X_0, $(X_0)^* = X$. Let Y be a Banach space and the conjugate of a given space Y_0, $(Y_0)^* = Y$. Let A be a continuous linear operator which maps X into Y. The operator A is continuous in the weak-*-topologies (i.e., A is regarded as a mapping acting between the space X with a weak-*-topology and the space Y with a weak-*-topology) if and only if A is the conjugate of a continuous linear operator A_0 which maps Y_0 into X_0.*

Proof. The conjugate operator A^* of the operator A is a continuous linear operator if and only if A is continuous. By the definition of weak *-topology and by Theorem 4.4.2, the operator A is continuous if and only if its dual operator A' maps Y_0 into X_0. Precisely, this means that A' maps $n(Y_0)$ into $n(X_0)$, where $n(Z)$ means as before the canonical image of Z in Z^{**}. We identify $n(Y_0)$ with Y_0 and $n(X_0)$ with X_0. Denoting by A_0 the restriction of the operator A' to Y_0, we obtain $A = A_0$. ∎

THEOREM 4.4.5. *Let Γ be a linear space and let X be the dual space of Γ, $X = \Gamma''$. Consider X with the Γ-topology. Let $c(\gamma)$ be an arbitrary function defined on Γ. Then the set*

$$K = \{x \in X : \gamma(x) \leq c(\gamma) \text{ for all } \gamma \in \Gamma\}$$

is compact in the Γ-topology.

Proof. Let

$$I(\gamma) = \{a : a \text{ is a number}, |a| \leq c(\gamma)\}$$

and let I be the Tikhonov product of the sets $I(\gamma)$,

$$I = \underset{\gamma \in \Gamma}{\times} I(\gamma)$$

(see Section 4.2). In virtue of the Tikhonov theorem (Theorem 4.2.13), the set I is a compact topological space.

Let T be a mapping of the set K into I given by the formula

$$T(x) = \{x_\gamma\},$$

where $x_\gamma = \gamma(x)$.

By the definition of Γ-topology, T is a homeomorphism. To complete the proof it is enough to show that the image of K under T is a closed set.

Now we take advantage of the fact that Γ is a linear set. Let

$$A(f, g) = \{\{x_\gamma\} \in I : x_{f+g} = x_f + x_g, f, g \in \Gamma\}$$

and

$$B(a, f) = \{\{x_\gamma\} \in I : x_{af} = a \cdot x_f, a\text{---number}, f \in \Gamma\}.$$

Since the mapping $\{x_\gamma\} \to x_f$, $f \in \Gamma$, is continuous in the Γ-topology, all the sets $A(f, g)$ and $B(a, f)$ are closed. This implies that the set

$$T(K) = [\underset{f, g \in \Gamma}{\bigcap} A(f, g)] \cap [\underset{\substack{f \in \Gamma \\ a\text{---number}}}{\bigcap} B(a, f)]$$

is closed. ■

As a corollary we obtain

THEOREM 4.4.6 (Alaoglu, 1940). *Let X be the conjugate space of a Banach space Y, $X = Y^*$. Then the closed unit ball in X is compact in the weak-$*$-topology.*

Proof. $K = \{x \in X: \|x\| \leqslant 1\} = \{x \in X: |x(y)| \leqslant \|y\|$ for all $y \in Y\}$. ∎

Let X be the conjugate space of a Banach space Y^*. A set $A \subset X$ is compact in the weak-$*$-topology if and only if A is bounded and closed in the weak-$*$-topology. This is an immediate consequence of the Alaoglu theorem.

So far we have considered weak topologies in Banach spaces. Now we shall investigate weak convergences.

The following theorem is a simple consequence of the Banach–Steinhaus theorem (Theorem 3.1.1).

THEOREM 4.4.7. *Let a Banach space X be the conjugate of a certain Banach space X_0. Then any sequence $\{x_n\}$, $x_n \in X$, which converges in the weak-$*$-topology is bounded.*

COROLLARY 4.4.8. *Each weakly convergent sequence of elements of a Banach space X is bounded.*

Proof. All elements of X should be regarded as elements of the second conjugate space. Then the assertion of the corollary follows from Theorem 4.4.7. ∎

THEOREM 4.4.9. *Let X and Y be Banach spaces. Suppose that X is the conjugate space of a space X_0 and hence the weak-$*$-topology is induced in X. Let $\{A_n\}$ be a sequence of operators which converges in the operator norm to a linear operator A. If all the operators A_n (regarded as mappings of the space X with the weak-$*$-topology into the space Y with the norm topology) are sequentially continuous, then the limit operator A is also sequentially continuous provided that A is regarded as a mapping of the space X with the weak-$*$-topology into the space Y with the norm topology.*

Proof. Let $\{x_n\}$ be a sequence of elements of the space X which converges to a certain x_0 in the weak-∗-topology. Then, by Theorem 4.4.7, the sequence $\{x_n\}$ is bounded. Write

$$M = \sup_n ||x_n||.$$

Let ε be an arbitrary positive number. Since the sequence $\{A_n\}$ converges in the operator norm to the operator A, there exists an index N such that

$$||A_N - A|| < \frac{\varepsilon}{3M}.$$

Since the operator A_N is sequentially continuous, there exists an n_0 such that for $n > n_0$

$$||A_N(x_n - x_0)|| < \frac{\varepsilon}{3}.$$

Hence, for $n > n_0$

$$||A(x_n) - A(x_0)|| \leq ||A(x_n) - A_N(x_n)|| + ||A_N(x_n) - A_N(x_0)|| + \\ + ||A_N(x_0) - A(x_0)||$$

$$\leq M \cdot \frac{\varepsilon}{3M} + \frac{\varepsilon}{3} + M \cdot \frac{\varepsilon}{3M} = \varepsilon.$$

This implies that $A(x_n) \to A(x_0)$ in the norm topology. ∎

COROLLARY 4.4.10. *Let a sequence $\{A_n\}$ of linear operators mapping a Banach space X into a Banach space Y converge in the operator norm to a linear operator A. If each of the operators A_n transforms weakly convergent sequences into sequences convergent in the norm topology, then the limit operator A transforms weakly convergent sequences into norm-convergent sequences.*

Proof. Consider the elements of the space Y as elements of the second conjugate space and apply Theorem 4.4.9. ∎

If we replace sequential continuity by weak-∗-continuity, then Theorem 4.4.9 does not hold. Indeed, let X and Y be two Banach spaces and let X be the conjugate space of a Banach space X_0. It is easy to observe that each finite-dimensional continuous operator K maps X with the weak-*-topology into Y with the norm topology in a continuous way. It follows from the fact that $K^{-1}(0)$ is a subspace of finite codimension. If an operator K is not finite-dimensional, then $K^{-1}(0)$ is a subspace of infinite codimension. Thus, for any bounded open set $U \subset Y$, the set $K^{-1}(U)$ does not contain any neighbourhood in the weak-*-topology. Therefore, in this case, K is not a continuous operator mapping X with the weak-*-topology into Y with the norm topology.

Suppose that A_0 is the limit of a sequence of finite-dimensional operators A_n which are continuous in the operator norm topology and A_0 is not finite-dimensional. Then, by the above arguments, A_0 is not continuous as an operator acting from X with the weak-*-topology into Y with the norm topology.

Observe that, by Theorem 4.4.9, A_0 is sequentially continuous. Thus, we have an example of an operator which is sequentially continuous but not continuous.

Now we shall show that closed balls in the conjugate space X^* are sequentially compact, provided that X is separable. This fact is a direct consequence of the Alaoglu theorem (Theorem 4.4.6). We present here another straightforward proof which does not exploit the Alaoglu theorem.

THEOREM 4.4.11. *Let X be a separable Banach space. Then the closed unit ball in the conjugate space X^* is sequentially weakly-*-compact.*

Proof. Let $\{f_n\}$ be an arbitrary sequence of functionals such that $\|f_n\| \leqslant 1$, $n = 1, 2, \ldots$ Since the space X is separable, there exists a sequence $\{x_m\}$ which is dense in X. Let us consider now a sequence $\{f_k^i\}$ of sequences such that: $f_k^0 = f_k$ and

(1) $\{f_k^i\}$ is a subsequence of the sequence $\{f_k^{i-1}\}$, $i = 1, 2 \ldots$,

(2) $\{f_k^i(x_i)\}$ is a convergent sequence, $i = 1, 2, \ldots$

Such a countable family of sequences can be easily constructed by induction. Observe that properties (1) and (2) imply the convergence of the sequence $\{f_k^i(x_m)\}$ for all x_m, $1 \leqslant m \leqslant i$. Consider the sequence $\{f_k^k\}$. This is the sequence $\{f_{n_k}\}$ which we seek. Clearly, $\{f_k^k\}$

is a subsequence of $\{f_n\}$. Since, for each m, almost all elements of $\{f_k^k\}$ belong to $\{f_k^m\}$, the sequence $\{f_{n_k}(x_m)\}$ converges for all m.

We shall show now that the sequence $\{f_{n_k}(x)\}$ converges for all $x \in X$. Take an arbitrary positive ε. Since the sequence $\{x_m\}$ is dense in X, there exists an index m_0 such that $||x - x_{m_0}|| < \frac{1}{3}\varepsilon$. Since the sequence $\{f_{n_k}\}$ converges for all x_m, there exists a natural number K such that for $k > K$

$$|f_{n_k}(x) - f_{n_k}(x_{m_0})| < \tfrac{1}{3}\varepsilon.$$

Hence, for $k, k' > K$

$$|f_{n_k}(x) - f_{n_{k'}}(x)| \leqslant |f_{n_k}(x) - f_{n_k}(x_{m_0})| + |f_{n_k}(x_{m_0}) - f_{n_{k'}}(x_{m_0})| +$$
$$+ f_{n_{k'}}(x_{m_0})| - f_{n_{k'}}(x)| < \varepsilon.$$

Since ε is arbitrary, the sequence $\{f_{n_k}(x)\}$ is fundamental for every $x \in X$. Hence, it converges to a certain $f(x)$. By the Banach–Steinhaus theorem, $f(x)$ is a continuous linear functional. Obviously, $||f|| \leqslant 1$ since for every $x \in X$

$$|f(x)| \leqslant \sup_k |f_{n_k}(x)| \leqslant ||x||.$$

The convergence of the sequence $\{f_{n_k}(x)\}$ to $f(x)$ for all $x \in X$ entails, by definition, convergence in the weak-$*$-topology. ∎

4.5. Reflexive spaces and weak compactness

Let X be a Banach space. Let X^* denote the conjugate space of X and let $X^{**} = (X^*)^*$ denote the second conjugate space. As we have shown in Section 2.5, to each element $x \in X$ we can assign a certain element $n(x) \in X^{**}$. This mapping, called the *canonical injection*, is a linear isometry.

If the image of the space X under the canonical injection coincides with the space X^{**}, $n(X) = X^{**}$, then the space X is said to be *reflexive*.

EXAMPLE 4.5.1. The space l^2 is reflexive since, by the results of Section 2.6, $(l^2)^{**} = (l^2)^* = (l^2)$.

4.5. Reflexive spaces and weak compactness

EXAMPLE 4.5.2. Every finite-dimensional space is reflexive.

EXAMPLE 4.5.3. The space $L^2[0, 1]$ is reflexive.

EXAMPLE 4.5.4. The space c_0 is not reflexive since
$$(c_0)^{**} = (l)^* = m \neq c_0.$$

EXAMPLE 4.5.5. The space $L^1[0, 1]$ is not reflexive since $(L^1[0, 1])^*$ $= M[0, 1]$ is a nonseparable space. By Theorem 2.5.13, $(L^1[0, 1])^{**}$ $= (M[0, 1])^*$ is also nonseparable and hence it cannot coincide with $L^1[0, 1]$, which is separable.

THEOREM 4.5.1 (Goldstine, 1938). *Let X be a Banach space over the field of real (complex) numbers. Let X^{**} denote the second conjugate space of X. Denote by K and K^{**} the closed unit balls in the spaces X and X^{**}, respectively. Then the set $n(K)$ is dense in K^{**} in the X^*-topology.*

Proof. Let K_1 denote the closure of $n(K)$ in the X^*-topology. By the Alaoglu theorem (Theorem 4.4.6), K^{**} is closed in the X^*-topology. This implies that $K_1 \subset K^{**}$. The set K_1 is convex as the closure of the convex set $n(K)$. Suppose that $K_1 \neq K^{**}$. Then there exists an element $x^{**} \in K^{**}$ which does not belong to K_1.

Applying the separation theorem (Theorem 4.3.4) we can deduce that there exist a linear functional f which is continuous in the X^*-topology, a number c and a positive number ε such that
$$f(x) \leqslant c \quad \text{for} \quad x \in K_1 \quad \text{and} \quad f(x^{**}) \geqslant c+\varepsilon \qquad (4.5.1)$$
(in the complex case
$$\operatorname{Re} f(x) \leqslant c \quad \text{for} \quad x \in K_1 \quad \text{and} \quad \operatorname{Re} f(x^{**}) \geqslant c+\varepsilon, \qquad (4.5.2)$$
respectively).

Since the functional f is continuous in the X^*-topology, by Theorem 4.4.1 the functional f is of the form
$$f(\tilde{x}) = \tilde{x}(x^*) \quad \text{for} \quad \tilde{x} \in X^{**}.$$
But $n(K) \subset K_1$. Thus, $f(x)$ (or $\operatorname{Re} f(x)$) is not greater than c for

$x \in n(K)$. This implies that $|x(x^*)| \leq c$ for all $x \in K$. Hence, by the definition of the norm of functionals, $\|\tilde{x}\| \leq c$, which contradicts inequalities (4.5.1) ((4.5.2)). ∎

THEOREM 4.5.2. *A Banach space X over the field of real (complex) numbers is reflexive if and only if the closed unit ball K is weakly compact (i.e., compact in the X^*-topology).*

Proof. Necessity. If X is a reflexive space, then $n(K) = K^{**}$. By the Alaoglu theorem (Theorem 4.4.6), K^{**} is compact in the X^*-topology and hence K is compact in the X^*-topology.

Sufficiency. Assume that K is weakly compact in the X^*-topology. The set $n(K)$ is compact in X^*-topology because it is a subset of X. By Theorem 4.5.1, $n(K)$ is dense in K^{**}. Hence, $n(K) = K^{**}$. ∎

COROLLARY 4.5.3. *A Banach space X over the field of real (complex) numbers is reflexive if and only if every bounded weakly closed set A is weakly compact.*

THEOREM 4.5.4. *Let X be a reflexive Banach space over the field of real (complex) numbers. Let A be a closed bounded convex set in X. Then the set A is weakly compact.*

Proof. By Corollary 4.5.3, it is enough to show that the set A is weakly closed. Let $p \notin A$. By the separation theorem (Theorem 4.3.4), there exist a continuous linear functional f, a constant c and a positive number ε such that

$$f(x) \leq c \quad \text{for} \quad x \in A,$$
$$f(p) \geq c + \varepsilon$$

($\operatorname{Re} f(x) \leq c$ for $x \in A$, $\operatorname{Re} f(p) \geq c + \varepsilon$). This implies that p does not belong to the weak closure of the set A. ∎

COROLLARY 4.5.5. *A subspace of a reflexive space is reflexive.*

Proof. Let X be a reflexive space and let X_0 be a subspace of X. Let A be an arbitrary closed bounded convex and compact set in X_0. Then A is also closed in X, and hence A is compact in the weak topology. ∎

As we mentioned earlier, the notions of compactness and sequential compactness do not coincide in arbitrary topology. However, in weak topologies the following theorem holds.

THEOREM 4.5.6 (Eberlein, 1950; Šmulian, 1940). *Let X be a Banach space. A set A contained in X is compact in the weak topology if and only if A is weakly sequentially compact in that topology.*

The proof of this theorem can be found in textbooks of functional analysis (see e.g. Dunford, Schwartz, 1952, Chapter V, Sec. 6; Rolewicz, 1986, Sec. 5.3). We omit this proof here since we shall use more frequently the following simpler result.

THEOREM 4.5.6'. *Let X be a reflexive Banach space. Every bounded closed convex set A contained in X is sequentially compact.*

Proof. First of all, observe that the set A is closed in the weak topology. Let $\{x_n\}$ be a sequence of elements of the set A. Let X_0 be the space spanned by the sequence $\{x_n\}$, $X_0 = \overline{\operatorname{lin}\{x_n\}}$. The space X_0 is separable and, as a subspace of the reflexive space X, it is reflexive. Hence, X_0 is the conjugate space of the space $(X_0)^*$. By Theorem 2.5.13, the space $(X_0)^*$ is separable. By Theorem 4.4.11, the sequence $\{x_n\}$ contains a subsequence $\{x_{n_k}\}$ which converges to a certain $x_0 \in X$ in the weak-∗-topology of the space X_0. By reflexivity of the space, the weak-∗-topology and the weak topology coincide, and hence the sequence $\{x_{n_k}\}$ converges to x_0 in the weak topology. By the closedness of A in the weak topology, we obtain $x_0 \in A$. ∎

4.6. EXTREME POINTS

Let X be a linear topological space. Let A be a convex set contained in X. A subset B of the set A is called an *extremal subset* if $x, y \in A$ and $tx+(1-t)y \in B$ for some t, $0 < t < 1$, imply $x, y \in B$. If an extremal set consists of a single element, then that element is called an *extreme point*. We denote by $E(A)$ the set of all extreme points.

PROPOSITION 4.6.1. *Let X be a linear space. Let A be a convex set. Let B be an extremal subset of A. Let C be an extremal subset of B. Then C is an extremal subset of A.*

Proof. Let x, y be arbitrary points belonging to A. Suppose that for a certain t, $0 < t < 1$, $tx+(1-t)y \in C$. Since $C \subset B$, we have $tx+(1-t)y \in B$. The set B is an extremal subset of A. Hence, $x, y \in B$. Since $tx+(1-t)y \in C$ and C is an extremal subset of B, $x, y \in C$. ∎

PROPOSITION 4.6.2. *Let X be a linear space. Let A be a convex set in X. Let A_α, $\alpha \in \mathfrak{A}$, be a family of extremal subsets of A. If $A_0 = \bigcap_{\alpha \in \mathfrak{A}} A_\alpha$ is nonempty, then A_0 is an extremal subset of A.*

Proof. Let x, y be arbitrary elements of A. Suppose that there is a $t, 0 < t < 1$, such that $tx+(1-t)y \in A_0$. This means that for each $\alpha \in \mathfrak{A}$, $tx+(1-t)y \in A_\alpha$. Since the sets A_α are extremal subsets, $x, y \in A_\alpha$ for all $\alpha \in \mathfrak{A}$. Thus, $x, y, \in A_0$. ∎

Let X, Y be two linear spaces. Let A be a convex set in X. Let T be a linear operator which maps X into Y. The image $T(B)$ of an extremal subset B need not be an extremal subset of $T(A)$, as is shown by the following example.

EXAMPLE 4.6.1. Let $X = R^3$, $Y = R^2$ with the Euclidean norms in both spaces. Let A be the closed unit ball in X. Let T be the natural projection, $T(x_1, x_2, x_3) = (x_1, x_2)$. It is easy to see that each point x, $||x|| = 1$, is an extreme point of A. Since $T(\{x: ||x|| = 1\}) = \{y: ||y|| \leq 1\}$, there are extreme points x of A such that the images $T(x)$ are interior points of $T(A)$. Hence, they are not extreme points.

However, we have

PROPOSITION 4.6.3. *Let X, Y be linear spaces. Let A be a closed set in X. Let T be a linear operator mapping X into Y. If B is an extremal subset of $T(A)$, then $T^{-1}(B) \cap A$ is an extremal subset of A.*

Proof. Take arbitrary $x, y \in A$. Suppose that there exists a $t, 0 < t < 1$, such that $tx+(1-t)y \in T^{-1}(B) \cap A$. This implies that $T(x), T(y)$

4.6. Extreme points

$\in T(A)$ and $t \cdot T(x)+(1-t) \cdot T(y) \in B$. Since B is an extremal subset of $T(A)$, $T(x), T(y) \in B$. Thus, $x, y \in T^{-1}(B) \cap A$. ∎

An extremal set consisting of more than one point is called a *face*. This definition differs from what we intuitively mean by a face. According to our definition, an edge of a polyhedron is also a face.

We say that a set U *has a finite* (*or countable*) *number of faces* if it contains only a finite (or countable) number of different faces.

In finite-dimensional spaces every bounded convex and closed set A has extreme points and moreover

$$A = \operatorname{conv} E(A), \tag{4.6.1}$$

where for an arbitrary set B, conv B denotes the smallest convex set containing B,

$$\operatorname{conv} B = \{y\colon y = a_1 x_1 + \ldots + a_n x_n, x_1, \ldots, x_n \in B, a_1, \ldots, a_n \text{ are nonnegative numbers such that } a_1 + \ldots + a_n = 1\}.$$

The set conv B is called the *convex hull* of the set B. The set $\overline{\operatorname{conv} B}$ is the *closed convex hull of* B.

In infinite-dimensional spaces it may happen that a closed bounded set has no extreme points.

EXAMPLE 4.6.2. Let $X = L^1[0, 1]$. Let A be the closed unit ball, $A = \{x \in L^1[0, 1]\colon ||x|| \leqslant 1\}$. The set A has no extreme points. Indeed, let x be an arbitrary element of the ball A. If $||x|| < 1$, i.e., x is an interior point of the set A, then x is not an extreme point. Let x be an arbitrary element of A such that $||x|| = 1$. Let

$$D = \{t \in [0, 1]\colon |x(t)| > 1/2\}.$$

Since $||x|| = 1$, the measure of D is positive. We split the set D into two disjoint sets D_1 and D_2 with equal measures. Let

$$x_1(t) = \begin{cases} \dfrac{1}{2} \dfrac{x(t)}{|x(t)|} & \text{for } t \in D_1, \\ -\dfrac{1}{2} \dfrac{x(t)}{|x(t)|} & \text{for } t \in D_2, \\ 0 & \text{for } t \notin D, \end{cases}$$

and $x_2 = -x_1$. It is easy to verify that $\|x+x_1\| = \|x+x_2\| = 1$ and $x = \frac{1}{2}(x_1+x_2)$, which means that x is not an extreme point.

The following theorem is the main result of the present section.

THEOREM 4.6.4 (Krein and Milman, 1940). *Let X be a locally convex space over the field of real (complex) numbers. Let A be a convex compact set contained in X. Then the set A has extreme points and A is equal to the closed convex hull of extreme points,*

$$A = \overline{\operatorname{conv} E(A)}. \tag{4.6.2}$$

Proof. Let $\mathscr{E}(A)$ denote the family of all closed extremal subsets of A. We can order the family $\mathscr{E}(A)$ by inclusion, i.e., for $E_1, E_2 \in \mathscr{E}(A)$, $E_1 \prec E_2$ if $E_1 \subset E_2$. We shall show that every totally ordered subfamily contains a minimal element, i.e., if $E_\alpha \prec E_\beta$, $\alpha, \beta \in \mathfrak{A}$, then $E_0 = \bigcap_{\alpha \in \mathfrak{A}} E_\alpha$
$\neq \emptyset$ and E_0 is an extremal subset of A.

The set E_0 is nonempty since the set A is compact. Moreover, E_0 is extremal since the intersection of an arbitrary family of extremal subsets is an extremal subset (see Proposition 4.6.2). Now, in virtue of the Kuratowski–Zorn lemma, there exists a minimal element in the set $E(A)$. We shall show that this minimal element, say $E_{\min}$, is a one-point set, i.e., $E_{\min}$ is an extreme point. Indeed, suppose that $E_{\min}$ contains at least two different points. Then, by the separation theorem (Theorem 4.3.3), there exists a continuous linear functional f which is not constant on $E_{\min}$ (Re f is not constant on $E_{\min}$). Let

$$E'_{\min} = \{x \in E_{\min}: f(x) = \min_{y \in E_{\min}} f(y)\}$$

$$(E'_{\min} = \{x \in E_{\min}: \operatorname{Re} f(x) = \min_{y \in E_{\min}} \operatorname{Re} f(y)\}).$$

Since the set $E_{\min}$ is nonempty, the set $E'_{\min}$ is nonempty. Moreover, $E'_{\min}$ is a proper subset of $E_{\min}$, since f (Re f) is not constant on $E_{\min}$. Furthermore, $E'_{\min}$ is an extremal set. Indeed, let $k = tk_1 + (1-t)k_2 \in E'_{\min}$, $0 < t < 1$. Then $k_1, k_2 \in E_{\min}$, since the set $E_{\min}$ is extremal. Suppose that

$$f(k_1) > \min_{y \in E_{\min}} f(y)$$

$$(\operatorname{Re} f(k_1) > \min_{y \in E_{\min}} \operatorname{Re} f(y)).$$

Then

$$f(k) > \min_{y \in E_{\min}} f(y)$$

$$(\operatorname{Re} f(k) > \min_{y \in E_{\min}} \operatorname{Re} f(y)).$$

Hence, we conclude that $k \in E'_{\min}$ entails $k_1, k_2 \in E'_{\min}$. This means that the set $E'_{\min}$ is extremal, which contradicts the minimality of $E_{\min}$. Thus, $E_{\min}$ is a one-point set, in other words, the set A has at least one extreme point.

Now we shall prove the second part of the theorem. Suppose that $A \neq \overline{\operatorname{conv} E(A)}$. This means that there exists an element x of A such that $x \notin \overline{\operatorname{conv} E(A)}$. Hence, by the separation theorem (Theorem 4.3.4), there exist a continuous linear functional f, a constant c and a positive number ε such that

$$f(x) \leqslant c, \quad f(y) \geqslant c+\varepsilon \quad \text{for} \quad y \in \overline{\operatorname{conv} E(A)}$$

$$(\operatorname{Re} f(x) \leqslant c, \operatorname{Re} f(y) \geqslant c+\varepsilon \quad \text{for} \quad y \in \overline{\operatorname{conv} E(A)}).$$

Let

$$E = \{x \in A : f(x) = \min_{y \in A} f(y)\}$$

$$(E = \{x \in A : \operatorname{Re} f(x) = \min_{y \in A} \operatorname{Re} f(y)\}).$$

The continuity of f implies that the set E is closed. As before, we can verify that the set E is extremal, and hence E contains an extreme point, say x_0. By Proposition 4.6.1, x_0 is also an extreme point of the set A. This leads to a contradiction, since $f(x_0) \leqslant c$ ($\operatorname{Re} f(x_0) \leqslant c$) entails $x_0 \notin E(A)$. ∎

The set of extreme points need not be closed even in finite-dimensional spaces as is shown by the following example.

EXAMPLE 4.6.3. Let $X = R^3$. Let

$$A_0 = \{(x_1, x_2, x_3) : |x_i| \leqslant 1, \; i = 1, 2, 3\},$$
$$A_1 = \{(x_1, x_2, x_3) : x_3 = 0, \; x_1^2 + (x_2-1)^2 = 1\}.$$

Let $A = \mathrm{conv}(A_0 \cup A_1)$. The set of extremal points of A is equal to the set

$$\{(\varepsilon_1, \varepsilon_2, \varepsilon_3): \varepsilon_i = \pm 1\} \cup$$
$$\cup \{(x_1, x_2, x_3): x_1^2 + (x_2-1)^2 = 1, x_3 = 0, x_2 > 1\}$$

and it is not closed (see Fig. 4.6.1).

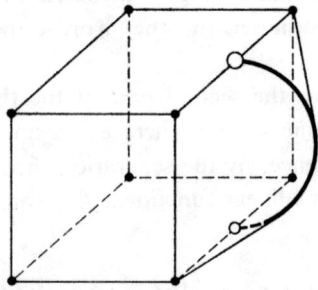

FIG. 4.6.1. •—extreme points, ○—points which are not extreme but belong to the closure of the set of extreme points.

In formula (4.6.2) the closed convex hull of $E(A)$ cannot be replaced by the convex hull of $E(A)$.

EXAMPLE 4.6.4. Let $X = l^1$. Let $A = \left\{x = \{x_n\}: 0 \leqslant x_n \leqslant \dfrac{1}{2^n}\right\}$. The set A is compact. The set $E(A)$ contains a sequence $\{y_n\}_{n=0}^{\infty}$ such that $y_0 = 0$, $y_n = \left\{0, ..., 0, \dfrac{1}{2^n}, 0, ...,\right\}$. Thus, $\mathrm{conv}\, E(A)$ contains only those sequences in which only finitely many elements are different from zero.

Chapter 5

Optimization and observability of linear systems

5.1. LINEAR SYSTEMS

By a *linear system* we mean a five-tuple consisting of three real Banach spaces X, $\square$, Y and two continuous linear operators: an operator A which maps the space X into the space $\square$ and an operator B which maps the space $\square$ into the space Y. A linear system will be denoted by

$$(X \underset{A}{\rightarrow} \square \underset{B}{\rightarrow} Y). \tag{5.1.1}$$

The spaces X, Y, $\square$ will be refferred to as the *input, output* and *trajectory* spaces, respectively. A will be referred to as the *input operator* and B as the *output operator*. When the trajectory space $\square$ does not play any role in our considerations, we shall use notation (5.1.1) in the abbreviated form

$$(X \underset{C}{\rightarrow} Y), \tag{5.1.2}$$

where $C = BA$ is a continuous linear operator which maps the space X into Y.

EXAMPLE 5.1.1. Given a time interval $[t_0, T]$. Let us consider a certain system in the interval $[t_0, T]$ which is not influenced by any external factor. Assume that the system is governed by a linear differential equation

$$\frac{dx}{dt} = A(t)x(t), \tag{5.1.3}$$

where $x(t)$ is an absolutely continuous n-dimensional vector function. $A(t)$ is an $n \times n$ matrix-valued function which is locally integrable.

Moreover, we require that equation (5.1.3) should be satisfied almost everywhere.

There exist numerous physical systems whose behaviour is described by equation (5.1.3). Moreover, those physical systems can be regarded as systems in the sense of our definition. The input space X is an n-dimensional space of initial values x_0. The trajectory space $\square$ is the space $C_n[t_0, T]$. The input operator A is an operator which assigns to each initial value x_0 a solution of equation (5.1.3), i.e., a solution $x(t)$ such that $x(t_0) = x_0$. The operator A can be expressed explicitly as

$$A(x_0) = \Phi(t)x_0,$$

where $\Phi(t)$ is the fundamental matrix of the equation (5.1.3) satisfying the initial condition $\Phi(t_0) = I$.

The output space and the output operator can be chosen in several different ways. For instance, they can be chosen as

(a) the space $\square$, $B = I$, if we are interested in trajectories on the whole interval $[t_0, T]$;

(b) an n-dimensional space, $B(x) = x(T)$, if we are interested in the final value of a trajectory (which is often the case).

If a system described by equation (5.1.3) is influenced by an external factor called a *control*, then the behaviour of the system may usually by described (not always, however) by a differential equation of the form

$$\frac{dx}{dt} = A(t)x(t) + B(t)u(t), \tag{5.1.4}$$

where $A(t)$, $x(t)$ are defined as before, $u(t)$ is an n-dimensional measurable vector-valued function, and $B(t)$ is an $n \times m$ measurable matrix-valued function. Equation (5.1.4) is solvable under additional conditions ensuring the integrability of the function $B(t)u(t)$. For instance, we can admit one of the following conditions:

(1) $B(t)$ bounded, $u(t)$ integrable;
(2) $B(t)$ integrable, $u(t)$ bounded;
(3) $B(t)$, $u(t)$ square integrable.

The space of controls U is an arbitrary Banach space of m-dimensional measurable functions $u(t)$ such that $B(t)u(t)$ is integrable. In particular, in case (1) the space U can be chosen as the space of m-dimensional absolutely integrable functions $L_m^1[t_0, T]$. In case (2) U can be regarded as the space of m-dimensional measurable essentially bounded functions $M_m[t_0, T]$. Finally, in case (3), U can be chosen as the space of m-dimensional square integrable functions $L_m^2[t_0, T]$. The input space X is the Cartesian product of the space of initial values R^n and the space of controls U. The trajectory space is, as before, the space $C_n[t_0, T]$. The input operator A is given by the formula

$$A(x_0, u) = x(t) = \Phi(t)x_0 + \Phi(t)\int_{t_0}^{t} \Phi^{-1}(s)B(s)u(s)\,ds. \tag{5.1.5}$$

The output space Y and the operator B can be chosen as before.

EXAMPLE 5.1.2. Given a system described by a linear equation with delays

$$\frac{dx}{dt} = \sum_{i=0}^{m} A_i(t)x(t-h_i), \tag{5.1.6}$$

where $0 = h_0 < h_1 < ... < h_m$. The functions $A_i(t)$, $x(t)$ satisfy the same conditions as in Example 5.1.1. We consider equation (5.1.6) in a given interval $[t_0, T]$ and we assume that $h_m < T-t_0$. As in Example 5.1.1, the trajectory space is the space $C_n[t_0, T]$, the input space X is the space $C_n[t_0-h_m, t_0]$, $X = C_n[t_0-h_m, t_0]$. The input operator A is given by the formula

$$A(\varphi) = x(t) = \int_{t_0-h_m}^{t_0} \varphi(s)\,d_s K(t, s), \tag{5.1.7}$$

where $K(t, s)$ is a matrix-valued function of bounded variation with respect to s and continuous with respect to t, $\varphi(s)$ is an initial function. The output space Y and the operator B can be chosen as in cases (a) or (b) of Example 5.1.1. Another interesting case is

(c) $Y = C[T-h_m, T]$ with the operator B defined as

$$B(x) = x(t)|_{[T-h_m, T]},$$

where, as usual, $f(t)|_A$ is the restriction of the function $f(t)$ to the set A.

If the above system is driven by a control $u(t)$, then it can often be described by the equation

$$\frac{dx}{dt} = \sum_{i=1}^{m} A_i(t)x(t-h_i) + B(t)u(t),$$

where $A_i(t)$, $x(t)$, $B(t)$, $u(t)$ are as in Example 5.1.1. The input space X is the Cartesian product of the space of initial functions $C[t_0-h_m, t_0]$ and the space of controls U (which is a certain space of functions defined on the interval $[t_0, T]$). The operator A is given by the formula

$$A(\varphi, u) = x(t) = \int_{t_0-h_m}^{t_0} \varphi(s)\, d_s K(t,s) + \int_{t_0}^{t} M(t,s)B(s)u(s)\, ds,$$

(5.1.8)

where $M(t,s)$ is a certain continuous matrix-valued function, which plays a similar role to that of the matrix-valued function $\Phi(t)\Phi^{-1}(s)$ in ordinary equations. The output space Y and the output operator B can be chosen as before.

There is a growing interest in differential-difference equations in automatics. Equations of this kind describe mainly those processes in which transportation delay appears, for instance, chemical processes.

Systems governed by partial differential equations, in other words; *systems with distributed parameters*, provide other important examples of systems. A detailed description of those examples in a linear case and the proof that they lead to systems in the sense of our definition require additional assumptions and results. Some applications of the theoretical results of the present chapter to systems with distributed parameters are presented in Chapter 7. To the reader who is more interested in systems with distributed parameters we recommend the books by Lions (1968), and by Butkovski (1965, 1975).

Let a linear system

$$(X \underset{A}{\rightarrow} \square \underset{B}{\rightarrow} Y)$$

be given. An element $y_0 \in Y$ is called a *point of controllability* if $y_0 \in BA(X)$. If $BA(X) = Y$, then the system is said to be *controllable*.

In the language of classical functional analysis a linear operator C which maps X onto Y is called a *surjection* (or C is said to be *surjective*).

5.1. Linear systems

In this language a linear system is controllable if and only if the operator BA is surjective.

Now we shall consider the problem how the perturbations of the operators A and B influence the controllability of a linear system. Let

$$(X \underset{A}{\to} \square \underset{B}{\to} Y) \tag{5.1.9}$$

be a controllable system. It can be shown (Gohberg–Krein, 1957) that there is a positive number δ_1 such that the system

$$(X \underset{A_1}{\to} \square \underset{B_1}{\to} Y) \tag{5.1.10}$$

is controllable, provided that $\|BA - B_1 A_1\| < \delta_1$. However, in many engineering examples this class of admissible perturbations is too narrow.

EXAMPLE 5.1.3. Let $X = \square = Y = C_0[0, 2\pi]$ be the space of 2π-periodic continuous functions defined on the real line. Let $A = I$ and $B_t x(s) = x(s-t)$. It is easy to verify that $\|B_t - B_{t_1}\| = 2$ for $t \ne t_1$. On the other hand, from the engineering point of view, the operators B_{t_1} and B_t are close to each other if t_1 and t are close.

Observe that in Example 5.1.3

$$\lim_{t_n \to t_0} B_{t_n}(x) = B_{t_0}(x) \quad \text{for all } x \in X. \tag{5.1.11}$$

We say that a family of operators B_t is *continuous in the strong operator topology* if it satisfies condition (5.1.11).

The following example shows that the strong operator topology is not an adequate tool to investigate the controllability of perturbed systems.

EXAMPLE 5.1.4. Let $X = \square = Y = L^1[0, 1]$. Let $A = I$ and let $B_t x(s) = x(s)\chi_{[t, 1]}(s)$ for $0 < t \le 1$. Of course, B_t, $0 \le t \le 1$, is continuous in the strong operator topology. It can be verified that $B_t A$ is not controllable for $0 < t \le 1$ but it is controllable for $t = 0$.

For that reason we introduce the notion of image continuity.

Let E be a linear topological space. Let (T, ϱ) be a metric space. A mapping which assigns to each $t \in T$ a subset $E(t) \subset E$ is called a *multifunction*. A multifunction $E(t)$ is called *linearly upper semicon-*

tinuous at t_0 if for each open neighbourhood V of zero there is a $\delta > 0$ such that

$$E(t) \subset E(t_0) + V \tag{5.1.12}$$

for all t satisfying the condition $\varrho(t, t_0) < \delta$. A multifunction $E(t)$ is called *linearly lower semicontinuous at* t_0 if for each neighbourhood V of zero there is a $\delta > 0$ such that

$$E(t_0) \subset E(t) + V \tag{5.1.13}$$

for all t satisfying the condition $\varrho(t, t_0) < \delta$. If a multifunction $E(t)$ is linearly upper semicontinuous and linearly lower semicontinuous, then we say that it is *linearly continuous at* t_0. If a multifunction $E(t)$ is linearly continuous (or linearly lower semicontinuous or linearly upper semicontinuous) at each $t_0 \in T$, we say that it is *linearly continuous* (*linearly lower semicontinuous, linearly upper semicontinuous*). A multifunction $E(t)$ such that the sets $E(t)$ are closed (compact) for all $t \in T$ will be called *closed-valued* (*compact-valued*).

Let $(X, \|\ \|_X)$, $(Y, \|\ \|_Y)$ be Banach spaces. Let T be a metric space. A family of operators C_t mapping X into Y is *image continuous* (*image lower semicontinuous, image upper semicontinuous*) if there is an open convex bounded set $U \subset X$ such that the multifunction $C(t) = C_t(U)$ is linearly continuous (linearly lower semicontinuous, linearly upper semicontinuous).

Observe that the family of operators C_t, $C_t = B_t A$, defined in Example 5.1.3 is image continuous. To verify this it is enough to take as the set U the unit ball in $C_0[0, 2\pi]$.

THEOREM 5.1.1. *Let C_t be a family of linear operators image lower semicontinuous at t_0. If the system*

$$(X \underset{C_t}{\rightarrow} Y) \tag{5.1.14}_t$$

is controllable for $t = t_0$, *then there is a* $\delta > 0$ *such that the system* $(5.1.14)_t$ *is controllable for all* t, $\varrho(t, t_0) < \delta$.

The proof of the theorem is based on the following lemmas.

LEMMA 5.1.2 (Rådström, 1952). *Let A, B, C be sets in a normed space X. Suppose that B is closed and convex and C is bounded. If*
$$A+C \subset B+C, \tag{5.1.15}$$
then
$$A \subset B. \tag{5.1.16}$$

Proof. Let a be an arbitrary element of A. By (5.1.15), there are $c_1, c_2 \in C$ and $b_1 \in B$ such that
$$a+c_1 = b_1+c_2. \tag{5.1.17}$$
Applying again (5.1.15), we obtain
$$a+c_2 = b_2+c_3 \tag{5.1.18}$$
for some $b_2 \in B$ and $c_3 \in C$. Repeating this procedure, we obtain by induction sequences $\{b_n\} \subset B$ and $\{c_n\} \subset C$ such that
$$a+c_n = b_n+c_{n+1} \tag{5.1.19}$$
for $n = 1, 2, \ldots$ Summing both sides of $(5.1.19)_n$ up to m, we obtain
$$ma + \sum_{i=1}^{m} c_i = \sum_{i=1}^{m} b_i + \sum_{i=2}^{m+1} c_i. \tag{5.1.20}$$
Hence,
$$ma = \sum_{i=1}^{m} b_i + c_{m+1} - c_1 \tag{5.1.21}$$
and
$$a = \frac{1}{m} \sum_{i=1}^{m} b_i + \frac{c_{m+1}}{m} - \frac{c_1}{m}. \tag{5.1.22}$$
Since B is convex, $d_m = \frac{1}{m} \sum_{i=1}^{m} b_i \in B$. Since C is bounded, the sequences $\left\{\frac{c_{m+1}}{m}\right\}, \left\{\frac{c_1}{m}\right\}$ tend to zero. Hence, $a \in B$ since B is closed. ∎

As an immediate consequence of Lemma 5.1.2 we obtain

LEMMA 5.1.3 (Rådström, 1952). *Let A, B be closed convex sets in a normed space X. Let C be a bounded set in X. If $A+C = B+C$, then $A = B$.*

LEMMA 5.1.4. *Let $K_r(x_0) = \{x: \|x-x_0\| < r\}$ be a ball of radius r and centre x_0 in a normed space X. Let Γ be an arbitrary closed convex set with the empty interior in X. Then $K_r(x_0)$ is not contained in the set $\Gamma + K_{r_0}(0)$ for any $r_0 < r$.*

Proof. Let r_1 be an arbitrary number such that $r_0 < r_1 < r$. Of course, $K_{r_1}(x_0) \subset K_r(x_0)$. Then

$$K_{r_0}(0) + \overline{K_{r_1-r_0}(x_0)} = K_{r_1}(x_0) \subset \Gamma + K_{r_0}(0). \tag{5.1.23}$$

Hence, by Lemma 5.1.2,

$$\overline{K_{r_1-r_0}(x_0)} \subset \Gamma \tag{5.1.24}$$

which leads to a contradiction since the interior of Γ is empty. ∎

LEMMA 5.1.5. *Let X be a normed space and let $E(t)$ be a multifunction with values in X. Suppose that $E(t)$ is linearly lower semicontinuous at t_0 and the sets $E(t)$ are closed and convex. If $E(t_0)$ has a nonempty interior, then there exists an $\eta > 0$ such that the sets $E(t)$ have nonempty interiors for all t satisfying $\varrho(t, t_0) < \eta$.*

Proof. Since $E(t_0)$ has a nonempty interior, there is a certain ball of radius ε and centre x_0, $K_\varepsilon(x_0)$, which is contained in $E(t_0)$. The multifunction $E(t)$ is linearly lower semicontinuous at t_0. Hence, there is an $\eta > 0$ such that

$$E(t_0) \subset E(t) + K_{\varepsilon/2}(0) \tag{5.1.25}$$

for all t, $\varrho(t, t_0) < \eta$. Therefore, by Lemma 5.1.4, the sets $E(t)$ have nonempty interiors. ∎

Proof of Theorem 5.1.1. By hypothesis, the family C_t is image lower semicontinuous at t_0. By the definition of image lower semicontinuity, there is an open convex bounded set U such that the multifunction $C(t) = C_t(U)$ is linearly lower semicontinuous at t_0. In virtue of the

Banach theorem (Theorem 3.2.1), since system $(5.1.14)_{t_0}$ is controllable, the set $C_{t_0}(U)$ has a nonempty interior. Therefore, by Lemma 5.1.5, the sets $C_t(U)$ have nonempty interiors, provided that t is close enough to t_0. Thus, system $(5.1.14)_t$ is controllable for all t sufficiently close to t_0. ∎

COROLLARY 5.1.6. *Let X and Y be normed spaces. Let (T, ϱ) be a metric space. Let C_t be a family of linear operators defined on X. Suppose that the family C_t is continuous at t_0 in the operator norm topology. If C_{t_0} is surjective (i.e., the system $(X \underset{C_{t_0}}{\to} Y)$ is controllable), then there is a $\delta > 0$ such that the operator C_t is surjective for all t, $\varrho(t, t_0) < \delta$.*

Proof. Let U be the unit ball. Then the multifunction $C(t) = C_t(U)$ is linearly continuous at t_0. Thus, it is linearly lower semicontinuous at t_0 and the assertion of the corollary follows from Theorem 5.1.1. ∎

There is no explicit relation between image continuity and strong continuity. As follows from Theorem 5.1.1, the family of operators defined in Example 5.1.4 is continuous in the strong operator topology and it is not image continuous. On the other hand, the following examples show that there are image continuous families of operators which are not continuous in strong topology.

EXAMPLE 5.1.5. Let $X = Y$ be the complex plane C. Let C_t be a family of continuous linear operators given by the formula

$$C_t(z) = \begin{cases} z & \text{for } t = 0, \\ ze^{i/t} & \text{for } 0 < t \leqslant 1. \end{cases} \tag{5.1.26}$$

It is easy to verify that C_t is image continuous and is not strong continuous.

Image continuity has several disadvantages. Namely, the superposition, the sum, the conjugate of the image continuous families of operators need not be image continuous.

EXAMPLE 5.1.6. Let $X = Y = R^2$ and let $\tilde{C}_t$ be the family of operators induced by the family C_t defined in Example 5.1.5 (we consider C as

R^2). Let $P_t = P$ be the orthogonal projection on the first coordinate. The families $\tilde{C}_t$ and P_t are both image continuous but the superposition $\tilde{C}_t P$ is not image continuous.

EXAMPLE 5.1.7. Let X, Y, $\tilde{C}_t$, $P_t = P$ be as in Example 5.1.6. As we have observed above, the family $\tilde{C}_t P$ is not image continuous. But the conjugate family $P\tilde{C}_t^*$ is image continuous.

EXAMPLE 5.1.8. Let $X = R^4$, $Y = R^2$, $A(x_1, x_2, x_3, x_4) = (x_1, x_2)$, $B(x_1, x_2, x_3, x_4) = (x_3, 0)$. Let

$$U = \{(x_1, x_2, x_3, x_4): (|x_1|+|x_3|)^2+x_2^2+x_4^2 < 1\}. \quad (5.1.27)$$

It is easy to verify that $A(U) = (A+B)(U) = \{(x_1, x_2): x_1^2+x_2^2 < 1\}$ and $B(U) = \{(x, 0): |x| < 1\}$. Let $\tilde{C}_t$ be defined as in Example 5.1.6. The families $\tilde{C}_t A$, $\tilde{C}_t(A+B)$ are image continuous but the family $\tilde{C}_t B$, $\tilde{C}_t B = \tilde{C}_t(A+B) - \tilde{C}_t A$ is not image continuous.

PROBLEM 5.1.1. Can we introduce a topology τ in the space of linear operators which map a given X into a given Y, such that

(1) Theorem 5.1.1 holds for the topology τ,

(2) the family of operators from Example 5.1.3 is continuous in the topology τ,

(3) the sum, the superposition and the conjugate of image continuous families in τ-topology are image continuous in τ-topology?

Now we return to the definition of a point of controllability. Namely, a point y_0 is a *point of controllability* if there exists an $x \in X$ such that $BA(x) = y_0$. However, in many applications the problem which interests us is whether there exists an x belonging to a certain linear manifold M, $M \subset X$, such that $BA(x) = y_0$. For instance, it is so in Example 5.1.1, provided we fix an initial value, and in Example 5.1.2, provided we fix an initial function. In both examples linear manifolds M are of the form $M = (x^0, U)$, where x^0 is a fixed initial value (or initial function) and U is the space of controls.

Thus we come to the concept of *relative controllability*. Let a linear system

$$(X \underset{A}{\rightarrow} \square \underset{B}{\rightarrow} Y) \quad (5.1.28)$$

5.1. Linear systems

be given. Suppose that the space X is the Cartesian product of two Banach spaces $X_0, X_1, X = X_0 \times X_1$. We say that y_0 is a *point of controllability of system* (5.1.28) *under the condition* $x_0 = x^0$, *where* x^0 *is a fixed element of the space* X_0, if there exists an $x_1 \in X_1$ such that $BA(x^0, x_1) = y_0$. Note that then

$$BA(0, x_1) = y_0 - BA(x^0, 0).$$

In other words, y_0 is a point of controllability of system (5.1.28) under the condition $x_0 = x^0$ if and only if $y_0 - BA(x^0, 0)$ is a point of controllability (without any conditions) of the system

$$(X_1 \underset{A_1}{\to} \square \underset{B}{\to} Y),$$

where $A_1(x_1) = A(0, x_1)$. This observation allows us to reduce the problem of relative controllability to controllability (without any conditions).

The notion of a linear system was extended to locally convex spaces $X, Y, \square$ by Singer (1980), who investigated some fundamental properties of such systems.

Let a linear system

$$(X \underset{A}{\to} \square \underset{B}{\to} Y)$$

be given. A point $y_0 \in Y$ is called a *point of approximative controllability* if there is a sequence $\{x_n\}$ of elements of X such that $BA(x_n) \to y_0$, in other words $y_0 \in \overline{BA(X)}$.

A point $y_0 \in Y$ is called a *point of bounded controllability* (Dolecki, 1977) if there is a bounded sequence $\{x_n\}$ of elements of X such that $BA(x_n) \to y_0$, in other words $y_0 \in \bigcup_{m=1}^{\infty} m \, \overline{BA(\{x: \|x\| \leq 1\})}$.

The set of all points of bounded controllability is called the *aureola* of the operator BA and is denoted by $\alpha(BA)$. (Dolecki, 1977).

Of course

$$BA(X) \subset \alpha(BA) \subset \overline{BA(X)}. \tag{5.1.29}$$

The following example shows that the aureola may differ from the range of the operator BA.

EXAMPLE 5.1.9 (Siemaszko, see Dolecki, 1977). Let $X = \square = C[0,1]$. Let $Y = L^2[0,1]$. Let A be the identity, $A = I$, and let B be the natural embedding of the space $C[0,1]$ into the space $L^2[0,1]$. It is easy to verify that the aureola $\alpha(BA)$ of the operator BA is the set of essentially bounded functions.

PROPOSITION 5.1.7. *If the image of the unit ball $\Gamma = BA(\{x: ||x|| \leq 1\})$ is closed, then a point y_0 is a point of bounded controllability if and only if it is a point of controllability.*

Proof. If Γ is closed, then

$$BA(X) = \bigcup_{m=1}^{\infty} BA(\{x: ||x|| \leq m\})$$

$$= \bigcup_{m=1}^{\infty} mBA(\{x: ||x|| \leq 1\}) = \alpha(BA). \blacksquare \quad (5.1.30)$$

COROLLARY 5.1.8 *If the input space X is reflexive, then each point of bounded controllability is a point of controllability.*

PROPOSITION 5.1.9 (Dolecki, 1977). *If the aureola $\alpha(BA)$ of the operator BA is equal to the whole space Y, $\alpha(BA) = Y$, then the linear system is controllable.*

Proof. By the Baire category theorem (Theorem 1.7.3), the set $\overline{BA(\{x: ||x|| \leq 1\})}$ contains an open ball. Then, by the repetition of the proof of the Banach theorem on open mappings (Theorem 3.2.1), the set $BA(\{x: ||x|| \leq 2\})$ contains an open ball. Thus $BA(X) = Y$. $\blacksquare$

By the definition, the aureola coincides with the range of the operator BA, $\alpha(BA) = BA(X)$, if and only if

$$\overline{BA(\{x: ||x|| \leq 1\})} \subset BA(X) = \bigcup_{m=1}^{\infty} mBA(\{x: ||x|| \leq 1\}).$$

(5.1.31)

Formula (5.1.31) can be replaced by the stronger formula (5.1.32).

THEOREM 5.1.10. (Dolecki, 1977). *The aureola $\alpha(BA)$ coincides with the range, $\alpha(BA) = BA(X)$, if and only if there is a $K > 0$ such that*

$$\overline{BA(\{x\colon ||x|| \leqslant 1\})} \subset K\overline{BA(\{x\colon ||x|| \leqslant 1\})}. \tag{5.1.32}$$

The proof of Theorem 5.1.10 is based on the following notion. Let y be a point of aureola, $y \in \alpha(BA)$. By $m(y)$ we shall denote the infimum of those positive numbers m, that there is a sequence $\{x_n\}$ of elements of X such that $||x_n|| \leqslant m$ and $BA(x_n) \to y$. If y belongs to Y and does not belong to the aureola we simply put $m(y) = +\infty$.

It is easy to verify the following properties of the function $m(y)$

$$m(ty) = |t|m(y) \quad \text{for each scalar } t, \tag{5.1.33}$$

$$m(y_1 + y_2) \leqslant m(y_1) + m(y_2), \tag{5.1.34}$$

$$m(y) = 0 \quad \text{if and only if} \quad y = 0. \tag{5.1.35}$$

PROPOSITION 5.1.11 (Dolecki, 1977). *There is a $k > 0$ such that*

$$m(y) \geqslant k||y||. \tag{5.1.36}$$

Proof. Suppose that (5.1.36) does not hold. Then there is a sequence of elements $\{y_n\}$ of norm one, $||y_n|| = 1$, such that

$$m(y_n) \leqslant \frac{1}{n}, \quad n = 1, 2, \ldots \tag{5.1.37}$$

By the definition of $m(y_n)$, there is an element $x_n \in X$ such that

$$||x_n|| < m(y_n) + \frac{1}{n}, \tag{5.1.38}$$

and

$$||BA(x_n) - y_n|| < \frac{1}{n}. \tag{5.1.39}$$

By (5.1.37), $m(y_n) \to 0$. Thus, by (5.1.38), $||x_n|| \to 0$. Therefore, by the continuity of BA, $||BA(x_n)|| \to 0$. Hence by (5.1.39)

$$1 = ||y_n|| \leqslant ||y_n - BA(x_n)|| + ||BA(x_n)|| \to 0,$$

which leads to a contradiction. ∎

Proof of Theorem 5.1.10. Formula (5.1.32) is a trivial consequence of formula (5.1.36) with $K = 1/k$. ∎

The following example shows that the number K in Theorem 5.1.10 cannot be uniformly bounded.

EXAMPLE 5.1.10. Let $X = l$ and $\Box = Y = c_0$ (or l, or l^2). Let B be the identity operator, $B = I$. Now we shall define the operator A. Let $\{y_n\}$ be a sequence of strongly linearly independent elements of the space Y, i.e., the convergence of the series $\sum_{n=0}^{\infty} t_n y_n$ to 0 implies that $t_n = 0$, $n = 0, 1, 2, \ldots$ Moreover we require that $\lim y_n = y_0$ and $\|y_0\| = 1$.

In the considered spaces such sequences exist, for example the sequence

$$y_0 = \{1, 0, 0, \ldots\}, \quad y_n = \{1, 0, \ldots, 0, \underbrace{\tfrac{1}{n}}_{n\text{-th position}}, 0, \ldots\}$$

satisifes the requested condition. Generally it can be shown, that such sequences exist in each infinite dimensional Banach space. The proof is complicated and requesting new definitions. Hence we omit it here and we restrict our considerations to the spaces mentioned above. Let $t = \{t_n\}$ be an arbitrary element of the space X. Let K be an arbitrary positive number. Let $A(t)$ be defined as follows

$$A(t) = \frac{1}{K} t_0 y_0 + \sum_{i=1}^{\infty} t_i y_i.$$

Since the sequence $\{y_n\}$ is bounded the operator A is well defined linear and continuous. By the strong linear independence of the elements of the sequence $\{y_n\}$ the operator A, and thus BA are one-to-one. Since $y_n \to y_0$, $y_0 \in \overline{BA(\{x: \|x\| \leq 1\})}$. On the other hand,

$$y_0 \notin mBA(\{x: \|x\| \leq 1\})$$

if $m < K$.

5.2. Optimization of a linear system with a fixed input operator

Let a linear system

$$(X \underset{A}{\rightarrow} \square \underset{B}{\rightarrow} Y) \tag{5.1.1}$$

be given. Let V be a subset of the product space $X \times \square \times Y$ and let F be a linear functional defined on V. The functional F is called a *performance functional*. Now we can formulate the following optimization problem:

find an element $v_0 = (x_0, \eta_0, y_0) \in V$ such that

$$A(x_0) = \eta_0,$$
$$B(\eta_0) = y_0,$$
$$F(v_0) = \inf\{F(u): u = (x, \eta, y) \in V, A(x) = \eta, B(\eta) = y\}.$$

Such a formulation covers a large number of problems related to optimization of linear systems. However, this formulation does not cover the minimum-time problem, which will be considered separately in Sections 5.4 and 5.5.

We say that a functional f defined on a linear space L (over the field of real or complex numbers) is *convex* if f is a real-valued functional and, for every pair of elements $x, y \in L$ and every real number t, $0 < t < 1$,

$$f(tx+(1-t)y) \leqslant tf(x)+(1-t)f(y). \tag{5.2.1}$$

Consider now the product space $R \times L$. Let

$$W = \{(t, x) \in R \times L: t \geqslant f(x)\}. \tag{5.2.2}$$

The set W is convex. To see this, let (r_1, w_1), $(r_2, w_2) \in W$. Then, for every real t, $0 < t < 1$, we have

$$r = tr_1+(1-t)r_2 \geqslant tf(x_1)+(1-t)f(x_2) \geqslant f(tx_1+(1-t)x_2)$$
$$= f(x), \tag{5.2.3}$$

where $x = tx_1+(1-t)x_2$. This implies that $(r, x) \in W$, which means that W is convex.

The set W is called the *epigraph* of the functional f. By the same arguments as above, the set $\{(t, x) \in W: t \leqslant \alpha\}$ is convex for each α. Since the projection of a convex set is convex, the set

$$K_\alpha = \{x \in L: f(x) \leqslant \alpha\} \tag{5.2.4}$$

is convex for each α. The sum of two convex functionals F, G is a convex functional since for arbitrary $x, y \in L$ and $0 < t < 1$,

$$\begin{aligned}F\bigl(tx+(1-t)y\bigr)&+G\bigl(tx+(1-t)y\bigr)\\&\leqslant tF(x)+(1-t)F(y)+tG(x)+(1-t)G(y)\\&= t\bigl(F(x)+G(x)\bigr)+(1-t)\bigl(F(y)+G(y)\bigr).\end{aligned}$$

Let Z be a Banach space and let f be a convex continuous functional defined on Z. The continuity of f implies that the epigraph W defined by (5.2.2) and the set K_α defined by (5.2.4) are both closed sets.

It is relatively easier to find the infimum of a convex functional over the whole space Z than over subsets of the space Z. In control theory these subsets are frequently of a particular form, namely, they are closed linear manifolds. Let us recall that a subset V of a linear space X (over the field of real or complex numbers) is called a linear manifold if for an arbitrary pair of elements $x, y \in V$ and arbitrary numbers a, b, $a+b = 1$, we have $ax+by \in V$.

By Theorem 2.3.1, every linear manifold is of the form

$$V = x_0 + M, \tag{5.2.5}$$

where x_0 is an arbitrary fixed element of V and M is a linear subset of the space X.

Since we have assumed that V is a closed set, the set M is a closed linear subset, whence M is a subspace.

In the sequel we shall consider exclusively linear manifolds which are closed. Therefore, by a *linear manifold* we mean a closed linear manifold.

Now we formulate the following problem: for a given linear manifold V expressed by formula (5.2.5) and a continuous convex functional F, determine the number r, $r = \inf_{v \in V} F(v)$.

By formula (5.2.2), the functional F defines in the space $R \times Z$ an epigraph W. By the continuity of F, the epigraph W has a nonempty

5.2. Optimization of linear systems with fixed input

interior. To see this, let $(t_0, \tilde{x}_0) \in W$ and $t_0 > F(\tilde{x}_0) + 2\varepsilon$, where ε is an arbitrary positive number. Since F is continuous, there exists a $\delta > 0$ such that for $\|x - \tilde{x}_0\| < \delta$ we have $F(x) < F(\tilde{x}_0) + \varepsilon$. This implies that $(t, x) \in W$ for $|t - t_0| < \varepsilon$ and $\|x - \tilde{x}_0\| < \delta$. This means that (t_0, x_0) is an interior point of the set W.

Let V' be a linear manifold in the space $R \times Z$ defined as

$$V' = \{(t, x) \in R \times Z : t = r, x \in V\}.$$

It immediately follows from the definition of the number α that V' has no common points with the interior of the set W. In virtue of the separation theorem (Theorem 2.5.9) there exists a continuous linear functional $\tilde{f}$, $\tilde{f} \neq 0$, defined on the space $R \times Z$ and such that

$$\sup \{\tilde{f}(t, x) : (t, x) \in V'\} \leqslant \inf \{\tilde{f}(t, x) : (t, x) \in W\}. \quad (5.2.6)$$

The set V' is a linear manifold. If a certain functional f is bounded (from above or from below) on a linear manifold then f must be constant on that manifold. Otherwise, the image of the manifold under f contains a line, which contradicts the boundedness of the functional f.

Hence,

$$\tilde{f}(t, x) = c \quad \text{for} \quad (t, x) \in V',$$

where

$$c = \sup \{\tilde{f}(t, x) : (t, x) \in V'\}.$$

Furthermore, $\tilde{f}(t, x) = \alpha t + f_0(x)$, where α is a certain constant and f_0 is a continuous linear functional defined on Z.

We claim that formula (5.2.6) entails $\alpha > 0$. On the contrary, suppose that $\alpha \leqslant 0$. Then the definition of the epigraph W implies

$$\inf \{\tilde{f}(t, x) : (t, x) \in W\} = -\infty,$$

which contradicts (5.2.6).

Observe that multiplication by positive numbers does not affect inequality (5.2.6). Hence, without loss of generality we can assume $\alpha = 1$. Thus, by the definition of V', we have

$$c = r + f_0(x_0),$$

and inequality (5.2.6) implies that

$$t + f_0(x) \geqslant r + f_0(x_0)$$

for $t \geqslant F(x)$. In particular, for $t = F(x)$ we have
$$F(x)+f_0(x)-f_0(x_0) \geqslant r. \qquad (5.2.7)$$
Since $f(t, x)$ is constant on V', we have $f_0(x) = 0$ for $x \in M$.

The set of linear functionals satisfying this condition will be denoted by $M^\perp$, i.e.,
$$M^\perp = \{f \in Z^* : f(x) = 0 \text{ for } x \in M\}.$$

Now we prove

THEOREM 5.2.1 (The method of Lagrange multipliers). *Let Z be a Banach space. Let V be a linear manifold in Z, $V = x_0 + M$, where M is a subspace. Let F be a continuous convex functional defined on Z. Then*
$$\inf_{v \in V} F(v) = \sup_{f \in M^\perp} \inf_{x \in Z} [F(x)+f(x)-f(x_0)]$$
and there exists a functional $f_0 \in M^\perp$ such that
$$\inf_{v \in V} F(v) = \inf_{x \in Z} [F(x)+f_0(x)-f_0(x_0)]. \qquad (5.2.8)$$

Proof. For every $f \in M^\perp$ and every $x \in V$ we have $f(x) = f(x_0)$. This implies that
$$r = \inf_{v \in V} F(v) \geqslant \sup_{f \in M^\perp} \inf_{x \in Z} [F(x)+f(x)-f(x_0)].$$
The opposite inequality and formula (5.2.8) follow immediately from formula (5.2.7). ∎

It happens frequently that V appearing in the formulation of the problem is of the form
$$V = \{(x, \xi, y) \in X \times \square \times Y : A(x) = \xi, B(\xi) = y_0\}, \qquad (5.2.9)$$
where y_0 is a fixed element of the output space Y and the performance functional F depends on inputs only, i.e.,
$$F(x, \xi, y) = F_0(x).$$

Problems in which $F_0(x)$ is either a norm or a seminorm in the space X are called *minimum-norm problems*. Problems of this type appear frequently in control theory.

5.2. Optimization of linear systems with fixed input

Now we analyse such problems by the method of Lagrange multipliers. By formula (5.2.5), $V = x_0 + M$, where x_0 denotes an arbitrary element of the set V and

$$M = \{(x, \xi, y): A(x) = \xi, B(\xi) = 0\}.$$

To determine $M^\perp$, observe that every continuous linear functional $f(x, \xi, y)$ defined on the space $X \times \square \times Y$ is of the form

$$f(x, \xi, y) = f_1(x) + f_2(\xi) + f_3(y),$$

where f_1, f_2, f_3 are continuous linear functionals defined on the spaces X, $\square$, Y, respectively. Since the element y_0, $y_0 = 0$, is fixed in the definition of M, f_3 can be chosen arbitrarily. This does not apply to f_1 and f_2. The functional f_1 (f_2) has to be chosen in such a way that the condition $BA(x) = 0$ ($B(\xi) = 0$) implies $f_1(x) = 0$ ($f_2(\xi) = 0$). The set of all continuous linear functionals which have this property will be denoted by V_1 (V_2).

An example of such a set of functionals is provided by functionals of the form

$$f_1(x) = \varphi(BA(x)), \quad \varphi \in Y^*$$

or

$$f_2(x) = \varphi(B(x)), \quad \varphi \in Y^*.$$

In other words,

$$A^*B^*(Y^*) \subset V_1$$

and respectively $B^*(Y^*) \subset V_2$. The question arises when the equality

$$A^*B^*(Y^*) = V_1$$

and respectively $B^*(Y^*) = V_2$ holds.

THEOREM 5.2.2. $A^*B^*(Y^*) = V_1$ ($B^*(Y^*) = V_2$) if and only if $BA(X)$ ($B(\square)$) is a subspace, i.e., a closed linear subset of the space Y.

Proof. We shall only prove the version of the theorem which is not in brackets. The proof of the version in brackets is the same.

Sufficiency. Consider the quotient space $X_0 = X/(BA)^{-1}(0)$. Every continuous linear functional f belonging to V_1 can be extended to a certain continuous linear functional $\tilde{f}$ defined on X_0 which is given by the formula $\tilde{f}([x]) = f(x)$. Conversely, every continuous linear functional

$\tilde{f}$ defined on the space X_0 can be extended to the functional f, $f(x) = \tilde{f}([x])$, which belongs to V_1. It is easy to see that the above correspondence is a linear isometry. The operator BA generates the operator $\widetilde{BA}$ by the formula

$$\widetilde{BA}([x]) = BA(x),$$

where $[x]$ denotes the equivalence class containing x. The operator $\widetilde{BA}$ is a one-to-one correspondence which maps X_0 onto the Banach space $BA(X)$. In virtue of the Banach theorem on the continuity of the inverse operator (Corollary 3.2.2), the operator $\widetilde{BA}$ is an isomorphism. Thus, every continuous linear functional defined on the space X_0 is of the form $B^*A^*\varphi$, where φ is a certain linear functional defined on $BA(X)$. By the Hahn–Banach theorem (Theorem 2.5.3), every continuous linear functional defined on $BA(X)$ can be extended to a continuous linear functional defined on the whole space Y. On the other hand, there exists a one-to-one correspondence between continuous linear functionals defined on X_0 and continuous linear functionals defined on V_1.

Hence, every functional $f \in V_1$ is of the form

$$A^*B^*(\varphi) \quad \text{where} \quad \varphi \in Y^*.$$

Necessity. In the same way as above we can prove that there exists an isomorphism between V_1 and X_0. The assumption $V_1 = A^*B^*(Y^*)$ implies that A^*B^* is an isomorphism of $(AB(X))^\perp$ onto X_0^*. Hence, $\widetilde{BA}$ is an isomorphism which maps X_0 onto $BA(X)$. Thus, $BA(X)$ is a closed set. ∎

COROLLARY 5.2.3. *If either X or Y are finite-dimensional spaces, then* $A^*B^*(Y^*) = V_1$.

Proof. If either X or Y are finite-dimensional spaces, then $BA(X)$ is a finite-dimensional subspace and thus closed. ∎

COROLLARY 5.2.4. *If either X or Y are finite-dimensional spaces, then there exists a $\varphi_0 \in Y^*$ such that*

$$\inf\{F(x): BA(x) = y_0\} = \inf\{F(x) + \varphi_0(BA(x)) - \varphi_0(y_0)\}.$$

5.2. Optimization of linear systems with fixed input

Corollary 5.2.4 can be regarded as a formulation of the maximum principle for linear systems.

Consider now the case where the performance functional F depends only on the input variable x. Then $f_2 = f_3 = 0$ since otherwise
$$\inf \{F(x)+f(x, \xi, y)-f(x_0, \xi_0, y_0)\} = -\infty.$$
Let V be a linear manifold of the form
$$V = \{(x, \xi, y) \in X \times \square \times Y: BA(x) = y_0\}$$
and let the performance functional F depend only on x. Then
$$\inf_{v \in V} F(v) = \inf_{x \in (BA)^{-1}(y_0)} F(x)$$
$$= \sup_{f \in (BA^{-1}(0))^\perp} \inf_{x \in X} [F(x)+f(x)-f(x_0)], \qquad (5.2.10)$$
where x_0 is an arbitrary element of $(BA)^{-1}(y_0)$.

One of the disadvantages of formula (5.2.10) is that it makes use of functionals f belonging to $(BA^{-1}(0))^\perp$. Moreover, in general $(BA^{-1}(0))^\perp$ cannot be replaced by $A^*B^*(Y^*)$ (cf. Theorem 5.2.2). This is a disadvantage. Indeed, the spaces X, $(BA)^{-1}(0)$ are determined uniquely by the problem itself. But in the space Y we are interested in one element only, namely y_0. We can use different norms in Y provided that they preserve the continuity of B. Thus, we can choose a norm $\| \ \|$ in such way that the space Y^* has a simple structure.

THEOREM 5.2.5. *Let the sets*
$$\Gamma_r = BA(\{x \in X: F(x) \leqslant r\})$$
be closed for each r. Then
$$\inf_{x \in (BA)^{-1}(y_0)} F(x) = \sup_{f \in A^*B^*(Y^*)} \inf_{x \in X} [F(x)+f(x)-f(x_0)]. \qquad (5.2.11)$$

Proof. Put
$$a = \inf_{x \in (BA)^{-1}(y_0)} F(x),$$
$$b = \sup_{f \in A^*B^*(Y^*)} \inf_{x \in X} [F(x)+f(x)-f(x_0)].$$
Since for every functional f belonging to $A^*B^*(Y^*)$ we have
$$f(x)-f(x_0) = \varphi(y_0)-\varphi(y_0) = 0$$

for $x \in (BA)^{-1}(y_0)$, it follows that
$$b \leqslant a.$$
Suppose that $b < a$. By assumption, the set $\Gamma_b = BA(\{x \in X: F(x) \leqslant b\})$ is closed. The inequality $b < a$ entails $y_0 \notin \Gamma_b$. As before, consider now the product space $R \times Y$. By $\tilde{W}$ we denote the set
$$\tilde{W} = \{(t, y): y \in BA(X), \ t \geqslant \inf F((BA)^{-1}(y))\}.$$
Since $y_0 \notin \Gamma_b$, we have $(b, y_0) \notin \tilde{W}$. Hence, there exists a continuous linear functional $\tilde{f}$ separating (b, y_0) and $\tilde{W}$. Without loss of generality we can assume that $\tilde{f}$ is of the form $\tilde{f}[(t, x)] = t + \varphi(x)$, where φ is a continuous linear functional defined on Y. This implies that the set
$$W = \{(t, x) \in R \times X: t \geqslant F(x)\}$$
can be separated in the space $R \times X$ from the set $(b, (BA)^{-1}(y_0))$ by a functional of the form $t + \bar{f}(x)$, where $\bar{f} \in A^*B^*(Y^*)$. From this point the proof is analogous to the proof of Theorem 5.2.1. ∎

If formula (5.2.11) holds, we say that we have *weak duality*. Consider now the case where V is of the form
$$V = \{x \in Z: f(x) = c\}.$$
Then, by the method of Lagrange multipliers,
$$\inf_{v \in V} F(v) = \sup_{\alpha \in R} \inf_{x \in Z} [F(x) + \alpha f(x) - \alpha c].$$
In particular, when $V \subset X$ is of the form
$$V = \{x \in X: \varphi(BA(x)) = \varphi(y_0)\}, \quad \text{where} \quad \varphi \in Y^*,$$
we have
$$\inf \{F(x): x \in X, \varphi(BA(x)) = \varphi(y_0)\}$$
$$= \sup_{\alpha \in R} \inf_{x \in X} [F(x) + \alpha \cdot \varphi(BA(x)) - \alpha \varphi(y_0)]. \qquad (5.2.12)$$
The set Y^* is linear. Taking the supremum over all functionals $\varphi \in Y^*$ for both sides of equality (5.2.12), we obtain
$$\sup_{\varphi \in Y^*} \inf \{F(x): x \in X, \varphi(BA(x)) = \varphi(y_0)\}$$
$$= \sup_{\varphi \in Y^*} \inf_{x \in X} [F(x) + \varphi(BA(x)) - \varphi(y_0)]. \qquad (5.2.13)$$
Hence, by Theorem 5.2.5, we obtain

5.2. Optimization of linear systems with fixed input

Theorem 5.2.6. *Let the set*
$$\Gamma_r = BA(\{x \in X: F(x) \leq r\})$$
be closed for each r. Then
$$\inf\{F(x): x \in X, BA(x) = y_0\}$$
$$= \sup_{\varphi \in Y^*} \inf\{F(x): \varphi(BA(x)) = \varphi(y_0)\}. \qquad (5.2.14)$$

There exists also a direct proof of Theorem 5.2.6.

Second proof. Let
$$a = \inf\{F(x): BA(x) = y_0\},$$
$$b = \sup_{\varphi \in Y^*} \inf\{F(x): \varphi(BA(x)) = \varphi(y_0)\}.$$

Since the equality $BA(x) = y_0$ implies $\varphi(BA(x)) = \varphi(y_0)$, we have
$$b \leq a.$$
We shall show that $b = a$. Suppose that the equality does not hold. Hence
$$b < a.$$
This implies that $y_0 \notin \Gamma_b$. By the assumption, the set Γ_b is closed. Hence, there exist a continuous linear functional $\varphi_0 \in Y^*$, a constant c and a positive number ε such that
$$\varphi_0(y_0) = c$$
and
$$\varphi_0(y) < c - \varepsilon \quad \text{for} \quad y \in \Gamma_b.$$
This means that
$$\inf\{F(x): \varphi_0(BA(x)) = \varphi_0(y_0)\} \geq b \frac{c}{c-\varepsilon} > b.$$
This contradicts the definition of the number b. ∎

Observe that in the case where $F(x)$ is a norm or a seminorm $\|\ \|$, the problem of finding the number a,
$$a = \inf\{\|x\|: \varphi_0(BA(x)) = \varphi_0(y_0)\}$$

amounts to the evaluation of the norm of the continuous linear functional $A^*B^*(\varphi_0)$. Indeed, if we write $\tilde{f} = \dfrac{1}{\varphi_0(y_0)} A^*B^*(\varphi_0)$ and $c = \varphi_0(y_0)$, then, in virtue of the considerations of Section 2.3, the hyperplane $H_{\tilde{f}}$ is a supporting hyperplane of the ball K_a of radius

$$a = \frac{1}{\|\tilde{f}\|} = \frac{|\varphi_0(y_0)|}{\|A^*B^*(\varphi_0)\|},$$

and in this way we find a connection between a, $|\varphi_0(y_0)|$ and $\|A^*B^*(\varphi_0)\|$.

Without the closedness of the sets Γ_r, Theorem 5.2.6 is not valid, as is shown by the following example.

EXAMPLE 5.2.1 (see also Singer, 1973). Let $X = l$ and $\square = Y = c_0$ (or l or l^2). Let B be the identity operator. Now we shall define the operator A. Choose a sequence $\{y_n\}$ of elements of the space Y which are strongly linearly independent, i.e., the convergence of the series $\sum\limits_{n=0}^{\infty} t_n y_n$ to zero implies that all t are equal to zero for $n = 0, 1, \ldots$ Moreover, we require that $\lim\limits_{n\to\infty} y_n = y_0$ and $\|y_0\| = 1$.

Such sequences exist in the space c_0 (l, l^2). For instance, the sequence

$$y_0 = \{1, 0, 0, \ldots\}, \quad y_n = \{\underbrace{1, 0, 0, \ldots, 0, \tfrac{1}{n}}_{(n-1)\text{-th position}}, 0, 0, \ldots\},$$

satisfies the above conditions. Generally, it can be shown that in each infinite-dimensional Banach space Y there exists a sequence satisfying the above conditions. The proof of this assertion is complicated and requires some new definitions. Hence, we omit it here and restrict our considerations to the spaces mentioned above. Let $\{t_n\} = t$ be an arbitrary element of the space l and let

$$A(t) = \frac{1}{2} t_1 y_0 + \sum_{i=1}^{\infty} t_{i+1} y_i.$$

Since the sequence $\{y_n\}$ is bounded, the operator A is well defined, linear and bounded. By the strong linear independence of the elements of the sequence $\{y_n\}$, the operator A is a one-to-one mapping.

5.2. Optimization of linear systems with fixed input

Let $F(x) = \|x\|$ be a norm of x in the space l. Since A is a one-to-one mapping, the set $\Gamma_1 = BA(\{x: \|x\| \leq 1\})$ contains the element $\tfrac{1}{2}y_0$, which is the image of the element $\{1, 0, \ldots\}$ under A. By the same argument, the set Γ_1 does not contain any element of the form by_0 for $b > \tfrac{1}{2}$. This immediately implies that

$$\inf\{\|u\|: BA(u) = y_0\} = 2.$$

Consider now an arbitrary functional $\varphi \in Y^*$. By the continuity of φ, the sequence $\{c_n\}$, $c_n = \varphi(y_n)$, tends to $c = \varphi(y_0)$. Let

$$y'_n = \frac{c}{c_n} \cdot y_n.$$

Then $\varphi(y'_n) = c$. Observe that, by the definition of the operator A,
$$y'_n = A(u_n),$$
where

$$u_n = \Big\{0, 0, \ldots, 0, \underbrace{\frac{c}{c_n}}_{(n+1)\text{-th position}}, 0, \ldots\Big\}.$$

Since $\|u_n\| = \dfrac{c}{c_n} \to 1$, we have

$$\inf\{\|u\|: \varphi(BA(u)) = \varphi(y_0)\} = 1$$

for all continuous linear functionals $\varphi \in Y^*$.

Dolecki (1977), fully explained when we have weak duality. Namely, he proved that weak duality holds for all $y_0 \in Y$ if and only if

$$\bigcap_{\varepsilon > 0} BA(\{x: F(x) < a+\varepsilon\}) = \bigcap_{\varepsilon > 0} \overline{BA(\{x: F(x) < a+\varepsilon\})}.$$

The proof of this fact will be given in the Appendix since it exploits the theory of multifunctions developed in the context of optimization by Dolecki and Kurcyusz (1978).

Corollary 5.2.4, Theorem 5.2.1 and Theorem 5.2.6 immediately imply

COROLLARY 5.2.7. *If $BA(X)$ is a closed subspace of the space Y (in particular, if the input space X or the output space Y are finite-dimensional), then there exists a $\varphi_0 \in Y^*$ such that*

$$\inf\{F(x): \varphi_0(BA(x)) = \varphi_0(y_0)\}$$
$$= \sup_{\varphi \in Y^*} \inf\{F(x): \varphi(BA(x)) = \varphi(y_0)\}.$$

Corollary 5.2.7 can be regarded as a certain analogue of the Pontryagin maximum principle for linear systems with a fixed final state. If the assertion of Corollary 5.2.7 holds we often say that we have *strong duality*.

The following theorem states the converse result to that of Corollary 5.2.7.

THEOREM 5.2.8. *Let a linear system*

$$(X \underset{A}{\rightarrow} \square \underset{B}{\rightarrow} Y)$$

be given and let F be a continuous convex functional defined on X. Assume that for each r the set $\Gamma_r = BA(\{x\colon F(x) \leqslant r\})$ is closed. If $BA(X)$ is not a subspace (i.e., $BA(X)$ is not closed), then there exists a $y_0 \in BA(X)$ such that for every functional $\varphi \in Y^*$

$$\inf\{F(x)\colon BA(x) = y_0\} > \inf\{F(x)\colon \varphi(BA(x)) = \varphi(y_0)\}.$$

The proof of this fact is based on a number of auxiliary notions and a lemma proved by Wojtaszczyk (1973).

Let X be a Banach space. Let K be a closed convex set in X. By the *algebraic interior* of the set K, $\text{Int}_a K$, we mean the set of all points $p \in K$ which have the following property: for each line L passing through p either $\{p\} = K \cap L$ or $K \cap L$ is equal to a certain interval I which contains p in its interior (the interior is taken with respect to the topology on L). By the *algebraic boundary* of the set K we mean the set $\text{Fr}_a K = K \setminus \text{Int}_a K$. Points belonging to the algebraic interior of a set will be called *algebraic interior points*. Points belonging to the algebraic boundary of a set will be called *algebraic boundary points*.

Let α be an arbitrary algebraic interior point of the set K and let p be an arbitrary algebraic boundary point of K. The convexity of the set K implies that the line L passing through α and p has a common interval with K. We shall show that p is one of the end-points of that interval. Suppose on the contrary that the above does not hold. Then $L \cap K$ is an interval $[c, d]$ and the points α, p belong to the interior of this interval. Without loss of generality we can assume that α belongs to the interior of the interval $[c, p]$ and p lies in the interior of the in-

5.2. Optimization of linear systems with fixed input

terval $[\alpha, d]$. Hence, there exist two numbers $t_\alpha, t_p, 0 < t_\alpha, t_p < 1$ such that

$$\alpha = t_\alpha p + (1 - t_\alpha) c, \tag{5.2.15}$$
$$p = t_p \alpha + (1 - t_p) d. \tag{5.2.16}$$

Let x be an arbitrary element of the space X such that $p + x \in K$. By the convexity of K and formula (5.2.15)

$$t_\alpha x + \alpha = t_\alpha (x + p) + (1 - t_\alpha) c \in K.$$

Since, by hypothesis, α is an algebraic interior point of the set K there exists an $\varepsilon > 0$ such that for all $\alpha, |\alpha| < \varepsilon$,

$$\alpha \cdot t_\alpha x + \alpha \in K.$$

By formula (5.2.16) and by the convexity of K, we obtain

$$p + \alpha t_\alpha t_p x = t_p(\alpha + \alpha t_\alpha x) + (1 - t_p) d \in K.$$

Since x is arbitrary, p belongs to the algebraic interior of K, which contradicts the assumption.

Observe that if the interior of K is nonempty then, for each point p of K, p belongs to the algebraic interior of K if and only if p belongs to the interior of K, and p belongs to the algebraic boundary of K if and only if p belongs to the boundary of K.

We say that a continuous linear functional f *supports* a convex set K at a point $x_0 \in K$ if $f(x_0) = 1$, $f(x) \leqslant 1$ for $x \in K$. A point $x_0 \in K$ with the property that there exists a continuous linear functional f which supports K at x_0 is called a *support point*.

LEMMA 5.2.9 (Wojtaszczyk, 1973). *Let* $(X, \| \ \|)$ *be a normed Banach space over the field of real numbers. Let K be a closed convex set containing 0 in its algebraic interior. Assume that the linear hull of the set K, lin K, is dense in X. Moreover, assume that each algebraic boundary point is a support point. Then the interior of K is nonempty.*

Proof. Suppose that 0 is not an interior point of K. Then we shall prove that there exists a sequence of algebraic boundary points which tends to 0. Suppose on the contrary that such a sequence does not exist. This means that there exists a $\delta > 0$ such that $\|p\| > \delta$ for all algebraic boundary points p of K.

Let x be an arbitrary point whose norm is not greater than δ, $||x|| \leq \delta$. Since, by hypothesis, the linear hull of K is dense in X and K is convex, there exists a sequence $\{x_n\}$ of points belonging to K and such that

$$\lim_{n \to \infty} \frac{x_n}{||x_n||} = \frac{x}{||x||}. \tag{5.2.17}$$

Let L_n denote the line passing through 0 and the point x_n. Let p_n be an algebraic boundary point lying on the half-line of L_n passing through x_n. By the assumptions, $||p_n|| > \delta$ and $||x|| \leq \delta$. Hence, by (5.2.17), there exists a sequence of numbers $\{\alpha_n\}$, $0 < \alpha_n < 1$, such that

$$\lim_{n \to \infty} \alpha_n p_n = x. \tag{5.2.18}$$

By convexity of K, $\alpha_n p_n \in K$. By formula (5.2.18) and closedness of K, $x \in K$. Since x is arbitrary, 0 is an interior point of K, which leads to a contradiction.

We have shown that if 0 is not an interior point of the set K then there exists a sequence of algebraic boundary points of K which converges to 0. The same reasoning applied to the set $K-a$, where $a \in K$, shows that if the interior of the set K is empty then the algebraic boundary $\mathrm{Fr}_a K$ is dense in K.

Assume now that K has an algebraic interior point, say x_0. Let p be an arbitrary algebraic boundary point. As we have shown, p is one of the end-points of the interval which is contained in K and lies on the line L passing through x_0 and p. Hence, in an arbitrary neighbourhood of p there are algebraic interior points. Thus, if the algebraic interior of K is nonempty, the set of algebraic interior points $\mathrm{Int}_a K$ is dense in K.

Consider now a set K satisfying all the assumptions of the lemma. By hypothesis, every algebraic boundary point of K is a support point. Let A_n denote the set of all algebraic boundary points such that the norms of the corresponding supporting functionals are not greater than n. The sequence $\{A_n\}$ is a decreasing sequence of sets. By hypothesis, the algebraic boundary $\mathrm{Fr}_a K$ can be expressed as the union of the sets A_n:

$$\mathrm{Fr}_a K = \bigcup_{n=1}^{\infty} A_n. \tag{5.2.19}$$

5.2. Optimization of linear systems with fixed input

We claim that the sets A_n are closed. To see this, let us consider a sequence $\{x_m\}$ such that $x_m \in A_n$, $m = 1, 2, \ldots$ and $x_m \to x_0$. By the definition of the set A_n, there exist continuous linear functionals f_m such that $f_m(x_m) = 1$, $f_m(x) \leqslant 1$ for all $x \in K$ and $\|f_m\| \leqslant n$.

In virtue of the Alaoglu theorem (Theorem 4.4.6), the set

$$Ac^*\{f_m\} = \bigcap_{N=1}^{\infty} \overline{\bigcup_{m=N}^{\infty} \{f_m\}},$$

where the closure is taken in the weak-∗-topology, is not empty. Let $f_0 \in Ac^*\{f_m\}$. Then there is a subsequence $\{f_{m_i}\}$ which converges to f_0 in the weak-∗-topology. It is easy to observe that $\|f_0\| \leqslant n$.

Of course, for all $x \in K$

$$f_0(x) = \lim_{i\to\infty} f_{m_i}(x) \leqslant 1.$$

Moreover, since $x_{m_i} \to x_0$ and $\|f_{m_i}\| \leqslant n$, we have

$$f_0(x_0) = \lim_{i\to\infty} f_{m_i}(x_0) = \lim_{i\to\infty} [f_{m_i}(x_0 - x_{m_i}) + f_{m_i}(x_{m_i})]$$
$$= \lim_{i\to\infty} f_{m_i}(x_{m_i}) = 1. \tag{5.2.20}$$

Hence, $x_0 \in A_n$ and the set A_n is closed.

This proves that under the assumptions of the lemma the algebraic boundary of K, $\mathrm{Fr}_a K$, is the union of the closed sets A_n. Let

$$K_n = \left(1 - \frac{1}{n}\right) K.$$

The sets K_n are closed and $\mathrm{Int}_a K = \bigcup_{n=1}^{\infty} K_n$. Hence, the set K is the union of a countable number of closed sets. Namely,

$$K = \left[\bigcup_{n=1}^{\infty} K_n\right] \cup \left[\bigcup_{n=1}^{\infty} A_n\right]. \tag{5.2.21}$$

The set K as a closed subset of a complete metric space can be regarded as a complete metric space. By the Baire theorem (Theorem 1.7.3), the set K is in itself of the second category. This means that one of the sets K_n, A_n, $n = 1, 2, \ldots$ is of the second category. Denote this set by B.

Since B is of the second category and closed, B contains a ball, i.e., there exist an $x_0 \in B$ and a positive number r such that

$$\{x \in K: \|x-x_0\| < r\} \subset B. \tag{5.2.22}$$

Recall that the algebraic interior of K, $\text{Int}_a K$, and the algebraic boundary of K, $\text{Fr}_a K$, are disjoint. Moreover, $\text{Int}_a K$ is dense in K. Hence, the set B cannot coincide with any of the sets A_n.

If the set K had the empty interior (in the sense of the space X), then, by the previous arguments, the algebraic boundary of K, $\text{Fr}_a K$, would be dense in K. Hence, there would exist no set $B = K_{n_0}$ satisfying (5.2.22). In this way we have proved that the interior of the set K is nonempty. ∎

By Lemma 5.2.9 we can prove Theorem 5.2.8.

Proof of Theorem 5.2.8. Let

$$r > \inf F(x)$$

and let x_0 be an arbitrary element such that $F(x_0) < r$. By the continuity of F, x_0 is an interior point and hence an algebraic interior point of the set

$$K_r = \{x \in X: F(x) \leq r\}.$$

Since the property of being an algebraic interior point is an invariant of linear (not necessarily continuous) mappings, the point $BA(x_0)$ is an algebraic interior point of the set

$$\Gamma_r = BA(K_r).$$

Hence, 0 is an algebraic interior point of the set $\Gamma_r'' = \Gamma_r - BA(x_0)$. Assuming strong duality, this would mean that $y_0 - BA(x_0)$ is a support point of the set Γ_r''. By Lemma 5.2.9, it would imply in turn that the set Γ_r'' (and also Γ_r) regarded as a subset of the space $Y_0 = \overline{\text{lin}\,\Gamma_r}$ has a nonempty interior.

Hence, $Y_0 = \text{lin}\,\Gamma_r$ and

$$BA(X) = BA(\text{lin}\,K_r) = \text{lin}\,BA(K_r) = \text{lin}\,\Gamma_r$$

is closed. ∎

5.2. Optimization of linear systems with fixed input

S. Kurcyusz (private comunication, 1975) observed that the hypothesis that Γ_r are closed is essential for the validity of Theorem 5.2.8, as is shown by the following example.

EXAMPLE 5.2.2. Let $X = Y = c_0$, $A = I$ and let B be given by the formula

$$B(\{x_n\}) = \left\{\frac{1}{n} \cdot x_n\right\}.$$

Let $F(\{x_n\}) = |x_1|$. Clearly, the sets

$$\Gamma_r = \{y \in c_0 \colon |y_1| \leqslant r, \lim_{n \to \infty} ny_n = 0\}$$

are not closed.

Since A and B are one-to-one mappings, for an arbitrary $y^0 \in BA(X)$ there is a unique x^0 such that $BA(x^0) = y^0$. Moreover,

$$F(x^0) = |x_1^0| = |y_1^0|.$$

Let $\varphi_0(F(x)) = x_1$. Then

$$\inf\{F(x)\colon BA(x) = y^0\} = F(x) = |x_1^0| = |y_1^0|$$
$$= \inf\{F(x)\colon \varphi_0(BA(x)) = \varphi_0(y^0)\}$$

and the strong duality holds.

Theorem 5.2.6 is especially useful in the case where $F(x) = \|x\|$ is a norm or a seminorm in the space X since then

$$\inf\{\|x\|\colon \varphi(BA(x)) = \varphi(y_0)\} = \frac{\varphi(y_0)}{\|A^*B^*(\varphi)\|}$$

and the optimization problem reduces to the calculation of the norm of the functional $A^*B^*(\varphi)$. Recall that norms of functionals in different spaces were considered in Section 4.6.

To apply Theorem 5.2.6 we should have the sets Γ'_b closed.

COROLLARY 5.2.10. *Let X be a reflexive space. Let F be a functional such that*

$$\lim_{\|x\| \to \infty} F(x) = +\infty.$$

Then the weak duality holds.

196   5. Optimization and observability of linear systems

Proof. By hypothesis, the sets $K_b = \{x: F(x) \leqslant b\}$ are bounded. By the convexity and the continuity of F, the sets K_b are closed and convex. Since X is reflexive, K_b are weakly compact (Theorem 4.5.4). The images of weakly compact sets under continuous linear operators are weakly compact (Corollary 4.4.3). Thus, $\Gamma_b = BA(K_b)$ are all weakly compact, whence they are closed and bounded. ∎

THEOREM 5.2.11. *Let X, $\Box$, Y be the conjugate spaces of spaces X_0, $\Box_0$, Y_0, respectively. Let $F(x) = \|x\|$ be a norm (or seminorm) of functionals corresponding to a norm (seminorm) $\|x\|_0$ in the space X_0. Let the input operator A and the output operator B be the conjugate operators of operators A_0, B_0 which map $\Box_0$ into X_0 and Y_0 into $\Box_0$, respectively. Then the following formula holds*

$$\inf\{\|x\|: BA(x) = y_0\} = \sup_{\varphi \in Y_0} \inf\{\|x\|: x(A_0 B_0(\varphi)) = y_0(\varphi)\}.$$

Proof. Let
$$a = \inf\{\|x\|: BA(x) = y_0\}$$
and
$$b = \sup_{\varphi \in Y_0} \inf\{\|x\|: x(A_0 B_0(\varphi)) = y_0(\varphi)\}.$$

Since $(A_0 B_0)^* = BA$, every x satisfying the equation $BA(x) = y_0$ satisfies also the equation $x(A_0 B_0(\varphi)) = y_0(\varphi)$ for every $\varphi \in Y_0$. Hence, $b \leqslant a$.

We shall prove that $b = a$. Suppose that the equality does not hold and $b < a$. This means that $y_0 \notin \Gamma_b = BA(K_b)$, where
$$K_b = \{x \in X: \|x\| \leqslant b\}.$$

By the Alaoglu theorem (Theorem 4.4.6), the sets K_b are weakly-∗-compact. Since B and A, the conjugate operators of B_0 and A_0, are weakly-∗-continuous (Corollary 4.4.4), the sets Γ_b are weakly-∗-compact and consequently weakly-∗-closed. By the separation theorems (Theorems 4.3.2 and 4.3.3), there exist a linear functional f continuous in the weak-∗-topology, a constant c, and a positive number ε such that

$$f(y_0) = c \qquad (5.2.23)$$

and
$$f(x) < c - \varepsilon \quad \text{for} \quad x \in \Gamma_b. \tag{5.2.24}$$

By the continuity of the functional f in the weak-$*$-topology, f is of the form $f(x) = x(\varphi_0)$ for some $\varphi_0 \in Y_0$.

Formulae (5.2.23) and (5.2.24) imply that
$$\inf\{\|x\|: x(A_0 B_0(\varphi_0)) = x(\varphi_0)\} > b,$$
which contradicts the definition of b.

So far we have considered problems in which the output was a single fixed element. However, in many engineering applications we relax this condition by admitting that the outputs belong to a certain set G called a *target set*. More precisely, we seek the value of
$$\inf\{F(x): BA(x) \in G\} \tag{5.2.25}$$
and an x_0 which realizes this infimum.

We shall restrict ourselves to the case where the target set G is closed and convex. Frequently, G is a closed ball of radius m and centre y_0,
$$G = \{y \in Y: \|y - y_0\| \leqslant m\}.$$

THEOREM 5.2.12 (Singer, 1980). *Consider the linear system* (5.1.1)
$$(X \underset{A}{\rightarrow} \square \underset{B}{\rightarrow} Y).$$

Let G be a closed convex set. Suppose that
$$\inf_{x \in X} F(x) < r = \inf\{F(x): BA(x) \in G\}. \tag{5.2.26}$$

Moreover, suppose that either the set G or the set $\Gamma_r = \{y: y \in BA(x), F(x) \leqslant r\}$ has a nonempty interior. Then there is a continuous linear functional $\varphi_0 \in Y^*$ such that
$$\inf\{F(x): BA(x) \in G\} = \inf\{F(x): \varphi_0(BA(x)) = c\}, \tag{5.2.27}$$
where
$$c = \inf_{y \in G} \varphi_0(y). \tag{5.2.28}$$

Proof. Take an arbitrary element x_0 such that $F(x_0) < r$. According to (5.2.26) such an element exists. By the convexity of F, the function

$f(t) = F(x_0+t(x-x_0))$ of the real argument t is strictly increasing in a neighbourhood of $t = 1$.

We shall show that the intersection Int $G \cap \Gamma_r$ is empty. Suppose that it is not empty. Take an arbitrary $x \in$ Int $G \cap \Gamma_r$. For t sufficiently close to 1 we have $x_t = x_0+t(x-x_0) \in G$. Moreover, $F(x_t) = f(t) < r$ for $t < 1$ and we obtain a contradiction with the definition of the number r.

Now we shall show that the intersection Int $\Gamma_r \cap G$ is empty. Suppose on the contrary that there exists an x belonging to the intersection $G \cap$ Int Γ_r. For $t > 1$ and sufficiently close to 1 we have $x_t = x_0 + t(x-x_0) \in \Gamma_r$ and $F(x_t) > F(x)$. By the definition of Γ_r, $F(x_t) \leq r$ and $F(x) < r$. This leads to a contradiction with the definition of r.

Thus, the sets Γ_r and G can be separated by a continuous linear functional $\varphi_0 \in Y^*$. Hence, there exists a constant c_1 such that

$$\varphi_0(y) \geq c_1 \quad \text{for} \quad y \in G \tag{5.2.29}$$

and

$$\varphi_0(y) \leq c_1 \quad \text{for} \quad y \in \Gamma_r. \tag{5.2.30}$$

Let

$$H_1 = \{y: \varphi_0(y) \geq c_1\}$$

and

$$H_2 = \{y: \varphi_0(y) \leq c_1\}.$$

By (5.2.29) and (5.2.30),

$$G \subset H_1, \quad \Gamma_r \subset H_2. \tag{5.2.31}$$

By the convexity of F,

$$\inf\{F(x): \varphi_0(BA(x)) \geq c_1\} = \inf\{F(x): \varphi_0(BA(x)) = c_1\} \tag{5.2.32}$$

and, by (5.2.31),

$$r = \inf\{F(x): BA(x) \in G\} \geq \inf\{F(x): \varphi_0(BA(x)) = c_1\} \geq r.$$

To complete the proof it is enough to show that $c_1 = c$. Take an x_0 such that $F(x_0) < r$. By the convexity of the functional F,

$$F(x) \geq c_1 + \varepsilon \frac{r - F(x_0)}{c - \varphi_0(BA(x_0))} \tag{5.2.33}$$

provided that $\varphi_0(BA(x)) \geq c_1 + \varepsilon$. If $G \subset \{y: \varphi_0(y) \geq c_1+\varepsilon\}$, then,

by (5.2.33), we obtain a contradiction with formula (5.2.30) and the definition of r. Thus,
$$c_1 \geqslant c = \inf\{\varphi_0(y): y \in G\}$$
and, by (5.2.31) and (5.2.29), $c = c_1$. ∎

I. Singer (1980) proved Theorem 5.2.12 in locally convex spaces under weaker assumptions on F. However, we shall restrict ourselves to Banach spaces and continuous convex functionals.

In the case $G = \{y: \|y - y_0\| \leqslant a\}$ Theorem 5.2.12 was proved in the Polish edition of this book.

Let Ω be a closed convex cone [1] in Y. We say that an element $y \in Y$ is *nonnegative* and we write $y \geqslant 0$ if $y \in \Omega$. We write $y_1 \geqslant y$ if $y_1 - y \geqslant 0$. We say that y is *nonpositive*, $y \leqslant 0$, if $-y \geqslant 0$. Let $G = -\Omega$ be the set of all nonpositive elements. By using notation introduced above, we can rewrite (5.2.27) in the form

$$\inf\{F(x): BA(x) \leqslant 0\} = \inf\{F(x): -\varphi_0(BA(x)) = 0\}$$
$$= \inf\{F(x): -\varphi_0(BA(x)) \geqslant 0\} = \inf\{F(x): \varphi_0(BA(x)) \leqslant 0\},$$
$$(5.2.34)$$

where the functional φ_0 is chosen so that (5.2.29) and (5.2.30) hold. Since G is a cone,
$$c = \inf\{\varphi_0: y \in G\} = 0$$
and, by (5.2.29),
$$\varphi_0(y) \leqslant 0 \quad \text{for} \quad y \in G = -\Omega.$$
Hence,
$$\varphi_0(y) \geqslant 0 \quad \text{for} \quad y \geqslant 0. \qquad (5.2.35)$$

The functionals satisfying (5.2.35) are called *nonnegative*. Thus, we can reformulate Theorem 5.2.12 as follows.

THEOREM 5.2.12'. *Let a linear system*
$$(X \underset{A}{\to} \square \underset{B}{\to} Y)$$

---

[1] We recall that Ω is a *cone* if $t\Omega \subset \Omega$, for $t > 0$.

be given. Let Ω be a closed convex order generating cone in the space Y and let F be a continuous convex functional defined on X. Assume that either the cone Ω or the set $\Gamma_r = BA(\{x: F(x) \leqslant r\})$, where $r = \inf\{F(x): BA(x) \leqslant 0\}$, has a nonempty interior. If

$$\inf_{x \in X} F(x) < r = \inf\{F(x): BA(x) \leqslant 0\},$$

then there is a nonnegative linear functional $\varphi_0 \in Y^*$ such that

$$\inf\{F(x): BA(x) \leqslant 0\} = \inf\{F(x): \varphi_0(BA(x)) \leqslant 0\}. \quad (5.2.36)$$

I. Singer (1980b) observed that in Theorem 5.2.12′ one can replace linear systems by so called *convex systems*.

We say that an operator C which maps a linear space X into a linear space Y with an order $\leqslant$ defined by a cone Ω is *convex* if

$$C(tx_1 + (1-t)x_2) \leqslant tC(x_1) + (1-t)C(x_2) \quad (5.2.37)$$

for all $x_1, x_2 \in X$ and $0 < t < 1$, i.e.,

$$tC(x_1) + (1-t)C(x_2) \in C(tx_1 + (1-t)x_2) + \Omega. \quad (5.2.38)$$

A convex operator C need not transform convex sets into convex sets. However, we have

PROPOSITION 5.2.13. *Let A be a convex set and let C be a convex operator. Then the set $C(A) + \Omega$ is convex.*

Proof. Let $w_1 = C(x_1) + z_1$, $w_2 = C(x_2) + z_2$, $x_1, x_2 \in A$, $z_1, z_2 \in \Omega$. Then, by (5.2.38),

$$tC(x_1) + (1-t)C(x_2) = C(tx_1 + (1-t)x_2) + z_3, \quad z_3 \in \Omega$$

for all t, $0 < t < 1$, and

$$\begin{aligned} tw_1 + (1-t)w_2 &= t(C(x_1) + z_1) + (1-t)(C(x_2) + z_2) \\ &= tC(x_1) + (1-t)C(x_2) + tz_1 + (1-t)z_2 \\ &= C(tx_1 + (1-t)x_2) + z_3 + tz_1 + (1-t)z_2 \\ &\in C(A) + \Omega. \quad \blacksquare \end{aligned}$$

Using Proposition 5.2.13, we can prove

THEOREM 5.2.14 (Singer, 1980). *Let X, $\square$, Y be Banach spaces and let Y be ordered by a closed convex cone Ω. Let A be a continuous operator which maps X into $\square$ and let B be a continuous operator which maps $\square$ into Y. Assume that BA is a convex operator and F is a continuous convex operator defined on X. Moreover,*

$$\inf_{x \in X} F(x) < r = \inf\{F(x): BA(x) \leqslant 0\}.$$

If either the cone Ω or the set $\Gamma_r = \{x: BA(x) \leqslant r\}$ has a nonempty interior, then there is a nonnegative continuous linear functional $\varphi_0 \in Y^*$, $\varphi_0 \geqslant 0$, such that

$$\inf\{F(x): BA(x) \leqslant 0\} = \inf\{F(x): \varphi_0(BA(x)) \leqslant 0\}.$$

Proof. The set $\Gamma_r + \Omega$ and the cone Ω have no common interior points. Hence, both sets can be separated by a continuous linear functional φ_0. The remainder of the proof is the same as the proof of Theorem 5.2.12. ∎

Problems of similar type with additional conditions on X were considered by Rolewicz (1979), and Phan Quoc Khanh (1985).

5.3. SUFFICIENT CONDITIONS FOR THE EXISTENCE OF AN OPTIMAL INPUT

In the previous section we considered the problem

$$\inf\{F(x, A(x), BA(x)): (x, A(x), BA(x)) \in V\},$$

where $F(x, \xi, y)$ is a convex continuous functional defined on the space $X \times \square \times Y$ and V is a closed convex subset of this space.

The aim of this section is to investigate the problem of existence of an element x_0 which realizes the above infimum, i.e., $(x_0, A(x_0), BA(x_0)) \in V$ and

$$F(x_0, A(x_0), BA(x_0)) =$$
$$= \inf\{F(x, A(x), BA(x)): (x, A(x), BA(x)) \in V\}.$$

Such an element x_0 is called an *optimal element* or an *optimal input*.

THEOREM 5.3.1. *Let V be a closed convex set contained in the product space $X \times \square \times Y$. Let*
$$V_0 = \{(x, \xi, y) \in V : A(x) = \xi, \ B(\xi) = y\}.$$
Then the set V_0 is closed and convex.

Proof. First we prove that the set V_0 is convex. Take (x_1, ξ_1, y_1), $(x_2, \xi_2, y_2) \in V_0$. Clearly, these points belong to V and
$$A(x_i) = \xi_i, \quad B(\xi_i) = y_i \quad \text{for} \quad i = 1, 2.$$
Therefore,
$$A(tx_1 + (1-t)x_2) = t\xi_1 + (1-t)\xi_2 \tag{5.3.1}$$
and
$$B(t\xi_1 + (1-t)\xi_2) = ty_1 + (1-t)y_2. \tag{5.3.2}$$

By the convexity of the set V and by formulae (5.3.1) and (5.3.2) we obtain
$$(tx_1 + (1-t)x_2, t\xi_1 + (1-t)\xi_2, ty_1 + (1-t)y_2) \in V_0,$$
which proves the convexity of the set V_0. To show the closedness of V_0, consider an arbitrary sequence $\{(x_n, A(x_n), BA(x_n))\}$ of elements of the set V_0, which converges to a certain element (x_0, ξ_0, y_0). Since V_0 is a subset of V and V is closed, we have $(x_0, \xi_0, y_0) \in V$. By the continuity of the operators A and B,
$$\xi_0 = \lim_{n \to \infty} A(x_n) = A(x_0), \quad y_0 = \lim_{n \to \infty} BA(x_n) = BA(x_0).$$
Hence, $(x_0, \xi_0, y_0) \in V_0$, which proves the closedness of V_0. ∎

Note that the set V_0 is fully determined by the set: $U_0 = \{x \in X : (x, A(x), BA(x)) \in V_0\}$. The convexity of the set V_0 entails the convexity of the set U_0. Moreover, the closedness and the convexity of a set U entail the closedness and the convexity of the set $\{(x, A(x), BA(x)) : x \in U\}$ (cf. the proof of Theorem 5.3.1). Hence, in the present section we restrict our considerations to the problem of finding the infimum of a functional of a single variable x,
$$\inf_{x \in U_0} F(x, A(x), BA(x)),$$
where F is defined on the input space and U_0 is a closed convex set.

5.3. Existence of optimal input

THEOREM 5.3.2. *Let $F(x, \xi, y)$ be a continuous convex functional defined on the product space $X \times \square \times Y$. Then the functional $F(x, A(x), BA(x))$ is a continuous convex functional defined on the input space X.*

Proof. Since the operators A and B and the performance functional F are continuous, the functional $F(x, A(x), BA(x))$ is also continuous as the superposition of continuous functions. The convexity of $F(x, A(x), BA(x))$ follows immediately, since

$$F(tx+(1-t)x', A(tx+(1-t)x'), BA(tx+(1-t)x'))$$
$$= F(tx+(1-t)x', tA(x)+(1-t)A(x'), tBA(x)+$$
$$+(1-t)BA(x'))$$
$$\leqslant tF(x, A(x), BA(x))+(1-t)F(x', A(x'), BA(x'))$$

for all t, $0 < t < 1$. ∎

Theorem 5.3.2 shows that the above problem of existence of an optimal input can be reduced to the investigation of functionals defined on the input space X.

THEOREM 5.3.3. *Let X be a Banach space and let G be a continuous convex functional defined on X. Then the functional G is lower semicontinuous in the weak topology, i.e., for every $x_0 \in X$ and an arbitrary positive number ε there exists a neighbourhood W of x_0 in the weak topology such that $\inf_{x \in W} G(x) \geqslant G(x_0) - \varepsilon$.*

Proof. Take an arbitrary $\varepsilon > 0$. Write $b = G(x_0) - \varepsilon$. Let $K_b = \{x \in X: G(x) \leqslant b\}$. By the continuity and the convexity of the functional G, the set K_b is closed and convex. Clearly, $x_0 \notin K_b$. Hence, there are a continuous linear functional f, a constant c, and a number $\delta > 0$ such that

$$f(x_0) = c$$

and

$$f(x) < c - \delta \quad \text{for} \quad x \in K_b. \tag{5.3.3}$$

Let

$$W = \{x \in X: |f(x) - f(x_0)| < \delta\}.$$

The set W is a neighbourhood of x_0 in the weak topology and, by formula (5.3.3), W is disjoint with K_b. This implies that
$$\inf_{x \in W} G(x) \geq b = G(x_0) - \varepsilon. \blacksquare$$

THEOREM 5.3.4. *If the input space X is reflexive and the set U_0 is closed and bounded, then there exists an optimal input.*

Proof. The set U_0 is weakly compact since it is convex and bounded. Write
$$U_b = U_0 \cap \{x \in X: F(x, A(x), BA(x)) \leq b\}, \quad (5.3.4)$$
where b is an arbitrary number greater than α, $\alpha = \inf_{x \in U_0} F(x, A(x), BA(x))$. The sets U_b are convex, closed and bounded. Hence, the sets U_b are compact in the weak topology. Moreover, the family of sets U_b is ordered by inclusion, i.e., $b_1 < b_2$ entails $U_{b_1} \subset U_{b_2}$. This implies, by the compactness of U_b, that
$$\tilde{U}_\alpha = \bigcap_{b > \alpha} U_b$$
is nonempty.

Let $x_0 \in \tilde{U}_\alpha$. Then $x_0 \in U_b$ for all $b > \alpha$ and, by the definition of U_b, $F(x_0, A(x_0), BA(x_0)) \leq b$. Thus, $F(x_0, A(x_0), BA(x_0)) = \alpha$. $\blacksquare$

THEOREM 5.3.5. *If the input space X is reflexive and the performance functional $F(x, A(x), BA(x))$ has the property that the set U_b defined by formula (5.3.4) is bounded for some b, then there exists an optimal input.*

The proof is similar to that of Theorem 5.3.4.

If we consider topologies other than weak topologies, then some additional requirements should be imposed on the set U_0 and the performance functional.

THEOREM 5.3.6. *Let X be a Banach space and let G be a convex functional defined on X. Assume that a certain Γ-topology τ is given in X such that for every b the sets*
$$K_b = \{x \in X: G(x) \leq b\}$$

are closed in that topology. Then the functional G is lower semicontinuous in the topology τ.

The proof proceeds in exactly the same way as the proof of Theorem 5.3.3 with only one difference. Namely, since K_b is closed in the topology τ, a functional f separating x_0 and K_b can be chosen so as to be continuous in the topology τ. Hence, f belongs to Γ.

THEOREM 5.3.7. *Let a certain topology τ be given in the input space X such that all the sets U_b defined by formula (5.3.4) are compact (sequentially compact) in this topology. Then there exists an optimal input.*

Proof. Let
$$\tilde{U}_\alpha = \bigcap_{n=1}^{\infty} U_{(\alpha+1/n)}.$$
The set $\tilde{U}_\alpha$ is nonempty. The remainder of the proof is exactly the same as the proof of Theorem 5.3.4. ∎

EXAMPLE 5.3.1. Let X be the conjugate space of a Banach space (or a seminormed complete space) X_0. Let the performance functional F be defined as the norm (or the seminorm) $||x||$ of functionals defined on the space X_0, $F(x) = ||x||$. Let U_0 be an arbitrary closed set in the weak-∗-topology. Then there exists an optimal input. In fact, the sets
$$K_b = \{x \in X: ||x|| \leq b\}$$
are all compact and sequentially compact in the weak-∗-topology. Hence, the sets
$$U_b = U_0 \cap K_b$$
are compact and sequentially compact in the weak-∗-topology. Thus, by Theorem 5.3.7, there exists an optimal input.

The set U_0 appearing in Example 5.3.1 frequently takes the form
$$U_\Xi = \{x \in X: x(\xi) = \alpha_\xi, \; \xi \in \Xi \subset X_0\}.$$
Sometimes the set U_0 is of the form
$$U_0 = \{x \in X: BA(x) = y_0\},$$
where y_0 is a fixed element of the output space Y.

Theorem 5.3.8. *Let $y_0 \in BA(X)$ and*
$$\alpha = \inf\{F(x): BA(x) = y_0\}.$$
Assume that the set $\Gamma_\alpha = BA(\{x \in X: F(x) \leqslant \alpha\})$ is closed. If there exists an element x_0 such that $F(x_0) \leqslant \alpha$, then there exists an x_1, $x_1 \in X$, such that $F(x_1) = \alpha$ and $BA(x_1) = y_0$.

Proof. By hypothesis, the set
$$K_\alpha = \{x \in X: F(x) \leqslant \alpha\}$$
is nonempty. Hence, the set
$$\Gamma_\alpha = BA(K_\alpha)$$
is also nonempty. Let x_0 be an arbitrary element of the set K_α. Without loss of generality we can assume additionally that $\alpha \geqslant 0$ and $F(0) = 0$. If it is not so, the functional F can be replaced by the functional F', $F'(x) = F(x - x_0) - F(x_0)$, and y_0 by $y_0' = y_0 - BA(x_0)$. Clearly, under these additional assumptions, $0 \in K_\alpha$.

We claim that $y_0 \in \overline{\Gamma}_\alpha$. Suppose on the contrary that $y_0 \notin \overline{\Gamma}_\alpha$. Let
$$c = \inf\left\{t > 0: \frac{y_0}{t} \in \Gamma_\alpha\right\}.$$
Since $y_0 \notin \Gamma_\alpha$, we have $c > 1$. Let d be an arbitrary number such that $1 < d < c$. By the convexity of the functional F,
$$F(x) \leqslant \frac{1}{d} F(dx) + \left(1 - \frac{1}{d}\right) F(0)$$
and
$$F(dx) \geqslant d F(x),$$
since $F(0) = 0$. Consequently,
$$K_{d\alpha} \subset d K_\alpha$$
and
$$\Gamma_{d\alpha} \subset d \Gamma_\alpha.$$
By the definition of the number d, $y_0 \notin \overline{d\Gamma_\alpha}$ and hence, $y_0 \notin \Gamma_{d\alpha}$. By the definition of $\Gamma_{d\alpha}$ we get
$$\inf\{F(x): BA(x) = y_0\} \geqslant d\alpha > \alpha$$

contrary to the definition of the number α. By the closedness of the set Γ_α, $y_0 \in \Gamma_\alpha$, i.e., there exists an x_1 such that $F(x_1) = \alpha$ and $BA(x_1) = y_0$. ∎

THEOREM 5.3.9. *Let F be a continuous convex functional defined on the input space X and, for every r, let the sets*

$$\Gamma_r = BA(\{x \in X : F(x) \leqslant r\})$$

be closed. Assume that there exist a functional $\varphi_0 \in Y^*$ such that

$$\inf \{F(x) : x \in X, BA(x) = y_0\}$$
$$= \inf \{F(x) : x \in X, \varphi_0(BA(x)) = \varphi_0(y_0)\} \quad (5.3.5)$$

and exactly one element x_0 such that

$$F(x_0) = \inf \{F(x) : x \in X, \varphi_0(BA(x)) = \varphi_0(y_0)\}. \quad (5.3.6)$$

Then

$$BA(x_0) = y_0. \quad (5.3.7)$$

Proof. As before, we put

$$\alpha = \inf \{F(x) : BA(x) = y_0\}.$$

By the closedness of the sets Γ_r, the set Γ_α is closed. Hence, by Theorem 5.3.8, there is an x' such that $F(x') \leqslant \alpha$ and $BA(x') = y_0$. By the definition of α, $F(x') = \alpha$. Moreover, $\varphi_0(BA(x')) = \varphi_0(y_0)$. But, by hypothesis, x_0 is the only element satisfying this equality and $F(x_0) = \alpha$. Hence, $x' = x_0$. ∎

If Γ_α is not closed, Theorem 5.3.8 need not be valid, as is shown by the following example.

EXAMPLE 5.3.2. Let $X = \square$ be the subspace of the space $L^1[0, 1]$ consisting of all functions which are constant on the interval $(1/4, 3/4)$. Let A be the identity operator and $Y = R^2$. Let B be given by the formula

$$B(x) = \{F_1(x), F_2(x)\},$$

where $F_i(x)$, $i = 1, 2$, are linear functionals of the form

$$F_i(x) = \int_0^1 f_i(t) x(t) \, dt, \quad i = 1, 2,$$

$$f_1(t) = \begin{cases} 4t & \text{for} \quad 0 \leq t < 1/4, \\ 1/2 & \text{for} \quad 1/4 \leq t < 3/4, \\ 0 & \text{for} \quad 3/4 \leq t \leq 1 \end{cases}$$

and

$$f_2(t) = f_1(1-t).$$

Let $F(x) = ||x||$, where $||x||$ denotes the norm in $L^1[0, 1]$. Consider the set

$$\Gamma_1 = BA(\{x: F(x) \leq 1\}).$$

Observe that

$$|F_1(x)| + |F_2(x)| = \int_0^{1/4} 4t |x(t)| \, dt + 2 \int_{1/4}^{3/4} \frac{1}{2} |x(t)| \, dt +$$

$$+ \int_{3/4}^1 4(1-t) |x(t)| \, dt.$$

This implies that $|F_1(x)| + |F_2(x)| \leq 1$. Moreover, the inequality is strong if $x(t)$ differs from zero on the set $(0, 1/4) \cup (3/4, 1)$ and it becomes an equality if the interval $(1/4, 3/4)$ is the support of $x(t)$. Hence,

$$\Gamma_1 = \{(x, y) \in R^2: |x| + |y| < 1\} \cup (\pm 1/2, \pm 1/2).$$

Let

$$y_0 = (1/3, 2/3).$$

Then the functional φ_0 satisfying equality (5.3.5) is of the form

$$\varphi_0(x, y) = x + y.$$

With the above φ_0 equation (5.3.6) has the unique solution $(1/2, 1/2) \neq (1/3, 2/3)$.

5.4. Minimum-time problem

So far we have considered problems with a fixed output operator. In the present section we deal with linear systems in which there is a family of output operators

$$(X \underset{A}{\to} \square \underset{B_t}{\to} Y),$$

where the output operator B_t depends on the real argument $t, 0 \leqslant t \leqslant \mathscr{T}$, which is traditionally called *time*.

We have an example of such a system in a system governed by ordinary differential equations or differential equations with a delay (see Examples 5.1.1 and 5.1.2) without control in which the input is defined by an initial value and the output is defined as the position of an object at the moment t (or as the part of the trajectory corresponding to the time interval $(t-h_m, t)$).

The minimum-time problem can be formulated as follows. Suppose we are given a certain convex closed set U in the space $X \times \square$. Let $Y(t)$ be a multifunction, i.e., a mapping assigning to each $t, 0 \leqslant t \leqslant \mathscr{T}$, a subset $Y(t) \subset Y$. We shall assume that $Y(t)$ is a closed-valued multifunction, i.e., the sets $Y(t)$ are closed for all t. Now the minimum-time problem is to find

$$T = \inf\{t: \text{ there exists an } x \in X \text{ such that } (x, A(x)) \in U$$
$$\text{ and } B_t A(x) \in Y(t)\}$$

and to verify the existence of an element $x \in X$ such that $(x, A(x)) \in U$ and $B_T A(x) \in Y(T)$.

THEOREM 5.4.1. *Let* $U_0 = \{x \in X: (x, A(x)) \in U\}$. *Then the set* U_0 *is convex.*

The proof proceeds in the same way as the proof of Theorem 5.3.1.

In the sequel we shall consider the sets U_0 which are closed.

THEOREM 5.4.2. *Let the input space X be reflexive and let $U \subset X$ be a closed convex bounded set. Let B_t be a family of continuous linear operators, continuous with respect to t in the topology generated by the operator norm. Assume that a closed-valued multifunction $Y(t)$ is linearly upper semicontinuous and there is a t such that $B_t A(U) \cap Y(t) \neq \emptyset$. Then there exists an x such that $B_T A(x) \in Y(T)$, where, as before,*

$$T = \inf\{t: B_t A(U) \cap Y(t) \neq \emptyset\}.$$

Proof. Let δ be an arbitrary positive number. Let
$$V_\delta = \overline{\operatorname{conv} \bigcup_{T < t \leqslant T+\delta} (B_t A)^{-1}(Y(t)) \cap U}.$$
By the definition of T, the sets V_δ are nonempty. Since V_δ are closed convex subsets of the bounded set U, all V_δ are weakly compact. The family V_δ is ordered by inclusion. Hence,
$$V_0 = \bigcap_{\delta > 0} V_\delta$$
is nonempty. We shall prove that $B_T A(V_0) \subset Y(T)$. Choose any $\varepsilon > 0$. By the continuity of the family B_t with respect to t in the operator norm topology and by the linear upper semicontinuity of the multifunction $Y(t)$, there exists a $\delta > 0$ such that, for all t, $T \leqslant t \leqslant T+\delta$,
$$\|(B_t - B_T) A(v)\| < \varepsilon \quad \text{for all } v \in U \tag{5.4.1}$$
and
$$Y(t) \subset Y(T) + \varepsilon K, \tag{5.4.2}$$
where K denotes the unit ball in Y, $K = \{y : \|y\| < 1\}$. Let x be an arbitrary element of the set V_0. By the definition of the sets V_δ, there exist a system of indices $t_1, \ldots, t_n$, $T \leqslant t_i \leqslant T+\delta$, $i = 1, \ldots, n$, a system of nonnegative numbers $a_1, \ldots, a_n$, $a_1 + \ldots + a_n = 1$, and a system of elements x_i, $i = 1, \ldots, n$, $x_i \in (B_{t_i} A)^{-1}(Y(t_i)) \cap U$, such that
$$\|x - (a_1 x_1 + \ldots + a_n x_n)\| < \varepsilon. \tag{5.4.3}$$
Since $B_{t_i} A(x_i) \in Y(t_i)$, we have, by (5.4.1) and (5.4.2),
$$B_T A(x_i) \in Y(t_i) + \varepsilon K \subset Y(T) + 2\varepsilon K.$$
Hence, by the convexity of the set $Y(T)$,
$$B_T A(a_1 x_1 + \ldots + a_n x_n) \subset Y(T) + 2\varepsilon K,$$
and according to formula (5.4.3) we conclude that
$$B_T A(x) \in Y(T) + (2 + \|B_T\| \|A\|) \cdot \varepsilon K.$$
Since ε is arbitrary and the set $Y(T)$ is closed, $B_T A(x) \in Y(T)$. ∎

THEOREM 5.4.3. *Suppose that the output space Y is reflexive and the family B_t is continuous with respect to t in the operator norm topology. Moreover, suppose that the multifunction $Y(t)$ is closed-valued and linearly upper* semicontinuous. If there exist a t and an x, $x \in U_0$, such that $B_t A(x) \in Y(t)$ and $B_T A(U_0)$ is closed and bounded, then there exists a v, $v \in U_0$, such that $B_T A(v) \in Y(T)$.

Proof. Choose any positive ε. By the continuity of the family B_t in the operator norm topology and by the linear upper semicontinuity of the multifunction $Y(t)$, there exists a $\delta > 0$ such that for all t, $|t-T| < \delta$, we have

$$B_t A(U_0) \subset B_T A(U_0) + \varepsilon \bar{K}, \tag{5.4.4}$$

where $\bar{K} = \{y \in Y: \|y\| \leqslant 1\}$ and

$$Y(t) \subset Y(T) + \varepsilon \bar{K}. \tag{5.4.5}$$

Write

$$W = B_T A(U_0) + \varepsilon \bar{K}.$$

The set W is convex and bounded. Let

$$V_\alpha = \overline{\mathrm{conv} \bigcup_{T < t \leqslant T+\alpha} [Y(t) \cap B_t A(U_0)]} \cap W.$$

The sets V_α are convex, bounded and closed. Hence, the sets V_α are also weakly compact since Y is reflexive. The family of sets V_α is ordered by inclusion and hence the set

$$V_0 = \bigcap_{\alpha > 0} V_\alpha$$

is nonempty. We shall prove that

$$V_0 = Y(T) \cap B_T A(U_0). \tag{5.4.6}$$

Let x be an arbitrary element of the set V_0. Take the same value of δ as that chosen at the beginning of the proof. Clearly $V_0 \subset V_\delta$ and, by the definition of the set V_0, for every $x \in V_0$ there exist a system of indices $t_1, \ldots, t_n$, $0 < t_i \leqslant \delta$, $i = 1, \ldots, n$, a system of nonnegative numbers $a_1, \ldots, a_n$, $a_1 + \ldots + a_n = 1$, and a system of elements y_i, $i = 1, \ldots, n$, satisfying

$$y_i \in Y(t_i) \cap B_{t_i} A(U_0), \tag{5.4.7}$$

and such that

$$\|x - (a_1 y_1 + \ldots + a_n y_n)\| < \varepsilon.$$

According to formulae (5.4.4) and (5.4.5) we have
$$y_i \in \big(Y(T)+\varepsilon \bar{K}\big) \cap \big(B_T A(U_0)+\varepsilon \bar{K}\big), \quad i=1,\ldots,n. \qquad (5.4.8)$$
The set $B_T A(U_0)+K$ is convex, whence
$$(a_1 y_1 + \ldots + a_n y_n) \in \big(Y(T)+\varepsilon \bar{K}\big) \cap \big(B_T A(U_0)+\varepsilon \bar{K}\big) \qquad (5.4.9)$$
and consequently
$$x \in \big(Y(T)+\varepsilon \bar{K}\big) \cap \big(B_T A(U_0)+\varepsilon \bar{K}\big)+\varepsilon \bar{K}. \qquad (5.4.10)$$
Since ε is arbitrary and the sets $Y(t)$, $B_T A(U_0)$ are closed, we obtain formula (5.4.6). This in turn directly implies the assertion of the theorem. ∎

Without assumptions which guarantee the compactness of the set $B_T A(U)$ in the corresponding topologies, Theorems 5.4.3 and 5.4.2 are not valid, even if the family B_t is continuous in the operator norm topology. This can be visualized by the following examples.

EXAMPLE 5.4.1. Let $X = L^1[0,1]$ and $U_0 = \{x \in X: \|x\| \leqslant 1\}$. Clearly, the set U_0 is convex and bounded. Let $\square = X$ and $A = I$. Let Y be a one-dimensional space. Then the family of operators B_t is a family of functionals. Let
$$B_t(x) = \int_0^1 b_t(s) x(s) \, ds,$$
where, for all t, $0 < t < 1$,
$$b_t(s) = \begin{cases} s & \text{for } 0 < s < 1-t, \\ 1 & \text{for } 1-t \leqslant s \leqslant 1 \end{cases}$$
and $b_0(s) = s$. As can easily be verified, $b_t(s)$ tends uniformly to $b_0(s)$ as $t \to 0$. Let $Y(t)$ consist of a single number 1 for all t, $0 \leqslant t < 1$.

Observe that $B_t A(U_0) = [-1,1]$ for all t, $0 < t < 1$. Hence, there exists an $x \in U_0$ such that $B_t A(x) \in Y(t)$ and, by the definition of T, $T = 0$. On the other hand, $B_0 A(U_0) = (-1,1)$ and
$$B_0 A(U_0) \cap Y(0) = \emptyset.$$

In Example 5.4.1 the set $B_T A(U_0)$ is not closed. Now, we give a similar example in which the set $B_T A(U_0)$ is closed and bounded. Clearly, in this example the space Y cannot be a reflexive space (see Theorem 5.4.3).

5.4. Minimum-time problem

EXAMPLE 5.4.2. Let X, $\Box$ and A be as in Example 5.4.1. Let $Y = R \times X$ be the product space of the one-dimensional space R and the space X. Define a family of operators $\tilde{B}_t$ as

$$\tilde{B}_t(x) = (x, B_t(x)),$$

where the operators B_t are defined as in Example 5.4.1.

By the continuity of the functionals B_t with respect to t in the operator norm topology we immediately obtain the continuity of the operators $\tilde{B}_t$ with respect to t in the operator norm topology. Moreover, $\tilde{B}_0 A(U_0)$ is closed. To see this, take an arbitrary sequence $\{(B_0 A(x_n), x_n)\}$ of elements of the set $\tilde{B}_0 A(U_0)$ which converges to an element (r, x_0). Then $x_n \to x_0$. By the closedness of the set U_0 we have $x_0 \in U_0$. Since the functionals B_0, A are continuous, $r = B_0 A(x_0)$. Consequently, $(r, x_0) \in \tilde{B}_0 A(U_0)$.

Let

$$Y(t) = \{(1, x) \colon x \in U_0\}$$

for all t, $0 \leq t < 1$. Then we obtain, as in the previous example,

$$\tilde{B}_t A(U_0) \cap Y(t) \neq \emptyset \quad \text{and} \quad \tilde{B}_0 A(U_0) \cap Y(0) = \emptyset$$

for all t, $0 < t < 1$.

Now we shall show that if $Y(t)$ is a closed-valued linearly upper semicontinuous multifunction then for every $\delta_0 > 0$ such that $\mathcal{T} - t \geq \delta_0$ we have

$$Y(t) = \bigcap_{0 < \delta < \delta_0} \bigcup_{t \leq \tau \leq t+\delta} Y(\tau). \tag{5.4.11}$$

The inclusion

$$Y(t) \subset \bigcap_{0 < \delta < \delta_0} \bigcup_{t \leq \tau \leq t+\delta} Y(\tau)$$

is evident. Now we shall prove the opposite inclusion. Take an arbitrary neighbourhood V of zero. Since the multifunction $Y(t)$ is linearly upper semicontinuous, there exists a δ, $0 < \delta < \delta_0$, such that for all τ, $t \leq \tau \leq t+\delta$,

$$Y(\tau) \subset Y(t) + V. \tag{5.4.12}$$

According to (5.4.12) we obtain

$$\bigcap_{0 < \delta < \delta_0} \bigcup_{t \leq \tau \leq t+\delta} Y(\tau) \subset Y(t) + V.$$

Since V is arbitrary and all $Y(t)$ are closed, we obtain the inclusion and formula (5.4.11).

THEOREM 5.4.4. *Let σ be a topology in the output space Y such that Y equipped with the topology σ is a linear topological space. Moreover, let multifunctions $Y(t)$ and $B_t A(U_0)$ be linearly upper semicontinuous in this topology. If the sets*
$$Y_\delta = \bigcup_{T < t \leqslant T+\delta} Y(t)$$
are closed in the topology σ for all δ, $0 < \delta < \delta_0$ and the sets
$$W_\delta = \bigcup_{T \leqslant t \leqslant T+\delta} B_t A(U_0)$$
are compact (sequentially compact) in the topology σ for all δ, $0 < \delta < \delta_0$, then
$$B_T A(U_0) \cap Y(T) \neq \emptyset$$
where, as before,
$$T = \inf\{t:\ B_t A(U_0) \cap Y(t) \neq \emptyset\}.$$

Proof. The sets
$$V_\delta = W_\delta \cap Y_\delta$$
are compact (sequentially compact) in the topology σ. This implies that the set
$$V_0 = \bigcap_{n=1}^\infty V_{1/n}$$
is nonempty as the intersection of a descending sequence of compact (sequentially compact) sets. The multifunctions $B_t A(U_0)$ and $Y(t)$ are both linearly upper semicontinuous and closed-valued in the topology σ. Hence,
$$\bigcap_{n=1}^\infty W_{1/n} = B_T A(U_0)$$
and
$$\bigcap_{n=1}^\infty Y_{1/n} = Y(T).$$

Finally,
$$V_0 = (\bigcap_{n=1}^{\infty} W_{1/n}) \cap (\bigcap_{n=1}^{\infty} Y_{1/n}) = B_T A(U_0) \cap Y(T). \blacksquare$$

THEOREM 5.4.5. *Theorem 5.4.4 remains true if the compactness (sequential compactness) of the sets W_δ is replaced by the closedness of the sets W_δ, and the sets Y_δ are assumed to be compact (sequentially compact).*

Proof. The proof proceeds in the same way as the proof of Theorem 5.4.4. ∎

Now we formulate various conditions ensuring the closedness and the compactness of the sets Y_δ.

THEOREM 5.4.6. *Let Y be a linear topological space and let $Y(t)$ be a family of ascending closed (compact, sequentially compact) sets. Then the sets*
$$Y_\delta = \bigcup_{T \leq t \leq T+\delta} Y(t)$$
are closed (compact, sequentially compact) for all δ.

Proof. By the assumption, the family $Y(t)$ is ascending and hence
$$Y_\delta = Y(T+\delta)$$
is evidently closed (compact, sequentially compact). ∎

THEOREM 5.4.7. *Let Y be a linear topological space. If $Y(t)$ is a closed-valued linearly upper semicontinuous multifunction, then the sets*
$$Y_{\delta, T} = \bigcup_{T \leq t \leq T+\delta} Y(t)$$
are closed for all T, $0 \leq T < \mathcal{T}$, and all positive δ such that $T+\delta \leq \mathcal{T}$.

Proof. Let $y \in \overline{Y_{\delta, T}}$. This means that for every neighbourhood U of zero there exists a t such that
$$T \leq t \leq T+\delta, \quad y \in Y(t)+U. \qquad (5.4.13)$$

Denote by E_U the closure of the set of all t satisfying formula (5.4.13). Clearly, E_U is compact. If a neighbourhood W is contained in the intersection of neighbourhoods U and V, $W \subset V \cap U$, then $E_W \subset E_V \cap E_U$. Hence, for any finite number of neighbourhoods $U_1, \ldots, U_n$, the intersection $E_{U_1} \cap E_{U_2} \cap \ldots \cap E_{U_n}$ contains $E_{U_1 \cap U_2 \cap \ldots \cap U_n}$ which is nonempty. Let U_α be a basis of neighbourhoods of zero. Let

$$E_0 = \bigcap_\alpha E_{U_\alpha}.$$

Since the sets E_U are compact and the intersection of every finite number of these sets is nonempty, the set E_0 is nonempty. Let t_0 be an arbitrary element of E_0. Let U be an arbitrary neighbourhood of zero. By the linear upper semicontinuity of the multifunction $Y(t)$, there exists a $\delta' > 0$ such that, for all t, $|t-t_0| < \delta'$, we have

$$Y(t) \subset Y(t_0) + U.$$

Next, by the definition of the set E_0, $t_0 \in E_U$ and hence there exists a t arbitrarily close to t_0 (in particular, $|t-t_0| < \delta'$) such that (5.4.13) holds. This implies that

$$y \in Y(t_0) + U + U.$$

Since U is arbitrary and $Y(t_0)$ is closed,

$$y \in Y(t_0) \subset Y. \blacksquare$$

COROLLARY 5.4.8. *Let Y be a Banach space and let $Y(t)$ be a multifunction which is closed-valued in the weak (weak-∗-) topology and linearly upper semicontinuous in the norm topology. Then the sets*

$$Y_{\delta, T} = \bigcup_{T \leqslant t \leqslant T+\delta} Y(t)$$

are closed in the weak (weak-∗-) topology.

Proof. Linear upper semicontinuity in the norm topology entails linear upper semicontinuity in the weak topology and in the weak-∗-topology. Now the corollary follows immediately from Theorem 5.4.7. ∎

THEOREM 5.4.9. *Let Y be a linear topological space and let $Y(t)$, $0 \leqslant t \leqslant \mathcal{T}$, be a compact-valued linearly upper semicontinuous multifunction. Then*

5.4. Minimum-time problem

the set
$$Y_{\delta,T} = \bigcup_{T \leq t \leq T+\delta} Y(t)$$
is compact for each T, $0 \leq T < \mathscr{T}$, and every positive $\delta \leq \mathscr{T} - T$.

Proof. Take an arbitrary covering W_α of the set $Y_{\delta,T}$. Denote by W_α^t the family of those neighbourhoods of W_α which have common points with $Y(t)$. Clearly, W_α^t is a covering of the set $Y(t)$. Since, by assumption, $Y(t)$ are compact, the covering W_α^t contains a finite covering $W_1^t, \ldots, W_{n_t}^t$. Write
$$V^t = \bigcup_{i=1}^{n_t} W_i^t.$$
The set V^t is an open set containing $Y(t)$. Hence, for every $y \in Y(t)$, there exists a neighbourhood U_y of zero such that $y + U_y \subset V^t$. Let M_y be a neighbourhood of zero such that $M_y + M_y \subset U_y$. Note that for every $y' \in y + M_y$
$$y' + M_y \subset y + M_y + M_y \subset y + U \subset V^t. \tag{5.4.14}$$

The sets $y + M_y$ form a covering of the set $Y(t)$. Hence, this covering contains a finite covering, say $y_1 + M_{y_1}, \ldots, y_m + M_{y_m}$. Let $M = \bigcap_{i=1}^{m} M_{y_i}$. Let y_0 be an arbitrary element of the set $Y(t)$. Then there exists an index i such that $y_0 \in y_i + M_{y_i}$. Hence, by (5.4.14), we get
$$y_0 + M \subset y_i + M_{y_i} + M \subset y_i + M_{y_i} + M_{y_i} \subset V^t. \tag{5.4.15}$$
Since $Y(t)$ is linearly upper semicontinuous, formula (5.4.15) implies that there exists a $\delta_t > 0$ such that for $t' \in (t - \delta_t, t + \delta_t) \cap [0, \mathscr{T}]$ we have
$$Y(t') \subset Y(t) + M \subset V^t. \tag{5.4.16}$$
The sets
$$(t - \delta_t, t + \delta_t) \cap [0, \mathscr{T}] \tag{5.4.17}$$
form a covering of the interval $[0, \mathscr{T}]$ which is compact. Hence, we can choose a finite covering which consists of the sets of the form (5.4.17) corresponding to some t_n. Then, by (5.4.16),
$$Y_{\delta,T} \subset \bigcup_{n=1}^{k} V^{t_n} \subset \bigcup_{n=1}^{k} \bigcup_{i=1}^{n_t} W_i^{t_n}, \tag{5.4.18}$$
which completes the proof. ∎

COROLLARY 5.4.10. *Let Y be a linear metric space. Let $Y(t)$ be a linearly upper semicontinuous multifunction. If all the sets $Y(t)$ are sequentially compact, then the sets*

$$Y_{\delta,T} = \bigcup_{T \leq t \leq T+\delta} Y(t)$$

are sequentially compact.

Proof. This is an immediate consequence of Theorem 5.4.9 and the fact that in metric spaces sequential compactness and compactness are equivalent. ∎

In Theorem 5.4.4 we required the multifunction $B_t A(U_0)$ to be linearly upper semicontinuous in a given topology. Now we formulate sufficient conditions for linear upper semicontinuity in some particular cases.

THEOREM 5.4.11. *Let X and Y be Banach spaces. Let U_0 be a bounded set contained in the space X. Let C_t be a family of continuous linear operators which map X into Y. Suppose that for every $f \in Y^*$ the function $C_t^* f$ (where C_t^* denotes the adjoint operator of C_t) defined on the interval $[0, \mathcal{T}]$, with values in X^*, is continuous at t_0 (on the whole interval $[0, \mathcal{T}]$). Then the multifunction $C_t(U_0)$ is linearly continuous at t_0 (on the whole interval $[0, \mathcal{T}]$) in the weak topology.*

Proof. Let
$$M = \sup\{\|x\|: x \in U_0\}.$$

Let V be an arbitrary neighbourhood of zero in the weak topology of the form
$$V = \{x \in Y: |f_i(x)| < \varepsilon, \ i = 1, 2, \ldots, n\}.$$

By assumption, there exists a number $\delta > 0$ such that for all t, $|t - t_0| < \delta$,

$$\|C_t^* f_i - C_{t_0}^* f_i\| < \frac{\varepsilon}{2M}, \quad i = 1, 2, \ldots, n. \tag{5.4.19}$$

Now, take an arbitrary element $x \in U_0$. It follows from (5.4.19) that

$$|f_i(C_t(x) - C_{t_0}(x))| < \varepsilon, \quad i = 1, 2, \ldots, n. \tag{5.4.20}$$

5.4. Minimum-time problem

Hence

$$C_t(x) \in C_{t_0}(x) + V \qquad (5.4.21)$$

and

$$C_{t_0}(x) \in C_t(x) + V. \qquad (5.4.22)$$

Since x is an arbitrary element of U_0, formulae (5.4.21) and (5.4.22) imply the linear continuity of the multifunction $C_t(U_0)$ at t_0.

If C_t^*f is a function which is continuous on the whole interval $[0, \mathcal{T}]$, then, by the above arguments, the multifunction $C_t(U_0)$ is linearly continuous on the whole interval $[0, \mathcal{T}]$. ∎

THEOREM 5.4.12. *Let X and Y be Banach spaces. Let C_t be a family of continuous linear operators which map the space X into the space Y. Assume that for every $x \in X$ the function $C_t(x)$ is continuous at t_0 (on the whole interval $[0, \mathcal{T}]$). Let U_0 be a bounded subset of the space Y^* which is closed in the weak-*-topology. Then the multifunction $C_t(U_0)$ is linearly continuous at t_0 (on the whole interval $[0, \mathcal{T}]$) in the weak-*-topology.*

Proof. By Corollary 4.4.7, the set U_0 is compact in the weak-*-topology. In virtue of Theorem 4.4.2, the operators C_t are continuous in the weak-*-topology. Hence, the sets $C_t(U_0)$ are compact in the weak-*-topology.

Let V be a neighbourhood of zero in the weak-*-topology of the form

$$V = \{f \in X^* : |f(x_i)| < \varepsilon, \; i = 1, \ldots, n, \; x_i \in X\}.$$

By the continuity of the functions $C_t(x)$ at t_0, there exists a $\delta > 0$ such that for all t, $|t - t_0| < \delta$, we have

$$\|C_t(x_i) - C_{t_0}(x_i)\| < \frac{\varepsilon}{M}; \quad i = 1, \ldots, n, \qquad (5.4.23)$$

where

$$M = \sup\{\|f\| : f \in U_0\}. \qquad (5.4.24)$$

It follows from (5.4.23) that for all $\varphi \in U_0$

$$|\varphi(C_t(x_i) - C_{t_0}(x_i))| < \varepsilon, \quad i = 1, \ldots, n. \qquad (5.4.25)$$

Formula (5.4.25) implies that

$$C_t^*(\varphi) \in C_{t_0}^*(\varphi) + V \qquad (5.4.26)$$

and

$$C_{t_0}^*(\varphi) \in C_t^*(\varphi) + V. \qquad (5.4.27)$$

Since V is arbitrary, formulae (5.4.26) and (5.4.27) imply that the multifunction $C_t(U_0)$ is linearly continuous at t_0 in the weak-$*$-topology.

The linear continuity of the multifunction $C_t(U_0)$ on the whole interval $[0, \mathcal{T}]$ is a consequence of the linear continuity of $C_t(U_0)$ at each point of the interval $[0, \mathcal{T}]$. But, as we have shown, the linear continuity at a given point follows from the continuity of $C_t(x)$ at this point. ∎

The continuity of the function $C_t(x)$ for every $x \in X$ need not imply the continuity of $C_t^*(f)$, as is shown by the example below.

EXAMPLE 5.4.3. Let $X = C[0, 1]$ and $Y = R$. Let C_t be a family of functionals which assign to a function $x(\cdot)$ a number $x(t)$. Clearly, each C_t maps X into R and $C_t(x) = x(t)$ is a continuous function.

The adjoint operators C_t^* map R into X^*. In particular, the operator C_t^* assigns to the number 1 a functional defined as follows:

$$C_t^*(1)(x(t)) = x(t).$$

The family of functionals $C_t^*(1)$ is not continuous at any point. Moreover, if K denotes the closed unit ball in X, then

$$C_t(K) = [-1, 1]$$

and $C_t(K)$ does not depend on t. Hence, the multifunction $C_t(K)$ is continuous in the norm topology.

This example illustrates also the fact that the condition given in Theorem 5.4.11 is sufficient but not necessary.

Example 5.4.3 leads to the following theorem.

THEOREM 5.4.13. *For every separable Banach space X there exists a family of continuous linear functionals C_t which map X into the real line R such that $C_t(x)$ is a continuous function for every $x \in X$ and $C_t^*(1)$ is not continuous.*

5.4. Minimum-time problem

Proof. Since X is a separable space, there is a sequence of functionals $\{f_n\}$ such that $\|f_n\| = 1$ and $\{f_n\}$ converges to zero in the weak-$*$-topology.
Define a family of functionals g_t as follows.

$$g_t = \begin{cases} 2^n\left(\dfrac{1}{2^{n-1}} - t\right)f_{n-1} + 2^n\left(t - \dfrac{1}{2^n}\right)f_n & \text{for } \dfrac{1}{2^n} \leq t \leq \dfrac{1}{2^{n-1}}, \\ 0 & \text{for } t = 0. \end{cases}$$

It can readily be checked that the family of functionals g_t is continuous with respect to t on the interval $(0, 1]$ in the norm topology. This family is not continuous at 0 in the norm topology but it is continuous at 0 in the weak-$*$-topology.

Clearly, the family g_t can be regarded as a family C_t of operators which map the space X into R. By the weak-$*$-continuity of this family, $C_t(x)$ is continuous on the whole interval $[0, 1]$ for all $x \in X$. The adjoint operators C_t^* map R into X^*. The value of the operator C_t^* at 1, $C_t^*(1)$, is the family g_t, which is not continuous at $t = 0$ in the norm topology. ■

If X is a reflexive space, the family of operators constructed in the proof of Theorem 5.4.13 has the following property: for $t \neq 0$, $C_t(\bar{K}) = [-1, 1]$, where K denotes the unit ball,

$$\bar{K} = \{x \in X : \|x\| \leq 1\}.$$

Note that $C_0(\bar{K}) = 0$. Hence, the multifunction $C_t(\bar{K})$ is not linearly upper semicontinuous.

As a consequence of Theorem 5.4.13 we obtain

COROLLARY 5.4.14. *Theorem 5.4.13 remains true if R is replaced by any Banach space.*

Proof. We define g_t as before. Let y be an arbitrary nonzero element of the space Y. Let $C_t(x) = g_t(x) \cdot y$. This family has the required property. ■

The following theorem appears to be useful in many applications.

THEOREM 5.4.15. *Let C_t be a family of continuous linear operators which* map a Banach space X into a Banach space Y. Let K be a compact set contained in the space X. Suppose that for every $x \in X$ the function $C_t(x)$ is continuous on the whole interval $[0, \mathcal{T}]$. Then the multifunction $C_t(K)$ is linearly continuous on the interval $[0, \mathcal{T}]$.

Proof. In virtue of the Banach–Steinhaus theorem (Theorem 3.1.1), the norms of the operators C_t are uniformly bounded,

$$\sup_{0 \leqslant t \leqslant \mathcal{T}} ||C_t|| = M.$$

Let ε be an arbitrary positive number. Since K is a compact subset of a metric space, there exists an $\frac{\varepsilon}{3M}$-net, say $\{x_1, ..., x_n\}$, in K. By the continuity of the function $C_t(x)$ for every $x \in X$ there is a $\delta > 0$ such that for all t, $|t-t'| < \delta$, we have

$$||C_t(x_i) - C_{t'}(x_i)|| < \frac{\varepsilon}{3}, \quad i = 1, 2, ..., n.$$

Take an arbitrary element x of K. Since $\{x_1, ..., x_n\}$ is an $\frac{\varepsilon}{3M}$-net, there is an index i such that

$$||x - x_i|| < \frac{\varepsilon}{3M}$$

and hence

$$||C_t(x) - C_{t'}(x)|| \leqslant ||C_t(x - x_i)|| + ||(C_t - C_{t'})(x_i)|| + ||C_{t'}(x_i - x)||$$

$$\leqslant \frac{\varepsilon}{3M} M + \frac{\varepsilon}{3} + \frac{\varepsilon}{3M} M = \varepsilon.$$

This proves the theorem. ∎

A linear continuous operator A is called *compact* if $\overline{A(\{x: ||x|| < 1\})}$ is compact.

COROLLARY 5.4.16. *Let X, $\square$, Y be Banach spaces. Let A be a compact operator which maps X into $\square$ and let B_t be a family of continuous linear operators which map $\square$ into Y. If, for every $y \in \square$, $B_t(y)$ is continuous, then the multifunction $B_t A(U_0)$ is linearly continuous in the norm topology for every bounded set U_0.*

5.4. Minimum-time problem

Suppose we are given a linear time depending system
$$(X \underset{A}{\to} \square \underset{B_t}{\to} Y) \quad 0 \leqslant t < +\infty. \tag{5.4.28}$$

Let X be the Cartesian product of two Banach spaces X_1, X_2,
$$X = X_1 \times X_2. \tag{5.4.29}$$

We say that the system (5.4.28) is *controllable* (is *controllable to zero* or *null controllable*) if for all $x_1 \in X_1$, $y \in Y$ ($x_1 \in X_1$) there are a moment t and an element $x_2^t \in X_2$ such that
$$B_t A(x_1, x_2^t) = y \tag{5.4.30}$$
$$(B_t A(x_1, x_2^t) = 0). \tag{5.4.30'}$$

Let
$$\tilde{X}_t = \{x_1 \in X_1 : 0 \in B_t A(x_1, X_2)\}. \tag{5.4.31}$$

THEOREM 5.4.17 (Rolewicz, 1976). *Suppose that*
$$\tilde{X}_t \subset \tilde{X}_{t_1} \quad \text{for} \quad t < t_1. \tag{5.4.32}$$
If system (5.4.28) *is controllable to zero, then there is a universal time t_u such that for all $x_1 \in X_1$ there is an x_2, $x_2 \in X_2$, such that $B_{t_u} A(x_1, x_2) = 0$.*

Proof. For each t, the set
$$W_t = \{(x_1, x_2) \in X_1 \times X_2 : B_t A(x_1, x_2) = 0\} \tag{5.4.33}$$
is closed as the inverse image of zero under a continuous operator.

Let P be the natural projection of $X_1 \times X_2$ onto X_1, $P(x_1, x_2) = x_1$. Observe that $\tilde{X}_t = P(W_t)$. By the Banach theorem (Theorem 5.2.1), either $\tilde{X}_t = X_1$ or $\tilde{X}_t$ is of the first category. Since system (5.4.28) is controllable to zero, by (5.4.33), we get
$$X_1 = \bigcup_{n=1}^{\infty} \tilde{X}_n. \tag{5.4.34}$$

Hence, there is an index n_0 such that $\tilde{X}_{n_0} = X_1$. Putting $t_u = n_0$ we obtain the assertion of the theorem. ∎

Let
$$C_t = B_t A(0, X_2). \tag{5.4.35}$$

THEOREM 5.4.18 (Rolewicz, 1976). *Let*
$$C_t \subset C_{t_1} \quad \text{for} \quad t < t_1. \tag{5.4.36}$$
If for each $y \in Y$ there are a moment t and an element x_2^t such that
$$B_t A(0, x_2^t) = y, \tag{5.4.37}$$
then there is an element x_2, $x_2 \in X_2$, such that
$$B_{t_u} A(0, x_2) = y. \tag{5.4.38}$$

Proof. Observe first that the operators $B_t A(0, \cdot)$ do not depend on x_1. Putting $Y = X_1$ and
$$B_t(y, x) = y - B_t A(0, x),$$
we apply Theorem 5.4.17, which completes the proof. ∎

COROLLARY 5.4.19 (Rolewicz, 1976). *If (5.4.28) and (5.4.36) hold, then there is a universal time t_u such that for all $x_1 \in X_1$, $y \in Y$, there is an element $x_2 \in X_2$ such that $B_{t_u} A(x_1, x_2) = y$.*

Proof. Let t_u be defined as in Theorem 5.4.18. Let $\tilde{y} = y - B_{t_u} A(x_1, 0)$. Applying Theorem 5.4.18, we obtain the assertion of the theorem. ∎

We say that system (5.4.28) is *approximately controllable* (*approximately controllable to zero*) if for all $x_1 \in X_1$, $y \in Y$ ($x_1 \in X_1$), there is a moment t such that for an arbitrary positive ε there is an element x_2, $x_2 \in X_2$, such that $\|B_t A(x_1, x_2) - y\| < \varepsilon$ ($\|B_t A(x_1, x_2)\| < \varepsilon$).
Let $X_t^0 = \{x_1 \in X_1: 0 \in \overline{B_t A(x_1, X_2)}\}$ for $0 \leq t < +\infty$.

LEMMA 5.4.20. *The sets X_t^0, $0 \leq t < +\infty$, are closed.*

Proof. Let $\{x_1^n\} \subset X_1$ be a sequence converging to x_1^0. Let ε be an arbitrary positive number. Since $\{x_1^n\}$ tends to x_1^0, there is an index N such that
$$\|x_1^N - x_1^0\| < \frac{\varepsilon}{2\|B_t\| \cdot \|A\|}. \tag{5.4.39}$$
Since $x_1^N \in X_t^0$, there is an element x_2, $x_2 \in X_2$, such that
$$\|B_t A(x_1^N, x_2)\| < \varepsilon/2. \tag{5.4.40}$$

Hence, by (5.4.39) and (5.4.40),

$$\|B_t A(x_1^0, x_2)\| \leq \|B_t A(x_1^N, x_2)\| + \|B_t A(x_1^N - x_1^0, 0)\|$$
$$< \frac{\varepsilon}{2} + \frac{\varepsilon}{2} = \varepsilon.$$

Since ε is arbitrary, X_t^0 is closed. ∎

THEOREM 5.4.21 (cf. Dolecki, 1976). *If*

$$X_t^0 \subset X_{t_1}^0 \quad \text{for} \quad t < t_1 \tag{5.4.41}$$

and system (5.4.28) *is approximately controllable, then there is a universal time t_u such that $X_{t_u}^0 = X_1$.*

Proof. The proof proceeds in the same way as the proof of Theorem 5.4.17. ∎

THEOREM 5.4.22 (cf. Dolecki, 1976). *If* (5.4.36) *holds and system* (5.4.28) *is approximately controllable, then there is a universal time t_u such that, for all $x_1 \in X_1$, $y \in Y$ and all positive ε, there is an element x_2, $x_2 \in X_2$, such that*

$$\|B_{t_u} A(x_1, x_2) - y\| < \varepsilon. \tag{5.4.42}$$

Proof. The proof proceeds in the same way as the proof of Theorem 5.4.18 and Corollary 5.4.19 with the only difference that C_t should be replaced by $\bar{C}_t$. ∎

5.5. REDUCTION OF THE MINIMUM-TIME PROBLEM TO THE MINIMUM-NORM PROBLEM

In the previous section we considered the minimum-time problem. By the relevant theorems, there exists (under appropriate assumptions) an element $x \in U_0$ such that $B_T A(x) \in Y(T)$. However, those theorems do not make it clear how to find that element.

In the present section, applying the results of Section 5.2 and Section 5.3, we propose a method for finding such an element.

Assume that the set U_0 appearing in the formulation of the minimum-time problem is closed, convex and bounded, and that the interior of U_0 is nonempty. Without loss of generality we can assume that 0 belongs to the interior of U_0. To see this, take an arbitrary point x_0, $x_0 \in \text{Int } U_0$. Let $U_0' = U_0 - x_0$. Clearly, $0 \in \text{Int } U_0'$. Observe that the existence of $x \in U_0$ such that $B_t A(x) \in Y(t)$ is equivalent to the existence of $x' \in U_0'$ such that $B_t A(x') \in Y'(t)$, where $Y'(t) = Y(t) + B_t A(x_0)$. It can be verified that the family $Y'(t)$ is linearly upper semicontinuous, provided that the family $Y(t)$ is linearly upper semicontinuous and the function $B_t A(x_0)$ is continuous. Hence, in view of the assumptions made so far, we can assume that $0 \in \text{Int } U_0$. Since U_0 is additionally assumed to be closed, convex and bounded, there exists a seminorm $||x||$ induced by U_0,

$$||x|| = \inf\left\{t > 0: \frac{x}{t} \in U_0\right\}.$$

Clearly, $x \in U_0$ if and only if $||x|| \leqslant 1$. Let

$$\varphi(t) = \inf\{||x||: x \in X, B_t A(x) \in Y(t)\}$$

and

$$\tilde{T} = \inf\{t > 0: \varphi(t) \leqslant 1\}. \tag{5.5.1}$$

Evidently, $\tilde{T} \leqslant T$, where, according to the notation introduced in the previous section,

$$T = \inf\{t > 0: B_t A(U_0) \cap Y(t) \neq \emptyset\}.$$

Under appropriate assumptions guaranteeing the existence of a solution of the minimum-norm problem for each t, i.e., the existence of an element x_t such that $||x_t|| = \varphi(t)$ and $B_t A(x_t) \in Y(t)$, the numbers T and $\tilde{T}$ are equal.

In further considerations we shall assume the existence of a solution of the minimum-norm problem for all t.

The following method is proposed for solving the minimum-time problems:

(1) Determine the function $\varphi(t)$ and the number $\tilde{T}$.

(2) Examine the existence of solutions of the corresponding minimum-norm problem.

(3) Check the inequality $\varphi(\tilde{T}) \leqslant 1$.

5.5. Minimum-time problem and minimum-norm problem

(1) and (2) were considered in detail in Sections 5.2 and 5.3. Now we investigate the properties of the function $\varphi(t)$. In particular, we formulate sufficient conditions ensuring the lower semicontinuity and the continuity of $\varphi(t)$.

The above method is called the *method of reduction of the minimum-time problem to the minimum-norm problem*.

THEOREM 5.5.1 (Rolewicz, 1976). *Let U_0 be a closed convex bounded set. Let the multifunctions $Y(t)$ and $B_t A(U_0)$ be linearly continuous at $t = t_0$. If $B_{t_0} A(U_0)$ has a nonempty interior, then $\varphi(t)$ is continuous at t_0.*

Proof. By Lemma 5.1.5, there is an $\eta > 0$ such that for all t, $|t-t_0| < \eta$ the sets $B_t A(U_0)$ have nonempty interiors. This implies that
$$\overline{B_t A(U_0)} \subset \text{Int}(1+\varepsilon) B_t A(U_0) \tag{5.5.2}$$
for all positive ε. Fix an $\varepsilon > 0$. By (5.5.1) and since the multifunction $B_t A(U_0)$ is linearly continuous with respect to t, there exists a $\delta_0 > 0$ such that for t', $|t'-t| < \delta_0$, we have
$$\overline{B_{t'} A(U_0)} \subset \text{Int}(1+\varepsilon) B_t A(U_0). \tag{5.5.3}$$
By the definition of $\varphi(t')$,
$$Y(t') \cap \overline{\varphi(t') B_{t'} A(U_0)} \neq \emptyset, \tag{5.5.4}$$
and hence, in view of (5.5.3), we have
$$Y(t') \cap \text{Int}(1+\varepsilon) \varphi(t') B_t A(U_0) \neq \emptyset. \tag{5.5.5}$$
So far we have not taken into account the fact that the multifunction $Y(t)$ is linearly continuous at t_0. Put $t = t_0$. By (5.5.5) and the linear continuity of $Y(t)$, there exists a number $\delta_1 > 0$ such that
$$Y(t) \cap \text{Int}(1+\varepsilon) \varphi(t_0) B_t A(U_0) \neq \emptyset \tag{5.5.6}$$
for all t, $|t-t_0| < \delta_1$. By (5.5.6) and the definition of $\varphi(t)$,
$$\varphi(t) \leq (1+\varepsilon) \varphi(t_0). \tag{5.5.7}$$
Next, by substituting $t = t_0$ and $t' = t$ in (5.5.5) and repeating the reasoning we obtain
$$\varphi(t_0) \leq (1+\varepsilon) \varphi(t). \tag{5.5.8}$$
Since ε is arbitrary, we immediately infer that $\varphi(t)$ is continuous at t_0. ∎

If there exists an x belonging to U_0 such that $B_T A(x) \in Y(T)$ and the function $\varphi(t)$ is continuous at $t = T$, then $||x|| = \varphi(T) = \lim_{t \to T} \varphi(t)$
$= 1$. This means that x lies on the boundary of U_0.

Now we give examples which show that the function $\varphi(t)$ need not be continuous, even if B_t is continuous with respect to t in the operator norm topology.

EXAMPLE 5.5.1. Let $X = \square = Y$ be the two-dimensional space R^2. Let U_0 be the unit disc
$$U_0 = \{(x_1, x_2): x_1^2 + x_2^2 \leqslant 1\}.$$
Let $A = I$ be the identity operator. Let B_t be defined by the matrix
$$B_t = \begin{bmatrix} t & 0 \\ 0 & 1 \end{bmatrix}.$$
The family B_t is continuous with respect to t in the norm topology. Let $Y(t)$ be a one-point set
$$Y(t) = \{(t, 1)\}.$$
It can readily be checked that $T = 0$ and $\varphi(T) = 1$. But $\varphi(t) = \sqrt{2}$ for $t > 0$. To see this, it is enough to observe that $\varphi(1) = \sqrt{2}$ and $B_t A(Y(1)) = Y(t)$ for all $t > 0$. This implies that $B_t A(x) \in Y(t)$ if and only if $B_1 A(x) \in Y(1)$.

Let $f(t)$ be a function defined on the interval $[a, b]$. We say that the function $f(t)$ is *lower semicontinuous on the right* at $t_0 \in [a, b]$ if
$$f(t_0) \leqslant \liminf_{\substack{t \to t_0 \\ t > t_0}} f(t).$$
Clearly, in Example 5.2.1 the function $\varphi(t)$ is lower semicontinuous on the right at $t = 0$.

THEOREM 5.5.2. *Under the assumptions of Theorems 5.4.2–5.4.4 the function $\varphi(t)$ is lower semicontinuous on the right.*

Proof. Let
$$\alpha = \lim_{\substack{t \to t_0 \\ t > t_0}} \varphi(t).$$

Let $U_\alpha = \alpha U_0$. In view of the assumptions of Theorems 5.4.2–5.4.4, there exists an x_α, $x_\alpha \in U_\alpha$ such that $B_{t_0} A(x_\alpha) \in Y(t_0)$. Hence, $\varphi(t_0) \leqslant \alpha$. ∎

If the above assumptions are not satisfied, the function $\varphi(t)$ need not be lower semicontinuous, as is shown by the following example.

EXAMPLE 5.5.2. As in Example 5.2.1, let $X = l$, $\square = Y = c_0$ (or l or l^2). Let B be the identity operator and let A be given by the formula

$$A(t) = \frac{1}{2} t_1 y_0 + \sum_{i=1}^{\infty} t_{i+1} y_i,$$

where $\{y_i\}$ is a sequence of strongly linearly independent elements which converges to y_0.

Define a function $y(t)$ by the formula

$$y(t) = \begin{cases} y_0 & \text{for } t = 0, \\ (n(n+1)t-n)y_n + (n+1-n(n+1)t)y_{n+1} & \text{for } \\ & \frac{1}{n+1} \leqslant t \leqslant \frac{1}{n}. \end{cases}$$

The function $y(t)$ is linear on the intervals $\left[\frac{1}{n+1}, \frac{1}{n}\right]$ and $y\left(\frac{1}{n}\right) = y_n$. Clearly, $y(t)$ is continuous. Hence, the multifunction consisting of the one-point set $\{y(t)\}$ for each t is continuous.

Moreover, in view of Example 5.2.1, $\varphi(0) = 2$. On the other hand, $y_n \in \Gamma_1$, which implies that $\varphi\left(\frac{1}{n}\right) = 1$. This proves that the function $\varphi(t)$ is not lower semicontinuous at 0.

In the space l the closed unit ball is the closed convex hull of the points $e_n = \{0, ..., 0, 1, 0, ...\}$ and $A(e_n) = y_{n-1}$. Thus, in this case
$\underbrace{\phantom{\{0,...,0,1,0,...\}}}_{n\text{-th position}}$
we can show even more, namely that $\varphi(t) = 1$ for $t > 0$.

The above example can be modified in such a way that $Y(t)$ is a fixed element x_0 and B_t varies.

It is a natural question to ask when condition (3), $\varphi(T) \leqslant 1$, can be replaced by the stronger condition $\varphi(T) = 1$.

THEOREM 5.5.3. *Let the multifunction $B_t A(U)$ be linearly lower semicontinuous at T and let the set $B_T A(U)$ have a nonempty interior. If $Y(t)$ is linearly lower semicontinuous at T, then $\varphi(t)$ is upper semicontinuous at T.*

Proof. Let ε be an arbitrary positive number. Since $B_T A(U)$ has a nonempty interior and $B_t A(U)$ is linearly lower semicontinuous at T, there is a $\delta_0 > 0$ such that $|t - T| < \delta_0$ implies

$$B_T A(U) \subset B_t A(U) + \varepsilon B_T A(U).$$

By the Rådström lemma (Lemma 5.1.3),

$$(1 - \varepsilon) B_T A(U) \subset B_t A(U).$$

Since $Y(t)$ is lower semicontinuous at T, there exists a $\delta_1 > 0$ such that

$$Y(T) \subset Y(t) + \varepsilon B_T A(U)$$

for all $t, |t - T| < \delta_1$. Thus, for all $t, |t - T| < \delta = \min(\delta_0, \delta_1)$ we have

$$\varphi(T) = \inf\{s : s B_T A(U) \cap Y(T) \neq \emptyset\}$$

$$\geq \inf\left\{s : s \frac{1}{1 - \varepsilon} B_t A(U) \cap Y(T) \neq \emptyset\right\}$$

$$\geq \inf\left\{s : \frac{s}{1 - \varepsilon} B_t A(U) \cap \bigl(Y(t) + \varepsilon B_T A(U)\bigr) \neq \emptyset\right\}$$

$$\geq \inf\left\{s : \left(\frac{s}{1 - \varepsilon} - \varepsilon\right) B_t A(U) \cap Y(t) \neq \emptyset\right\}$$

$$= \inf\{u(1 - \varepsilon) + \varepsilon(1 - \varepsilon) : u B_t A(U) \cap Y(t) \neq \emptyset\}$$

$$= (1 - \varepsilon) \varphi(t) + \varepsilon(1 - \varepsilon).$$

Since ε is arbitrary, the function $\varphi(t)$ is upper semicontinuous at T. ∎

COROLLARY 5.5.4. *If the family of operators $B_t A$ is continuous in the operator norm topology and the set $B_T A(U)$ has a nonempty interior, then $\varphi(t)$ is upper semicontinuous at T.*

5.6. OBSERVABILITY OF LINEAR SYSTEMS

Before we pass to the main topic of this section, we explain a certain ambiguity caused by different uses of the terms *observability, observable*, etc.

These terms have different meanings. We shall explain this on a simple example.

Suppose we are given a linear system

$$(X \underset{A}{\to} \square \underset{B}{\to} Y),$$

where X is a space of initial values (finite- or infinite-dimensional), and $\square$ is the space of all continuous functions defined on the interval $[0, T]$ with values in X,

$$\square = C_X[0, T].$$

Some authors (cf. Lions, 1968) use the term "controllability" in the case where $X = Y$ and the operator B has a special structure, namely, B assigns to a continuous function $x(t)$ the value $x(T)$. In other cases they apply the term "observability".

Observe that in the sense of our definitions (see Section 5.1) both cases concern the problem of controllability.

Other authors (cf. Kalman, 1960, 1968; Kalman, Falb, Arib, 1969; Krassowski, 1968) apply the term observability when the problem is to derive the attributes of input from the attributes of output.

In the sequel we shall use the term observability in the latter sense. Since we investigate linear systems, we shall focus our attention on the "linear attributes" of the objects under consideration.

Suppose we have a linear system

$$(X \underset{A}{\to} \square \underset{B}{\to} Y).$$

Attributes of this system (which can be interpreted as physical quantities) assign numbers to elements of X. Therefore, they are functionals defined on the space X. To preserve the continuity of physical quantities, only continuous functionals $f(x)$ should be considered.

It frequently happens that a direct observation of the input space X is not possible and $f(x)$ cannot be determined directly. However, the outputs can be observed by measurements or calculations of $\varphi(y)$ for every $y \in Y$ and every $\varphi \in Y^*$.

For instance, it is so when we want to determine, at a moment t, the velocity of an object which is described by a differential equation of order n by observing the trajectory of the object in a certain time interval.

The question arises whether it is possible to choose a $\varphi \in Y^*$ such that

$$f(x) = \varphi(BA(x)) \tag{5.6.1}$$

for all $x \in X$. If the answer is affirmative, we say that the continuous linear functional f is *observable*. A continuous linear functional φ satisfying (5.6.1) is sometimes called (cf. Krassowski, 1968) a *method of observation*.

If we pass to conjugate spaces, equation (5.6.1) takes the form

$$f = A^*B^*\varphi. \tag{5.6.2}$$

THEOREM 5.6.1. *A continuous linear functional f is observable if and only if*

$$0 \notin \overline{BA(\{x \in X: f(x) = 1\})}. \tag{5.6.3}$$

Proof. Necessity. Suppose that formula (5.6.3) does not hold; then there exists a sequence $\{x_n\}$, $x_n \in X$, such that $f(x_n) = 1$ and $BA(x_n) \to 0$. Let φ be an arbitrary continuous linear functional defined on Y. Then $\varphi(BA(x_n)) \to 0$. On the other hand, $\varphi(BA(x_n)) = f(x_n) = 1$. This leads to a contradiction.

Sufficiency. Assume that (5.6.3) holds. Then, by the separation theorem (Theorem 2.5.7), there exist a continuous linear functional $\varphi \in Y^*$ and a constant c such that

$$\varphi(y) > c \quad \text{for} \quad y \in \overline{BA(\{x \in X: f(x) = 1\})}.$$

But the set $BA(\{x \in X: f(x) = 1\})$ is a linear manifold, whence we can show as before that the functional φ is constant on this set and is equal to c'. Without loss of generality (multiplying, if necessary, by $1/c'$) we can assume that $c' = 1$. Hence, if $f(x) = 1$, then $\varphi(BA(x)) = 1$. By homogeneity of functionals, we immediately obtain formula (5.6.1). ∎

5.6. Observability of linear systems

Clearly, if a functional f is observable, then there may exists many different $\varphi \in Y^*$ which satisfy equation (5.6.1) and hence (5.6.2). Thus, we can choose φ in different ways.

Now we can formulate the following problem. Suppose the output $y = BA(x)$ is measured with some error. Derive a continuous linear functional (a method of observation) such that $f(x)$ can be determined with a minimal error. First of all, we must know the error of direct measurements of y. In many practical applications we can state a priori that $y_r - y_0 \in W$, where y_r denotes a real output, y_0 denotes an observed (measured) output and W is a certain bounded closed convex set which contains 0 in its interior.

EXAMPLE 5.6.1. Let the output of a system be given by an n-dimensional continuous vector-valued function $y(t) = [y_1(t), ..., y_n(t)]$. Assume that each component function $y_i(t)$, $i = 1, ..., n$, can be measured independently and the result of measurement does not depend on time. Then

$$W = \{y(t) \in C_n[a, b]: |y_i(t)| \leqslant a_i, \ i = 1, 2, ..., n\},$$

where a_i denotes the accuracy of measurement of the i-th component function.

EXAMPLE 5.6.2. Let the output of a system be as in Example 5.6.1. Assume that each component can be measured with a given accuracy a_i, $i = 1, ..., n$, and 'moreover' the sum of the components can be measured with a given accuracy b, $b < a_1 + ... + a_n$. Then

$$W_1 = \{y(t) \in C_n[a, b]: |y_i(t)| \leqslant a_i, \ i = 1, ..., n,$$
$$|y_1(t) + ... + y_n(t)| \leqslant b\}.$$

Since $b < a_1 + ... + a_n$, the set W_1 is a proper subset of W (i.e., W_1 does not coincide with W).

EXAMPLE 5.6.3. Suppose we are given two systems: the first supplies a pattern signal, which is the time-varying voltage, and the second follows up that signal. Suppose that between these two systems there is a unit resistor. Then, the quantity of heat dissipated on that resistor is the measure of the accuracy with which the second system follows up signals generated by the first system.

In this example, the output space Y is equal to $L^2[0, T]$ and the set W is given by the formula

$$W = \left\{y(t) \in L^2[0, T]: \int_0^T |y(t)|^2 dt \leq M^2\right\}.$$

Clearly, by previous considerations, each bounded closed convex set which contains 0 in its interior induces a seminorm on the space Y by the formula

$$\|y\| = \inf\left\{t > 0: \frac{x}{t} \in W\right\}.$$

In most practical applications the set W is symmetric and hence the above seminorm is a norm.

Since $\|y\| \leq 1$ if and only if $y \in W$, we can say that the accuracy of available instruments generates a certain seminorm (or a norm) $\|\ \|$ such that $\|y_r - y_0\| \leq 1$.

Because of possible units changes, we assume that the accuracy of observations is given by the formula

$$\|y_r - y_0\| < \varepsilon.$$

The problem of optimal observability consists in determining a continuous linear functional φ, $\varphi \in Y^*$, such that $\varphi(y_0 - y_r)$ is minimal whenever $\|y_r - y_0\| \leq \varepsilon$.

In other words, of all the possible φ's, $\varphi \in Y^*$, we want to choose the one with the minimal norm. Thus, we should solve the minimum-norm problem for a system

$$(Y^* \underset{B^*}{\to} \square^* \underset{A^*}{\to} X^*)$$

which is the conjugate of the given system. We say that a continuous linear functional f is *optimally observable* if there exists a $\varphi_0 \in Y^*$ such that $\|\varphi_0\| = \inf\{\|\varphi\|: A^*B^*\varphi = f\}$. The fact that the optimal observability problem induces a minimum-norm problem for a conjugate system is called the *duality principle*.

5.6. Observability of linear systems

Since the weak-$*$-topology can be used to investigate conjugate systems, we can show without any additional assumptions, the following theorem.

THEOREM 5.6.2. *Every observable functional f is optimally observable.*

Proof. Put
$$r = \inf\{\|\varphi\|: \varphi \in Y^*, f = A^*B^*\varphi\}.$$
The set $V = \{\varphi \in Y^*: A^*B^*\varphi = f\}$ is closed in the weak-$*$-topology. Hence, by the Alaoglu theorem (Theorem 4.4.6), the sets
$$M_n = \left\{\varphi \in V: \|\varphi\| \leqslant r + \frac{1}{n}\right\}$$
are compact in the weak-$*$-topology. Observe that, by the definition of the number r, the sets M_n are nonempty. Hence, the intersection V_0 of the sets M_n, $V_0 = \bigcap_{n=1}^{\infty} M_n$, is nonempty. By the definition of M_n, $\|\varphi\| \leqslant r$ for every $\varphi \in V_0$. On the other hand, $V_0 \subset V$ and hence $\|\varphi\| \geqslant r$ for $\varphi \in V_0$. Thus, $\|\varphi\| = r$. Clearly, all elements of V_0, $V_0 \subset V$, satisfy equation (5.6.2). ∎

Observe that, in some sense, the minimal norm does not depend on the space Y. Speaking more precisely, for an arbitrary Banach space Y_1 which contains Y as a subspace we have
$$\inf\{\|\varphi\|: f(x) = \varphi(BA(x)), \varphi \in Y^*\}$$
$$= \inf\{\|\varphi\|: f(x) = \varphi(BA(x)), \varphi \in Y_1^*\}. \tag{5.6.4}$$
This is an immediate consequence of the fact that every functional φ, $\varphi \in Y^*$, can be extended to the functional φ_1, $\varphi_1 \in Y_1^*$, in such a way that $\|\varphi\| = \|\varphi_1\|$ (Theorem 2.5.2). Clearly, $\varphi_1(BA(x)) = \varphi(BA(x)) = f(x)$ since $BA(X) \subset Y \subset Y_1$.

Consider Banach spaces Y_1, Y_2 such that on the intersection $Y = Y_1 \cap Y_2$ the norms of those spaces are equivalent and $BA(X) \subset Y_1 \cap Y_2$. Applying formula (5.6.2) to the spaces Y, Y_1 and to the spaces Y, Y_2 we obtain formula (5.6.2) for the spaces Y_1, Y_2.

Similarly to Theorem 5.2.4, the following theorem holds for the problem of observability.

THEOREM 5.6.3. *The following equality holds*
$$\inf\{||\varphi||: f = A^*B^*\varphi\} = \sup_{x \in X} \inf\{||\varphi||: f(x) = \varphi(BA(x))\}.$$

Proof. Put
$$\alpha = \inf\{||\varphi||: f = A^*B^*\varphi\}$$
and
$$b = \sup_{x \in X} \inf\{||\varphi||: f(x) = \varphi(BA(x))\}.$$

Every functional φ satisfying the equation $f = A^*B^*\varphi$ satisfies also the equation $f(x) = \varphi(BA(x))$ for every $x \in X$, we have $b \leqslant \alpha$.

Suppose $b < \alpha$. Put
$$\Gamma_b = \{g: g = A^*B^*\varphi, \ ||\varphi|| \leqslant b\}.$$

By hypothesis, $b < \alpha$. Hence $f \notin \Gamma_b$. Observe that the set Γ_b is closed in the weak-*-topology since, as we have shown, Γ_b is compact in that topology. Hence, by the separation theorem (Theorem 4.3.3), there exist a constant c, a positive number ε and an element $x \in X$ such that
$$g(x) < c - \varepsilon \quad \text{for} \quad g \in \Gamma_b,$$
$$f(x) \geqslant c.$$

This means that for all φ such that $||\varphi|| \leqslant b$ we have $\varphi(BA(x)) < f(x) - \varepsilon$. This contradicts the definition of the number b. ∎

Suppose now that X is a finite-dimensional space. Then X is reflexive and can be considered as the conjugate of X^*. The set Γ_α is closed and f lies on the boundary of Γ_α. Hence, there exists an $x_0 \in X$ such that
$$f(x_0) = \sup\{g(x_0): g \in \Gamma_\alpha\}.$$

Therefore, we obtain

COROLLARY 5.6.4. *If X is a finite-dimensional space, then there exists an $x_0 \in X$ such that*
$$\inf\{||\varphi||: f = A^*B^*\varphi\} = \inf\{||\varphi||: f(x_0) = \varphi(BA(x_0))\}.$$

THEOREM 5.6.5. *Let a linear system*
$$(X \underset{A}{\to} \Box \underset{B}{\to} Y) \tag{5.1.1}$$
be given. Suppose that the input space X is n-dimensional, $\dim X = n$. Then there exists a solution φ of the equation
$$f = A^*B^*\varphi, \tag{5.6.2}$$
which is a convex combination,
$$\varphi = a_1\varphi_1 + \ldots + a_n\varphi_n$$
$(a_i \geqslant 0,\ a_1 + \ldots + a_n = 1)$, *of extreme points $\varphi_1, \ldots, \varphi_n$ of the ball*
$$K_r^* = \{\varphi \in Y^* \colon ||\varphi|| \leqslant r\},$$
where
$$r = \inf\{||\varphi|| \colon A^*B^*\varphi = f\}.$$

The proof is based on the following lemma.

LEMMA 5.6.6 (Carathéodory, 1921). *Let X be a Banach space. Let U be a closed bounded and convex set. Suppose that the dimension of the affine hull of U is equal to n, $\dim \mathrm{aff}(U) = n$. Then each $x \in U$ can be represented as a convex combination of $n+1$ extreme points of U.*

Proof. The proof will proceed by induction with respect to n. For $n = 1$, U is a closed interval and the representation is trivial. Suppose that the assertion of the lemma is true for a positive integer n. Take U such that $\dim \mathrm{aff}(U) = n+1$. Let u be an arbitrary point of U. Let e_0 be an arbitrary extreme point of U. By the Krein–Milman theorem (Theorem 4.6.4), such an extreme point exists. If $e_0 = u$, the proof is completed. If $e_0 \neq u$, we take the line passing through e_0 and u. The intersection of this line and the set U forms a closed interval $[e_0, w]$ $= \{te_0 + (1-t)w \colon 0 \leqslant t \leqslant 1\}$. The point w belongs to the boundary of U. Thus, there is an extremal subset W of U, $W \neq U$, such that $w \in W$. Of course, $\dim \mathrm{aff}(W) \leqslant n$. By using the inductive assumption, we can find extreme points $e_1, \ldots, e_n \in W$ such that w is a convex combination of $e_1, \ldots, e_n$,
$$w = a_1e_1 + \ldots + a_ne_n, \quad a_1, \ldots, a_n \geqslant 0, \quad a_1 + \ldots + a_n = 1.$$

Thus, each point of the closed interval $[e_0, w]$ is a convex combination of elements $e_0, e_1, \ldots, e_n$ of the form
$$te_0 + (1-t)a_1 e_1 + \ldots + (1-t)a_n e_n$$
and
$$t + (1-t)a_1 + \ldots + (1-t)a_n = 1.$$
Since $u \in [e_0, w]$, u is a convex combination of elements $e_0, \ldots, e_n$. By Proposition 4.6.1, $e_1, \ldots, e_n$ are extreme points of U. ■

COROLLARY 5.6.7. *Let U be a closed bounded and convex set in a Banach space X. Let $\dim \mathrm{aff}(U) = n$. If $u \in U$ and u is not an algebraic interior point, then u can be represented as a convex combination of n extreme elements $e_1, \ldots, e_n$ of U.*

Proof. Since u does not belong to the algebraic interior of U, there is an extremal subset W, $W \subset U$, such that $u \in W$, $\mathrm{aff}(W) \leq n-1$. Then we can apply Lemma 5.6.6 to the set W. ■

Proof of Theorem 5.6.5 Let $\bar{K}_r^*$ denote the closed ball of radius r in the space Y^*, $\bar{K}_r^* = \{\varphi \in Y^*: \|\varphi\| \leq r\}$. By the Alaoglu theorem (Theorem 4.4.6), the ball $\bar{K}_r^*$ is weakly-*-compact. Since the operator A^*B^* is continuous in the weak-*-topology (Corollary 4.4.4), the set $\Gamma_r^* = A^*B^*(\bar{K}_r^*)$ is compact in the weak-*-topology. Since X^* is n-dimensional, the weak-*-topology coincides with the norm topology. Thus, Γ_r^* is compact. The functional f which appears in the definition of r belongs to the algebraic boundary of Γ_r^*. Hence, by Corollary 5.6.7, f is a convex combination of n extreme elements of Γ_r^*, $f_1, \ldots, f_n$,
$$f = a_1 f_1 + \ldots + a_n f_n, \quad a_1, \ldots, a_n \geq 0, \quad a_1 + \ldots + a_n = 1.$$
The sets $(A^*B^*)^{-1}(f_i) \cap K_r^*$, $i = 1, \ldots, n$, are compact in the weak-*-topology. By the Krein–Milman theorem (Theorem 4.6.4), those sets contain extreme points φ_i. By Proposition 4.6.1, φ_i are extreme points of K_r^*. Of course, $A^*B^*\varphi_i = f_i$. Hence,
$$A^*B^*\varphi = a_1 A^*B^*\varphi_1 + \ldots + a_n A^*B^*\varphi_n = a_1 f_1 + \ldots + a_n f_n = f$$
for $\varphi = a_1 \varphi_1 + \ldots + a_n \varphi_n$. ■

5.6. Observability of linear systems

So far we have considered the problem of observability of a single continuous linear functional $f \in X^*$. Now we shall be interested in the problem of observability of an n-tuple F of linear functionals defined on X, $F = (f_1, \ldots, f_n)$. Without loss of generality we can assume that the functionals $f_1, \ldots, f_n$ are linearly independent.

We say that the system F is *observable* if all the functionals f_i, $i = 1, \ldots, n$ are observable.

THEOREM 5.6.8. *The system F is observable if and only if every linear combination of functionals $f_1, \ldots, f_n$,*

$$a_1 f_1 + \ldots + a_n f_n$$

is observable.

Proof. Sufficiency. If every linear combination is observable, then, in particular, the functionals $f_1, \ldots, f_n$ are observable.

Necessity. Assume that the functionals $f_1, \ldots, f_n$ are observable. This means that there exist functionals $\varphi_1, \ldots, \varphi_n \in Y^*$ such that

$$f_i = A^* B^* \varphi_i, \quad i = 1, \ldots, n. \tag{5.6.5}$$

Multiplying both sides of each of equation (5.6.5) by a_i, $i = 1, \ldots, n$, and summing the resulting equations, we obtain

$$a_1 f_1 + \ldots + a_n f_n = A^* B^* (a_1 \varphi_1 + \ldots + a_n \varphi_n).$$

This means that the linear combination $a_1 f_1 + \ldots + a_n f_n$ is observable. ∎

Clearly, for a given system $F = (f_1, \ldots, f_n)$ there may be many different systems $\Phi = (\varphi_1, \ldots, \varphi_n)$ which satisfy equation (5.6.5). The problem consists in selecting those systems which are in some sense optimal. To do this, we should specify the measure of deviation of points in n dimensional spaces.

If we are interested in maximal deviation on each component, then we use the norm $\|y\| = \sup\limits_{1 \leq i \leq n} |y_i|$. If we are interested in deviation measured by the Euclidean distance, then we use the Euclidean distance.

In many applications it is enough to assume that the difference between a real vector and that obtained by measurements belongs to a certain convex set W. Then the distance can be measured by the corresponding norm (or seminorm), $||\ ||$, generated by W.

Let E denote an n-dimensional space with the norm $||\ ||$ which measures deviations.

The system $F = (f_1, ..., f_n)$ can be regarded as a linear mapping F of the space X onto the space E, defined by the formula

$$F(x) = (f_1(x), ..., f_n(x)).$$

In this framework the term *observability* means that there exists such an operator Φ which maps the space Y into E that

$$F = \Phi BA. \tag{5.6.6}$$

Introducing the diagram

$$\begin{array}{c} X \xrightarrow{A} \square \xrightarrow{B} Y \\ {}_F\searrow \quad \swarrow_\Phi \\ E \end{array} \tag{5.6.7}$$

we can say that the existence of an operator Φ which satisfies (5.6.6) results from the existence of an operator Φ such that the diagram (5.6.7) commutes. In other words, the operator F can be factorized by the operator BA. With these terms, Theorem 5.6.8 can be formulated in a different way.

THEOREM 5.6.9. *An operator F is observable if and only if for each continuous linear functional $f \in E^*$ the functional $\varphi = F^*f$ is observable.*

Passing to the conjugate system we obtain

THEOREM 5.6.10. *An operator F is observable if and only if*

$$F^*(E^*) \subset A^*B^*(Y^*),$$

i.e., the range of F^*, R_{F^*}, is contained in the range of A^*B^*, $R_{A^*B^*}$,

$$R_{F^*} \subset R_{A^*B^*}.$$

The problem of optimal observability is the problem of determining of an operator Φ_0 satisfying (5.6.6) and such that

$$||\Phi_0|| = \inf\{||\Phi||: \Phi \in B(Y \to E),\ F = \Phi BA\}. \tag{5.6.8}$$

If such an operator exists, we say that the system F is *optimally observable*.

THEOREM 5.6.11. *Every observable system is optimally observable.*

The proof of this theorem is based on a generalization of the Alaoglu theorem (Theorem 4.4.6).

Let E and E_1 be Banach spaces. Assume that in the space E there is a linear topology τ such that the closed ball $\overline{K}_r = \{x \in E_1: \|x\| \leqslant r\}$ is compact in the topology τ for all r.

In the space of continuous linear operators $B(E \to E_1)$ we introduce a topology $\tilde{\tau}$ in the following way. U is a neighbourhood of a given operator $A_0 \in B(E \to E_1)$ if U is of the form

$$U = \{A \in B(E \to E_1): A(x_i) - A_0(x_i) \in V_i, i = 1, \ldots, n\},$$
(5.6.9,

where $x_1, \ldots, x_n$ is a finite system of elements of X and V_i, $i = 1, \ldots, n$) are neighbourhoods of zero in the space E_1 in the topology τ.

THEOREM 5.6.12. *Under the above assumptions the set*

$$K_M = \{A: A \in B(E \to E_1), \|A\| \leqslant M\} \quad (5.6.10)$$

is compact in the topology $\tilde{\tau}$.

Proof. Let I be the Tikhonov product space of the sets

$$I(x) = \{y \in E_1: \|y\| \leqslant M \|x\|\}. \quad (5.6.11)$$

By the assumptions, the sets $I(x)$ are compact in the topology τ. By the Tikhonov theorem (Theorem 4.2.13), the set

$$I = \underset{x \in E}{\times} I(x)$$

is compact in the product topology. Clearly, the set I can be regarded as the set of all functions $f(x)$ defined on E with values in E_1 which satisfy the condition $\|f(x)\| \leqslant M \|x\|$.

The set K_M is a subset of the set I. Moreover, K_M is homeomorphically imbedded in I. This follows from the fact that the topology $\tilde{\tau}$ and the product topology define the same neighbourhoods on K_M. To complete

the proof, we show the closedness of the set K_M in I in the same way as in the proof of the Alaoglu theorem (Theorem 4.4.6). Namely, observe that the sets
$$A(x, y) = \{T \in I: T(x+y) = T(x)+T(y)\}$$
and
$$B(\alpha, x) = \{T \in I: T(\alpha x) = \alpha T(x), x \in X, \alpha\text{—number}\}$$
are closed in the product topology since the projection on a coordinate is a continuous operator. Hence, the set
$$K_M = \bigcap_{x,y \in E} A(x, y) \cap \bigcap_{\substack{x \in E \\ \alpha\text{—number}}} B(\alpha, x)$$
is closed in the product topology. ∎

Proof of Theorem 5.6.11. Let
$$r = \inf\{\|\Phi\|: F = \Phi BA\}.$$
By Theorem 5.6.12, the sets
$$K_{r+1/m} = \left\{A \in B(Y \to E): \|A\| \leq r + \frac{1}{m}\right\}$$
are compact in the strong operator topology, i.e., in the topology generated by neighbourhoods U of a given operator A_0,
$$U = \{A \in B(Y \to E): A(x_i) - A_0(x_i) \in V_i, i = 1, ..., n\},$$
where $x_1, ..., x_n$ denotes a system of elements and $V_1, ..., V_n$ denotes a system of neighbourhoods of zero in E in the norm topology. This is a consequence of the fact that in the finite-dimensional space E all closed balls are compact. The set
$$V = \{\Phi \in B(Y \to E): F = \Phi BA\}$$
is closed in the strong operator topology. This implies that the sets
$$V_m = V \cap K_{r+1/m}$$
form a descending sequence of sets which are compact in the strong operator topology. Thus, the intersection
$$V_0 = \bigcap_{m=1}^{\infty} V_m$$
is nonempty. Let $\Phi_0 \in V_0$. Then $\Phi_0 \in V$ and $\|\Phi_0\| \geq r$. On the other

hand, $\Phi_0 \in K_{r+1/m}$ for all m and $\|\Phi_0\| \leq r$. Finally, $\|\Phi_0\| = r$. ∎

If E is of dimension greater than 1, the minimal norm may depend heavily upon the space Y.

If a norm in the space E is given by the formula

$$\|(t_1, \ldots, t_n)\| = \sup_{1 \leq i \leq n} a_i |t_i|, \quad a_i > 0,$$

then by applying componentwise the Hahn–Banach theorem (Theorem 2.5.2) to every subspace Y_1 which contains Y as a subspace we can extend each operator acting from the space Y into E to an operator defined on Y_1. In this case an analogue of formula (5.6.4) holds. In other cases the situation is more complicated.

If a given space E is not isometric to the space just considered, then there exist Banach spaces Y and Y_1, where Y is a subspace of the space Y_1, and an operator T which maps Y into E and is such that every extension T_1 of the operator T, which maps the space Y_1 into E satisfies $\|T_1\| > \|T\|$ (Goodner, 1950; Kelley, 1952; Nachbin, 1950).

Recall that if $\dim X = n$ and $\dim E = m$ then $\dim B(X \to E) = nm$. The linear system

$$(X \underset{A}{\to} \square \underset{B}{\to} Y)$$

induces an operator $(BA)^+$ which maps the space $B(Y \to E)$ into $B(X \to E)$ and is defined by the formula

$$((BA)^+ G)(x) = G(BA(x)).$$

Since the space $B(X \to E)$ is nm-dimensional, by repeating the argumentation of the proof of Theorem 5.6.5 we obtain

THEOREM 5.6.13. *Suppose we are given a linear system*

$$(X \underset{A}{\to} \square \underset{B}{\to} Y) \tag{5.1.1}$$

and $\dim X = n$. *Let E be a Banach space and* $\dim E = m$. *Let F be an arbitrary linear operator which maps X into E. If F is observable, then there is a minimum-norm solution G_0 of the equation*

$$F = (BA)^+ G$$

such that

$$G_0 = \sum_{i=1}^{nm} \alpha_i G_i,$$

G_0 *is a convex combination of elements* G_i, $i = 1, \ldots, nm$ *which are extreme points of the ball in the space* $B(Y \to E)$ *of radius* r,

$$r = \inf \{\|G\|: G \in B(Y \to E), (BA)^+ G = F\}.$$

We can reduce nm to a smaller number if we consider a subspace Φ of the space $B(X \to E)$ of smaller dimension. Thus, we have the following

THEOREM 5.6.14. *Suppose we are given a linear system*

$$(X \underset{A}{\to} \square \underset{B}{\to} Y).$$

Let E *be a finite-dimensional Banach space. Let* Φ *be a subspace of the space* $B(X \to E)$. *Assume that* $\dim \Phi = k$. *Let* Ψ *be a subspace of the space* $B(Y \to E)$. *Suppose that*

$$\Phi = (BA)^+(\Psi).$$

Then for each $F \in \Phi$ *there is a minimum-norm solution* G_0 *of the equation*

$$(BA)^+ G = F, \quad G \in \Psi.$$

Moreover, G_0 *is a convex combination,*

$$G_0 = \sum_{i=1}^{k} \alpha_i G_i,$$

where G_i, $i = 1, \ldots, k$ *are extreme points of the ball in the space* Ψ *of radius* r,

$$r = \inf \{\|G\|: G \in \Psi, (BA)^+ G = F\}.$$

The method of reduction of the number of extreme points, which is presented in Theorem 5.6.14, has certain disadvantages. Namely, by diminishing the space Ψ one may cause the increase of the norm of the optimal solution. On the other hand, in some spaces it is relatively easy to find extreme points of balls.

If the space X is finite-dimensional, then by putting $X = E$, $F = I$, we find by Theorem 5.6.11 that there exists a continuous linear functional Φ such that $x = \Phi(BA(x))$ and the norm of Φ is minimal. We often have to solve the minimum-norm problems when the input space X is infinite-dimensional, for instance when systems with distributed parameters or systems with delays are considered. Thus, it may be of interest to investigate diagram (5.6.7) in the case where E is an infinite-dimensional Banach space.

More precisely, consider a linear system
$$(X \underset{A}{\rightarrow} \square \underset{B}{\rightarrow} Y)$$
and a continuous linear operator F which maps X into a Banach space E. We want to determine a continuous linear functional Φ which maps Y into E and is such that
$$F = \Phi BA \qquad (5.6.6')$$
or, in other words, we want to find a Φ such that the diagram (5.6.7) is commutative. If such a Φ exists, we say that the operator F is *observable* and Φ is called a *method of observation*. An observable operator F will be called *optimally observable* if there exists a Φ_0 satisfying (5.6.6') and such that
$$\|\Phi_0\| = \inf\{\|\Phi\|: F = \Phi BA, \Phi \in B(Y \to E)\}.$$

Let Z be a Banach space. We recall that an operator P which maps Z into itself is called a projection if $P^2 = P$. Put $Z_0 = P(Z)$. If P is a projection, then P is the identity operator on Z_0. In the sequel we shall consider only projections which are continuous and linear. Hence, by a projection we shall mean a continuous linear projection.

We say that a subspace Z' is a *projection of the space* Z if there exists a projection P such that $P(Z) = Z'$.

THEOREM 5.6.15. *Every finite-dimensional subspace Z' is a projection of Z.*

Proof. Let $e_1, \ldots, e_n$ be a basis in Z'. This means that each element $x \in Z'$ can be written in the form
$$x = f_1(x)e_1 + \ldots + f_n(x)e_n,$$

where $f_1, ..., f_n$ are continuous linear functionals defined on Z'. In virtue of the Hahn-Banach theorem (Theorem 2.5.2), functionals $f_1, ..., f_n$ can be extended to continuous linear functionals $\tilde{f}_1, ..., \tilde{f}_n$ which are defined on the whole Z. The operator P given by the formula

$$P(x) = \tilde{f}_1(x)e_1 + ... + \tilde{f}_n(x)e_n$$

is a projection operator which maps Z onto Z'. ∎

THEOREM 5.6.16. *If*
  (I) *$BA(X)$ is a projection of Y,*
  (II) *BA is invertible on $BA(X)$,*
then for every Banach space E, all operators F, $F \in B(X \to E)$, are observable.

Proof. By (I), there exists a projection operator Q which maps Y onto $BA(X)$. By (II) the operator $(BA)^{-1}$ is well defined on $BA(X)$ and by the Banach theorem (Corollary 3.2.2) $(BA)^{-1}$ is continuous. Hence, the operator $\Phi = F(BA)^{-1}Q$ is a continuous linear operator which maps Y into E. It can be readily checked that

$$F = \Phi BA. \blacksquare$$

THEOREM 5.6.17. *If $E = X$ and $F = I$ (I denotes the identity mapping), then conditions (I) and (II) of Theorem 5.6.16 are also necessary.*

Proof. Let Φ be an operator satisfying (5.6.6′). Then the operator $BA\Phi$ is a projection operator which maps Y onto $BA(X)$. ∎

If the space E is infinite-dimensional, there exist continuous linear functionals which are observable but not optimally observable. This fact can be illustrated by the following example.

EXAMPLE 5.6.4 (cf. Isbell, Semadeni, 1963). Let

$$\square = Y = L^1[0, 1]$$

and let B be the identity operator, $B = I$. Let

$$X = E = \{x \in Y: f(x) = 0\},$$

5.6. Observability of linear systems

where f is a functional given by the formula

$$f(x) = \int_0^1 tx(t)\,dt. \tag{5.6.12}$$

Let A be the operator of natural injection of X into $\square$. Let F be the identity operator.

We shall show that F is observable. It is enough to prove that there exists a projection operator of Y onto X. Take an arbitrary element x_0, $x_0 \notin X$. Let

$$P(x) = x - \frac{f(x)}{f(x_0)} x_0, \tag{5.6.13}$$

where f is defined by (5.6.12). It can be shown that $f(P(x)) = 0$, which means that P maps Y onto X. Moreover, $P(x) = x$ for $x \in X$. This proves that P is the required projection operator. Now determine the norm of P. Firstly,

$$||P(x)|| \leqslant ||x|| + \frac{1}{|f(x_0)|} ||x||\,||x_0||.$$

Consider a sequence

$$x_n(t) = \begin{cases} 0 & \text{for } 0 \leqslant t < 1-1/n, \\ n & \text{for } 1-1/n \leqslant t \leqslant 1. \end{cases}$$

For each n the norm of x_n is equal to 1. Hence,

$$||P(x_n)|| = \int_0^1 \left| x_n(t) - \frac{f(x_n)}{f(x_0)} x_0(t) \right| dt$$

$$\geqslant \left| \int_0^{1-1/n} \frac{f(x_n)}{f(x_0)} x_0(t)\,dt \right| + \left(\int_{1-1/n}^1 |x_n(t)|\,dt - \left| \int_{1-1/n}^1 \frac{f(x_n)}{f(x_0)} x_0(t)\,dt \right| \right)$$

$$\to \frac{1}{|f(x_0)|} ||x_0|| + 1.$$

This implies that

$$||P|| = \frac{1}{|f(x_0)|} ||x_0|| + 1.$$

Observe that the definition of the operator F implies that every operator Φ satisfying (5.6.6) must be a projection operator on X and every projection operator on X is of the form (5.6.13).

Let ε be an arbitrary positive number. Since the norm of the functional f is equal to 1, there exists an element x_0, $\|x_0\| = 1$, such that $\dfrac{1}{|f(x_0)|}$ < $1+\varepsilon$. This implies that the norm of the corresponding projection operator is not greater than $2+\varepsilon$. On the other hand, there is no element x_0, $\|x_0\| = 1$, such that $f(x_0) = 1$. Hence, there is no projection operator projecting Y onto X, with the norm equal to 2.

Theorem 5.6.12 implies the following generalization of Theorem 5.6.11.

THEOREM 5.6.18. *Suppose that we are given a linear topology τ on a space E such that balls are compact in that topology. Then every observable operator F is optimally observable.*

Proof. We show that the set
$$V = \{\Phi \in B(Y \to E): F = \Phi BA\}$$
is closed in the topology $\tilde{\tau}$ defined in the proof of Theorem 5.6.12. To see this, suppose that a certain operator T does not belong to V. This means that there is an x_0, $x_0 \in X$, such that $F(x_0) \neq TBA(x_0)$. Write $BA(x_0) = y$. Consider a neighbourhood U of the point $F(x_0)$ in the topology τ on the space Y, such that $y \notin U$. Let
$$\tilde{U} = \{S \in B(Y \to E): S(y) \in U\}.$$
$\tilde{U}$ is a neighbourhood of the operator T in the topology $\tilde{\tau}$ and has no common points with the set V. Hence, the set V is closed. The remainder of the proof is the same as in the proof of Theorem 5.6.11. ∎

COROLLARY 5.6.19. *If E is either a reflexive space or the conjugate of a given Banach space, then every observable operator F is optimally observable.*

Proof. The weak topology (the weak-∗-topology) has the property that every ball is compact in this topology. ∎

The considerations of this section can also be applied to systems with incomplete information (Krassowski, 1976, 1979b; Kurzhanskii, 1975, 1977).

Namely, we consider a linear system

$$(X \xrightarrow[A]{} \Box \xrightarrow[B]{} Y),$$

where inputs and trajectories are perturbed. In other words,

$\eta = A(x+u),$
$y = B(\eta+v),$

where u, v are unknown elements of U and V respectively. U and V are closed convex and bounded sets.

The idea is to consider the perturbations in X and $\Box$ as errors of measurement since under the conditions described above

$$y_r - y_0 \in W,$$

where

$$W = \{y: \|y\| < \varepsilon\} + B(V) + BA(U)$$

is a bounded convex set with a nonempty interior. Therefore, the set W induces a seminorm $\|x\|_1 = \inf\left\{t > 0: \dfrac{x}{t} \in W\right\}$ which is equivalent to the seminorm $\|\ \|$. Hence, the problem of optimal observability of system (5.1.1) with incomplete information can be reduced to the problem of optimal observability of system (5.1.1) with complete information with $\|\ \|_1$ as a measure of error in the output space Y.

This means that we seek a minimum-norm solution of the problem

$$f = A^*B^*\varphi,$$

where the norm of functionals is that induced by the norm $\|\ \|_1$.

5.7. MINIMUM-TIME PROBLEMS OF OBSERVABILITY THEORY

Minimum-time problems can be formulated also in the context of optimal observability.

EXAMPLE 5.7.1. Suppose that we want to measure the velocity of an object by observing its trajectory in a certain interval. Since no measuring

250   5. Optimization and observability of linear systems

is exact, the longer the time interval the more accurate the velocity measurement.

EXAMPLE 5.7.2. We want to find the temperature distribution in a bar by measuring the temperature at a fixed point during a certain time interval. If the bar is insulated, then, as we shall show later, the temperature distribution can be determined. However, the question arises whether by considering longer time intervals one can diminish the inaccuracy of observations.

Just as in the minimum-time problems of control theory, we consider a family of linear systems

$$(X \underset{A}{\to} \square \underset{B_t}{\to} Y), \tag{5.7.1}$$

where the family of operators B_t depends upon a real parameter t, $0 \leqslant t \leqslant \mathcal{T}$, which is called *time*.

Suppose we are given a continuous linear functional f defined on the input space X. The considerations of the previous section imply that for every t there exists a continuous linear functional $\varphi_t \in Y^*$ (a method of observation) such that

$$||\varphi_t|| = \inf\{||\varphi||: \varphi(B_t A(x)) = f(x)\}.$$

Recall that the norm $||\ ||$ can be regarded as a measure of accuracy of an observation method when the accuracy of the measuring instruments and the direct measurement technique are predetermined (see Section 5.6). Now the question is whether it is possible to find for a given number M a parameter t such that $||\varphi_t|| \leqslant M$ and $||\varphi_T|| \leqslant M$, where $T = \inf\{t > 0: ||\varphi_t|| \leqslant M\}$.

We say that the minimum-time problem has a solution if there exists a φ_0, $\varphi_0 \in Y^*$, $||\varphi_0|| \leqslant M$, such that $f(x) = \varphi_0(B_T A(x))$.

In other words, this is the minimum-time problem for the system

$$(Y^* \underset{B_t^*}{\to} \square^* \underset{A^*}{\to} X^*).$$

Now, we can apply the theorems of Sections 5.4 and 5.5. In those theorems the operators have the form $B_t A$ and here the corresponding operators are of the form $A^* B_t^*$. But this difference is of no importance

5.7. Minimum-time problems of observability theory

since we do not exploit the space $\square^*$ and we do not consider A and B_t separately.

The problem in question is related mainly to a norm in the space Y^*; thus, the method developed in Section 5.5 is applicable.

Write, as before,

$$\varphi(t) = \inf\{\|y^*\|: y^* \in Y^*, f = A^*B_t^*y^*\}. \tag{5.7.2}$$

THEOREM 5.7.1. *In minimum-time problems of observability theory the function $\varphi(t)$ is lower semicontinuous provided that the function $B_t A(x)$ is continuous with respect to t for every $x \in X$.*

Proof. Let
$$M = \liminf_{\substack{t \to T \\ t > T}} \varphi(t).$$

The ball
$$K_M = \{y^* \in Y^*: \|y^*\| \leqslant M\}$$

is compact in the weak-$*$-topology, in view of the Alaoglu theorem (Theorem 4.4.6). By the assumption and by Theorem 5.4.12 the multifunction $A^*B_t^*(K_M)$ is continuous in the weak-$*$-topology. Hence, by Theorem 5.4.6, the sets

$$W_{\delta, T} = \bigcup_{T \leqslant t \leqslant T+\delta} A^*B_t^*(K_M)$$

are compact in the weak-$*$-topology. In view of Theorem 5.4.4, this entails the existence of a solution of the minimum-time problem. This means that the function $\varphi(t)$ is lower semicontinuous. ∎

Theorem 5.7.1 can be proved also for systems F of n functionals, $F = (f_1, \ldots, f_n)$ and for some more general problems of observability. Suppose we are given a general problem of observability

$$\begin{array}{c} X \xrightarrow{A} \square \xrightarrow{B_t} Y \\ {\scriptstyle F} \searrow \quad \swarrow {\scriptstyle \Phi} \\ E \end{array} \tag{5.7.3}$$

Assume that the operator F is observable, i.e., that there exist a t_0 and an operator $\Phi \in B(Y \to E)$ such that diagram (5.7.3) commutes.

Write
$$\varphi(t) = \inf\{||\Phi||: \Phi \in B(Y \to E), F = \Phi B_t A\}.$$

Theorem 5.7.2. *Let τ be a topology in the space E such that the balls $K_r = \{x \in E: ||x|| \leq r\}$ are compact in that topology for each r. Then the function $\varphi(t)$ is lower semicontinuous provided that $B_t A(x)$ is continuous with respect to t for every $x \in X$.*

Proof. Write, as before,
$$M = \liminf_{\substack{t \to T \\ t > T}} \varphi(t).$$

Let $K_M = \{\Phi \in B(Y \to E): ||\Phi|| \leq M\}$. In virtue of Theorem 5.6.5, the set K_M is compact in the topology $\tilde{\tau}_Y$ which is generated by the topology τ as described in Section 5.6. In the same way, the topology τ generates the topology $\tilde{\tau}_X$ on the space $B(X \to E)$.

We shall prove that the mapping $\mathscr{B}_t \mathscr{A}$ which maps the space $B(Y \to E)$ with the topology $\tilde{\tau}_Y$ into the space $B(X \to E)$ with the topology $\tilde{\tau}_X$, $\mathscr{B}_t \mathscr{A} \Phi = \Phi B_t A$, is continuous for every t.

Take an arbitrary element F_0 such that $F_0 = \Phi_0 B_t A$ for some $\Phi_0 \in B(Y \to E)$. Consider a neighbourhood U of the element F_0,
$$U = \{F \in B(X \to E): F(x_i) - F_0(x_i) \in V_i, \ i = 1, 2, \ldots, n\},$$
where $x_1, \ldots, x_n \in X$, and $V_1, \ldots, V_n$ are neighbourhoods of zero on the space E and the topology τ.

Let
$$W = \{\Phi \in B(Y \to E): \Phi(y_i) - \Phi_0(y_i) \in V_i, \ i = 1, 2, \ldots, n\},$$
where $y_i = B_t A(x_i)$. Clearly, by definition, $\mathscr{B}_t \mathscr{A}(W) = U$. This means that the mapping $\mathscr{B}_t \mathscr{A}$ is continuous. Hence, the set $\mathscr{B}_t \mathscr{A}(K_M)$ is compact in the topology $\tilde{\tau}_X$.

We shall show that the multifunction $\mathscr{B}_t \mathscr{A}(K_M)$ is continuous in the topology $\tilde{\tau}_X$. Let U be an arbitrary neighbourhood of zero of the form
$$U = \{\psi \in B(X \to E): \psi(x_i) \in V_i, \ i = 1, \ldots, n\},$$
where $x_1, \ldots, x_n \in X$, and $V_1, \ldots, V_n$ are neighbourhoods of zero in the

topology τ. Since the topology τ is weaker than the norm topology, there exist $\varepsilon_i > 0$, $i = 1, \ldots, n$ such that

$$U \supset U_0 = \{\psi \in B(X \to E): \|\psi(x_i)\| \leqslant \varepsilon_i, \ i = 1, \ldots, n\}.$$

By the continuity of $B_t A(x)$ for every $x \in X$, there exists a $\delta > 0$ such that for all t, $|t - t_0| < \delta$, we have

$$\|B_t A(x_i) - B_{t_0} A(x_i)\| \leqslant \frac{\varepsilon_i}{2M}, \quad i = 1, \ldots, n.$$

Hence, for $\psi \in K_M$,

$$\|\psi(B_t A(x_i)) - \psi(B_{t_0} A(x_i))\| \leqslant \varepsilon_i.$$

This implies that

$$\psi(B_t A(x_i)) - \psi(B_{t_0} A(x_i)) \in V_i, \quad i = 1, \ldots, n,$$

and therefore, for $\psi \in K_M$,

$$\mathscr{B}_t \mathscr{A} \psi - \mathscr{B}_{t_0} \mathscr{A} \psi \in U.$$

This entails the continuity of the multifunction $\mathscr{B}_t \mathscr{A}(K_M)$ in the topology $\tilde{\tau}_X$.

Further, we proceed as in the proof of Theorem 5.7.1. In view of Theorem 5.4.7, the set

$$W_{\delta, T} = \bigcup_{T \leqslant t \leqslant T + \delta} \mathscr{B}_t \mathscr{A}(K_M)$$

is compact for each positive δ. The one-element set $\{F\}$ is also compact and hence, by Theorem 5.4.4, the minimum-time problem is solvable and the function $\varphi(t)$ is lower semicontinuous. ∎

COROLLARY 5.7.3. *If E is a finite-dimensional space, or a reflexive space, or the conjugate space of a certain Banach space, then the function $\varphi(t)$ defined by the formula (5.7.4) is lower semicontinuous.*

5.8. NONCONTINUOUS LINEAR SYSTEMS. ESTIMABILITY, OBSERVABILITY, CONTROLLABILITY

In the previous section we considered linear systems $(X \underset{A}{\to} \square \underset{B}{\to} Y)$, where the linear operators A and B were assumed to be continuous. This continuity assumption is important for the minimization of continuous functionals, since it ensures that for each continuous functional $F(x, \eta, y)$, $(x, \eta, y) \in X \times \square \times Y$, the functional $F(x, A(x), BA(x))$ is continuous on X. On the other hand, if we consider problems of controllability or observability, this continuity assumption is not so important. What is more, there are some models for which the continuity assumption is too restrictive.

A more general approach is proposed by Dolecki (1977), and by Dolecki and Russell (1977), but with a different definition of observability. We shall present here their approach. Our presentation, however, will be consistent with our definition of observability.

To begin with, we observe that it is more convenient to consider the composition BA as a single operator C, $C = BA$. In this way we overcome the necessity of giving a precise definition of the composition of noncontinuous operators.

We consider a linear system
$$X \supset D_C \underset{C}{\to} Y, \tag{5.8.1}$$
where X and Y are Banach spaces and C is a closed operator which maps the domain D_C into Y. The domain D_C is assumed to be dense. Let E be a Banach space and let F be a continuous linear operator which maps X into E.

We say that F is *observable* if there is a continuous linear operator Φ, called a *method of observation*, such that
$$F(x) = \Phi(C(x)) \quad \text{for all } x \in D_C. \tag{5.8.2}$$

In other words, there is a continuous operator Φ such that the diagram
$$\begin{array}{c} X \supset D_C \underset{C}{\to} Y \\ {}_F\searrow \quad \swarrow_\Phi \\ E \end{array} \tag{5.8.3}$$
is commutative. Of course, if F is observable, then
$$||F(x)|| \leqslant ||\Phi|| \, ||C(x)||. \tag{5.8.4}$$
This means that if F is observable, then there is a constant $K > 0$ such that
$$||F(x)|| \leqslant K ||C(x)||. \tag{5.8.5}$$

5.8. Noncontinuous linear systems

Inequality (5.8.5) represents a necessary condition for the observability of F. However, this condition is not sufficient. The operators F satisfying (5.8.5) will be called *C-estimable*. In particular, if the identity operator I is C-estimable (of course, in this case we assume that $X = E$), we say that the operator C is *estimating*. If C is estimating, then for an arbitrary Banach space E and an arbitrary continuous linear operator F, the operator F is C-estimable.

The above considerations fit in with the framework elaborated in the previous sections. Namely, we shall introduce a new topology on D_C by a norm $\| \ \|_C$, $\|x\|_C = \|x\| + \|C(x)\|$. The space D_C with the norm $\| \ \|_C$ is a Banach space. If F is a continuous operator acting from X into E, then it is also a continuous operator acting from D_C into E. Hence, we can replace system (5.8.1) by the continuous linear system

$$(D_C \underset{C_1}{\to} Y), \tag{5.8.6}$$

where C_1 is a continuous linear operator which maps $(D_C, \| \ \|_C)$ into Y and is induced by C.

However, if together with a given system we consider the conjugate system, systems (5.8.1) and (5.8.6) differ significantly. Namely, by taking the conjugate of system (5.8.1) we obtain the system

$$D_{C^*} \subset Y^* \to X^*, \tag{5.8.7}$$

where D_{C^*} is the set of continuous linear functionals $\eta \in Y^*$ such that the functionals $(C^*\eta)(x) = \eta(C(x))$ are continuous linear functionals defined on X, $C^*\eta \in X^*$. The conjugate operator C^* is closed (see Proposition 3.4.3).

The conjugate of system (5.8.6) is a linear system,

$$(Y^* \underset{C_1^*}{\to} D_{C^*}),$$

where C_1^* is a continuous linear operator, the conjugate of C_1.

The operators C^* and C_1^* differ significantly. They have different domains and different ranges.

As in the case of continuous linear systems, we say that system (5.8.1) (or (5.8.7)) is *controllable* if

$$C(D_C) \supset Y \quad (C^*(D_{C^*}) \supset X^*).$$

Rewriting Theorems 3.4.4–3.4.6 in terms of controllability and estimability, we obtain the following theorems.

THEOREM 5.8.1 (Dolecki, Russell, 1977). *The operator C is estimating if and only if the conjugate system* (5.8.7) *is controllable.*

THEOREM 5.8.2 (Dolecki, Russell, 1977). *System* (5.8.1) *is controllable if and only if the conjugate operator C^* is estimating.*

We say that the conjugate system (5.8.1) is F^*-controllable if

$$R_{C^*} = C^*(D_{C^*}) \supset F^*(E^*) = R_{F^*}. \tag{5.8.8}$$

THEOREM 5.8.3 (Dolecki, Russell, 1977). *Let a linear system*

$$(X \supset D_C \underset{C}{\to} Y) \tag{5.8.1}$$

be given. A continuous linear operator F which maps X into a Banach space E is C-estimable if and only if the conjugate system (5.8.7) *is F^*-controllable.*

Proof. Necessity. Suppose that the range of F^* is contained in the range of C^*, $R_{F^*} \subset R_{C^*}$. Let $Y_1 = \overline{R_{C^*}}$. We consider the system

$$\begin{array}{c} X \supset D_C \underset{C}{\to} Y_1 \\ {}_F \searrow \\ E \end{array} \tag{5.8.9}$$

and the conjugate system

$$\begin{array}{c} D_{C^*} \subset Y_1^* \underset{C^*}{\to} X^* \\ \nearrow {}_{F^*} \\ E^* \end{array} \tag{5.8.10}$$

Since $Y_1 = \overline{R_{C^*}}$ and $\ker C^*|_{Y_1^*} = \{0\}$, by the Douglas theorem (Theorem 3.3.11) there is a continuous linear operator G such that $F^* = C^*G$.

Thus, $F^{**} = G^*C^{**}$. However, $F^{**}(x) = F(x)$ and $C^{**}(x) = C(x)$ for $x \in X$ and finally

$$||F(x)|| \leqslant ||G^*|| \, ||C(x)|| = ||G|| \, ||C(x)||. \tag{5.8.11}$$

To complete this part of the proof it is enough to observe that the range of the operator C^*, R_{C^*}, coincides with the range of the operator C^* restricted to Y_1, $R_{C^*} = R_{C^*}|_{Y_1^*}$.

5.8. Noncontinuous linear systems

Sufficiency. Suppose that F is C-estimable, i.e., there is a $K > 0$ such that for $x \in D_C$ we have

$$\|F(x)\| \leqslant K \cdot \|C(x)\|. \tag{5.8.12}$$

Take any functional $f \in R_{F^*}$, i.e., an f of the form $f(x) = \varphi(F(x))$, where $\varphi \in E^*$. Let $H = \{x\colon f(x) = 1\}$. Since $1 = |f(x)| \leqslant \|\varphi\| \|F(x)\|$, we have $\inf\{\|F(x)\|\colon x \in H\} \geqslant \dfrac{1}{\|\varphi\|} > 0$. By (5.8.12) we have $\inf\{\|C(x)\|\colon x \in H\} \geqslant \dfrac{1}{K \cdot \|\varphi\|} > 0$ and $0 \notin \mathrm{cl}(C(H))$. By the same arguments as in the proof of Theorem 5.6.1, we conclude that there is a ψ belonging to Y^* such that $\psi(C(x)) = f(x)$ for $x \in D_C$. Since f is continuous, $\psi \in D_{C^*}$ and $f \in R_{C^*}$. Since f is arbitrary, we get

$$R_{F^*} \subset R_{C^*}. \blacksquare$$

Theorem 5.8.3 gives a relation between the C-estimability of an operator F and the F^*-controllability of the conjugate system.

Consider a system

$$X \supset D_C \underset{C}{\to} Y \tag{5.8.1}$$

and a continuous linear operator F which maps a Banach space E into Y. We say that system (5.8.1) is *F-controllable* if the range of F is contained in the range of C,

$$R_F \subset R_C.$$

Now the following question arises. Is it true that system (5.8.1) is F-controllable if and only if the conjugate operator F^* is C^*-estimable? In the sequel we shall study this problem.

THEOREM 5.8.4 (Dolecki, Russell, 1977). *If system* (5.8.1) *is F-controllable, then F^* is C^*-estimable.*

Proof. Replace X by the quotient space $X_1 = Y/\ker C$. The operator C induces an operator C_1 acting from X_1 into Y by the formula $C_1([x]) = C(x)$, where $[x]$ denotes the equivalence class containing x. It is easy to see that $||C^*f|| = ||C_1^*f||$. Now consider the system

$$X_1 \supset D_{C_1} \xrightarrow[C_1]{} Y$$
$$E \xrightarrow{F}$$
(5.8.13)

By the Douglas theorem (Theorem 3.3.11), there is a continuous linear operator G mapping E into X_1 and such that

$$F(x) = C_1 G(x) \quad \text{for} \quad x \in E.$$

Take any $f \in Y^*$. We get

$$f(F(x)) = f(C_1 G(x))$$

and

$$(F^*f)(x) = (C_1^*f)(G(x)).$$

Hence

$$||F^*f|| = \sup_{||x|| \leq 1} |(F^*f)(x)|$$
$$\leq \sup_{||z|| \leq ||G||} |(C_1^*f)(z)| \leq ||G|| \cdot ||C_1^*f|| = ||G|| \cdot ||C^*f||. \blacksquare$$

The proof of the converse implication is more complicated and requires additional assumptions.

THEOREM 5.8.5 (Dolecki, Russell, 1977). *Let* $(X, || \ ||_X)$ *and* $(Y, || \ ||_Y)$ *be Banach spaces. Consider a linear system*

$$(X \supset D_C \xrightarrow[C]{} Y),$$
(5.8.1)

where C is a closed linear operator. Let F be a continuous linear operator which maps a Banach space $(E, || \ ||_E)$ *into* $(Y, || \ ||_Y)$. *Assume that the range of the conjugate operator C^* is dense in* $(\ker C)^\perp = \{f \in X^*: f(x) = 0$ *for* $x \in \ker C\}$. *Moreover, assume that either $F^{-1}(R_C)$ is dense in E and the range R_F is dense in R_C, $\overline{R_F} \supset R_C$, or $C^{-1}(R_F)$ is dense in X. Then the C^*-estimability of F^* implies the F-controllability of system* (5.8.1), $R_F \subset R_C$.

Proof. Without loss of generality we can assume that the operators C and F are one-to-one mappings. To see this, consider the quotient

5.8. Noncontinuous linear systems

spaces $X_1 = X/\ker C$, $E_1 = E/\ker F$. The norms $\| \, \|_X$ and $\| \, \|_E$ induce the norms $\| \, \|_X^1$ and $\| \, \|_E^1$ in the spaces $X_1 = X/\ker C$, $E_1 = E/\ker F$. Of course, C and F induce continuous linear one-to-one operators C_1 and F_1 which map the spaces X_1 and E_1 into Y, $C_1([x]) = C(x)$, $F_1([e]) = F(e)$, where $[x]$ ($[e]$) denote the equivalence class containing x (e).

It is easily shown that the ranges of operators C and C_1 (F and F_1) coincide, $R_C = R_{C_1}$ ($R_F = R_{F_1}$).

Let $\eta \in Y^*$. We shall calculate $\|C_1^* \eta\|_{X_1^*}$. Namely,

$$\|C_1^* \eta\|_{X_1^*} = \sup_{[x] \neq 0} \frac{|\eta(C_1([x]))|}{\|[x]\|_X^1} = \sup_{[x] \neq 0} \frac{|\eta(C(x))|}{\inf_{x \in [x]} \|x\|_X}$$

$$= \sup_{[x] \neq 0} \sup_{x \in [x]} \frac{|\eta(C(x))|}{\|x\|_X} = \sup_{x \neq 0} \frac{|\eta(C(x))|}{\|x\|}$$

$$= \|C^* \eta\|_{X^*}, \qquad (5.8.14)$$

where $\| \, \|_{X^*}$ ($\| \, \|_{X_1^*}$) denotes the norm induced in X^* (X_1^*) by the norm $\| \, \|_X$ ($\| \, \|_X^1$).

In a similar way we obtain

$$\|F_1^* \eta\|_{E_1^*} = \|F \eta\|_{E^*} \qquad (5.8.15)$$

where $\| \, \|_{E^*}$ ($\| \, \|_{E_1^*}$) denotes the norm induced in the space E^* (E_1^*) by the norm $\| \, \|_E$ ($\| \, \|_{E_1}$).

By formulae (5.8.14) and (5.8.15) we find that F^* is C^*-estimable if and only if F_1^* is C_1^*-estimable. Thus, as was mentioned at the beginning of the proof, we may assume that the operators C and F are both one-to-one mappings.

Now we introduce a new norm, $\| \, \|_{R_F}$, on the range of F, R_F, by the formula

$$\|y\|_{R_F} = \|F^{-1}(y)\|_E$$

and a new norm, $\| \, \|_{R_C}$, on the range of the operator C, R_C, by the formula

$$\|y\|_{R_C} = \|C^{-1}(y)\|_X.$$

In general, the topologies induced by these norms are noncomparable.

Observe that $(R_F, \| \, \|_{R_F})$ is a complete space because it is the image of the space E under an isometric mapping.

Let
$$X_0 = \overline{C^{-1}(R_F)} \subseteq X$$
and let C_0 be the restriction of C to X_0. We shall show that
$$C_0(X_0) \supset F(E) = R_F. \qquad (5.8.16)$$
By the definition, the operator C_0 is densely defined on X_0. It is also a closed operator from X_0 into $(R_F, \|\ \|_{R_F})$, since $\|\ \|_{R_F}$ is stronger than $\|\ \|_Y$ (see Proposition 3.4.4). By Theorems 3.4.4 and 3.4.5, C_0 maps $(X_0, \|\ \|_X)$ onto $(R_F, \|\ \|_{R_F})$ if and only if there is a constant $K_0 > 0$ such that
$$\|\eta\|_{R_F^*} \leq K_0 \cdot \|C_0\eta\|_{X_0^*}, \qquad (5.8.17)$$
where $\|\ \|_{R_F^*}$ ($\|\ \|_{X_0^*}$) denotes the norm induced by $\|\ \|_{R_F}$ ($\|\ \|_X$) and η is an arbitrary continuous linear functional defined on $(R_F, \|\ \|_{R_F})$.

To complete the proof it is enough to prove that the C^*-estimability of F^* implies (5.8.17).

To see this, take an arbitrary continuous linear functional $\hat{\eta}$ defined on $(Y, \|\ \|_Y)$. By η we denote the restriction of $\hat{\eta}$ to R_F, $\eta = \hat{\eta}|_{R_F}$. Thus,
$$\|F^*\eta\|_{E^*} = \sup_{\substack{z \in E \\ z \neq 0}} \frac{|\eta(F(z))|}{\|z\|_E} = \sup_{\substack{y \in R_F \\ y \neq 0}} \frac{|\eta(y)|}{\|y\|_{R_F}} = \|\eta\|_{R_F^*}.$$

We have two different norms on R_F, $\|\ \|_Y$ and $\|\ \|_{R_F}$. Hence, the operator C_0 induces two conjugate operators, C_0^* and C_0^F. The operators C_0^* and C_0^F have different domains. However, if η belongs to the intersection of those domains, then
$$C_0^*\eta = C_0^F\eta.$$

Since $\|\ \|_{R_F}$ is stronger than $\|\ \|_Y$, we have $D_{C_0^F} \supset D_{C_0^*}$. Suppose that η belongs to $D_{C_0^*}$. Thus, since η is the restriction of $\hat{\eta}$, we obtain
$$\|C_0^F\eta\|_{X_0^*} = \sup_{\substack{x \in X_0 \\ x \neq 0}} |\hat{\eta}(C(x))| \leq \sup_{x \in D_C} |\hat{\eta}(C(x))|$$
$$\leq \|C^*\hat{\eta}\|_{X^*}. \qquad (5.8.18)$$

By the assumption that either R_F is dense in R_C or $C^{-1}(R_F)$ is dense

5.8. Noncontinuous linear systems

in X and by the continuity of $\hat{\eta}$, inequality (5.8.18) reduces to the equality

$$||C_0^F \eta||_{X_0^*} = ||C_0^* \eta||_{X_0^*} = ||C^* \hat{\eta}||_X \qquad (5.8.19)$$

and, by (5.8.17),

$$||\eta||_{R_{F^*}} \leqslant K \cdot ||C_0^* \eta||_{X_0^*} \qquad (5.8.20)$$

for all η which are continuous on $(R_F, ||\ ||_Y)$. Of course, there are also functionals η which are continuous on $(R_F, ||\ ||_{R_F})$ and not continuous on $(R_F, ||\ ||_Y)$. We claim that for such η formula (5.8.20) holds.

To see this, take an arbitrary η which is continuous on $(R_F, ||\ ||_{R_F})$ and belongs to the domain C_0^F, i.e., $\eta(C(x))$ is continuous on $(X, ||\ ||_X)$. Observe that $(C_0^F \eta) \in (\ker C)^\perp$.

We have assumed that the range R_{C^*} is dense in $(\ker C)^\perp$. On the other hand, by the definition of C_0 the range of C_0^* is dense in $(X_0^*, ||\ ||_{X_0^*})$. Thus, there is a sequence $\{\eta_k\} \subset D_{C_0^*}$ such that

$$\lim_{k \to \infty} ||C_0^* \eta_k - C_0^F \eta||_{X_0^*} = 0. \qquad (5.8.21)$$

By (5.8.20), the sequence $\{\eta_k\}$ is fundamental in the norm $||\ ||_{R_{F^*}}$. Hence, $\{\eta_k\}$ converges in the norm $||\ ||_{R_{F^*}}$ to a certain linear functional η_0 which is continuous on $(R_F, ||\ ||_{R_F})$. The operator C_0^F is closed. Since $C_0^* \eta_k = C_0^F \eta_k$, the limit $C_0^F \eta_k$ exists. The range $R(C_0)$ is dense in X_0. Hence, $\eta_0 = \eta$.

Since (5.8.21) holds for all η_k, $k = 1, 2, \ldots$, then, passing to the limit, we find that (5.8.21) holds for all η which are continuous on $(R_F, ||\ ||_{R_F})$. This completes the proof. ∎

Chapter 6

Systems governed by ordinary differential equations

6.1. PROBLEMS OF MINIMIZATION OF CONVEX FUNCTIONALS RELATED TO SYSTEMS GOVERNED BY ORDINARY DIFFERENTIAL EQUATIONS

In optimization problems we often encounter objects governed by systems of ordinary linear differential equations.

Suppose we are given a differential equation

$$\frac{dx}{dt} = A(t)x + S(t)u(t), \quad 0 \leqslant t \leqslant T, \tag{6.1.1}$$

where $x(t)$ is an n-dimensional vector-valued function, $u(t)$ is an m-dimensional vector-valued function, and $A(t)$, $S(t)$ are matrix-valued functions of dimensions $n \times n$ and $n \times m$, respectively.

To investigate equation (6.1.1) one should make some assumptions concerning continuity and measurability of functions $x(t)$, $u(t)$, $A(t)$, $S(t)$. The usual assumption about the continuity of both sides of equation (6.1.1) and the assumption that the equation is satisfied at each point have two disadvantages. Firstly, from the mathematical point of view, the space $C[0, T]$ is not the conjugate space of any Banach space, whence, the weak-$*$-topology cannot be introduced in it. The second disadvantage, which is of a technical nature, is that in many cases controls $u(t)$ do not change continuously but with jumps. For these two reasons we shall consider $x(t)$, $A(t)$, $S(t)$ and $u(t)$ satisfying the following assumptions:

(a) the function $x(t)$ has the derivative $x'(t)$ for almost all t and, moreover,

$$x(t) = x(0) + \int_0^t x'(t) dt;$$

6.1. Convex minimization in ODE* systems

(b) the function $A(t)$ is measurable and integrable (i.e., each term of the matrix $A(t)$, $a_{l,k}(t)$, is measurable and belongs to $L^1[0, T]$);

(c) the functions $S(t)$ and $u(t)$ are measurable and the function $S(t)u(t)$ is integrable.

Moreover, we require equation (6.1.1) to be satisfied for almost all t.

We can prove the existence and the form of solutions on the basis of the so called *fundamental matrix* $\Phi(t)$, which solves an operator equation of the form

$$\frac{d\Phi}{dt} = A(t)\Phi(t), \tag{6.1.2}$$

where $\Phi(t)$ is a continuous matrix-valued function (of dimension $n \times n$), differentiable almost everywhere and the derivative $\Phi'(t)$ satisfies equation (6.1.2) almost everywhere.

Of course, the matrix $\Phi(t)$ can be regarded as an n^2-dimensional vector-valued function. The right-hand side of equation (6.1.2) is a linear function. Hence, it satisfies the Lipschitz condition with a Lipschitz constant $L(t)$. $L(t)$ can be estimated by a linear combination of the modulus of the coefficients of the matrix $A(t)$ and hence, $L(t)$ is integrable.

Therefore, using the Bielecki method (cf. Section 1.6), we can prove the existence of a solution of equation (6.1.2), which corresponds to a given initial condition.

Let $\Phi(t)$ be solution of equation (6.1.2), which corresponds to a given initial condition

$$\Phi(0) = I. \tag{6.1.3}$$

Having determined $\Phi(t)$, we can determine the solution $x(t)$ of the homogeneous equation

$$\frac{dx}{dt} = A(t)x, \tag{6.1.4}$$

which satisfies the initial condition

$$x(0) = x_0. \tag{6.1.5}$$

---
* Ordinary differential equations

Namely, the solution $x(t)$ is of the form

$$x(t) = \Phi(t)x_0. \qquad (6.1.6)$$

Now we shall prove that the fundamental matrix is invertible. To do this, we shall introduce a time reversing variable s,

$$s = T-t. \qquad (6.1.7)$$

In this new variable equations (6.1.2) and (6.1.4) take the form

$$\frac{dy}{ds} = -\tilde{A}(s)y(s), \quad 0 \leq s \leq T, \qquad (6.1.8)$$

$$\frac{d\psi}{ds} = -\tilde{A}(s)\psi(s), \quad 0 \leq s \leq T, \qquad (6.1.9)$$

where $\tilde{A}(s) = A(T-s)$. Equations (6.1.8) and (6.1.9) satisfy the conditions ensuring the existence of solutions. Thus, for an arbitrary initial condition

$$y(0) = y_0 \qquad (6.1.10)$$

there is a unique solution of equations (6.1.8) and (6.1.9) corresponding to the initial condition (6.1.10). Returning to the variable t we find that for an arbitrary final condition

$$x(T) = x_0 \qquad (6.1.11)$$

there is a unique solution of equation (6.1.4) satisfying (6.1.11). This means that $\Phi(T)$ maps R^n onto R^n and consequently it is invertible. Let t_0 be an arbitrary point of the interval $[0, T]$. Replacing T by t_0 in the above argumentation we find that the matrix $\Phi(t_0)$ is invertible for all t_0, $0 \leq t_0 \leq T$.

The existence of the fundamental matrix, which we have just proved, does not imply that it can easily be calculated. If $A(t)$ does not depend on t, $A(t) \equiv A$, then

$$\Phi(t) = e^{tA} = \sum_{n=0}^{\infty} \frac{t^n A^n}{n!}. \qquad (6.1.12)$$

In the case where $A(t)$ depends on time, there are technical difficulties in finding the fundamental matrix $\Phi(t)$. In many cases the matrix $\Phi(t)$ cannot be formed by elementary functions.

6.1. Convex minimization in ODE* systems

However, in the sequel we shall assume the fundamental matrix to be known. This assumption allows us to prove some qualitative results.

Under the assumption that the fundamental matrix is known, we can express the solution of equation (6.1.1) which corresponds to an initial condition

$$x(0) = x_0 \tag{6.1.13}$$

in the form

$$x(t) = \Phi(t)x_0 + \int_0^t \Phi(t)\Phi^{-1}(s)S(s)u(s)\,ds. \tag{6.1.14}$$

Let $\hat{U}$ be a space of m-dimensional vector-valued functions $u(t)$. We shall consider mainly the following spaces:

(1) the space $M_m[0, T]$ of all m-dimensional measurable vector-valued functions with all components essentially bounded, with the norm

$$\|u\| = \max_{1 \leq i \leq m} \sup_{0 \leq t \leq T} |u_i(t)|;$$

(2) the space $L_m^1[0, T]$ of all m-dimensional measurable vector-valued functions with all components integrable, with the norm

$$\|u\| = \sum_{i=1}^{m} \int_0^T |u_i(t)|\,dt;$$

(3) the space $L_m^2[0, T]$ of all m-dimensional measurable vector-valued functions with all components square integrable, with the norm

$$\|u\| = \left(\sum_{i=1}^{m} \int_0^T |u_i(t)|^2\,dt\right)^{1/2}.$$

As in the case of $m = 1$, we can prove that the conjugate space of the space $L_m^2[0, T]$ can be identified with itself, and hence $L_m^2[0, T]$ is reflexive. The space $M_m[0, T]$ can be identified with the conjugate space of the space $L_m^1[0, T]$. This means that every continuous linear functional F defined on $L_m^2[0, T]$ ($L_m^1[0, T]$) is of the form

$$F(u) = \sum_{i=1}^{m} \int_0^T f_i(t)u_i(t)\,dt, \tag{6.1.15}$$

where $f = (f_1(t), \ldots, f_m(t))$ belongs to $L_m^2[0, T]$ ($L_m^1[0, T]$) and $\|F\| = \|f\|$.

Balls in $L_m^1[0, T]$ have no extreme points (cf. Example 4.6.1). Hence, by the Krein–Milman theorem (Theorem 4.6.1), on that space we cannot introduce a topology τ such that the unit ball is compact in τ. This implies, by the Alaoglu theorem (Theorem 4.4.6), that $L_m^1[0, T]$ is not isometric to any conjugate space.

Let the input space X be the product of the n-dimensional space R^n of initial states and the space $\hat{U}$. Let $\square$ be the space $C_n[0, T]$ and let $Y = R^n$. Then equation (6.1.14) defines the input operator A and the output operator B is defined as $B(x) = x(T)$.

THEOREM 6.1.1. *Let $\hat{U} = L_m^2[0, T]$. Let U be a closed convex set contained in $X \times \square \times Y$ and such that the set U_0, $U_0 = \{x \in X: (x, A(x), BA(x)) \in U\}$, is bounded and closed. Let $F(x, y, z)$ be a convex continuous functional defined on $X \times \square \times Y$. Then there exists an x_0 such that $(x_0, A(x_0), BA(x_0)) \in U$ and*

$$F(x_0, A(x_0), BA(x_0))$$
$$= \inf \{F(x, A(x), BA(x)): (x, A(x), BA(x)) \in U\}. \quad (6.1.16)$$

Proof. The proof is an immediate consequence of Theorem 5.3.4. ■

Here we present typical examples of sets U and functionals F.

Sets U:

$x(0) \in V_0$, $x(T) \in W_0$, $u \in U_0$, $x(t) \in \tilde{U}(t)$ for $0 \leqslant t \leqslant T$, where V_0, W_0, U_0 are convex closed sets, U_0, V_0 are bounded and the multifunction $\tilde{U}(t)$ is linearly upper semicontinuous with values being the closed convex bounded sets $U(t)$. In particular, it often happens that V_0 or W_0 (or both) are one-point sets and the set U_0 is of the form

$$U_0 = \{u: u(t) \in U(t)\},$$

where $U(t)$ is an linearly upper semicontinuous multifunction. In particular, $U(t)$ can be constant.

Functionals F:

(1) $F(u) = \|u\|$,

(2) $F(u) = \int_0^T [u(t), x(t)] K(t) \begin{bmatrix} u(t) \\ x(t) \end{bmatrix} dt + (x(T), f)^2$,

where $K(t)$ is a measurable matrix-valued function, the values of which are positive semidefinite matrices of dimension $(n+m)\times(n+m)$, $[u, x]$ is an $(n+m)$-dimensional row vector, $\begin{bmatrix} u \\ x \end{bmatrix}$ is an $(n+m)$-dimensional column vector, f is a given vector of an n-dimensional space and (a, b) is the inner product in an n-dimensional space.

Functionals of type (2) are called *quadratic performance functionals*.

6.2. CONTROLLABILITY OF FINITE-DIMENSIONAL SYSTEMS

Suppose we are given a system which is described by equation (6.1.1)

$$\frac{dx}{dt} = A(t)x(t)+S(t)u(t), \quad 0 \leqslant t \leqslant T,$$

where $x(t)$, $u(t)$, $A(t)$, $S(t)$ satisfy the conditions specified at the beginning of Section 6.1. We say that system (6.1.1) is controllable if for every pair of vectors $x_0, x_1 \in R^n$ there exists a control $u \in U$ such that the solution $x(t)$ corresponding to the control u and the initial condition $x(0) = x_0$ satisfies the final condition $x(T) = x_1$.

By formula (6.1.14), after simple transformations, we find that system (6.1.1) is controllable if and only if there exists a $u \in U$ such that

$$K(u) = \int_0^T \Phi^{-1}(s)S(s)U(s)ds = y_0, \qquad (6.2.1)$$

where $\Phi(s)$ denotes the fundamental matrix of equation (6.1.2) and

$$y_0 = \Phi^{-1}(T)x_1 - x_0. \qquad (6.2.2)$$

Since we seek a u satisfying (6.2.1) for all $x_0, x_1 \in R^n$, system (6.1.1) is controllable if and only if the operator K maps U onto the whole $Y = R^n$. Since the operator K is determined by the matrix $\Phi^{-1}(s)S(s)$ with n rows and m columns, K maps U onto Y if and only if the rows $K_1(t), \ldots, K_n(t)$ are linearly independent as vector-valued functions defined on the interval $[0, T]$. It is evident that, in the general case, controllability depends upon the length of the interval on which our system is defined. For instance, it may happen that for $T_0 < T$ the system is not controllable on the interval $[0, T_0]$ but it is controllable on the interval $[0, T]$. Clearly, if the system is controllable on the interval $[0, T_0]$, then it is also controllable on the interval $[0, T]$.

Consider now the case where $A(t)$ and $S(t)$ are constant, $A(t) = A$, $S(t) = S$. Then $\Phi(t) = e^{tA} = \sum_{i=0}^{\infty} \frac{1}{i!} t^i A^i$ and $\Phi(t) S = e^{-tA} S$.

Now we prove

THEOREM 6.2.1 (Kalman, 1960). *System* (6.1.1) *is controllable if and only if the matrix*
$$[S, AS, \ldots, A^{n-1}S] \tag{6.2.3}$$
with n rows and mn columns is of rank n.

Proof. Necessity. Assume that the matrix given by formula (6.2.3) is of rank less than n. This means that there exists an n-dimensional row vector v such that
$$vS = vAS = \ldots = vA^{n-1}S = 0. \tag{6.2.4}$$

But the matrix A satisfies its own characteristic equation
$$A^n = \sum_{i=0}^{n-1} c_i A^i, \tag{6.2.5}$$
and hence, by (6.2.4),
$$vA^n S = 0. \tag{6.2.6}$$

Multiplying both sides of (6.2.6) by A and using (6.2.4), we get
$$vA^{n+1}S = 0. \tag{6.2.7}$$

Continuing this process, we obtain by induction,
$$vA^m S = 0, \quad m = 1, 2, \ldots \tag{6.2.8}$$

Hence,
$$v \cdot \sum_{i=1}^{\infty} \frac{(-t)^i}{i!} A^i S = v e^{-tA} S = 0. \tag{6.2.9}$$

This proves that the system is not controllable.

6.3. Minimum-norm problems for sup type norms

Sufficiency. Suppose that system (6.1.1) is not controllable. This means that there exists an n-dimensional nonzero vector v such that for all u

$$v\int_0^T e^{-tA} Su(t)\,dt = 0. \tag{6.2.10}$$

Since u is arbitrary,

$$ve^{-tA}S = 0 \quad \text{for} \quad 0 \leqslant t \leqslant T. \tag{6.2.11}$$

Differentiating both sides of (6.2.11) n times and putting $t = 0$, we get

$$vS = vAS = \ldots = vA^{n-1}S = 0. \tag{6.2.12}$$

This implies that matrix (6.2.3) is of rank less than n since v is a nonzero vector. ∎

COROLLARY 6.2.2. *If system (6.1.1) with constant coefficients is controllable on a given interval* $[0, T_0]$, *then it is also controllable on every interval* $[0, T]$.

6.3. MINIMUM-NORM PROBLEMS FOR sup TYPE NORMS

Consider system (6.1.1),

$$\frac{dx}{dt} = A(t)x(t) + S(t)u(t), \quad 0 \leqslant t \leqslant T,$$

with the initial condition $x(0) = x_0$ and the final condition $x(T) = x_1$. The assumptions about the functions $x(t)$, $u(t)$ and the matrices $A(t)$, $S(t)$ are the same as in Section 6.1.

Moreover, we shall assume that u belongs to $M_m[0, T]$, i.e., that the components $u_i(t)$ are measurable and essentially bounded functions. We introduce a norm $\| \ \|$ in the space $M_m[0, T]$ in the following way. Suppose that in the m-dimensional space R^m we are given a certain norm (or seminorm) $\| \ \|_E$. The space R^m with this norm will be denoted by E. Now we define a norm (or seminorm) of $u(t)$ by the formula

$$\|u\| = \text{ess sup} \|u(t)\|_E. \tag{6.3.1}$$

Norms of this type often appear in practice. For instance, they arise when we solve a minimum-time problem under the assumption that $u(t) \in U$, where U is a closed, convex bounded set. Without loss of generality we can assume that 0 is an interior point of the set U (otherwise we can shift the set U). Then U generates a norm (if, additionally, U is symmetric) or a seminorm. Let us denote that norm (or seminorm) by $\| \ \|_E$. The condition $u(t) \in U$ can be restated as follows

$$\|u\| = \operatorname{ess\,sup} \|u(t)\|_E \leqslant 1.$$

As follows from the considerations of Section 5.5, if the image of a ball is closed, then the minimum-time problem can be reduced to the munimum-norm problem. We shall prove this fact later on.

However, minimum-norm problems may arise without any relation to minimum-time problems. For instance, minimum-norm problems arise if we want to determine the minimum power (intensity, voltage) of a source, which guarantees that our system can be driven from the position x_0 to the position x_1 in a time interval $[0, T]$.

As in Section 6.2, we can reduce the minimum-time problem corresponding to equation (6.1.1) to the minimum-norm problem corresponding to equation (6.2.1)

$$K(u) = \int_0^T \Phi^{-1}(s) S(s) u(s) \, ds = y_0$$

where, according to (6.2.2),

$$y_0 = \Phi^{-1}(T) x_1 - x_0,$$

and $\Phi(t)$ is the fundamental matrix of the homogeneous equation (6.1.2) corresponding to the initial condition (6.1.3).

The output space Y is finite-dimensional. Hence, in virtue of Corollary 5.2.6, there exists a functional $\varphi_0 \in Y^*$ such that

$$\inf \{\|u\| : K(u) = y_0\} = \inf \{\|u\| : \varphi_0(K(u)) = \varphi_0(y_0)\}. \quad (6.3.2)$$

Now we explain the meaning of the expression $\varphi_0(K(u))$. Since Y is m-dimensional, every functional $\varphi \in Y^*$ can be interpreted as a row vector. Hence,

$$\varphi_0(K(u)) = \int_0^T \varphi_0 \Phi^{-1}(s) S(s) u(s) \, ds = \int_0^T \tilde{\psi}(s) S(s) u(s) \, ds, \quad (6.3.3)$$

6.3. Minimum-norm problems for sup type norms

where $\tilde{\psi}(s) = \varphi_0 \Phi^{-1}(s)$. Let $\psi(s)$ be the transposition of the row vector $\tilde{\psi}(s)$. Then (6.3.3) can be rewritten as

$$\varphi_0(K(u)) = \int_0^T (\psi(s), S(s)u(s))\,ds, \tag{6.3.4}$$

where (a, b) denotes the inner product of vectors a and b.

By definition, $\psi(s) = (\Phi^{-1}(s))^{tr}\varphi_0^{tr}$, where $(C)^{tr}$ denotes the transposed matrix (vector) of a matrix (vector) C.

Now we prove that $\psi(s)$ satisfies the equation

$$\frac{d\psi}{ds} = -(A(s))^{tr}\psi(s), \tag{6.3.5}$$

which is called the *conjugate equation* of equation (6.1.2).

To prove this, it is enough to show that the matrix $(\Phi^{-1}(s))^{tr}$ is the fundamental matrix of equation (6.3.5).

Differentiating the matrix $\Phi^{-1}(s)$ we obtain

$$\begin{aligned}
\frac{d\Phi^{-1}(s)}{ds} &= \lim_{h\to 0} \frac{\Phi^{-1}(s+h) - \Phi^{-1}(s)}{h} \\
&= \lim_{h\to 0} \Phi^{-1}(s+h) \frac{\Phi(s) - \Phi(s+h)}{h} \Phi^{-1}(s) \\
&= -\Phi^{-1}(s)\frac{d\Phi}{ds}\Phi^{-1}(s) = -\Phi^{-1}(s)A\Phi(s)\Phi^{-1}(s) \\
&= -\Phi^{-1}(s)A, \tag{6.3.6}
\end{aligned}$$

because $\Phi(s)$ is the fundamental matrix of equation (6.1.1).

By transposing equation (6.3.6) we obtain

$$\frac{d(\Phi^{-1}(s))^{tr}}{ds} = -(A(s))^{tr}(\Phi^{-1}(s))^{tr}. \tag{6.3.7}$$

It immediately follows from (6.3.7) that $\psi(s)$ is a solution of equation (6.3.3), which is the conjugate of equation (6.1.1).

Consider now the problem of existence of an element u_0 which solves the minimum-norm problem, i.e., an element u_0 such that

$$\|u_0\| = \inf\{\|u\|: K(u) = y_0\}. \tag{6.3.8}$$

To solve this problem we shall apply Theorem 5.3.8 and we shall show that the space $M_E[0, T]$ is isometric to the conjugate space of a certain

Banach space; namely, this Banach space is the space $L^1_{E^*}[0, T]$ of all integrable m-dimensional vector-valued functions with the norm

$$||x|| = \int_0^T ||x(t)||_{E^*} dt, \qquad (6.3.9)$$

where E^* denotes the conjugate space of the space E and $||\ ||_{E^*}$ denotes the norm in the conjugate space. Observe that the space E is reflexive as a finite-dimensional space, and thus E can be regarded as the conjugate space of the space E^*.

Let $f \in M_E[0, T]$. Then f defines a continuous linear functional $\tilde{f}$ by the formula

$$\tilde{f}(x) = \int_0^T (f(t), x(t)) dt, \qquad (6.3.10)$$

where (a, b) denotes the inner product of a and b. Estimating

$$|\tilde{f}(x)| \leq \int_0^T ||f(t)||_E ||x(t)||_{E^*} dt \leq ||f||\ ||x||,$$

we obtain $||\tilde{f}|| \leq ||f||$. The opposite inequality and the fact that each continuous linear functional belonging to the space $M_E[0, T]$ is of the form (6.3.10) can be deduced from the arguments used in the corresponding proofs for the spaces $L[0, T]$ and $M[0, T]$ and from the fact that for every $f(t)$ there exists an $x(t)$ such that

$$(f(t), x(t)) = ||f(t)||_E \cdot ||x(t)||_{E^*}.$$

The operator K is determined by a system of continuous linear functionals which are generated by the rows of the matrix $\Phi^{-1}(s)S(s)$. By the assumptions which we have made about $S(s)$ and by the continuity of $\Phi^{-1}(s)$, the rows of the matrix $\Phi^{-1}(s)S(s)$ belong to the space $L^1_{E^*}[0, T]$. This implies that the operator K is continuous in the weak-$*$-topology (note that in the space E norm topologies, the weak topology and the weak-$*$-topology coincide since E is finite-dimensional). Hence, by Theorem 5.3.8, there exists an element u_0 satisfying formula (6.3.8).

Let u_0 be an element satisfying formula (6.3.8). By (6.3.2) and (6.3.4),

6.3. Minimum-norm problems for sup type norms

there exists a solution $\psi(t)$ of the conjugate equation (6.3.5) such that

$$\int_0^T (\psi(t), S(t)u(t))\,dt \leq \int_0^T (\psi(t), S(t)u_0(t))\,dt \qquad (6.3.11)$$

for all $u(t)$ such that $||u|| \leq ||u_0||$, i.e., by the definition of the norm in the space $M_E[0, T]$, for all $u(t)$ such that $||u(t)||_E \leq ||u_0||$ for almost all t.

It is easy to see that (6.3.11) holds if and only if

$$(\psi(t), S(t)u_0(t)) = \sup\{(\psi(t), S(t)u(t)) : ||u|| \leq ||u_0||\}. \qquad (6.3.12)$$

Formula (6.3.12) is called the *maximum principle for linear minimum-norm problems*.

Consider now a minimum-time problem. We want to find a minimal T such that there exists a control $u_0(t)$, $u_0(t) \in U$, which satisfies the condition

$$K(u_0) \in Y(T). \qquad (6.3.13)$$

Write

$$T = \inf\{t > 0: K_t\{u: u(t) \in U\} \cap Y(t) \neq \emptyset\}, \qquad (6.3.14)$$

where

$$K_t(u) = \int_0^t \Phi^{-1}(s)S(s)u(s)\,ds.$$

In view of our previous considerations, the sets $K_t\{u: u(t) \in U\}$ are compact. As we have already assumed (cf. Section 5.5), the multifunction $Y(t)$ is linearly upper semicontinuous. Hence, by Theorem 5.4.4, there exists an u_0 satisfying (6.3.13).

Let $y_0 = K(u_0)$. Now the problem is to find an $u_0(t)$ such that $u_0(t) \in U$ and $u_0(t)$ satisfies (6.2.1). Let v_0 be an arbitrary interior point of the set U. Let $V = U - v_0$. The set V contains zero in its interior and therefore the set V generates a seminorm $||\ ||_E$ (in particular, if V is symmetric, V generates a norm). It is an immediate consequence of the equality $u = v + v_0$ that problem (6.2.1) reduces to the minimum-norm problem:

$$\inf\{||v||_E : K(v) = y_0 - K(v_0)\}. \qquad (6.3.15)$$

Let $u_0(t)$ be an element which satisfies (6.3.13). Then the element v^0, $v^0 = u_0(t) - v_0$, realizes the infimum in (6.3.15).

This implies that there exists a solution $\psi(t)$ of the conjugate equation (6.3.5) such that
$$(\psi(t), S(t)v^0(t)) = \sup_{v \in V} (\psi(t), S(t)v).$$

By adding $(\psi(t), v_0)$ to both sides of the above formula we obtain

THEOREM 6.3.1 (Maximum principle for minimum-time problems). *If $u_0(t)$ solves a minimum-time problem, then there exists a solution $\psi(t)$ of the conjugate equation (6.3.5) such that*
$$(\psi(t), S(t)u_0(t)) = \sup_{u \in U} (\psi(t), S(t)u) \qquad (6.3.16)$$
for almost all t.

6.4. CRITERIA FOR THE UNIQUENESS OF OPTIMAL CONTROL

Consider the minimum-time problem for system (6.1.1) or the corresponding minimum-norm problem.

The maximum principle implies that if a solution $\psi(t)$ of the conjugate equation satisfies equality (6.3.16) and has the additional property that $(\psi(t), S(t)u)$ attains its maximum on U at exactly one point $u_0(t)$ for almost all t, then the maximum principle uniquely determines the optimal control $u_0(t)$.

The assumption that $(\psi(t), S(t)u)$ attains its maximum at exactly one point of the set U implies that that point is an extreme point of the set U. It is easy to verify that the set
$$V_t = \{u' : (\psi(t), S(t)u') = \sup_{u \in U} (\psi(t), S(t)u)\}$$
is an extremal subset of the set U. If the set V_t consists of one point, then this point is an extreme point.

A set W is said to be *strictly convex* if all its boundary points are extreme points. The only extremal subsets of strictly convex sets are extreme points. Hence, if the constraint set U is strictly convex, then the sets V_t are extremal subsets. Thus, in this case, an optimal control is determined uniquely.

6.4. Criteria for uniqueness of optimal control

Consider now a system in which matrices A and S do not depend on t (i.e., A and S are constant). Let U be a polyhedron. We say that U is in the *general position* if the matrix

$$[Sw, ASw, \ldots, A^{n-1}Sw]$$

is of rank n for every vector w which is parallel to an edge of the polyhedron U.

THEOREM 6.4.1 (Pontryagin, Boltyanskii, Gamkrelidze, Mishchenko, 1961). *If a polyhedron U is in the general position with respect to the system*

$$\frac{dx}{dt} = Ax + Su,$$

then the optimal control u_0 is uniquely determined.

Proof. In virtue of the maximum principle, there exists a nonzero solution $\psi(t)$ of the conjugate equation such that (6.3.16) holds, i.e.,

$$(\psi(t), Su_0(t)) = \sup_{u \in U} (\psi(t), Su).$$

Suppose that the optimal control $u_0(t)$ is not unique and $u_1(t)$ is another optimal control. Then, by the maximum principle, $u_1(t)$ also satisfies equality (6.3.16). This means that $u_0(t) \neq u_1(t)$ on a set D of positive measure and

$$(\psi(t), S[u_0(t) - u_1(t)]) = 0. \tag{6.4.1}$$

Let $W(t)$ denote the set

$$W(t) = \{u \in U: (\psi(t), Su) = \sup_{v \in U} (\psi(t), Sv)\}.$$

Clearly, the set $W(t)$ is a certain k-dimensional face of the polyhedron U. By (6.4.1), $\dim W(t) \geq 1$ on the set D.

Since the polyhedron U has only a finite number of faces (we mean faces of all dimensions), there exist a set D_1 of positive measure and a vector w parallel to a certain edge, such that

$$(\psi(t), Sw) = 0 \quad \text{for} \quad t \in D_1. \tag{6.4.2}$$

By the continuity of the function $\psi(t)$ we find that (6.4.1) holds on the closure of the set D_1. Observing that $\psi(t)$ is a solution of an

equation with constant coefficients, we conclude that $\psi(t)$ is an analytic function. Moreover, since $\bar{D}_1$ is an infinite compact set, we have
$$(\psi(t), Sw) = 0 \quad \text{for all } t, \quad 0 \leqslant t \leqslant T. \tag{6.4.3}$$
Differentiating each side of equality (6.4.3) $n-1$ times and using the fact that $\psi(t)$ is a solution of the conjugate equation (6.3.3), we obtain
$$-(A^{\text{tr}}\psi(t), Sw) = 0,$$
$$\cdots\cdots\cdots\cdots\cdots$$
$$(-1)^{n-1}((A^{\text{tr}})^{n-1}\psi(t), Sw) = 0.$$
Hence,
$$(\psi(t), Sw) = 0,$$
$$\cdots\cdots\cdots\cdots$$
$$(\psi(t), A^{n-1}Sw) = 0$$
and $\psi(t) = 0$ because U is in the general position. This contradicts the fact that $\psi(t)$ is nonzero. ∎

We say that a set U with a finite or countable number of faces *is in the general position with respect to the system*
$$\frac{dx}{dt} = Ax + Su,$$
if, for every nonzero vector w parallel to a face of U, the matrix
$$[Sw, ASw, \ldots, A^{n-1}Sw]$$
is of rank n.

As can easily be verified, Theorem 6.4.1 remains valid if U is an arbitrary set with a finite or countable number of faces. The proof of this fact proceeds in the same way as the proof of Theorem 6.4.1. However, if the number of faces is uncountable, Theorem 6.4.1 may be false, as is shown by the following example.

EXAMPLE 6.4.1. Let the control space and the input space be three-dimensional spaces of vector-valued functions. Let S be the identity mapping, $S = I$, and let A be a diagonal matrix,
$$A = \begin{bmatrix} a_1 & 0 & 0 \\ 0 & a_2 & 0 \\ 0 & 0 & a_3 \end{bmatrix},$$

where all numbers a_1, a_2, a_3 are different. It can be verified that every vector $w = (w_1, w_2, w_3)$, where $w_i \neq 0$, $i = 1, 2, 3$, is in the general position. Let $\psi(t)$ be an arbitrary solution of the conjugate equation such that all its components are nonzero. Observe two facts:

(1) if all components of the vector $\psi(t)$ are nonzero for a certain t, then so are the components of $\psi(t)$ for all other t;

(2) the same holds for the derivative $\Phi'(t)$.

Let

$$U = \{(x_1, x_2, x_3): x_1^2 + x_2^2 + x_3^2 \leq 4,$$
$$(\psi(t), x) \leq \sqrt{\psi_1^2(0) + \psi_2^2(0) + \psi_3^2(0)}, \ 0 \leq t \leq T\}.$$

(1) and (2) imply that the envelope of the family of planes,

$$H_t = \{(x_1, x_2, x_3): (\psi(t), x) = \sqrt{\psi_1^2(0) + \psi_2^2(0) + \psi_3^2(0)}\},$$

contains intervals but all components of those intervals are nonzero. Hence, every vector contained in the boundary of U is in the general position with respect to the system in question.

On the other hand, it can readily be checked that two different functions $u_0(t)$ and $u_1(t)$ can be chosen such that

$$(\psi(t), u_0(t) - u_1(t)) = 0,$$

and hence either $K(u_0) = K(u_1)$, which implies nonuniqueness of control, or the whole segment joining the points $K(u_0)$ and $K(u_1)$ lies on the boundary of the set U (since u_0 and u_1 satisfy (6.3.16)).

In the next section we shall prove that in the latter case there exist also nonunique optimal controls for the system in question.

6.5. BANG-BANG PRINCIPLE

Consider a linear system

$$\frac{dx}{dt} = A(t)x + S(t)u \tag{6.1.1}$$

on the interval $0 \leq t \leq T$ with the initial condition $x(0) = x_0$ and the final condition $x(T) = x_1$. Moreover, consider controls $u(t)$ such that $u(t) \in U$, where U is a closed convex set with a nonempty interior. Without loss of generality we can assume that $0 \in \text{Int } U$.

278   6. Systems governed by ordinary differential equations

Let $\|\ \|_E$ denote the seminorm generated by the set U (the norm when U is symmetric).

Take $M_E[0, T]$ as the control space and recall that $M_E[0, T]$ is isometric to the conjugate space of the space $L_E^1[0, T]$ (cf. Section 6.3). This implies that the closed unit ball in the space $M_E[0, T]$,

$$K = \{u \in M_E[0, T]: \|u\| \leqslant 1\} = \{u: u(t) \in U\},$$

is compact in the weak-∗-topology. In view of the considerations of Section 5.4, every minimum-norm (or minimum-time) control u_0 has the property that $u_0(t)$ is a boundary point of the set U for almost all t. Hence, if the set U is strictly convex, then $u_0(t)$ is an extreme point of the set U for almost all t.

Controls u such that the point $u(t)$ is an extreme point of the set U for almost all t are called *bang-bang controls*. Thus, if the set U is strictly convex then the minimum-norm (minimum-time) control is always a bang-bang control. If the set U is not strictly convex, there may exist optimal controls which are not bang-bang controls. However, the following theorem holds.

THEOREM 6.5.1 (Bang-bang principle). *Let U be a bounded convex closed set. Assume that U has at most a countable number of faces. If there exists a control u_0 such that $u_0(t) \in U$ almost everywhere and the solution of equation* (6.1.1) *corresponding to the control u_0 and to the initial condition $x(0) = x_0$ satisfies also the final condition $x(T) = x_1$, then there is a control u_1 satisfying the same conditions and such that $u_1(t)$ is an extreme point of U almost everywhere.*

Theorem 6.5.1 was proved by La Salle (1960) in the case where

$$U = \{(u_1, \ldots, u_n) \in R^n: |u_i| \leqslant 1,\ i = 1, \ldots, n\}.$$

The proof of Theorem 6.5.1 relies on a number of auxiliary results.

THEOREM 6.5.2. *Let U be a set with at most a countable number of faces. An element u_0 is an extreme element of the set*

$$\tilde{U} = \{u: u(t) \in U \text{ for almost all } t\}$$

if and only if $u_0(t)$ is an extreme point of the set U for almost all t.

6.5. Bang-bang principle

Proof. Sufficiency. Let $u_0(t)$ be an extreme point of the set U for almost all t. Suppose that $u_0 = ku_1 + (1-k)u_2$, where $u_1, u_2 \in \tilde{U}$ and $0 < k < 1$. This means that

$$u_0(t) = ku_1(t) + (1-k)u_2(t)$$

and $u_1(t), u_2(t) \in U$ almost everywhere. Since $u_0(t)$ is an extreme point of the set U almost everywhere, $u_1(t) = u_2(t)$ almost everywhere. Thus, by definition, $u_0(t)$ is an extreme point of the set $\tilde{U}$.

Necessity. Suppose that there is a set D of positive measure such that $u_0(t)$ is not an extreme point of the set D for $t \in D$. By assumption, the set U has at most a countable number of faces (faces of different dimensions). Hence, there exist a set D_1 of positive measure contained in D, $D_1 \subset D$, and a face W such that $u_0(t)$ is not an extreme point of the face W for $t \in D_1$. Moreover, we require W to have the least dimension of all the faces satisfying the above conditions. Now there is a set D_2, $D_2 \subset D_1$, of positive measure and a vector $r \neq 0$ such that $u_0(t) \pm r \in W$.

Putting

$$u_1(t) = \begin{cases} u_0(t) & \text{for } t \notin D_2, \\ u_0(t) + r & \text{for } t \in D_2, \end{cases}$$

and

$$u_2(t) = \begin{cases} u_0(t) & \text{for } t \notin D_2, \\ u_0(t) - r & \text{for } t \in D_2, \end{cases}$$

we obtain $u_1, u_2 \in U$, $u_1 \neq u_2$. But $u_0 = \tfrac{1}{2}(u_1 + u_2)$, which means that u_0 is not an extreme point of the set U. ∎

THEOREM 6.5.3. *Let U be a closed convex bounded set with at most a countable number of faces. As before, let*

$$\tilde{U} = \{u \in M_E[0, T]: u(t) \in U \text{ almost everywhere}\}.$$

Let $\Gamma = K(\tilde{U})$, where K is the operator given by formula (6.2.1),

$$K(u) = \int_0^T \Phi^{-1}(s) S(s) u(s) \, ds.$$

If an element $y_0 = \Phi^{-1}(T) x_1 - x_0$ is an extreme point of the set Γ, then there exists an optimal control $u_0(t)$ such that $u_0(t)$ is an extreme point of the set U for almost all t.

Proof. The set $\tilde{U}$ is compact in the weak-$*$-topology. By Proposition 4.6.3, the set $\tilde{U} \cap K^{-1}(y_0)$ is an extremal subset of $\tilde{U}$. Thus, by the Krein–Milman theorem (Theorem 4.6.4), the set $\tilde{U} \cap K^{-1}(y_0)$ contains an extreme point, say u_0. Of course, $K(u_0) = y_0$. By Theorem 6.5.2, u_0 has the property that $u_0(t)$ is an extreme point of U for almost all t. ∎

Now we shall prove Theorem 6.5.1 in the case where $m = 1$, i.e., U is a one-dimensional set and the control space is a one-dimensional space of vector functions. In this case the assertion of Theorem 6.5.1 is an immediate consequence of the following theorem.

THEOREM 6.5.4. *Let $U \subset R^1$ be defined as*
$$U = \{x\colon 0 \leqslant x \leqslant 1\}.$$
Write
$$\tilde{U} = \{u \in M[0, T]\colon 0 \leqslant u(t) \leqslant 1 \text{ for almost all } t\}$$
and
$$\tilde{U}_0 = \{u \in M[0, T]\colon u(t) \text{ is equal to } 0 \text{ or } 1 \text{ for almost all } t\}.$$

Suppose we are given a system of n measurable integrable functions $\psi_1(t), \ldots, \psi_n(t)$. Let
$$\psi(u) = \left\{\int_0^T \psi_i(t) u(t) \, dt\right\} \tag{6.5.1}$$
be an operator which maps the space $M[0, T]$ into the n-dimensional space R^n. Then
$$\psi(\tilde{U}_0) = \psi(\tilde{U}).$$

Proof. The set $\tilde{U} - u_0$, where $u_0(t) = \tfrac{1}{2}$, forms a ball in the space $M[0, T]$. Since this space is isometric to the conjugate space of the space $L^1[0, T]$, by the Alaoglu theorem (Theorem 4.4.6) the set $\tilde{U} - u_0$ is compact in the weak-$*$-topology and thereby the set $\tilde{U}$ is compact in

6.5. Bang-bang principle

the weak-∗-topology. Since $\psi_i(t) \in L[0, T]$, $i = 1, \ldots, n$, the operator ψ is continuous in the weak-∗-topology. Let y be an arbitrary element of the set $\psi(\tilde{U})$. The set $\psi^{-1}(y)$ is closed in the weak-∗-topology, and by the Krein–Milman theorem (Theorem 4.6.4) the set $\psi^{-1}(y) \cap \tilde{U}$ has extreme points. However, it is not evident that the extreme points of the set $\psi^{-1}(y) \cap \tilde{U}$ are the extreme points of the set $\tilde{U}$. We shall prove this fact by induction with respect to n.

Suppose that $u \in \tilde{U} \cap \psi^{-1}(y)$ and u is not an extreme point of the set $\tilde{U}$. We shall prove that in this case u is not an extreme point of the set $\tilde{U} \cap \psi^{-1}(y)$. If u is not an extreme point of the set $\tilde{U}$, then there exist a positive number ε and a set D of positive measure such that $\varepsilon < u(t) < 1-\varepsilon$ for $t \in D$.

We shall not prove the case $n = 1$ separately since the proof would proceed in exactly the same way as the proof of the inductive step. Assume that the operator ψ determined by $n-1$ functions belonging to $L[0, T]$ has the property that the images of $\tilde{U}$ and $\tilde{U}_0$ under the operator ψ coincide.

Now take in D two disjoint sets of positive measure, D_1 and D_2. By the inductive assumption, for the systems $(\psi_1 \chi_{D_1}, \ldots, \psi_{n-1} \chi_{D_1})$ and $(\psi_1 \chi_{D_2}, \ldots, \psi_{n-1} \chi_{D_2})$ there exist two sets F_1 and F_2 of positive measure such that $F_1 \subset D_1, F_2 \subset D_2$ and

$$\int_0^T \psi_i(t) \chi_{F_j} dt = \frac{1}{2} \int_0^T \psi_i(t) \chi_{D_j} dt, \quad i = 1, \ldots, n-1, \ j = 1, 2. \tag{6.5.2}$$

Let $h_j = 2\chi_{F_j} - \chi_{D_j}, j = 1, 2$. Then, by (6.5.2),

$$\int_0^T \psi_i(t) h_j(t) dt = 0, \quad i = 1, \ldots, n-1, \ j = 1, 2. \tag{6.5.3}$$

Note that the functions h_1 and h_2 have disjoint supports. Choose two numbers a and b which do not vanish simultaneously and are such that $|a| \leqslant \varepsilon, |b| \leqslant \varepsilon$ and $h = ah_1 + bh_2$ satisfies the equality

$$\int_0^T \psi_n(t) h(t) dt = a \int_0^T \psi_n(t) h_1(t) dt + b \int_0^T \psi_n(t) h_2(t) dt = 0. \tag{6.5.4}$$

Since $|a| \leqslant \varepsilon$ and $|b| \leqslant \varepsilon$ we have $u \pm h \in \tilde{U}$. On the other hand, by (6.5.3) and (6.5.4), $\psi(u \pm h) = \psi(u) = y$. This means that u is not an extreme point of the set $\tilde{U} \cap \psi^{-1}(y)$.

Consequently, each extreme point of the set $\tilde{U} \cap \psi^{-1}(y)$ belongs to $\tilde{U}_0$. This implies that $y \in \psi(\tilde{U}_0)$ and $\psi(\tilde{U}) = \psi(\tilde{U}_0)$. ∎

Since the functions $\psi_i(t)$ are integrable with respect to the Lebesgue measure, there exists a countably additive function defined for subsets E of the interval $[0, T]$ with values in the n-dimensional space R^n. This function is given by the formula

$$\mu(E) = \{\mu_i(E)\}, \tag{6.5.5}$$

where

$$\mu_i(E) = \int_E \psi_i(t) dt, \quad i = 1, \ldots, n.$$

Every countably additive function $\mu(E)$ with values in R^n is called a *vector-measure*. If all functions $\mu_i(E)$ are absolutely continuous with respect to the Lebesgue measure, then we say that the measure $\mu(E)$ is *absolutely continuous* with respect to the Lebesgue measure.

Clearly, by the Radon–Nikodym theorem, every vector-measure which is absolutely continuous with respect to the Lebesgue measure is of the form (6.5.5).

Now we can formulate Theorem 6.5.4 in a slightly different way.

THEOREM 6.5.4′. *Suppose we are given an n-dimensional vector-measure μ defined on measurable subsets of the interval $[0, T]$. Let μ be absolutely continuous with respect to the Lebesgue measure. Then the set of values of the measure μ,*

$$H = \{\mu(E): E \text{ is a measurable subset of the interval } [0, T]\},$$

is a convex compact set in the n-dimensional space R^n.

Proof. In virtue of the Radon–Nikodym theorem,

$$\mu(E) = \left\{ \int_0^T \psi_i(t) \chi_E dt \right\}. \tag{6.5.6}$$

Let ψ denote the operator generated by functions $(\psi_1, \ldots, \psi_n)$. By (6.5.6), we have $H = \psi(\tilde{U}_0)$. Now, by Theorem 6.5.5, $H = \psi(\tilde{U})$ and, as before, $\psi(\tilde{U})$ is a convex compact set. ∎

Theorem 6.5.4 is a weakened version of the Lyapunow theorem (cf. Lyapunow, 1940, 1948; Halmos, 1948) which states that the set of values of an n-dimensional, nonatomic and bounded vector-measure is a convex compact set. The method of proof presented here is due to Lindenstrauss (1966).

THEOREM 6.5.5 (Dvoretzky, Wald, Wolfowitz, 1951). *Suppose we are given a finite system of n-dimensional measures* $\mu_1(E), \ldots, \mu_m(E)$ *which are defined on measurable subsets of the interval* $[0, T]$ *and which are absolutely continuous with respect to the Lebesgue measure. Let* $a_1, \ldots, a_n$ *be nonnegative numbers such that*

$$a_1 + \ldots + a_n = 1. \tag{6.5.7}$$

Then every measurable set E can be represented as the union of disjoint measurable sets E_i, $E = E_1 \cup \ldots \cup E_m$ *such that*

$$\mu_i(E_i) = a_i \mu_i(E), \quad i = 1, \ldots, m. \tag{6.5.8}$$

Proof. We prove this theorem by induction with respect to the number of measures. We start with $m = 2$. Let a_1 be a number such that $0 \leqslant a_1 \leqslant 1$. Clearly, a_2 which satisfies (6.5.7) is equal to $1 - a_1$. We form a $2n$-dimensional vector-measure $(\mu_1(E), \mu_2(E))$, by putting first the components of the measure $\mu_1(E)$ and next the components of the measure $\mu_2(E)$. By Theorem 6.5.4', there exists a set E_1 such that

$$\mu_1(E_1) = a_1 \mu_1(E) \quad \text{and} \quad \mu_2(E_1) = a_1 \mu_2(E). \tag{6.5.9}$$

By taking $E_2 = E \setminus E_1$, we obtain the required partition of the set E.

Suppose that the assertion of the theorem holds for all systems of $m-1$ vector-measures which are absolutely continuous. We form an nm-dimensional vector-measure

$$(\mu_1(E), \ldots, \mu_m(E)),$$

by putting successively the components of the first measure $\mu_1(E)$, the second measure $\mu_2(E)$, etc.

By Theorem 6.5.4', there exists a measurable subset E_1 of the set E such that

$$\mu_i(E_1) = a_1 \mu_i(E), \quad i = 1, \ldots, m. \tag{6.5.10}$$

Write $E' = E \setminus E_1$. By the inductive assumption, the set E' can be represented as the union of disjoint sets, $E' = E_2 \cup \ldots \cup E_m$, such that

$$\mu_i(E_i) = \frac{a_i}{1-a_1} \mu_i(E') = \frac{a_i}{1-a_1} (1-a_1) \mu_i(E) = a_i \mu_i(E),$$

$$i = 2, \ldots, m, \quad (6.5.11)$$

since

$$\frac{a_2}{1-a_1} + \ldots + \frac{a_m}{1-a_1} = \frac{1-a_1}{1-a_1} = 1. \quad \blacksquare \tag{6.5.12}$$

THEOREM 6.5.1'. *Let U be a closed convex bounded set in an n-dimensional space. Suppose that U has at most a countable number of faces. If for a given y_0 there are minimum-time controls, then there is a minimum-time control $u(t)$ such that, for almost all t, the control $u(t)$ is an extremal point of the set U.*

Proof. Write

$$\tilde{U} = \{u \in M_m[0, T]: u(t) \in U \text{ for almost all } t\}.$$

Let $\Gamma = K(\tilde{U})$, where the operator K is given by formula (6.2.1),

$$K(u) = \int_0^T \Phi^{-1}(s) S(s) u(s) \, ds. \tag{6.5.13}$$

Let y be an arbitrary element of the set Γ. By the Carathéodory theorem (Lemma 5.6.6), the element y can be represented in the form

$$y = a_0 e_0 + \ldots + a_n e_n, \tag{6.5.14}$$

where $a_0, \ldots, a_n$ are nonnegative numbers such that

$$a_0 + \ldots + a_n = 1. \tag{6.5.15}$$

By Theorem 6.5.3, there exist controls $u_0, \ldots, u_m$ such that $K(u_i) = e_i$, $i = 0, 1, \ldots, n$, and, moreover, each $u_i(t)$, $i = 0, 1, \ldots, n$, is an extreme point of the set U for almost all t.

Consider now the system of $n+1$ vector-measures $\mu_0(E), \ldots, \mu_n(E)$ which are defined on measurable subsets of the interval $[0, T]$ by the formula

$$\mu_i(E) = \int_E \Phi^{-1}(s)S(s)u_i(s)\,ds, \quad i = 0, 1, \ldots, n. \tag{6.5.16}$$

Note that, by definition, $\mu_i([0, T]) = e_i$. By the Dvoretzky–Wald–Wolfowitz theorem (Theorem 6.5.5), the interval $[0, T]$ can be represented as the union of $n+1$ disjoint measurable sets, $E_0, \ldots, E_n$, $[0, T] = E_0 \cup \ldots \cup E_n$, such that

$$\mu_i(E_i) = a_i\mu_i([0, T]) = a_i e_i, \quad i = 0, 1, \ldots, n. \tag{6.5.17}$$

Let

$$u(t) = u_0(t)\chi_{E_0} + \ldots + u_n(t)\chi_{E_n}.$$

Since the sets $E_0, \ldots, E_n$ are disjoint and their union is equal to the interval $[0, T]$, $u(t)$ is an extreme point of the set U for almost all t. On the other hand, by (6.5.17) and (6.5.14), we have

$$K(u) = \int_0^T \Phi^{-1}(s)S(s)u(s)\,ds = \sum_{i=0}^n \int_{E_i} \Phi^{-1}(s)S(s)u_i(s)\,ds$$

$$= \sum_{i=0}^n \mu_i(E_i) = \sum_{i=0}^n a_i e_i = y. \blacksquare$$

6.6. MEASURABLE MULTIFUNCTIONS AND THEIR APPLICATIONS

In the previous section we proved Theorem 6.5.1 under the strong condition that the set U has at most a countable number of faces. The reason was that this condition appeared in the formulation of Theorem 6.5.2 which was applied in the proof of Theorem 6.5.1. However, Theorem 6.5.2 is valid without this condition; the proof of this fact is our main aim in the present section.

As a tool we shall use the theory of measurable multifunctions.

Let E be a set. A family Σ of subsets of E such that $\emptyset \in \Sigma$, $A \in \Sigma$ entails $E \setminus A \in \Sigma$ and, for an arbitrary sequence $\{E_n\}$, $E_n \in \Sigma$, $n = 1, 2, \ldots$, $\bigcup_{n=1}^{\infty} E_n \in \Sigma$, will be called a σ-algebra Σ.

In our considerations E will be the interval $[0, T]$ and Σ will be the σ-algebra of all measurable subsets of $[0, T]$.

Let (H, ϱ) be a metric space. Let F be a *multifunction*, i.e., a mapping which maps E into subsets of H. The set

$$F^{-1}(B) = \{e \in E: F(e) \cap B \neq \emptyset\}$$

is called the *inverse image* of the set $B \subset H$ under the multifunction F.

Let B_α, $\alpha \in \mathfrak{A}$, be a family of subsets of H. It can be shown that

$$F^{-1}\left(\bigcup_{\alpha \in \mathfrak{A}} B_\alpha\right) = \bigcup_{\alpha \in \mathfrak{A}} F^{-1}(B_\alpha). \tag{6.6.1}$$

However, it should be mentioned (Example 6.6.1) that a similar formula for the intersection of a family of subsets is not valid. Moreover, there is a significant difference between the definition of the inverse image of a set under a function and the definition of the inverse image of a set under a multifunction.

EXAMPLE 6.6.1. Let E be a one-point set, $E = \{e\}$. Let H consist of three points, $H = \{0, 1, 2\}$. Let $F(e) = \{1, 2\}$. Let $B_1 = \{0, 1\}$, $B_2 = \{0, 2\}$. Thus, $B_1 \cap B_2 = \{0\}$ and $F^{-1}(B_1 \cap B_2) = \emptyset$. On the other hand, $F^{-1}(B_i) = \{e\}$, $i = 1, 2$, and

$$F^{-1}(B_1) \cap F^{-1}(B_2) = \{e\} \neq F^{-1}(B_1 \cap B_2) = \emptyset.$$

We say that a multifunction F is *measurable* if for each closed set K the inverse image of K belongs to Σ,

$$F^{-1}(K) \in \Sigma.$$

PROPOSITION 6.6.1. *Let F be a measurable multifunction. Let G be a fixed closed set in H. Then the multifunction*

$$F_G(e) = F(e) \cap G$$

is measurable.

Proof. Let K be an arbitrary closed set. Then

$$F_G^{-1}(K) = \{e: F_G(e) \cap K \neq \emptyset\} = \{e: F(e) \cap K \cap G \neq \emptyset\} \in \Sigma,$$

since $G \cap K$ is closed. ∎

6.6. Measurable multifunctions

PROPOSITION 6.6.2. *Let F be a measurable multifunction. Let B be an open set. Then the inverse image of B belongs to Σ, $F^{-1}(B) \in \Sigma$.*

Proof. Let
$$K_n = \left\{x \in H: \varrho(x, y) \geq \frac{1}{n} \text{ for all } y \notin B\right\}$$
$$= \bigcap_{y \notin B} \left\{x \in H: \varrho(x, y) \geq \frac{1}{n}\right\}.$$

Of course, the sets K_n are closed and $\bigcup_{n=1}^{\infty} K_n = B$. Thus, by (6.6.1),
$$F^{-1}(B) = F^{-1}\left(\bigcup_{n=1}^{\infty} K_n\right) = \bigcup_{n=1}^{\infty} F^{-1}(K_n) \in \Sigma. \blacksquare$$

The converse result to that of Proposition 6.6.2 is not true.

EXAMPLE 6.6.2. Let $E = [0, 1]$ and let Σ be the algebra of all Lebesgue measurable subsets of $[0, 1]$. Let H be the real line R with the standard metric. Let E_0 be a nonmeasurable set. Let
$$F(t) = \begin{cases} [0, 1) & \text{for } t \notin E_0, \\ [0, 1] & \text{for } t \in E_0. \end{cases}$$
Let B be an arbitrary open set. Observe that if $1 \in B$ then there exists a t, $0 < t < 1$, such that $t \in B$. Therefore $F^{-1}(B)$ is either empty or equal to E. Thus, $F^{-1}(B) \in \Sigma$. On the other hand, $F^{-1}(\{1\}) = E_0 \notin \Sigma$.

We recall that a multifunction F is called *compact-valued* (*closed-valued*) if $F(e)$ is a compact (closed) set for all e.

PROPOSITION 6.6.3. *Let F be a compact-valued multifunction. Then F is measurable if and only if, for each open set B, $F^{-1}(B) \in \Sigma$.*

Proof. As follows from Proposition 6.6.2, if F is measurable then $F^{-1}(B) \in \Sigma$. To prove the converse implication, suppose that, for each open set B, $F^{-1}(B) \in \Sigma$. Let K be an arbitrary closed set in H. Let
$$B_n = \left\{x \in H: \varrho(x, K) < \frac{1}{n}\right\},$$

where $\varrho(x, K) = \inf\{\varrho(x, y): y \in K\}$. The sets B_n are open, $B_{n+1} \subset B_n$, and $\bigcap_{n=1}^{\infty} B_n = K$.

We claim that

$$F^{-1}(K) = \bigcap_{n=1}^{\infty} F^{-1}(B_n). \tag{6.6.2}$$

To see this, observe that, since $K \subset B_n$, $n = 1, 2, \ldots$, we have

$$F^{-1}(K) \subset \bigcap_{n=1}^{\infty} F^{-1}(B_n). \tag{6.6.3}$$

Suppose now that $e \in F^{-1}(B_n)$, $n = 1, 2, \ldots$ This means that there is an x_n, $x_n \in F(e) \cap B_n$, $n = 1, 2, \ldots$ The set $F(e)$ is compact, and therefore the sequence $\{x_n\}$ contains a subsequence $\{x_{n_k}\}$ which converges to x_0, $x_0 \in F(e)$. Since $\varrho(x_{n_k}, K) < \dfrac{1}{n_k}$ and the set K is closed, $x_0 \in K$, i.e., $x_0 \in F(e) \cap K$. Hence, $e \in F^{-1}(K)$. Since e is an arbitrary element of $\bigcap_{n=1}^{\infty} F^{-1}(B_n)$, we obtain the inclusion converse to that of (6.6.3). Thus, (6.6.2) holds. ∎

An open question is whether it is possible to replace the requirement that F should be compact-valued by the requirement that F should be closed-valued.

Now we shall give an example of a measurable multifunction.

EXAMPLE 6.6.3. Let $(H, \|\ \|)$ be a Banach space. Let U be a compact set in H. Let $u(t)$ be a measurable function defined on E with values in H. Then the multifunction

$$F(t) = u(t) + U$$

is measurable. To prove this, assume that K is an arbitrary closed set. Then

$$F^{-1}(K) = \{t: F(t) \cap K \neq \emptyset\} = \{t: u(t) \in K - U\} \in \Sigma,$$

since the set $K - U$ is closed.

Let $F(t)$ be a closed-valued multifunction. A function $f(t)$ is called a *selector* of $F(t)$ if $f(t) \in F(t)$ for all t.

THEOREM 6.6.4 (Kuratowski, Ryll-Nardzewski, 1965). *Let E be a set and let Σ be a σ-algebra of subsets of E. Let (H, ϱ_0) be a separable complete metric space. Let $F(e)$ be a measurable closed-valued multifunction from E into subsets of H. Then there is a measurable selector $f(e)$ of $F(e)$, i.e., a measurable function f such that $f(e) \in F(e)$ for all e.*

The proof is based on the following lemma.

LEMMA 6.6.5 (Kupka, private communication, 1984). *A limit $f(x)$ of a sequence of pointwise convergent measurable functions is a measurable function.*

Proof. Let K be an arbitrary closed set. Let
$$K_m = \left\{ x \in H: \varrho(x, K) \leq \frac{1}{m} \right\}.$$
The sets K_m are closed because they are inverse images of the closed intervals $\left[0, \dfrac{1}{m}\right]$ under a continuous function f_K, $f_K(x) = \varrho(x, K)$. By the assumption, $f(e) = \lim\limits_{n \to \infty} f_n(e)$. Thus, $f(e) \in K$ if and only if for each m there is an n_0 such that, for $n \geq n_0$, $f_n(e) \in K_m$. In other words,
$$f^{-1}(K) = \bigcap_{m=1}^{\infty} \bigcup_{n_0=1}^{\infty} \bigcap_{n=n_0}^{\infty} f_n^{-1}(K_m).$$
Since the right-hand side of this formula belongs to Σ, f is measurable. ∎

Lemma 6.6.5 was proved in a similar way by Kuratowski and Ryll-Nardzewski (1965), but instead of pointwise convergence they exploited uniform convergence.

Proof of Theorem 6.6.4. Let
$$\varrho(x, y) = \frac{1}{2} \frac{\varrho_0(x, y)}{1 + \varrho_0(x, y)}.$$

This metric is equivalent to the metric ϱ_0 and, moreover,

$$\delta(H) = \sup\{\varrho(x,y):\, x,y \in H\} \leqslant \frac{1}{2} < 1. \tag{6.6.4}$$

Let $\{r_i\}$ be a dense sequence in H. We define by induction a sequence of measurable functions $f_n(e)$, $n = 1, 2, \ldots$, such that

$$\varrho(f_n(e), F(e)) < \frac{1}{2^n} \tag{6.6.5}_n$$

and

$$\varrho(f_n(e), f_{n-1}(e)) \leqslant \frac{1}{2^{n-1}} \tag{6.6.6}_n$$

for all $e \in E$. Denote by f_0 an arbitrary measurable function. By (6.6.4), formula $(6.6.5)_0$ holds.

Suppose that functions $f_0, f_1, \ldots, f_{n-1}$ have been constructed. Let

$$\begin{aligned} C_i^n &= \left\{ e \in E:\, \varrho(r_i, F(e)) < \frac{1}{2^n} \right\} \\ &= \left\{ e \in E:\, F(e) \cap \left\{ y \in H:\, \varrho(r_i, y) < \frac{1}{2^n} \right\} \neq \emptyset \right\} \\ &= F^{-1}\left(\left\{ y \in H:\, \varrho(r_i, y) < \frac{1}{2^n} \right\}\right) \end{aligned} \tag{6.6.7}$$

and

$$\begin{aligned} D_i^n &= \left\{ e:\, \varrho(r_i, f_{n-1}(e)) < \frac{1}{2^{n-1}} \right\} \\ &= f_{n-1}^{-1}\left(\left\{ x:\, \varrho(r_i, x) < \frac{1}{2^{n-1}} \right\}\right). \end{aligned} \tag{6.6.8}$$

By the measurability of $F(e)$ (cf. Proposition 6.6.2), and the measurability of $f_{n-1}(e)$, the sets C_i^n, D_i^n, $i = 1, 2, \ldots$ belong to Σ. Thus,

$$A_i^n = C_i^n \cap D_i^n \in \Sigma, \quad i = 1, 2, \ldots \tag{6.6.9}$$

We shall prove that

$$E = \bigcup_{i=1}^{\infty} A_i^n. \tag{6.6.10}$$

Let e be an arbitrary element of E. By the inductive hypothesis,

$$\varrho\big(f_{n-1}(e), F(e)\big) < \frac{1}{2^{n-1}}. \tag{6.6.11}$$

The sequence $\{r_i\}$ is dense in H. Hence, we can find an r_i such that

$$\varrho\big(f_{n-1}(e), r_i\big) < \frac{1}{2^{n-1}} \tag{6.6.12}$$

and

$$\varrho\big(r_i, F(e)\big) < \frac{1}{2^n}. \tag{6.6.13}$$

Formulae (6.6.12) and (6.6.13) imply that $e \in A_i^n$. Let

$$f_n(e) = r_i \quad \text{for} \quad e \in A_i^n \setminus (A_1^n \cup \ldots \cup A_{i-1}^n). \tag{6.6.14}$$

By (6.6.10), $f_n(e)$ is defined on the whole E. The function $f_n(e)$ is measurable because $A_i^n \in \Sigma$. By (6.6.13), we obtain $(6.6.5)_n$. By (6.6.12), we obtain $(6.6.6)_n$. By $(6.6.6)_n$, the sequence $\{f_n(e)\}$ is uniformly convergent on the whole E. Since H is complete, $\{f_n(e)\}$ converges to a certain function $f(e)$ and, by Lemma 6.6.5, $f(e)$ is measurable. ∎

As a consequence of the Kuratowski–Ryll-Nardzewski theorem (Theorem 6.6.4) we obtain

THEOREM 6.6.6 (Castaing, 1967). *Let (H, ϱ) be a separable complete metric space. Let $F(e)$ be a closed-valued measurable multifunction. Then there is a countable family of measurable functions $\{u_n\}$ such that for all e, the sequence $\{u_n(e)\}$ is dense in $F(e)$, $\overline{\{u_n(e)\}} = F(e)$.*

Proof. Since $F(e)$ is closed-valued and measurable, there is a measurable selector u_0, $u_0(e) \in F(e)$, for all $e \in E$. Let $\{a_k\}$ be a dense sequence in H. By $\overline{B}(a_k, 1/m)$ we denote the closed ball with radius $1/m$ and centre a_k. The family of sets $\overline{B}(a_k, 1/m)$, $k, m = 1, 2, \ldots$, is countable and we can order it in a sequence, say $\{K_n\}$. Of course, $K_n \cap F(e)$ is a measurable multifunction (cf. Proposition 6.6.1). By Theorem 6.6.4, it has a measurable selector $\{\tilde{u}_n(e)\}$. Putting

$$u_n(e) = \begin{cases} \tilde{u}_n(e) & \text{if} \quad K_n \cap F(e) \neq \emptyset, \\ u_0(e) & \text{otherwise}, \end{cases}$$

we extend the function $\tilde{u}_n(e)$ to the function $u_n(e)$ defined on the whole E. The sequence $\{u_n(e)\}$ has the required properties. Indeed, let y be an arbitrary element of $F(e)$ and let ε be an arbitrary positive number. Take a_k and $1/m$ such that $1/m < \varepsilon$ and $y \in \bar{B}(a_k, 1/m) = K_n$. Thus,

$$\varrho(u_n(e), y) < \varrho(u_n(e), a_n) + \varrho(a_n, y) < 2\varepsilon.$$

Since ε is arbitrary, the proof is complete. ∎

The following theorem is a partially converse result to that of Theorem 6.6.6.

THEOREM 6.6.7. *Let (H, ϱ) be a metric space and let $\{u_n\}$ be a sequence of measurable functions. If the multifunction $F(e) = \overline{\{u_n(e)\}}$ is compact-valued, then F is measurable.*

Proof. Let K be an arbitrary closed set. Let ε be an arbitrary positive number. Let

$$D_\varepsilon = \bigcup_{n=1}^\infty \{e : \varrho(u_n(e), K) < \varepsilon\}$$

and $D_0 = \bigcap_{\varepsilon > 0} D_\varepsilon$. Of course, $D_0 \in \Sigma$. We shall prove that

$$F^{-1}(K) = D_0.$$

Indeed, if $e \in F^{-1}(K)$, then there exists an x, $x \in F(e) \cap K$. Since $\{u_n(e)\}$ is dense in $F(e)$, for an arbitrary $\varepsilon > 0$ there is an index n_0 such that

$$\varrho(u_{n_0}(e), x) < \varepsilon.$$

Thus, $\varrho(u_{n_0}(e), K) < \varepsilon$ and $e \in D_\varepsilon$. Since ε is an arbitrary positive number, $e \in D_0$.

To prove the opposite inclusion assume that $e \in D_0$. This means that $e \in D_{1/n}$. Thus, by the definition of $D_{1/n}$, there are indices i_n such that

$$\varrho(u_{i_n}(e), K) < \frac{1}{n}.$$

The set $\overline{\{u_n(e)\}}$ is compact. Hence, there is a subsequence $\{i_{n_k}\}$ of indices such that the sequence $\{u_{i_{n_k}}(e)\}$ converges to a certain x_0. Of course, $x_0 \in F(e) = \overline{\{u_n(e)\}}$. On the other hand,

$$\varrho(u_{i_{n_k}}(e), K) < \frac{1}{n_k} \to 0.$$

Thus, $x_0 \in K$ and $F(e) \cap K \neq \emptyset$. Therefore $e \in F^{-1}(K)$. ∎

By Theorem 6.6.7, we obtain the following example.

EXAMPLE 6.6.4. Let H be a separable normed space. Let $U(e)$, $V(e)$ be compact-valued measurable multifunctions. Then the multifunction
$$U(e) + V(e)$$
is a compact-valued measurable multifunction. To see this, take a sequence $\{u_n(e)\}$ of measurable functions which is dense in $U(e)$. Moreover, let $\{v_m(e)\}$ be a sequence of measurable functions, which is dense in $V(e)$. Then the sequence $\{u_n(e) + v_m(e)\}$ of measurable functions is dense in $U(e) + V(e)$. Since the set $U(e) + V(e)$ is compact, it follows by Theorem 6.6.7 that $U(e) + V(e)$ is a compact-valued measurable multifunction. In particular, taking a one-point set $\{v(e)\}$ as $V(e)$ we find that if $v(e)$ is a measurable function and $U(e)$ is a compact-valued measurable multifunction then $v(e) + U(e)$ is a compact-valued measurable multifunction.

In the proof of Theorem 6.6.4 and in the proof of Theorem 6.6.6 we do not assume measurability of F but we assume only that for each open set B the set $F^{-1}(B)$ belongs to Σ. Hence, we can also formulate

THEOREM 6.6.7'. *Let H be a separable metric space. Let $\{u_n\}$ be a sequence of measurable functions. Then the multifunction F, $F(e) = \overline{\{u_n(e)\}}$ is a closed valued multifunction such that, for each open set B, $F^{-1}(B) \subset \Sigma$.*

THEOREM 6.6.8. *Let H be a separable metric space. Let $F_1(e)$ and $F_2(e)$ be measurable multifunctions. Let $F_1(e)$ be compact-valued and let $F_2(e)$ be closed-valued. Then the intersection*
$$(F_1 \cap F_2)(e) = F_1(e) \cap F_2(e)$$
is a measurable compact-valued multifunction.

294   6. Systems governed by ordinary differential equations

Proof. Of course, $F_1 \cap F_2$ is compact-valued since $F_1(e) \cap F_2(e)$ is compact for each e. By the Castaing theorem (Theorem 6.6.6), there exist sequences of measurable functions $\{u_n\}$ and $\{v_n\}$ such that $\{u_n(e)\}$ is dense in $F_1(e)$, $\overline{\{u_n(e)\}} = F_1(e)$, and $\{v_n(e)\}$ is dense in $F_2(e)$, $\overline{\{v_n(e)\}} = F_2(e)$. Let K be an arbitrary closed set. Let

$$D_\varepsilon = \bigcup_{i,j=1}^\infty \{e: \varrho(u_i(e), v_j(e)) < \varepsilon,\ \varrho(u_i(e), K) < \varepsilon,\ \varrho(v_j(e), K) < \varepsilon\}.$$

where ε is any positive number. The sets

$$\{e: \varrho(u_i(e), v_j(e)) < \varepsilon,\ \varrho(u_i(e), K) < \varepsilon,\ \varrho(v_j(e), K) < \varepsilon\}$$

belong to Σ since u_i and v_j are measurable. Thus, $D_\varepsilon \in \Sigma$ and $D_0 = \bigcap_{n=1}^\infty D_{1/n} \in \Sigma$. We claim that

$$D_0 = (F_1 \cap F_2)^{-1}(K).$$

To see this, observe that if $e \in (F_1 \cap F_2)^{-1}(K)$ then there exists an x, $x \in F_1(e) \cap F_2(e) \cap K$. Let ε be a positive number. By the density of $\{u_i(e)\}$, $\{v_j(e)\}$ we can find indices i_0, j_0 such that

$$\varrho(x, u_{i_0}(e)) < \varepsilon/2,\quad \varrho(x, v_{j_0}(e)) < \varepsilon/2.$$

This implies that

$$e \in \{e: \varrho(u_{i_0}(e), v_{j_0}(e)) < \varepsilon,\ \varrho(u_{i_0}(e), K) < \varepsilon,\ \varrho(v_{j_0}(e), K) < \varepsilon\}$$

and $e \in D_\varepsilon$ for all $\varepsilon > 0$. Thus, $e \in D_0$. To prove the opposite inclusion, let $e \in D_0$. Then $e \in D_{1/n}$ for $n = 1, 2, \ldots$ and there exist indices i_n, j_n such that

$$\varrho(u_{i_n}(e), v_{j_n}(e)) < \frac{1}{n},$$

$$\varrho(u_{i_n}(e), K) < \frac{1}{n},$$

$$\varrho(v_{j_n}(e), K) < \frac{1}{n}.$$

The set $F_1(e)$ is compact. Thus, the sequence $\{u_{i_n}(e)\}$ contains a subse-

quence $\{u_{i_{n_k}}(e)\}$ which converges to a certain x_0, $x_0 \in F_1(e)$. Of course, $\varrho\left(u_{i_{n_k}}(e), K\right) \to 0$. Hence, since the set K is closed, we have $x_0 \in K$. The sequence $\{v_{j_{n_k}}(e)\}$ converges to x_0 and $x_0 \in F_2(e)$ since $F_2(e)$ is closed. Thus,
$$x_0 \in F_1(e) \cap F_2(e) \cap K$$
and
$$e \in (F_1 \cap F_2)^{-1}(K). \blacksquare$$

THEOREM 6.6.9. *Let E be an interval $[0, T]$ equipped with the Lebesgue measure. Let Σ be the σ-algebra of all Lebesgue measurable sets. Let $U(t)$ be a multifunction with values in R^n. We assume that $U(t)$ is closed-valued and uniformly bounded. Then the set*
$$\tilde{U} = \{u(\cdot)\colon u\text{—measurable, } u(t) \in U(t) \text{ almost everywhere}\}$$
is a subset of $M_m[0, T]$, compact in the weak-$*$-topology.

Proof. The multifunction $U(t)$ is uniformly bounded, i.e., there is a constant $M > 0$ such that
$$\sup_{0 \leqslant t \leqslant T} \sup_{x \in U(t)} \|x\| \leqslant M$$
and
$$\tilde{U} \subset \tilde{B}_M = \{u\colon u\text{—measurable, } \operatorname*{ess\,sup}_{0 \leqslant t \leqslant T} \|u(t)\| < M\}. \quad (6.6.15)$$
The set $\tilde{B}_M$ is compact in the weak-$*$-topology by the Alaoglu theorem (Theorem 4.4.6). We shall show that $\tilde{U}$ is closed in the weak-$*$-topology. On the contrary, suppose that $u_0 \notin \tilde{U}$. By the definition of $\tilde{U}$, there are $\varepsilon > 0$ and a set D_1 of positive measure such that
$$\inf_{x \in U(t)} \|u_0(t) - x\| < \varepsilon \quad \text{for} \quad t \in D_1. \quad (6.6.16)$$
Let
$$h_i = \{0, \ldots, 0, \underbrace{\chi_{D_1}}_{i\text{-th position}}, 0, \ldots, 0\}, \quad i = 1, 2, \ldots, m,$$
where χ_{D_1} denotes the characteristic function of the set D_1. Evidently, $h_i \in L_m^1[0, T]$, $i = 1, \ldots, m$. The set
$$V = \left\{v \in M_m[0, T]\colon \int_0^T (h_i(t), v(t) - u_0(t))dt < \frac{\varepsilon}{n}|D_1|\right\},$$

where $|D_1|$ denotes the Lebesgue measure of the set D_1, is a neighbourhood of u_0 in the weak-$*$-topology. By (6.6.10), V is disjoint with the set $\tilde{U}$. Therefore $\tilde{U}$ is closed in the weak-$*$-topology. ∎

To prove the general bang-bang principle we need the following lemma.

LEMMA 6.6.10. *Let X be a linear space. Let U be a convex set in X. A point $u \in U$ is an extreme point of the set U if and only if the set $U \cap (2u - U)$ contains only one point, namely the point u.*

Proof. Necessity. Since $u \in U$, we have $2u - u \in U$. Hence, $u \in U \cap (2u - U)$. Suppose that $U \cap (2u - U)$ contains also an element u_1, $u_1 \neq u$. Then $u_2 = 2u - u_1 \in U \cap (2u - U)$ and therefore $u_1, u_2 \in U$. The element u is not extreme since $u = \dfrac{u_1 + u_2}{2}$.

Sufficiency. Suppose that u is not an extreme point. This implies that there are a, $0 < a < 1$, and $u_1, u_2 \in U$ such that
$$u = au_1 + (1-a)u_2 \in U.$$
Without loss of generality we can assume that $a \geqslant \tfrac{1}{2}$. Let
$$u_3 = (2a-1)u_1 + 2(1-a)u_2.$$
Since $(2a-1) + 2(1-a) = 1$ and U is convex, we have $u_3 \in U$. By simple calculations we get
$$u = \frac{u_1 + u_3}{2}.$$
Thus, $u_3 = 2u - u_1$, $u_1 = 2u - u_3$ and $u_1, u_3 \in U \cap (2u - U)$. This contradicts the fact that $U \cap (2u - U)$ is a one-point set. ∎

On the basis of Lemma 6.6.10 we can prove the following extension of Theorem 5.6.2.

THEOREM 5.6.11. *Let $U(t)$ be a closed-valued measurable multifunction with values in subsets of R^n. Assume that $U(t)$ is uniformly bounded. Let*
$$\tilde{U} = \{u \in M_m[0, T]: u(t) \in U(t)\}.$$

6.6. Measurable multifunctions

A measurable function u_0 is an extreme point of $\tilde{U}$ if and only if $u_0(t)$ is an extreme point of $U(t)$ almost everywhere.

Proof. Sufficiency. If u_0 is not an extreme point of $\tilde{U}$, then
$$u_0 = au_1 + (1-a)u_2,$$
where $0 < a < 1$, $u_1, u_2 \in \tilde{U}$ and $u_1(t) \neq u_2(t)$ on a certain set D of positive measure. Thus, $u_0(t)$ is not an extreme point of $U(t)$ for $t \in D$.

Necessity. Suppose that u_0 is an extreme point of the set $\tilde{U}$. The multifunction $V(t) = U(t) \cap (2u_0(t) - U(t))$ is measurable and closed-valued (see Theorem 6.6.8 and Example 6.6.3). Clearly, $u_0(t)$ is a measurable selector of $V(t)$. By the Castaing theorem (Theorem 6.6.6), there is a sequence of measurable functions $v_n(t)$ such that $\overline{\{v_n(t)\}} = V(t)$. Let
$$D_i = \{t: v_i(t) \neq u_0(t)\} \quad (i = 1, 2, \ldots)$$
and let $D = \bigcup_{i=1}^{\infty} D_i$. If D is of nonzero measure, then some of the sets D_i is also of nonzero measure. Note that $u_0(t)$ is the centre of symmetry of $V(t)$. Thus, the functions
$$u_1 = \begin{cases} v_i(t) & \text{for } t \in D_i, \\ u_0(t) & \text{elsewhere}, \end{cases}$$
$$u_2 = \begin{cases} 2u_0(t) - v_i(t) & \text{for } t \in D_i, \\ u_0(t) & \text{elsewhere}, \end{cases}$$
belong to $\tilde{U}$ and $u_1 \neq u_2$, $u_0 = \dfrac{u_1 + u_2}{2}$. This leads to a contradiction with the assumption that u_0 is an extreme point of $\tilde{U}$.

Hence, D is a set of measure zero. Observe that $V(t) = \{u_0(t)\}$ for $t \notin D$. Thus, $u_0(t)$ is an extreme point of $U(t)$ almost everywhere. ∎

By applying Theorem 6.6.11, we can generalize Theorem 6.5.1.

THEOREM 6.6.12 (bang-bang principle). *Let $U(t)$, $0 \leqslant t \leqslant T$, be a measurable convex closed-valued multifunction with values in subsets of R^n.*

Assume that $U(t)$ is uniformly bounded, i.e., that there is a constant $M > 0$ such that

$$\sup_{0 \leqslant t \leqslant T} \sup \{\|u\|: u(t) \in U(t)\} \leqslant M.$$

If there is a control u_0, $u_0(t) \in U(t)$, such that the solution $x(t)$ of equation (6.1.1) corresponding to the control u_0 and to the initial condition $x(0) = x_0$ satisfies the final condition $x(T) = x_1$, then there is a control u_1 having the same properties as u_0 and such that $u_1(t)$ is an extreme point of $U(t)$ almost everywhere.

Proof. The proof proceeds in the same way as the proof of Theorem 6.5.1. Using the Carathéodory theorem (Lemma 5.6.6) we represent $y_0 = \Phi^{-1}(T)x_0 - x_1$ as a convex combination of n points y_i, $i = 1, \ldots, n$, belonging to the set $K(\tilde{U})$, where $\tilde{U} = \{u \in M_m[0, T]: u(t) \in U(t)\}$. Thus, there are controls u_i, $i = 1, \ldots, n$, such that $K(u_i) = y_i$ and u_i is an extreme point of $\tilde{U}$. By Theorem 6.6.11, $u_i(t)$, $i = 1, \ldots, n$, is an extreme point of $U(t)$ almost everywhere. The remainder of the proof is the same as the proof of Theorem 6.5.1. ∎

Theorem 6.6.12 was proved for polyhedra by Malanowski (1966) and, in general case, by Olech (1966), who used another technique, namely the so called "lexicographic order".

So far we have assumed that $U(t)$ are convex. The following result concerns a more general case.

PROPOSITION 6.6.13. *Let $U(t)$ be a compact-valued measurable multifunction whose values are subsets of a separable Banach space X. Then the multifunction $\overline{\text{conv}\,(U(t))}$ is compact-valued and measurable.*

Proof. It can be shown that the closed convex hull of a compact set is compact. Thus, $\overline{\text{conv}\,(U(t))}$ is compact-valued. By the Castaing theorem (Theorem 6.6.6), there exists a sequence of measurable functions $u_n(t)$ such that $\{u_n(t)\}$ is dense in $U(t)$. Consider all finite convex combinations with rational coefficients of the elements of the sequence $\{u_n(t)\}$. In this way we obtain a countable family of measurable functions $v_\alpha(t)$ such that $\{v_\alpha(t)\}$ is dense in $\overline{\text{conv}\,(U(t))}$ for each t. Thus, by Theorem 6.6.7, the multifunction $\overline{\text{conv}\,(U(t))}$ is measurable. ∎

Now we can prove Theorem 6.6.12 without the convexity assumption for the sets $U(t)$.

THEOREM 6.6.14. *Let $U(t)$, $0 \leq t \leq T$, be a compact-valued measurable multifunction with values in subsets of the space R^n. Let $U(t)$ be uniformly bounded. The set of points which can be reached from the point $x(0) = x_0$ by measurable controls u, $u(t) \in U(t)$, coincides with the set of points which can be reached from the point $x(0) = x_0$ by the application of measurable controls v, $v(t) \in \overline{\mathrm{conv}\,(U(t))}$.*

Proof. Let us denote by Γ_U the set of all points which can be reached by the application of controls u such that $u(t)$ belongs to $U(t)$. By Γ_V we denote the set of points which can be reached by the application of controls v, $v(t)$ belonging to $\overline{\mathrm{conv}\,(U(t))}$. Since $U(t) \subset \overline{\mathrm{conv}\,(U(t))}$, we have $\Gamma_U \subset \Gamma_V$. Let $x_1 \in \Gamma_V$. By Theorem 6.6.12, there exists a control v_1 such that x_1 can be reached by the application of v_1 and $v_1(t)$ is an extreme point of the set $\overline{\mathrm{conv}\,(U(t))}$ almost everywhere. Thus, $v_1(t) \in U(t)$ almost everywhere and $\Gamma_U = \Gamma_V$. ∎

By using the methods described in Section 5.5 we can reduce a given minimum-time problem to a minimum-norm problem. In these considerations, an essential role is played by the function

$$\varphi(t) = \inf\{||u||: B_t A(u) = y_0(t)\},$$

where

$$B_t A(u) = \int_0^t \Phi^{-1}(\tau) S(\tau) u(\tau) \, d\tau,$$

$$y(t) = \Phi^{-1}(t) x_1 - x_0,$$

and

$$||u|| = \inf\left\{s: \frac{u}{s} \in \tilde{U}\right\}.$$

We shall show that the function $\varphi(t)$ is lower semicontinuous on the right, i.e.,

$$\varphi(t_0) \leq \lim_{\delta \to 0} \inf_{t_0 < t < t_0 + \delta} \varphi(t).$$

Indeed, the multifunction $B_t A(\tilde{U})$ is linearly upper semicontinuous.

The set $U = \{x: \|x\| \leq 1\}$ is compact and the function $y(t)$ is continuous. Thus, by Theorem 5.5.1, $\varphi(t)$ is lower semicontinuous on the right.

If $B_T A(U)$ has empty interior, the function $\varphi(t)$ need not be continuous.

Indeed, observe that in Example 5.5.1 the operator B_t can be obtained from the equation

$$\frac{dx}{dt} = S(t)u, \quad -1 \leq t \leq 1,$$

where

$$S(t) = \begin{cases} \begin{bmatrix} 0 & 0 \\ 0 & 1 \end{bmatrix} & \text{for } -1 \leq t < 0, \\ \begin{bmatrix} 1 & 0 \\ 0 & 0 \end{bmatrix} & \text{for } 0 \leq t \leq 1. \end{cases}$$

On the basis of this example a similar example can be obtained for fixed y.

6.7. OBSERVABILITY OF SYSTEMS DESCRIBED BY ORDINARY DIFFERENTIAL EQUATIONS

Suppose we are given a system

$$\frac{dx}{dt} = A(t)x, \quad 0 \leq t \leq T, \tag{6.7.1}$$

where, as before, $x(t)$ is an n-dimensional vector-valued function, $A(t)$ is a matrix-valued function of dimension $n \times n$. The function $x(t)$ is absolutely continuous and $A(t)$ is locally integrable. We require equation (6.7.1) to be satisfied almost everywhere. By $\Phi(t)$ we denote the fundamental matrix of equation (6.7.1).

Apart from system (6.7.1), consider a continuous matrix $G(t)$ with n columns and m rows, called a *matrix of observations*. Let

$$y(t) = G(t)x(t). \tag{6.7.2}$$

We define the input, trajectory and output spaces in the following way. The input space X is the n-dimensional space R^n of final values.

6.7. Observability of ODE systems

The space $\square$ is the space $C_n[0, T]$ of all n-dimensional continuous functions. The output space Y is the space $C_m[0, T]$ of all m-dimensional continuous functions. The operator A is given by the formula

$$A(x_T)(t) = \Phi(t)\Phi^{-1}(T)x_T, \tag{6.7.3}$$

where x_T denotes an arbitrary vector, $x_T \in R^n$. The operator B is defined as

$$B(x(\cdot))(t) = G(t)x(t). \tag{6.7.4}$$

In this way we have defined a linear system of the form

$$(X \xrightarrow[A]{} \square \xrightarrow[B]{} Y). \tag{6.7.5}$$

Suppose we are given a linear functional f defined on the space of final values. According to the general scheme, the problem of optimal observability is to find a continuous linear functional $\varphi \in Y^*$ whose norm is as small as possible and such that

$$f = A^*B^*\varphi. \tag{6.7.6}$$

Basing ourselves on formula (6.3.7), we find that the minimum-norm observability problem corresponds to the minimum-norm control problem

$$\frac{dF}{dt} = -(A(t))^{tr}F + (G(t))^{tr}y$$

with the initial condition

$$F(0) = f$$

and the final condition

$$F(T) = 0.$$

Norms in the output space can be chosen in several ways. We shall consider the space $C_m[0, T]$ with the following norm. Suppose we are given a norm $\| \ \|_E$ in the space R^m. The space R^m with the norm $\| \ \|_E$ will be denoted by E. We define a norm $\| \ \|$ in $C_m[0, T]$ by the formula

$$\|x\| = \sup_{0 \leqslant t \leqslant T} \|x(t)\|_E. \tag{6.7.7}$$

The space $C_m[0, T]$ with the above norm will be denoted by $C_E[0, T]$.

Norms of type (6.7.7) frequently appear in observability problems. They appear for example if $x(t)$ can be measured at each moment t with an a priori given accuracy ε, $\|x_r - x_0\|_E < \varepsilon$, where x_r is the exact value while x_0 is the value of measurement.

If the observations of all components are independent, then
$$\|x\|_E = \sup_{1 \leq i \leq m} \frac{1}{a_i} \cdot |x_i|, \qquad (6.7.8)$$
where a_i denotes the accuracy of observation of the i-th component.

To solve the minimum-norm problem for equation (6.7.6) we should know the conjugate space of the space $C_E[0, T]$ and the norm of functionals defined on this space.

By the general form of functionals defined on the space $C[0, T]$, it follows immediately that every continuous linear functional F defined on the space $C_m[0, T]$ (i.e., also on the space $C_E[0, T]$) is of the form
$$F(x) = \int_0^T (x(t), dg(t)) = \sum_{i=1}^m \int_0^T x_i(t) \, dg_i(t), \qquad (6.7.9)$$
where $x = (x_1(t), \ldots, x_m(t)) \in C_m[0, T]$ and $g_1, \ldots, g_m$ are functions of bounded variation.

Let $g = (g_1(t), \ldots, g_m(t))$ be an m-dimensional vector-valued function of bounded variation. By $\mathrm{Var}_{E^*} g$ we denote the number
$$\mathrm{Var}_{E^*} g = \sup \sum_{i=1}^n \|g(t_i) - g(t_{i-1})\|_{E^*}, \qquad (6.7.10)$$
where the supremum is taken over all subdivisions of the interval $[0, T]$,
$$0 = t_0 < t_1 < \ldots < t_n = T$$
and $\| \ \|_{E^*}$ denotes the norm in the space E^*.

THEOREM 6.7.1. *Consider a functional F given by formula* (6.7.9). *Then*
$$\|F\| = \mathrm{Var}_{E^*} g.$$

Proof. We have
$$F(x) = \int_0^T \sum_{i=1}^n x_i(t) \, dg_i(t) = \lim \sum_{j=1}^n (x(t'_j), g(t_j) - g(t_{j-1})), \qquad (6.7.11)$$

6.7. Observability of ODE systems

where the limit is taken with respect to an arbitrary sequence of normal subdivisions, (a, b) denotes the inner product and t'_j is an arbitrary point such that $t_{j-1} < t'_j \leq t_j$.

We have the estimations

$$\sum_{j=1}^{n} \left(x(t'_j), g(t_j) - g(t_{j-1})\right) \leq \sum_{j=1}^{n} \|x(t'_j)\|_E \|g(t_j) - g(t_{j-1})\|_{E^*}$$

$$\leq \sup_{0 \leq t \leq T} \|x(t)\|_E \sum_{j=1}^{n} \|g(t_j) - g(t_{j-1})\|_{E^*} \leq \|x\| \cdot \mathrm{Var}_{E^*} g; \quad (6.7.12)$$

and hence

$$\|F\| \leq \mathrm{Var}_{E^*} g. \quad (6.7.13)$$

Before we prove the opposite inequality we show that every function of bounded variation is continuous everywhere except at most a countable number of points.

Write

$$w(t) = \limsup_{\substack{t_1, t_2 \to t \\ t_1 < t < t_2}} \|g(t_2) - g(t_1)\|_{E^*}.$$

A function $g(t)$ is continuous at t_0 if and only if $w(t_0) = 0$. Moreover,

$$\sum_{i=1}^{n} w(t_i) \leq \mathrm{Var}_{0}^{T} \|g(t)\|_{E^*},$$

where the summation is taken over an arbitrary n-tuple of points $t_1, \ldots, t_n$ which belong to the interval $[0, T]$. This immediately implies that $w(t) \neq 0$ for at most a countable number of points. Thus, every function of bounded variation is continuous except at most at a countable number of points.

If $g(t)$ is continuous at t', then, for an arbitrary positive ε, there exists a $\delta > 0$ such that the variation of $g(t)$ on the interval $[t' - \delta, t' + \delta]$ is less than ε. This implies that we can choose a subdivision

$$0 = t_0 < t_1 < \ldots < t_n = T \quad (6.7.14)$$

of the interval $[0, T]$ such that $g(t)$ is continuous at each point t_i, $i = 1, \ldots, n$, and

$$\sum_{j=1}^{n} \|g(t_j) - g(t_{j-1})\|_{E^*} > \mathrm{Var}_{E^*} g - \varepsilon. \quad (6.7.15)$$

Considering the fact that $g(t)$ is continuous at $t_1, \ldots, t_n$, we can choose a number $\delta > 0$ such that the variation of the function $g(t)$ on the interval $(t_j - \delta, t_j + \delta)$ is less than ε/n for $j = 1, 2, \ldots, n-1$.

Let x_j be an m-dimensional vector such that $\|x_j\|_E = 1$ and

$$(x_j, g(t_j) - g(t_{j-1})) = \|g(t_j) - g(t_{j-1})\|_{E^*}. \tag{6.7.16}$$

Let $x(t)$ be a continuous function such that $\|x(t)\|_E \leq 1$ and $x(t) = x_j$ for $t \in (t_{j-1}, t_j)$ and $|t - t_j| > \delta$ for $j = 1, \ldots, n-1$.

By formulae (6.7.15) and (6.7.16) and the definition of the number δ, we immediately obtain

$$\int_0^T (x(t), dg(t)) \geq \mathrm{Var}_{E^*} g - 2\varepsilon. \tag{6.7.17}$$

Since ε is arbitrary, we get

$$\|F\| \geq \mathrm{Var}_{E^*} g. \quad\blacksquare \tag{6.7.18}$$

The space of all functions of bounded variation with the norm given by formula (6.7.10) will be denoted by $\mathrm{Var}_{E^*}[0, T]$.

In subsequent considerations we shall use the following result.

THEOREM 6.7.2. *The extreme points of the unit ball in the space* $\mathrm{Var}_{E^*}[0, T]$ *are those functionals F,*

$$F(x) = \int_0^T (x(t), dg(t)), \tag{6.7.9}$$

for which the function $g(t)$ has the following properties: $g(t)$ *is continuous on the left*, $g(t)$ *is of bounded variation, there exists a point t_0 in the interval $[0, T]$ such that the function $g(t)$ is constant on the intervals $[0, t_0)$ and $(t_0, T]$, and $g(t_0 + 0) - g(t_0 - 0)$ is an extreme point of the unit ball K in the space E^*.*

Proof. Take an arbitrary function of bounded variation g, i.e., $g \in \mathrm{Var}_{E^*}[0, T]$, the norm of g is equal to 1. Suppose that there exists a point t' such that the function $g(t)$ is constant neither on the interval $[0, t')$ nor on the interval $(t', T]$. Since the number of points of discontinuity

6.7. Observability of ODE systems

of the function $g(t)$ is at most countable, we can assume without loss of generality that $g(t)$ is continuous at t'. Write

$$g_1(t) = \begin{cases} g(t) & \text{for } 0 \leq t \leq t', \\ g(t') & \text{for } t' \leq t \leq T, \end{cases}$$
$$g_2(t) = \begin{cases} g(t') & \text{for } 0 \leq t \leq t', \\ g(t) & \text{for } t' \leq t \leq T. \end{cases} \qquad (6.7.19)$$

It can readily be checked that

$$\int_0^T (x(t), dg(t)) = \int_0^T (x(t), dg_1(t)) + \int_0^T (x(t), dg_2(t)) \qquad (6.7.20)$$

for all $x \in C_E[0, T]$. It follows from (6.7.20) that

$$F = F_1 + F_2 \qquad (6.7.21)$$

and

$$||F|| = ||F_1|| + ||F_2||, \qquad (6.7.22)$$

where F_i is a continuous linear functional determined by the formula

$$F_i(x) = \int_0^T (x(t), dg_i(t)), \quad i = 1, 2. \qquad (6.7.23)$$

It follows from (6.7.21) and (6.7.22) that

$$F = \frac{||F_1||}{||F||} \cdot \frac{F_1}{||F_1||} + \frac{||F_2||}{||F||} \cdot \frac{F_2}{||F_2||} \qquad (6.7.24)$$

and

$$\left\| \frac{F_1}{||F_1||} \right\| = \left\| \frac{F_2}{||F_2||} \right\| = 1.$$

But

$$\frac{||F_1||}{||F||} + \frac{||F_2||}{||F||} = 1.$$

By (6.7.24), F is not an extreme point. Thus, we have shown that for F to be an extreme point it is necessary (but not sufficient) that there exists a t_0 such that the function $g(t)$ is constant on the intervals $[0, t_0]$ and $(t_0, T]$.

Suppose now that $g(t+0)-g(t-0)$ is not an extreme point of the unit ball in E. Then there exist vectors g_1 and g_2 in the space E^*, such that $\|g_1\|_{E^*} = \|g_2\|_{E^*} = 1$ and

$$g(t_0+0)-g(t_0-0) = ag_1+bg_2, \qquad (6.7.25)$$

where a and b are positive numbers, $a+b = 1$.

Let

$$g_1(t) = \begin{cases} 0 & \text{for } 0 \leqslant t < t_0, \\ g_1 & \text{for } t_0 \leqslant t \leqslant T, \end{cases} \qquad (6.7.26)$$

and

$$g_2(t) = \begin{cases} 0 & \text{for } 0 \leqslant t < t_0, \\ g_2 & \text{for } t_0 \leqslant t \leqslant T. \end{cases} \qquad (6.7.27)$$

Clearly, $g(t) = ag_1(t)+bg_2(t)$ and $\|g_1(t)\| = \|g_2(t)\| = 1$. Hence, $g(t)$ is not an extreme point of the unit ball in the space $\text{Var}_{E^*}[0, T]$.

Now we have to show that if g is of the above form then g is an extreme point of the unit ball in the space $\text{Var}_{E^*}[0, T]$. Therefore, assume that g can be represented as

$$g = ag_1+bg_2, \qquad (6.7.28)$$

where $\|g_1\| = \|g_2\| = 1$ and a, b are positive numbers, $a+b = 1$.

It follows from (6.7.28) that the functions $g_1(t)$ and $g_2(t)$ have at t_0 jumps v_1 and v_2, respectively. Since $\|g_1\| = \|g_2\| = 1$, we get $\|v_1\|_{E^*}, \|v_2\|_{E^*} \leqslant 1$.

By the assumption, the function $g(t)$ has a jump v at t_0, where v is an extreme point of the unit ball. From (6.7.28) we have

$$v = av_1+bv_2. \qquad (6.7.29)$$

Hence, the fact that v is an extreme point of the unit ball implies that $v = v_1 = v_2$. Thus, $\|v_1\|_{E^*} = \|v_2\|_{E^*} = 1$. Since $\|g_1\| = \|g_2\| = 1$, the functions $g_1(t)$ and $g_2(t)$ are constant on the intervals $[0, t_0]$ and $(t_0, T]$. ∎

As a consequence of Theorem 6.7.2 we obtain

THEOREM 6.7.3. *Consider the linear system* (6.7.5)

$$(X \xrightarrow[A]{} \square \xrightarrow[B]{} Y),$$

6.7. Observability of ODE systems

where X is a space of final values $x(T)$, $\square$ is the space $C_n[0, T]$, Y is the space $C_E[0, T]$. The operators A and B are given by formulae (6.7.3) and (6.7.4). Let f be an arbitrary observable functional defined on the input space X. Then there exists a minimum-norm solution of equation (6.7.6)

$$f = A^* B^* \varphi$$

and this solution has the form

$$\varphi(y(t)) = \sum_{i=1}^{n} (v_i, y(t_i)), \tag{6.7.30}$$

where n is the dimension of the space X, $t_1, \ldots, t_n$ are points which belong to the interval $[0, T]$, $v_1, \ldots, v_n$ are vectors in an m-dimensional space, and (a, b) denotes the inner product.

Proof. By Lemma 5.6.6, there are extreme points of the ball K_r, $\varphi_1, \ldots, \varphi_n$, such that

$$\varphi = \alpha_1 \varphi_1 + \ldots + \alpha_n \varphi_n$$

is an optimal observation. The form of φ_i follows from Theorem 6.7.2. Namely, each φ_i is determined by a certain function g_i with the property that there exists a point t_i such that $g_i(t)$ is constant on the intervals $[0, t_i]$ and $(t_i, T]$. Moreover, the function $g_i(t)$ has a jump v_i at the point t_i. Hence,

$$\varphi_i(x) = \int_0^T (x(t), \mathrm{d}g_i(t)) = (x(t_i), v_i) \tag{6.7.31}$$

and φ is a minimum-norm solution of problem (6.7.6) of the required form (6.7.30). ∎

In particular, in the case $m = 1$ there exists a minimum-norm solution of the form

$$\varphi(y(t)) = \sum_{i=1}^{n} v_i y(t_i). \tag{6.7.30'}$$

Functionals of forms (6.7.30) and (6.7.30') will be called *quantified functionals*.

For $m = 1$, Theorem 6.7.4 results directly from the following

308   6. Systems governed by ordinary differential equations

THEOREM 6.7.4 (Singer, 1957; Ptak, 1958). *Let L be an n-dimensional subspace of the space $C(\Omega)$, where Ω is a compact set. Each continuous linear functional $f \in L^*$ can be represented in the form*

$$f(x) = \sum_{i=1}^{n} \alpha_i x(t_i) \quad \text{for} \quad x \in L,$$

where $\{t_1, ..., t_n\}$ are fixed points and $\{\alpha_1, ..., \alpha_n\}$ are numbers such that

$$\sum_{i=1}^{n} |\alpha_i| = \|f\|.$$

Theorem 6.7.3 has an important physical interpretation. Suppose we are given a system described by a differential equation of order n. To determine an optimal method of measurement (an optimal method of observation) of a given linear parameter (attribute) of the system it suffices to perform at most n measurements at prescribed points. Clearly, since outputs belong to an m-dimensional space, a single measurement must determine all the m components of an output vector.

Theorem 6.7.3 can be generalized to the case of k linearly independent linear parameters (attributes) of a system (i.e., continuous linear functionals).

Recall that we can consider a system of functionals $f_1, ..., f_k$ as a mapping F of the input space X into a certain k-dimensional Banach space H. The problem is to find a continuous linear operator Φ, $\Phi \in B(Y \to H)$, such that the diagram

$$\begin{array}{ccccc} X & \xrightarrow{A} & \Box & \xrightarrow{B} & Y \\ & {}_F\searrow & & \swarrow{}_\Phi & \\ & & H & & \end{array} \tag{6.7.32}$$

is commutative and Φ is a solution of the minimum-norm problem corresponding to this diagram (see Section 5.6).

Since the space H is finite-dimensional, the existence of such a solution follows from the general theory (cf. Theorem 5.6.4). We shall determine the form of this solution by using the fact that the output space is $C_E[0, T]$.

Considering that each continuous linear functional $\tilde{G}$ which maps the space $C_E[0, T]$ into the k-dimensional space H can be regarded as a system of k continuous linear functionals defined on $C_E[0, T]$, we obtain the form of $\tilde{G}$. Namely, there exists a system of m-dimensional vector-valued functions of bounded variation $\{g_1(t), \ldots, g_k(t)\}$ such that

$$\tilde{G}(x) = \left\{ \int_0^T (x(t), dg_i(t)) \right\}. \tag{6.7.33}$$

The system of vector-valued functions $\{g_1(t), \ldots, g_k(t)\}$ can be treated as a matrix $G(t)$ whose elements are functions of bounded variation. Matrices of this type will be called *matrix-valued functions of bounded variation*.

Formula (6.7.33) can be rewritten in a different way as

$$\tilde{G}(x) = \int_0^T dG(t) x(t). \tag{6.7.34}$$

Clearly, as can be readily verified, we have

$$\int_0^T dG(t) x(t) = \lim \sum_{i=1}^n [G(t_i) - G(t_{i-1})] x(t_i'), \tag{6.7.35}$$

where lim is taken with respect to an arbitrary normal sequence of partitions Δ:

$$0 = t_0 < t_1 < \ldots < t_n = T, \tag{6.7.36}$$

and t_i' is an arbitrary point such that $t_{i-1} < t_i' \leq t_i$.

Denote by $\|\tilde{G}\|_{E,H}$ the norm of an operator $\tilde{G}$, $\tilde{G} \in B(E \to H)$. On the set of all matrices of bounded variation of dimension $m \times k$ we define the variation of a matrix $G(t)$ by the formula

$$\mathrm{Var}_{B(E \to H)} G(t) = \sup \sum_{i=1}^n \|G(t_i) - G(t_{i-1})\|_{E,H}, \tag{6.7.37}$$

where the supremum is taken over all partitions (6.7.36) of the interval $[0, T]$.

310   6. Systems governed by ordinary differential equations

It follows from formulae (6.7.34) and (6.7.35) that
$$\|\tilde{G}\| \leqslant \mathrm{Var}_{B(E \to H)} G(t). \qquad (6.7.38)$$
Repeating the argumentation of the proof of Theorem 6.7.1 without significant changes, we can prove the opposite inequality to inequality (6.7.38). In this way we obtain

THEOREM 6.7.5. *Each operator $\tilde{G}$, $\tilde{G} \in B(C_E[0, T] \to H)$, is given by formula* (6.7.34), *where $G(t)$ is a matrix-valued function of bounded variation. The norm of G is equal to*
$$\|\tilde{G}\| = \mathrm{Var}_{B(E \to H)} G(t). \qquad (6.7.39)$$

Clearly, the linear space of matrices with k rows and m columns is an mk-dimensional space. The operator norm $\|\tilde{G}\|_{E,H}$ can be regarded as a norm in an mk-dimensional space. Hence, we can apply Theorem 6.7.2 to the space $\mathrm{Var}_{B(E \to H)}[0, T]$ of all functions of bounded variation defined on the interval $[0, T]$ with the norm given by formula (6.7.4). Thus, we obtain

THEOREM 6.7.2'. *The extreme points of the unit ball in the space $\mathrm{Var}_{B(E \to H)}[0, T]$ are those matrix-valued functions $G(t)$ for which there exists a point t_0 such that $G(t)$ is constant on the intervals $[0, t_0]$ and $(t_0, T]$, and the operator*
$$V = \lim_{t \to t_0 + 0} G(t) - \lim_{t \to t_0 - 0} G(t),$$
called the jump *of the matrix-valued function $G(t)$ at t_0, is an extreme point of the unit ball in the space $B(E \to H)$.*

Theorem 6.7.2' implies the following result.

THEOREM 6.7.6. *Suppose we are given a diagram of form* (6.7.32) *in which X is an n-dimensional space, Y is the space $C_E[0, T]$, and H is a k-dimensional space. Then there exists an operator Φ with a minimal norm, such that diagram* (6.7.32) *is commutative and*
$$\Phi(y) = \sum_{i=1}^{nk} V_i y(t_i), \qquad (6.7.40)$$

where $\{t_1, \ldots, t_{nk}\}$ are points and $\{V_1, \ldots, V_{nk}\}$ are matrices with k rows and m columns.

Proof. The proof proceeds in exactly the same way as the proof of Theorem 6.7.3. Firstly, the space $B(X \to H)$ is an nk-dimensional space. Secondly, the operator F lies on the boundary of the image of a certain ball in the space $\mathrm{Var}_{B(E \to H)}[0, T]$. Hence, the operator Φ can be represented in the form

$$\Phi = \sum_{i=1}^{nk} \alpha_i \Phi_i, \qquad (6.7.41)$$

where Φ_i, $i = 1, \ldots, nk$, are extreme points of a certain ball in the space $\mathrm{Var}_{B(E \to H)}[0, T]$. Therefore, by Theorem 6.7.2,

$$\Phi_i(y) = \int_0^T dG_i(t) y(t), \qquad (6.7.42)$$

where the function $G_i(t)$ is constant everywhere except at a point t_i at which it has a jump V_i. Together with formula (6.7.41), this gives formula (6.7.40). ∎

Recall that a scalar equation of order n can be regarded as an n-dimensional equation of order 1.

EXAMPLE 6.7.1. Consider an object which moves rectilinearly and uniformly in a time interval $[0, T]$.

The problem is to determine, with minimal error, the velocity of the object by measuring its space coordinates. The accuracy of measurement is known and does not depend on time.

The motion law has the form

$$\frac{d^2 x}{dt^2} = 0, \qquad (6.7.43)$$

where x is a scalar-valued function. The space of final (or initial) conditions of equation (6.7.43) is two-dimensional. In virtue of Theorem 6.7.3, there exists an optimal method of observation which requires only two measurements, i.e., there exists a functional φ which is a minimum-norm solution of equation (6.7.6) and

$$\varphi(x) = ax(t_1) + bx(t_2), \qquad (6.7.44)$$

where $x(t) \in C[0, T]$. Recall that solutions of equation (6.7.43) are of the form $x(t) = ct + d$.

To determine the velocity c from formula (6.7.44) one should put

$$a = \frac{1}{t_1 - t_2}, \quad b = -a = \frac{-1}{t_1 - t_2}. \qquad (6.7.45)$$

By simple calculations,

$$\|\varphi\| = \frac{2}{t_2 - t_1} \qquad (6.7.46)$$

and clearly $\|\varphi\|$ is minimal when $T = t_2$, $t_1 = 0$, which agrees with our intuition.

EXAMPLE 6.7.2. Suppose we are given a harmonic oscillator described by the equation

$$\frac{d^2 x}{dt^2} + x = 0, \quad 0 \leqslant t \leqslant T, \qquad (6.7.47)$$

where x is a scalar-valued function. As in the previous example, the problem is to determine at a given t_0 the velocity of the oscillator by observing its position. We assume that the error of measurement is known and does not depend on time. Assume that $T - t_0 \geqslant \frac{1}{2}\pi$ or $t_0 \geqslant \frac{1}{2}\pi$. By Theorem 6.7.3, there exists a functional φ which is a minimum-norm solution of equation (6.7.6). Moreover, this functional can be represented in form (6.7.44). Note that the solution of equation (6.7.47) which corresponds to the velocity c and the position d at a moment t_0 can be represented as

$$x(t) = c\sin(t - t_0) + d\cos(t - t_0). \qquad (6.7.48)$$

Coefficients a and b in formula (6.7.44) must be chosen so as to ensure $\varphi(x) = c$ for each function x of the form (6.7.48). Hence,

$$a\bigl(c\sin(t_1 - t_0) + d\cos(t_1 - t_0)\bigr) + b\bigl(c\sin(t_2 - t_0) + d\cos(t_2 - t_0)\bigr)$$
$$= c, \qquad (6.7.49)$$

6.7. Observability of ODE systems

for all c and d. This implies that

$$a\sin(t_1-t_0)+b\sin(t_2-t_0) = 1,$$
$$a\cos(t_1-t_0)+b\cos(t_2-t_0) = 0. \qquad (6.7.50)$$

By solving equation (6.7.55) we obtain

$$a = \frac{\cos(t_2-t_0)}{\sin(t_1-t_0)\cos(t_2-t_0)-\sin(t_2-t_0)\cos(t_1-t_0)},$$
$$b = \frac{-\cos(t_1-t_0)}{\sin(t_1-t_0)\cos(t_2-t_0)-\sin(t_2-t_0)\cos(t_1-t_0)}. \qquad (6.7.51)$$

It can be verified that the norm of functional (6.7.44) is expressed by the formula

$$\|\varphi\| = |a|+|b|.$$

Observe that

$$|a|+|b| = \frac{|\cos(t_2-t_0)|+|\cos(t_1-t_0)|}{|\sin(t_1-t_0)\cos(t_2-t_0)-\sin(t_2-t_0)\cos(t_1-t_0)|}$$
$$\geq \frac{|\cos(t_2-t_0)|+|\cos(t_1-t_0)|}{|\sin(t_1-t_0)\cos(t_2-t_0)|+|\sin(t_2-t_0)\cos(t_1-t_0)|}$$
$$\geq \frac{|\cos(t_2-t_0)|+|\cos(t_1-t_0)|}{|\cos(t_2-t_0)|+|\cos(t_1-t_0)|} = 1. \qquad (6.7.52)$$

On the other hand, for $t_1 = t_0+\tfrac{1}{2}\pi$, we have $\cos(t_1-t_0) = 0$, $a = \pm 1$, $b = 0$. The functional (6.7.44) takes the form

$$\varphi(x) = \pm x(t_0 \pm \tfrac{1}{2}\pi) \qquad (6.7.53)$$

and $\|\varphi\| = 1$. It follows from (6.7.52) that (6.7.53) is a minimum-norm solution of equation (6.7.6).

On the other hand, the optimal measurement of the position at t_0 can be obtained by a direct measurement at t_0.

Let f be a continuous linear functional defined on the space of initial values R^1. The functional f is of the form

$$f(x) = ax(0)+bx'(0).$$

Suppose that $a, b > 0$. Let $t_0 = \cos^{-1}\dfrac{a}{\sqrt{a^2+b^2}}$. Then

$$\cos t_0 = \frac{a}{\sqrt{a^2+b^2}}, \quad \sin t_0 = \frac{b}{\sqrt{a^2+b^2}}$$

and
$$f(x) = \sqrt{a^2+b^2}\, x(t_0),$$
for every function $x(t)$ which solves equation (6.7.47). Indeed, $x(t) = c\cos t + d\sin t$ and
$$f(x) = \sqrt{a^2+b^2}\, x(t_0) = (ac+bd) = ax(0)+bx'(0).$$
If $a > 0$, $b < 0$, we put $t_0 = \dfrac{\pi}{2} - \sin^{-1}\dfrac{b}{\sqrt{a^2+b^2}}$. Then
$$\sin t_0 = \frac{a}{\sqrt{a^2+b^2}}, \quad \cos t_0 = \frac{-b}{\sqrt{a^2+b^2}}$$
and
$$f(x) = -\sqrt{a^2+b^2}\, x'(t_0).$$
Indeed, for $x(t) = c\cos t + d\sin t$,
$$-\sqrt{a^2+b^2}\, x'(t_0) = -\sqrt{a^2+b^2}\,(-c\sin t_0 + d\cos t_0)$$
$$= ac+db = f(x).$$

The cases $a < 0$, $b > 0$, and $a < 0$, $b < 0$, can be reduced to the cases just considered by changing the sign of the functional f. As a consequence of these considerations it follows that each continuous linear functional f defined on R^2 can be optimally observed by measuring either the position or the velocity at a point t_0, $0 \leqslant t_0 \leqslant T$, $T \geqslant \pi/2$. Moreover, measurement of the velocity at t_0 can be replaced by measurement of the position at $t_0 \pm \pi/2$. Finally, we find that there is always an optimal observation method based on one measurement of the position only.

The results of the present section can be extended to the case where the space E is not fixed. This happens when the accuracy of measurement depends continuously on time, i.e., when we deal with a multifunction of symmetric open sets $U(t)$, $0 \leqslant t \leqslant T$, which depends continuously on time t. For each set $U(t)$ we define a corresponding Banach space $E(t)$ in which the unit ball coincides with $U(t)$.

By $C_{E(t)}[0, T]$ we denote the space of all continuous functions defined on the interval $[0, T]$, with the norm $||x|| = \sup\limits_{0 \leqslant t \leqslant T} ||x(t)||_{E(t)}$, where $||\ ||_{E(t)}$ denotes a norm in the space $E(t)$. As can be verified, the space $C_{E(t)}[0, T]$ consists of the same functions as the space $C_m[0, T]$

6.7. Observability of ODE systems

and is equipped with the same topology. The conjugate space of the space $C_m[0, T]$ coincides with the conjugate space of the space $C_{E(t)}[0, T]$; the elements are the same and the topologies are equivalent but norms are given by the different formulae.

The norm of the functionals is

$$\|\tilde{G}\| = \text{Var}_{E(t)} G(t) = \sup \sum_{i=1}^{n} \|G(t_i) - G(t_{i-1})\|_{E(t_i')}, \quad (6.7.54)$$

where the supremum is taken over all subdivisions of the interval $[0, T]$,

$$0 = t_0 < t_1 < \ldots < t_n = T$$

and t_i' is an arbitrary point of the interval $[t_{i-1}, t_i]$.

For the space of functions with bounded variation and the norm given by formula (6.7.54) we can prove Theorem 6.7.2. As a consequence, Theorem 6.7.3 and Theorem 6.7.4 follow.

The above approach can be used to investigate the following system

$$\frac{dx}{dt} = A(t)x + S(t)u, \quad 0 \leq t \leq T.$$

Assume that $x(t) \in R^n$ and u belongs to a certain k-dimensional subspace U_0 of the control space U. The input space X is the product of the space of initial values R^n and the space U_0, $X = R^n \times U_0$. The trajectory space $\square$ is the space $C_n[a, b] \times U_0$, $A(x, u(t)) = (x(t), u(t))$, the output space Y is the space $C_m[a, b]$ and the output operator is given by the formula

$$y(t) = G(t)x(t) + H(t)u(t).$$

Let f be an arbitrary observable functional on the space $X = R^n \times U_0$. Since the dimension of the input space is $k+n$, there exists an optimal observation of the functional f, which is based on $m(n+k)$ measurements.

In the particular case where $m = 1$, the optimal observation is of the form

$$\varphi(y) = \sum_{i=1}^{n+k} a_i y(t_i).$$

6.8. OPTIMAL OBSERVATION OF STATIONARY SYSTEMS

Suppose that the matrix A appearing in equation (6.7.1) and the matrix G from formula (6.7.2) are constant.

Let U be a polytope, $U \subset R^m$, dim $U = m$, $m \geq 2$. We say that the polytope U *is in the dual general position with respect to the stationary system* (6.7.1) *and the stationary observation* (6.7.2) if for every vector w which is orthogonal to a face of dimension $m-1$ of U the matrix

$$[G^*w, A^*G^*w, ..., (A^*)^{n-1}G^*w] \tag{6.8.1}$$

is of order n.

In the above definition we have assumed that $m \geq 2$. For $m = 1$ the polytope U reduces to an interval. The boundary of U is zero-dimensional. In this sense each vector is orthogonal to the boundary of U. The interval U *is in the dual general position* if the matrix (6.8.1) is of order n for the unit vector $w \in R^1$.

THEOREM 6.8.1. *Consider a stationary system with a stationary observation. If the unit ball in the space E is in the dual general position, then every minimum-norm solution φ of equation* (6.7.6)

$$f = A^*B^*\varphi$$

is quantified, i.e., φ is of the form (6.7.30).

Proof. Let φ be a minimum-norm solution of equation (6.7.6). Clearly, φ is defined by a certain vector-valued function $g(t)$ of bounded variation.

By Corollary 5.6.4, there exists a nonzero element x_0 such that

$$\inf\{||\varphi||: \varphi \in Y^*, f = A^*B^*\varphi\}$$
$$= \inf\{||\varphi||: \varphi \in Y^*, f(x_0) = \varphi(BA(x_0))\}. \tag{6.8.2}$$

Write

$$D = \{t: ||x(t)||_E = \sup_{0 \leq t \leq T} ||x(t)||_E\},$$

where $x(t) = Ge^{tA}x_0$. The set D is closed. It follows from (6.8.2) that the function $g(t)$ is constant on the set $[0, T] \setminus D$.

We shall prove that the set D consists of a finite number of points.

6.8. Optimal observation of stationary systems

The surface of the set D is the union of a finite number of faces of dimension $m-1$. If the set D were infinite, there would exist a face W of dimension $m-1$ such that $x(t) \in W$ for an infinite number of points, $\{t_1, \ldots, t_n, \ldots\}$. But that is not possible. To see this, denote by w a vector which is orthogonal to W. By definition,

$$(w, Ge^{t_i A}x_0) = 0, \quad i = 1, \ldots$$

Since the function Ge^{tA} is analytic,

$$(w, Ge^{tA}x_0) = 0 \quad \text{for all } t. \tag{6.8.3}$$

Differentiating formula (6.8.3) $n-1$ times with respect to t and putting $t = 0$, we obtain

$$\begin{aligned}
(w, Gx_0) &= (G^*w, x_0) = 0, \\
(w, GAx_0) &= (A^*G^*w, x_0) = 0, \\
&\cdots \cdots \cdots \cdots \\
(w, GA^{n-1}x_0) &= ((A^*)^{n-1}G^*w, x_0) = 0.
\end{aligned} \tag{6.8.4}$$

This implies that $x_0 = 0$, which contradicts the assumption about the dual general position. ∎

Note, that the fact that every minimum-norm solution of equation (6.7.6) is quantified (i.e. is of the form (6.7.30)) does not entail the uniqueness of a solution. This can be observed in Example 6.7.2, where under the additional assumption

$$\tfrac{1}{2}\pi < t_0 < T - \tfrac{1}{2}\pi$$

the minimum-norm solution of equation (6.7.6) is of the form

$$\varphi(y) = ay\left(t_0 + \tfrac{1}{2}\pi\right) + (a-1)y\left(t_0 - \tfrac{1}{2}\pi\right),$$

for every a, $0 \leqslant a \leqslant 1$.

The following theorem is a generalization of Theorem 6.8.1.

THEOREM 6.8.2. *Suppose that the boundary of a set U can be represented as the union of a countable number of faces W_n of dimension i, $i = 1, \ldots$*

..., $m-1$. *Assign to each W_n a nonzero vector w_n which is orthogonal to W_n. If for each w_n the matrix*

$$[G^*w_n, A^*G^*w_n, \ldots, (A^*)^{n-1}G^*w_n]$$

is of order n, then the minimum-norm solution of equation (6.7.6) *can be expressed in the form*

$$\varphi(y) = \sum_{i=1}^{\infty} (a_i, y_i(t_i)), \qquad (6.8.5)$$

where $\{t_i\}$ *is a sequence of points belonging to the interval* $[0, T]$, *and* $\{a_i\}$ *is a sequence of vectors.*

Proof. By the same arguments as in the proof of Theorem 6.8.1, there exists an x_0 which solves equation (6.8.2). Now we shall prove that the set D defined in the proof of Theorem 6.8.1 is countable. Suppose that it is not so. Since the whole of the boundary of D can be represented as the union of a countable number of faces, there exist a face, say W_n, and an infinite sequence of points $\{t_i\}$ such that $x(t_i) \in W_n$. The remainder of the proof is the same as the proof of Theorem 6.8.1. ∎

6.9. Functionals observed by less than n measurements

We shall consider a linear system

$$\frac{dx}{dt} = A(t)x, \quad 0 \leqslant t \leqslant T, \qquad (6.7.1)$$

with an observation

$$y(t) = G(t)x(t). \qquad (6.7.2)$$

All assumptions about A, G and x are the same as in Section 6.7. The input space X is the space of initial values R^n, $X = R^n$. According to Theorem 6.7.3, each linear functional f defined on X, $f \in X^*$, which is observable is also optimally observable by at most n measurements,

$$f(x) = \sum_{i=1}^{n} (v_i, y(t_i)), \qquad (6.7.30)$$

where y is an output given by formulae (6.7.1) and (6.7.2).

6.9. Functionals observed by less than n measurements

The right-hand side of formula (6.7.30) defines a continuous linear functional φ, $\varphi \in C_E^*[0, T]$. The norm of this functional,

$$||\varphi|| = \sum_{i=1}^{n} ||v_i||_{E^*}, \tag{6.9.1}$$

is equal to the norm of the optimal method of observation.

We shall denote by U_r ($\tilde{U}_r$) those linear functionals defined on X which can be observed (optimally observed) by r measurements. Thus, such functionals can be written in the form

$$f(x) = \sum_{i=1}^{r} (v_i, y(t_i)) \tag{6.9.2}$$

$(f(x) = \sum_{i=1}^{r} (v_i, y(t_i))$ and $\sum_{i=1}^{r} ||v_i||_{E^*} = \inf\{||\varphi||: A^*B^*\varphi = f\})$.

Suppose that $f \in U_{r+s}$. This means that

$$f(x) = \sum_{i=1}^{r+s} (v_i, y(t_i)). \tag{6.9.3}$$

Hence,

$$f = f_1 + f_2, \tag{6.9.4}$$

where

$$f_1(x) = \sum_{i=1}^{r} (v_i, y(t_i)), \tag{6.9.5}$$

$$f_2(x) = \sum_{i=r+1}^{s} (v_i, y(t_i)). \tag{6.9.6}$$

Clearly, $f_1 \in U_r$ and $f_2 \in U_s$. Hence, by (6.9.4),

$$U_{r+s} \subset U_r + U_s.$$

Now we shall prove the converse inclusion. Take any $f_1 \in U_r, f_2 \in U_s$. The functionals f_1 and f_2 can be represented in the forms (6.9.5) and (6.9.6), respectively. By formula (6.9.4),

$$U_r + U_s \subset U_{r+s}.$$

Finally,
$$U_{r+s} = U_r + U_s. \tag{6.9.7}$$

Suppose now that $f \in \tilde{U}_{r+s}$. Thus, f can be represented in the form (6.9.3) and additionally
$$\sum_{i=1}^{r+s} ||v_i||_{E^*} = \inf\{||\varphi||: A^*B^*\varphi = f\}.$$

We can express f in the form (6.9.4), where f_1 is given by (6.9.5) and f_2 is given by (6.9.6).

Let
$$a_i = \inf\{||\varphi||: A^*B^*\varphi = f_i\}, \quad i = 1, 2. \tag{6.9.8}$$

In view of (6.9.4), $a_1 + a_2 \geq \inf\{||\varphi||: A^*B^*\varphi = f\}$. On the other hand, formulae (6.9.5) and (6.9.6) give rise to functionals φ_1, φ_2 such that
$$A^*B^*\varphi_i = f_i, \quad i = 1, 2,$$
and
$$||\varphi_1|| + ||\varphi_2|| = \inf\{||\varphi||: A^*B^*\varphi = f\}.$$

This implies that
$$a_i = ||\varphi_i||, \quad i = 1, 2$$
and the functionals φ_1, φ_2 define optimal observations. Thus,
$$\tilde{U}_{r+s} \subset \tilde{U}_r + \tilde{U}_s. \tag{6.9.9}$$

The converse implication is not true, as will be shown later.

THEOREM 6.9.1 (Phan Quoc Khanh, 1979).

(a) $\{0\} = U_0 \subsetneq U_1 \subsetneq U_2 \subsetneq \ldots \subsetneq U_r = U_{r+1} = U_{r+2} = \ldots = U_n = \mathscr{F}$,

(b) $\{0\} = \tilde{U}_0 \subsetneq \tilde{U}_1 \subsetneq \tilde{U}_2 \subsetneq \ldots \subsetneq \tilde{U}_s = \tilde{U}_{s+1} = \tilde{U}_{s+2} = \ldots = \tilde{U}_n = \mathscr{F}$,

where $\mathscr{F}$ denotes the set of all observable functionals.

Proof. Suppose that, for some k, $U_{k-1} = U_k$. Then, by (6.9.7),
$$U_{k+1} = U_k + U_1 = U_{k-1} + U_1 = U_k.$$

6.9. Functionals observed by less than n measurements

This proves (a). Suppose that $\tilde{U}_{k-1} = \tilde{U}_k$ and $f \in \tilde{U}_{k+1}$. Hence, by (6.9.9), $f = f_k + f_1$, where $f_k \in \tilde{U}_k$ and $f_1 \in \tilde{U}_1$. By using formulae (6.9.5) and (6.9.6) we can find φ_k and φ_1 such that

$$||\varphi_k|| + ||\varphi_1|| = \inf\{||\varphi||: A^*B^*\varphi = f\}$$

and

$$f^k = A^*B^*\varphi_k.$$

Under our hypothesis, there is a functional ψ, $||\psi|| = ||\varphi_k||$, such that $A^*B^*\psi = f$ and the functional ψ is of the form

$$\psi(y) = \sum_{i=1}^{k-1} (v'_i, y(t'_i)).$$

By taking points $t'_1, \ldots, t'_{k-1}, t_{k+1}$ and vectors $v'_1, \ldots, v'_{k-1}, v_{k+1}$ we construct a functional ψ_k such that

$$||\psi_k|| = \inf\{||\varphi||: A^*B^*\varphi = f\}.$$

Note that ψ_k is based on k measurements. Thus (b) holds. ∎

It is of interest to investigate the behaviour of U_s and $\tilde{U}_s$ under successive extensions of the time intervals on which the system is defined.

Let us consider two systems:

$$\frac{dx}{dt} = A(t)x, \quad 0 \leq t \leq T_1,$$

$$y(t) = G(t)x(t), \qquad (6.9.10)_1$$

and

$$\frac{dx}{dt} = A(t)x, \quad 0 \leq t \leq T_2,$$

$$y(t) = G(t)x(t), \qquad (6.9.10)_2$$

where $T_1 \leq T_2$. By $\tilde{U}_r^i$ (U_r^i) we denote the set of observable functionals (optimally observable functionals) by r measurements for systems $(6.9.10)_i$, $i = 1, 2$.

In general

$$U_r^1 \subset U_r^2. \qquad (6.9.11)$$

For optimally observable functionals there is no such relation as is shown by the following example.

EXAMPLE 6.9.1. Let
$$\frac{d^2x}{dt^2} = \begin{cases} -x, & 0 \leqslant t \leqslant \pi, \\ 0, & \pi < t \leqslant \pi+4, \end{cases}$$
$$y(t) = x(t).$$

Let $T_1 = \pi$, $T_2 = \pi+4$. By Example 6.7.2, $X^* = \tilde{U}_1^1 = U_1^1$ and by (6.9.11), $U_1^1 = U_1^2$. However, $\tilde{U}_2^2 \neq \tilde{U}_1^2$, since $x'(0) = -x'(\pi)$ and φ can be expressed in the form
$$\varphi(y) = -\frac{y(\pi+4)-y(\pi)}{4}.$$

The norm of this functional is equal to $\frac{1}{2}$. Hence, by measurement at two points we obtain a better result than by measurement at one point (in the latter case the norm is equal at least to 1). The example shows also that $U_1^2 \neq \tilde{U}_1^2$.

6.10. OPTIMAL INTEGRATION AND DIFFERENTIATION FORMULAE

In this section we present the results of Pallaschke (1976), who observed that the optimal observation approach proposed in this chapter can be used to derive some integration and differentiation formulae.

Let X be the space of polynomials of order $n-1$. X can be identified with the set of solutions of the differential equation
$$\frac{d^n x}{dt^n} = 0.$$

The output space Y is the space of all continuous scalar-valued functions, $Y = C[0, 1]$.

Let f be a functional of the form
$$f(x) = \int_0^1 x(t)\,d\mu,$$

6.10. Optimal integration and differentiation

where μ is a nonnegative measure. According to Theorem 6.7.4, the functional f can be represented in the form

$$f(x) = \int_0^1 x(t) \, d\mu = \sum_{i=1}^n \alpha_i x(t_i), \qquad (6.10.1)$$

where t_i are properly chosen points and

$$\sum_{i=1}^n \alpha_i = \mu([0, 1]). \qquad (6.10.2)$$

In the sequel we shall assume that the cardinality of the support of the measure μ is greater than n since otherwise the representation (6.10.1) would be trivial. Moreover, we assume that the measure μ is normalized, i.e., $\mu([0, 1]) = 1$.

We show how to choose the points t_i. Let Φ_n be a polynomial of order n such that

$$\int_0^1 S(t) \Phi_n(t) \, d\mu = 0, \qquad (6.10.3)$$

for each polynomial $S(t)$ of order $n-1$. The existence of the polynomial Φ_n follows from the fact that the space of polynomials of order n is $n+1$-dimensional and the space of polynomials of order $n-1$ is n-dimensional. The polynomial $\Phi_n(t)$ is uniquely determined up to a constant factor. Thus, its zeros $t_1, \ldots, t_n$ are also uniquely determined.

We claim that these zeros are different. To see this, suppose that

$$\Phi_n(t) = (t-t_0)^2 P(t),$$

where $P(t)$ is a polynomial of order $n-2$. Thus,

$$\int_0^1 P(t) \Phi_n(t) \, d\mu = \int_0^1 P^2(t) (t-t_0)^2 \, d\mu > 0, \qquad (6.10.4)$$

which contradicts (6.10.3). By using the zeros $t_1, \ldots, t_n$ of the polynomial $\Phi_n(t)$, we construct the Lagrange interpolation polynomials

$$L_k(t) = \prod_{\substack{j=1 \\ j \neq k}}^n \frac{(t-t_j)}{(t_k-t_j)}. \qquad (6.10.5)$$

We have
$$L_k(t_j) = \begin{cases} 0 & \text{for } j \neq k, \\ 1 & \text{for } j = k. \end{cases} \quad (6.10.6)$$

Thus, if $x(t)$ is a polynomial of order $n-1$ we have
$$x(t) = \sum_{k=1}^{n} x(t_k) L_k(t) \quad (6.10.7)$$

and
$$\int_0^1 x(t) \, d\mu = \sum_{k=1}^{n} x(t_k) A_k, \quad (6.10.8)$$

where
$$A_k = \int_0^1 L_k(t) \, d\mu, \quad k = 1, \ldots, n, \quad (6.10.9)$$

are called *Christoffel coefficients* (Christoffel, 1858).

Observe that
$$L_k(t) = \frac{\Phi_n(t)}{(t-t_k)\Phi_n'(t_k)}, \quad k = 1, \ldots, n. \quad (6.10.10)$$

and
$$A_k = \int_0^1 \frac{\Phi_n(t)}{(t-t_k)\Phi_n'(t_k)} \, d\mu, \quad k = 1, \ldots, n. \quad (6.10.11)$$

Putting $x(t) = 1$, we obtain by (6.10.8) and (6.10.9)
$$\sum_{k=1}^{n} A_k = 1. \quad (6.10.12)$$

Putting $x(t) = L_k^2(t)$, we obtain by (6.10.6) and (6.10.8)
$$A_k = \sum_{j=1}^{n} L_k^2(t_j) A_j = \int_0^1 L_k^2(t) \, d\mu \geq 0, \quad k = 1, \ldots, n. \quad (6.10.13)$$

By (6.10.12) and (6.10.13), the norm of the functional (6.10.8) equals 1. Since the norm of the integral is also equal to 1, formula (6.10.8) is optimal.

6.10. Optimal integration and differentiation

It is worth mentioning that representation (6.10.8) is also valid for the space of polynomials of order $2n-1$. Indeed, suppose that $x(t)$ is a polynomial of degree not greater than $2n-1$. By (6.10.7) and (6.10.10), we can write

$$W(t) = x(t) - \sum_{k=1}^{n} x(t_k) L_k(t)$$

$$= x(t) - \sum_{k=1}^{n} x(t_k) \left(L_k(t) - L_k(t_k)\right)$$

$$= x(t) - \sum_{k=1}^{n} x(t_k) \frac{\Phi_n(t) - \Phi_n(t_k)}{(t-t_k)\Phi_n'(t_k)} \cdot \qquad (6.10.14)$$

Observe that $W(t)$ has zeros at the points $t_1, \ldots, t_k$. Thus, $W(t)$ is divisible by $\Phi_n(t)$,

$$W(t) = S(t)\Phi_n(t), \qquad (6.10.15)$$

where the polynomial $S(t)$ is of degree not greater than $n-1$. Thus, by (6.10.3),

$$\int_0^1 W(t)\,d\mu = 0$$

and we get formula (6.10.8) for all polynomials of degree $2n-1$. Formula (6.10.8) was proved by Christoffel (1858). In the case where μ is a uniformly distributed measure the proof was given by Gauss (1814).

Let X be the space of polynomials of order $2m$. Of course, X is $2m+1$-dimensional. Consider a functional $x'(0)$ defined on X. We seek points

$$t_{-m} < t_{-m+1} < \ldots < t_0 < t_1 < \ldots < t_m$$

such that

$$x'(0) = \sum_{i=-m}^{m} a_i x(t_i), \qquad (6.10.16)$$

for all polynomials of order $2m-1$ and such that the norm of the functional defined by the right-hand side of formula (6.10.16) is minimal.

By using the Lagrange interpolation formula, we obtain

$$x(t) = \sum_{k=-m}^{m} x(t_k) L_k(t), \qquad (6.10.17)$$

where $L_k(t)$ are Lagrange polynomials constructed for the points $t_{-m}, \ldots, t_m$. Differentiating (6.10.17), we obtain in view of (6.10.16)

$$a_k = \frac{d}{dt} L_k(t)|_{t=0}. \qquad (6.10.18)$$

Assume that $t_{-k} = -t_k$ and $t_0 = 0$. Differentiating $L_k(t)$ we obtain

$$\frac{d}{dt} L_k(t)|_{t=0} = \prod_{\substack{j=-m \\ j \neq k}}^{m} \frac{1}{t_k - t_j} \sum_{\substack{i=-m \\ i \neq k}}^{m} \prod_{\substack{j=-m \\ j \neq i \\ j \neq k}}^{m} (t - t_j)|_{t=0}$$

$$= - \prod_{\substack{j=-m \\ j \neq k}}^{m} \frac{1}{t_k - t_j} \cdot \sum_{\substack{i=-m \\ i \neq k}}^{m} \prod_{\substack{j=-m \\ j \neq i \\ j \neq k}}^{m} t_j$$

$$= - \prod_{\substack{j=-m \\ j \neq k}}^{m} \left(\frac{1}{t_k - t_j} \right) \prod_{\substack{j=-m \\ j \neq 0 \\ j \neq k}}^{m} t_j, \qquad (6.10.19)$$

for $k = \pm 1, \ldots, \pm m$. The function $L_0(t)$ is even, hence, $\frac{d}{dt} L_0(t)|_{t=0} = 0$. By (6.10.19),

$$\operatorname{sgn} \frac{d}{dt} L_{-k}(t)|_{t=0} = \operatorname{sgn} \frac{d}{dt} L_k(t)|_{t=0} = - \frac{d}{dt} L_{k+1}(t)|_{t=0}. \qquad (6.10.20)$$

Now we consider the Tchebychev polynomial of order $2m-1$,

$$T(t) = \cos((2m-1)t).$$

Observe that for t, $|t| < 1$, $|T(t)| \leq 1$. Let

$$t_k = \cos \frac{2m-k}{2m-1} \pi, \quad k = 1, 2, \ldots, m.$$

Putting $t_{-k} = t_k$ we can verify that

$$|T(t_k)| = 1 \quad \text{for} \quad k = \pm 1, \ldots, \pm m \qquad (6.10.21)$$

6.10. Optimal integration and differentiation

and
$$T(t_k) = -T(t_{k+1}), \quad k = 1, 2, \ldots, m-1. \tag{6.10.22}$$
Put $t_0 = 0$. By (6.10.20), (6.10.21) and (6.10.22)
$$\sum_{j=-m}^{m} |a_j| = \left|\sum_{j=-m}^{m} a_j T(t_j)\right| \tag{6.10.23}$$
and by (6.10.16)
$$\sum_{j=-m}^{m} |a_j| = T'(0) = 2m-1.$$

On the other hand, the norm of the functional $f(x) = x'(0)$ is not smaller than $2m$ since
$$\|f\| = \sup_{\|x\|\leq 1} |x'(0)| \geq x'(0) = 2m-1.$$

Thus, f_k and a_k given above determine the optimal differentiation formulae.

Let $x(t)$ be a twice continuously differentiable function defined on the interval $[-1,1]$. Formula (6.10.16), where a_k are given by (6.10.18) and $t_{-m}, \ldots, t_m$ are given by the Tchebychev polynomial provides us with an approximation of the derivative $x'(0)$ as $m \to \infty$ (Pallaschke, 1977). The same holds for functions whose first derivative satisfy the Hölder condition and moreover, the convergence of the approximations is uniform on the interval $[-1,1]$ (Haverkamp, 1978, 1982).

Chapter 7

Systems with distributed parameters

In applied sciences, by *systems with distributed parameters* we mean systems described by partial differential equations. The *heat diffusion equation*

$$\frac{\partial Q}{\partial t} = a\left(\frac{\partial^2 Q}{\partial x_1^2} + \ldots + \frac{\partial^2 Q}{\partial x_n^2}\right) \qquad (7.0.1)$$

or the *wave equation*

$$\frac{\partial^2 Q}{\partial t^2} = a\left(\frac{\partial^2 Q}{\partial x_1^2} + \ldots + \frac{\partial^2 Q}{\partial x_n^2}\right) \qquad (7.0.2)$$

are examples of such systems.

In equations (7.0.1) and (7.0.2), the argument t belongs to an interval $[0, T]$ and $x = (x_1, \ldots, x_n)$ belongs to a fixed domain D, $D \subset R^n$. Some boundary and initial conditions are relevant to equations (7.0.1) and (7.0.2).

In the present chapter we shall apply the ideas developed in Chapter 5 to systems with distributed parameters.

The method proposed in this chapter represents a certain abstract setting of the Fourier method. The approach we present here is not a complicated one and, in certain simple cases, allows us to determine solutions of a problem in an explicit form. Therefore, we restrict our considerations to the simplest one-dimensional systems.

It should be noticed, however, that even in those simplest cases the results derived here are weaker than the corresponding results obtained by applying the general theory of differential equations. For instance, by using the proposed method, one can prove the existence of the generalized solution of the one-dimensional equation (7.0.2) with boundary value conditions which are not continuous. But the method fails in the proof of continuity of this solution regarded as a function of t which takes values in L^2. This fact can be proved by the general theory (cf. Lions, 1968; Lions and Magenes, 1968) and the reader who is interested in theory of equations with distributed parameters should be acquainted with the interesting book by Lions (1968).

The method presented in this chapter deals with equations

$$\frac{dx}{dt} = A(x) \quad \left(\text{or } \frac{d^2x}{dt^2} = A(x)\right) \tag{7.0.3}$$

regarded as differential equations in appropriately chosen Banach spaces. Those spaces are decomposed into one-dimensional spaces corresponding to the eigenvectors of the operator A.

By the derivative of a function $x(t)$ of a real argument t with values in a Banach space X we mean as usual the limit

$$\frac{dx}{dt} = \lim_{h \to 0} \frac{x(t+h) - x(t)}{h}.$$

Assume that A is a linear (not necessarily continuous) operator defined on a certain domain D which is dense in X, with values in X. The boundary conditions for equations (7.0.1) and (7.0.2) imply the requirement that solutions $x(t)$ of the equation (7.0.3) for each t should belong to a manifold V contained in X and determined by boundary conditions.

7.1. BASES IN BANACH SPACES

Let X be a Banach space. A sequence $\{e_n\}$ of elements of X is called a *Schauder basis* (or simply a *basis*) of the space X if each element x which belongs to X can be represented in the form of the sum of a series

$$x = \sum_{i=1}^{\infty} t_i e_i \tag{7.1.1}$$

and the representation is unique.

Clearly, each basis in a finite-dimensional space is a Schauder basis.

EXAMPLE 7.1.1. In the spaces l, l^2, c_0, a sequence $\{e_n\}$ defined as

$$e_n = \{0, \ldots, \underbrace{0, 1}_{n\text{-th position}}, 0, \ldots\}$$

is a basis.

EXAMPLE 7.1.2. In the space $L^2[-\pi, \pi]$ a sequence $\{e_n\}$, $n = 0, +1, -1, +2, -2, \ldots$ defined as follows:

(a) $e_n = e^{int}$ for the space over the field of complex numbers,

(b) $e_n = \begin{cases} \cos nt & \text{for } n \leq 0, \\ \sin nt & \text{for } n > 0, \end{cases}$

for the space over the field of real numbers, forms a basis.

We should add that in this example the summation of series is performed after rearranging all integers into the sequence $0, 1, -1, 2, -2, \ldots$

Now we prove some fundamental facts about bases.

Let $(X, \|\ \|)$ be a Banach space and let $\{e_n\}$ be a basis in X. Let

$$P_n(x) = \sum_{i=1}^{n} t_i e_i \quad \text{where} \quad x = \sum_{i=1}^{\infty} t_i e_i.$$

THEOREM 7.1.1 (Schauder, 1927). *The operators P_n are continuous and*

$$\sup_n \|P_n\| < +\infty.$$

Proof. Let $X_1 = \lim_\infty \{e_n\}$ denote the space of all sequences y, $y = \{\eta_i\}$, for which the series $\sum_{i=1}^{\infty} \eta_i e_i$ is convergent. The arithmetic properties of the limit imply that X_1 is a linear space. We define a norm in X_1 by the following formula

$$\|y\|^* = \sup_n \left\| \sum_{i=1}^{n} \eta_i e_i \right\|. \tag{7.1.2}$$

The space X_1 with the norm $\|\ \|^*$ is a normed space. We shall prove that X_1 is a Banach space, i.e. that X_1 is complete.

Consider a fundamental sequence $\{y^k\}$, $y^k = \{\eta_n^k\}$, in the space X_1. By the definition of a fundamental sequence and the definition of the norm $\|\ \|^*$, for every positive ε there exists a natural number m_0, such that

$$\|y^m - y^k\|^* = \sup_n \left\| \sum_{i=1}^{n} (\eta_i^m - \eta_i^k) e_i \right\| < \varepsilon \tag{7.1.3}$$

7.1. Bases in Banach spaces

for $m, k > m_0$. This implies that for all n

$$\|(\eta_n^m - \eta_n^k)e_n\| \le \left\|\sum_{i=1}^{n}(\eta_i^m - \eta_i^k)e_i\right\| + \left\|\sum_{i=1}^{n-1}(\eta_i^m - \eta_i^k)e_i\right\|$$
$$< 2\varepsilon. \tag{7.1.4}$$

Since ε is arbitrary, for each fixed n the sequence $\{\eta_n^m\}$ is a fundamental sequence of numbers. The field of scalars is complete and therefore there exists an η_n such that

$$\eta_n = \lim_{m \to \infty} \eta_n^m. \tag{7.1.5}$$

Passing to the limit in (7.1.3) with k tending to infinity, we obtain

$$\sup_n \left\|\sum_{i=1}^{n}(\eta_i^m - \eta_i)e_i\right\| \le \varepsilon \quad \text{for} \quad m > m_0. \tag{7.1.6}$$

Write

$$S_n^m = \sum_{i=1}^{n} \eta_i^m e_i \quad \text{and} \quad S_n = \sum_{i=1}^{n} \eta_i e_i;$$

by (7.1.6) we obtain

$$\|S_n - S_r\| \le \|S_n^m - S_r^m\| + 2\varepsilon \tag{7.1.7}$$

for $m > m_0$ and all n, r. Let us fix m_1, $m_1 > m_0$. The sequence $\{S_m^{m_1}\}$ is fundamental since $\{e_n\}$ is a basis. Hence, there exists a natural number N such that for $n, r > N$

$$\|S_n^{m_1} - S_r^{m_1}\| < \varepsilon. \tag{7.1.8}$$

Therefore, by (7.1.7),

$$\|S_n - S_r\| < 3\varepsilon. \tag{7.1.9}$$

This implies that the series $\sum_{i=1}^{\infty} \eta_i e_i$ is convergent. By the definition of X_1 we obtain $y = \{\eta_i\} \subset X_1$.

Inequality (7.1.6) entails $\|y^m - y\| < \varepsilon$ for $m > m_0$. This means that X_1 is complete.

Let A be an operator mapping X_1 onto X which is given by the formula

$$A(y) = \sum_{i=1}^{\infty} \eta_i e_i. \tag{7.1.10}$$

By the definition of X_1, the operator A is well-defined on the whole of X_1. Clearly, A is linear and continuous since

$$\|A(y)\| = \left\|\sum_{i=1}^{\infty} \eta_i e_i\right\| \leqslant \sup_n \left\|\sum_{i=1}^{n} \eta_i e_i\right\| = \|y\|^*. \tag{7.1.11}$$

It follows from the definition of a basis that A is a one-to-one mapping. By the Banach theorem on inverse operators (Corollary 3.2.2), the inverse operator A^{-1} is continuous. By (7.1.11), for every n and every x,
$x = \sum_{i=1}^{\infty} \eta_i e_i$,

$$\|P_n(x)\| = \left\|\sum_{i=1}^{n} \eta_i e_i\right\| \leqslant \|y\|^* = \|A^{-1}(x)\| \leqslant \|A^{-1}\|\|x\| \tag{7.1.12}$$

and hence

$$\|P_n\| \leqslant \|A^{-1}\|. \quad \blacksquare \tag{7.1.13}$$

COROLLARY 7.1.2. *Coefficients $\{t_i\}$ of representations with respect to a basis $\{e_i\}$ represent continuous linear functionals. They are called basic functionals.*

The following theorem can be regarded as converse to Theorem 7.1.1.

THEOREM 7.1.3. *Let X be a Banach space. Let $\{e_n\}$ be a sequence of linearly independent elements of X. Let $X_0 = \text{lin}\{e_n\}$ be a linear space spanned by the sequence $\{e_n\}$. Clearly, X_0 is a linear subset of X and X_0 can be considered a normed space. On the space X_0 we define operators P_n by the formula*

$$P_n\left(\sum_{i=1}^{m} t_i e_i\right) = \sum_{i=1}^{n} t_i e_i, \tag{7.1.14}$$

7.1. Bases in Banach spaces

where, for $m < n$, we assume $t_i = 0$, $i = m+1, \ldots, n$. If

$$\sup_n \|P_n\| < +\infty,$$

then the sequence $\{e_n\}$ forms a basis in the space $\bar{X}_0$ which is the completion of the space X_0'.

Proof. Denote by $X_1 = \lin_\infty \{e_n\}$ the set of all elements x which can be represented as the sum of a convergent series

$$x = \sum_{i=1}^{\infty} t_i e_i. \tag{7.1.15}$$

Now we shall prove that if the operators P_n are continuous then the sequence $\{e_n\}$ consists of strongly linearly independent elements, i.e., if the series converges to zero then $t_i = 0$, $i = 1, 2, \ldots$ Suppose on the contrary that the elements of the sequence $\{e_n\}$ are not strongly linearly independent. Then, by definition, there exists a sequence $\{t_i^0\}$, with elements not all equal to zero, such that the series $\sum_{i=1}^{\infty} t_i^0 e_i$ converges to zero. Since not all t_i^0 are zero, there exists an n such that for $m > n$

$$a = P_n\left(\sum_{i=1}^{m} t_i^0 e_i\right) \neq 0.$$

On the other hand, $\sum_{i=1}^{m} t_i^0 e_i$ tends to zero with m tending to infinity. This contradicts the continuity of the operator P_n.

By the strong linear independence of the elements of the sequence $\{e_n\}$, the operators P_n can be uniquely extended to $X_1 = \lin_\infty \{e_n\}$ by the formula

$$P_n\left(\sum_{i=1}^{\omega} t_i e_i\right) = \sum_{i=1}^{n} t_i e_i.$$

Define a norm in the space X_1 by the formula

$$\|x\|^* = \sup_n \left\|\sum_{i=1}^{n} t_i e_i\right\|, \tag{7.1.16}$$

334 7. Systems with distributed parameters

where, as before,

$$x = \sum_{i=1}^{\infty} t_i e_i.$$

Denote by X_2 the space X_1 with the norm given by formula (7.1.16). Observe that

$$||x|| \leq ||x||^*.$$

On the other hand, by the assumption, there exists a constant M such that for all n

$$||P_n|| \leq M. \tag{7.1.17}$$

From (7.1.17) we obtain

$$||x||^* \leq M||x||.$$

Hence, the spaces X_1 and X_2 are isomorphic. By the same arguments as in the proof of Theorem 7.1.1, the space X_2 is complete. Thus, the space X_1 is also complete. This means that $\overline{X_0} = X_1$. By the definition of the space X_1 and by the strong linear independence of the elements of the sequence $\{e_n\}$, the sequence $\{e_n\}$ is a basis in the space X_1. ∎

A sequence $\{e_n\}$ of elements of a Banach space X is called a *basic sequence* if it forms a basis in the space X_1 spanned by $\{e_n\}$,

$$X_1 = \overline{\text{lin } \{e_n\}}.$$

By using the above definition Theorem 7.1.3 can be restated in the following way.

THEOREM 7.1.3'. *If the operators P_n are continuous and*

$$\sup_n ||P_n|| < +\infty,$$

then the sequence $\{e_n\}$ is a basic sequence.

COROLLARY 7.1.4. *Let X be a Banach space. Let $\{e_n\}$ be a sequence of elements of X. The sequence $\{e_n\}$ is a basis if and only if*

   (a) *linear combinations of elements of the sequence $\{e_n\}$ are dense in X,*

(b) *there exists a constant M such that for all n, m, $m > n$,*

$$\left\|\sum_{i=1}^{n} t_i e_i\right\| \leq M \cdot \left\|\sum_{i=1}^{m} t_i e_i\right\|.$$

EXAMPLE 7.1.3. Let H be a Hilbert space. Let $\{e_n\}$ be an orthogonal sequence in H. Then the sequence $\{e_n\}$ is a basic sequence. Indeed, if $\{e_n\}$ is an orthogonal sequence, then for each n

$$\left\|\sum_{i=1}^{n} t_i e_i\right\| = \sqrt{\left(\sum_{i=1}^{n} t_i e_i, \sum_{i=1}^{n} t_i e_i\right)} = \sqrt{\sum_{i=1}^{n} |t_i|^2 \|e_i\|^2}$$

since $(e_i, e_j) = 0$ for $i \neq j$. Hence, for all m, $m > n$,

$$\left\|\sum_{i=1}^{n} t_i e_i\right\| \leq \left\|\sum_{i=1}^{m} t_i e_i\right\|.$$

If linear combinations of elements of the orthogonal sequence $\{e_n\}$ form a dense set in H, then the sequence $\{e_n\}$ forms a basis, called an *orthogonal basis*.

An orthogonal basis $\{e_n\}$ is called an *orthonormal basis* if $\|e_n\| = 1$. If $\{e_n\}$ is an orthonormal basis, then the coefficients $\{c_n\}$ of expansions with respect to this basis belong to l^2, and conversely, for each sequence $\{e_n\} \subset l^2$, the series $\sum_{i=1}^{\infty} c_i e_i$ converges to an element x, $x \in H$. Moreover,

$$\|x\|_H = \|\{c_n\}\|_{l^2}.$$

Let X and Y be Banach spaces. Let $\{e_n\}$ be a basis in the space X and let $\{f_n\}$ be a basis in the space Y. We say that the bases $\{e_n\}$ and $\{f_n\}$ are *equivalent* if the series $\sum_{i=1}^{\infty} t_i e_i$ converges if and only if the series $\sum_{i=1}^{\infty} t_i f_i$ converges.

THEOREM 7.1.5. *If a basis $\{e_n\}$ in a Banach space X and a basis $\{f_n\}$ in a Banach space Y are equivalent, then the spaces X and Y are isomorphic.*

Proof. Let X_1 (resp. Y_1) denote the space of all sequences $\xi = \{\xi_i\}$ (resp. $\eta = \{\eta_i\}$) such that the series $\sum_{i=1}^{n} \xi_i e_i$ (resp. $\sum_{i=1}^{n} \eta_i f_i$) converges. Introduce a norm in the space X_1 (resp. Y_1) by the formula

$$\|\xi\|_X = \sup_n \left\| \sum_{i=1}^{\infty} \xi_i e_i \right\| \quad \left(\text{resp. } \|\eta\|_Y = \sup_n \left\| \sum_{i=1}^{\infty} \eta_i f_i \right\| \right). \tag{7.1.18}$$

The spaces X_1 and Y_1 are isomorphic to the spaces X and Y, respectively, as follows from the proof of Theorem 7.1.1. The hypothesis that the bases $\{e_n\}$ and $\{f_n\}$ are equivalent implies that the spaces X and Y consist of the same sequences. By Corollary 7.1.2, the components of those sequences ξ_i (resp. η_i) are linear functionals which are continuous in both topologies.

Clearly, these functionals form a total family of functionals. Hence, by Corollary 3.2.4, the spaces X_1 and Y_1 are isomorphic. ∎

THEOREM 7.1.6 (Krein, Milman, Rutman, 1940). *Let $(X, \|\ \|)$ be a Banach space with a basis $\{e_n\}$ such that $\|e_n\| = 1$. Let $\{f_n\}$ be a sequence of strongly linearly independent elements such that the series*

$$\sum_{i=1}^{\infty} \|e_i - f_i\| \tag{7.1.19}$$

is convergent. Then $\{f_n\}$ forms a basis in the space X.

Proof. Let $M = \sup_n \|P_n\|$. Since the series (7.1.19) is convergent, there exists a number N such that

$$K_N = \sum_{i=N}^{\infty} \|f_i - e_i\| < \frac{1}{2M}. \tag{7.1.20}$$

Let X_1 be the space spanned by the elements $\{e_N, e_{N+1}, \ldots\}$ and let Y_1 be the space spanned by the elements $\{f_N, f_{N+1}, \ldots\}$.

7.1. Bases in Banach spaces

We shall prove that the set $\{f_N, f_{N+1}, \ldots\}$ forms a basis in the space Y_1 and this basis is equivalent to the basis $\{e_N, e_{N+1}, \ldots\}$ in the space X_1.

Indeed, by the triangle inequality and by formula (7.1.20), for all $n, n' > N$,

$$\left\|\sum_{i=n}^{n'} t_i e_i\right\| - \sum_{i=n}^{n'} |t_i| \, \|f_i - e_i\| \leq \left\|\sum_{i=n}^{n'} t_i f_i\right\|$$

$$\leq \left\|\sum_{i=n}^{n'} t_i e_i\right\| + \sum_{i=n}^{n'} |t_i| \cdot \|f_i - e_i\|. \tag{7.1.21}$$

Considering the fact that $\|e_i\| = 1$, we obtain by formula (7.1.4)

$$|t_i| \leq 2M \cdot \left\|\sum_{i=n}^{n'} t_i e_i\right\|, \quad i = n, n+1, \ldots, n', \tag{7.1.22}$$

and hence

$$\sum_{i=n}^{n'} |t_i| \cdot \|f_i - e_i\| \leq 2M K_N \left\|\sum_{i=n}^{n'} t_i e_i\right\|. \tag{7.1.23}$$

Thus,

$$(1-\delta)\left\|\sum_{i=n}^{n'} t_i e_i\right\| \leq \left\|\sum_{i=n}^{n'} t_i f_i\right\| \leq (1+\delta)\left\|\sum_{i=n}^{n'} t_i e_i\right\|, \tag{7.1.24}$$

where $\delta = K_N 2M < 1$. This implies that the series $\sum_{i=N}^{\infty} t_i f_i$ converges if and only if the series $\sum_{i=N}^{\infty} t_i e_i$ converges. Moreover, by formula (7.1.24) we have

$$\left\|\sum_{i=1}^{m} t_i f_i\right\| \leq (1+\delta)\left\|\sum_{i=N}^{n} t_i e_i\right\| \leq 2M(1+\delta)\left\|\sum_{i=N}^{m} t_i e_i\right\|$$

$$\leq \frac{2M(1+\delta)}{1-\delta}\left\|\sum_{i=N}^{m} t_i f_i\right\|, \tag{7.1.25}$$

for $N < n < m$. This implies in turn that the sequence $\{f_N, f_{N+1}, \ldots\}$ forms a basis in the space Y_1.

By hypothesis, the elements of the sequence $\{f_i\}$ are strongly linearly independent. This means that $f_i \notin Y_1$ for $i = 1, \ldots, N-1$. Since Y_1 is a closed subspace, there exist continuous linear functionals F_i, $i = 1, \ldots, N-1$, such that

$$F_i(f_j) = \begin{cases} 1 & \text{for } i = j, \\ 0 & \text{for } i \neq j, \end{cases} \tag{7.1.26}$$

for $i, j = 1, 2, \ldots, N-1$ and

$$F_i(y) = 0 \quad \text{for} \quad y \in Y_1. \tag{7.1.27}$$

Let Y_2 be a space of elements of the form

$$y = \sum_{i=1}^{\infty} t_i f_i.$$

We define an operator Q_n by the formula

$$Q_n\left(\sum_{i=1}^{\infty} t_i f_i\right) = \sum_{i=1}^{n} t_i f_i.$$

It follows from (7.1.26) and (7.1.27) that the operators Q_n are continuous for $n = 1, 2, \ldots, N-1$. By formula (7.1.25),

$$\|Q_n\| \leq \|Q_{N-1}\| + \frac{2M(1+\delta)}{(1-\delta)} \tag{7.1.28}$$

for $n > N$. Hence, by Corollary 7.1.4, $\{f_n\}$ is a basis in the space Y_2.

The elements of the sequence $\{f_n\}$ are strongly linearly independent. In order to prove that $\{f_n\}$ is a basis in X it is enough to show that the codimension of Y_1 in X is $N-1$. Indeed, if the codimension of the space Y_1 in X is $N-1$, then linear combinations of elements of $\{f_n\}$ are dense in X and, by Corollary 7.1.4, $\{f_n\}$ is a basis.

Thus, we shall prove that Y_1 is of codimension $N-1$ in X. Suppose on the contrary that it is not true. Then the space $Y_0 = \lin\{e_1, \ldots, e_{N-1}, Y_1\}$ does not coincide with X. Hence, there exists an element x_0, $x_0 \in X$, such that $\|x_0\| = 1$ and

$$\inf\{\|x_0 - y\| : y \in Y\} > \delta. \tag{7.1.29}$$

Since $\{e_n\}$ is a basis, x_0 can be represented in the form

$$x_0 = \sum_{i=1}^{\infty} t_i^0 e_i.$$

Let

$$y_0 = t_1^0 e_1 + \ldots + t_{N-1}^0 e_{N-1} + \sum_{i=N}^{\infty} t_i^0 f_i. \tag{7.1.30}$$

Clearly, $y_0 \in Y$ and

$$\|x_0 - y_0\| = \left\|\sum_{i=N}^{\infty} t_i^0 f_i\right\| \leq \delta \|x_0\| = \delta. \tag{7.1.31}$$

This contradicts (7.1.29). ∎

Observe that formula (7.1.24) implies strong linear independence of elements $\{f_N, f_{N+1}, \ldots\}$. Hence, if the sum (7.1.19) is sufficiently small, one can prove that the sequence $\{f_n\}$ is a basis, even if the elements of $\{f_n\}$ are not strongly linearly independent.

In Hilbert spaces we can formulate a result close to the Krein–Milman–Rutman theorem.

THEOREM 7.1.7 (Singer, 1970). *Let H be a Hilbert space. Let $\{e_n\}$ be an orthonormal basis in H. Let $\{f_n\}$ be a sequence of strongly linearly independent elements such that*

$$\sum_{n=1}^{\infty} \|e_n - f_n\|^2 < +\infty. \tag{7.1.32}$$

Then $\{f_n\}$ is a basis in H and it is equivalent to the basis $\{e_n\}$.

Proof. We shall prove that if the sum (7.1.32) is sufficiently small then the sequence $\{f_n\}$ is a basic sequence. Suppose that

$$\sum_{i=1}^{\infty} \|r_i\|^2 = C^2 < 1, \tag{7.1.33}$$

where

$$r_i = f_i - e_i, \quad i = 1, 2, \ldots$$

By the Schwarz inequality and by formula (7.1.33),

$$\left\|\sum_{i=1}^n t_i r_i\right\| \leq \sum_{i=1}^n |t_i|\,\|r_i\| \leq \Big(\sum_{i=1}^n |t_i|^2\Big)^{1/2}\Big(\sum_{i=1}^n \|r_i\|^2\Big)^{1/2}$$

$$\leq C\Big(\sum_{i=1}^n |t_i|^2\Big)^{1/2} = C\left\|\sum_{i=1}^n t_i e_i\right\|. \tag{7.1.34}$$

It follows from the triangle inequality and formula (7.1.34) that

$$(1-C)\left\|\sum_{i=1}^n t_i e_i\right\| \leq \left\|\sum_{i=1}^n t_i f_i\right\| \leq (1+C)\left\|\sum_{i=1}^n t_i e_i\right\|, \tag{7.1.35}$$

for all n and all n-tuples of numbers $\{t_1, \ldots, t_n\}$. From (7.1.35) we infer that for all m, $m > n$,

$$\left\|\sum_{i=1}^n t_i f_i\right\| \leq (1+C)\left\|\sum_{i=1}^n t_i e_i\right\| \leq (1+C)\left\|\sum_{i=1}^m t_i e_i\right\|$$

$$\leq \frac{1+C}{1-C}\left\|\sum_{i=1}^m t_i f_i\right\|.$$

By Theorem 7.1.3, this implies that the sequence $\{f_n\}$ is a basic sequence.

Consider now the general case, where the sum (7.1.32) is assumed to be finite. By (7.1.32) there exists an N such that

$$C_N = \sum_{i=N}^\infty \|r_i\|^2 < \frac{1}{2}. \tag{7.1.36}$$

By the above arguments, $\{f_N, f_{N+1}, \ldots\}$ is a basic sequence in H. The proof that linear combinations of elements of $\{f_n\}$ are dense in H proceeds in the same way as the corresponding part of the proof of Theorem 7.1.6. ∎

7.2. Eigenvalues and eigenvectors

Let X be a Banach space. Let A be a linear (not necessarily continuous) operator defined on a linear (not necessarily closed) subset D_A, called

7.2. Eigenvalues and eigenvectors

the *domain* of the operator A. The range of A, R_A, is contained in X. If there exist a number λ and a vector $x_\lambda \neq 0$ such that

$$A(x_\lambda) = \lambda x_\lambda, \qquad (7.2.1)$$

then the number λ is called an *eigenvalue* of the operator A and the vector x_λ is called an *eigenvector* of the operator A.

EXAMPLE 7.2.1. Let $X = C[0, T]$ and $A = \dfrac{d}{dt}$. The domain of the operator A is the space of all continuously differentiable functions. Every number λ is an eigenvalue of the operator A. The corresponding eigenvectors are x_λ, $x_\lambda = e^{\lambda t}$.

EXAMPLE 7.2.2. Let $X = C[0, T]$ and $A = \dfrac{d^2}{dt^2}$. The domain of the operator A is the space of all twice continuously differentiable functions. Every number λ is an eigenvalue of A. Two eigenvectors correspond to each λ: $e^{\pm\sqrt{\lambda}t}$ if X is complex, $e^{\pm\sqrt{\lambda}t}$ for $\lambda \geq 0$ and $\sin\sqrt{\lambda}t$, $\cos\sqrt{\lambda}t$ for $\lambda < 0$ if X is real.

EXAMPLE 7.2.3. Let X be a subspace of the space $C^{(1)}[0, T]$ consisting of real-valued functions such that $x(0) = x(T) = x'(0) = x'(T)$. As before, let A be the second derivative operator. The domain of the operator A is the set of all twice differentiable functions which belong to X. By easy calculations we can get the eigenvalues λ_n of the operator A. Namely,

$$\lambda_n = -\left(\frac{2n\pi}{T}\right)^2$$

and the eigenvectors which correspond to λ_n are

$$\sin\frac{2n\pi}{T}t, \quad \cos\frac{2n\pi}{T}t.$$

In particular, if $T = 2\pi$, then $\lambda_n = -n^2$ and the corresponding eigenvectors are $\sin nt$, $\cos nt$.

Let X be a Banach space and let A be a linear operator. Assume that a sequence $\{e_n\}$ of eigenvectors of A forms a basis in X. Then the operator A can be represented in the form

$$A\left(\sum_{i=1}^{\infty} t_i e_i\right) = \sum_{i=1}^{\infty} \lambda_i t_i e_i, \qquad (7.2.2)$$

where λ_i is the eigenvalue which corresponds to an eigenvector e_i.

The representation of the operator A in the form (7.2.2) is called the *diagonal representation*.

If the operator A can be represented in the diagonal form, then a formula which gives a solution of the differential equation (7.0.3) can be derived, provided that a solution exists.

To derive such a formula, denote by $\{f_n\}$ the basic functionals corresponding to the basis $\{e_n\}$. Suppose that $x(t)$ is a solution of equation (7.0.3),

$$\frac{d}{dt} x = A(x) \quad \left(\text{resp. } \frac{d^2}{dt^2} x = A(x)\right)$$

with the initial conditions

$$x(0) = x_0 \quad \left(\text{resp. } x(0) = x_0, \left.\frac{dx}{dt}\right|_0 = x_1\right). \qquad (7.2.3)$$

By applying the functionals f_i, $i = 1, 2, \ldots$, to both sides of equation (7.0.3) we obtain the infinite sequence of differential equations,

$$\frac{d\varphi_i}{dt} = \lambda_i \varphi_i \quad \left(\text{resp. } \frac{d^2\varphi_i}{dt^2} = \lambda_i \varphi_i\right) \qquad (7.2.4)$$

with the initial conditions

$$\varphi_i(0) = f_i(x_0) \quad \left(\text{resp. } \varphi_i(0) = f_i(x_0), \left.\frac{d}{dt}\varphi_1\right|_0 = f_i(x_1)\right),$$

where $\varphi_i(t) = f_i(x(t))$, $i = 1, 2, \ldots$ Hence, the solution, if it exists, is of the form

$$x(t) = \sum_{i=1}^{\infty} \varphi_i(t) e_i, \qquad (7.2.5)$$

where $\varphi_i(t)$ are solutions of equation (7.2.4).

7.2. Eigenvalues and eigenvectors

In this chapter we shall not be interested in general existence results. Instead, we shall investigate directly whether a solution $x(t)$ given by formula (7.2.5) belongs to the space X and whether $x(t)$ satisfies the initial condition (7.2.3). In many applications it happens that the eigenvectors of the operator A do not form a basis in the space X. However, they may form a basis in a certain Banach space Y which contains X as a linear subset and which induces on X a topology weaker than the original topology on X.

EXAMPLE 7.2.4. Let X and A be as in Example 7.2.3. For simplicity, take $T = 2\pi$. The sequence $\{e_n\}$ of eigenvectors of A,

$$e_n = \begin{cases} \cos \dfrac{n}{2} t & \text{for} \quad n = 2k, \quad k = 0, 1, \ldots, \\ \sin \dfrac{n-1}{2} t & \text{for} \quad n = 2k+1, \quad k = 1, 2, \ldots, \end{cases}$$

does not form a basis in the space $C[0, 2\pi]$ but $\{e_n\}$ forms an orthonormal basis in the space $L^2[0, 2\pi]$. Observe that $C[0, 2\pi]$ is a linear subset of $L^2[0, 2\pi]$ and the topology induced by the topology of $L^2[0, 2\pi]$ is clearly weaker than the original topology in $C[0, 2\pi]$.

In all such cases we proceed as follows. Observe that the basic functionals f_i, $f_i \in Y$, $i = 1, 2, \ldots$, are continuous on X. Thus, we can apply the functionals f_i to both sides of equation (7.0.3). Hence, we obtain the infinite sequence of equations (7.2.4) with the corresponding initial conditions (7.2.3).

If the series (7.2.5) converges in X for each t, the series

$$\sum_{i=1}^{\infty} \varphi_i'(t) e_i \quad \left(\text{or} \sum_{i=1}^{\infty} \varphi_i'(t) e_i \quad \text{and} \quad \sum_{i=1}^{\infty} \varphi_i''(t) e_i \right) \tag{7.2.6}$$

converges (converge) in Y and the sum (sums) belongs (belong) to the domain of the operator A, D_A, then the series (7.2.5) represents a solution of equation (7.0.3) which corresponds to the initial condition (conditions) (7.2.3).

It may happen that series (7.2.6) do not converge in X but in Y. Moreover, also series (7.2.5) may be convergent not in X but in Y. In all such cases we say that equation (7.0.3) has a generalized solution.

Hence, of all the possible definitions of generalized solutions we prefer to adopt the least restrictive one. Therefore, by a *generalized solution* of equation (7.0.3) we mean the limit of series (7.2.5) in Y, whenever it exists.

So far we have considered the homogeneous equation (7.0.3). However, in many applications we encounter equations which are not homogeneous. Namely,

$$\frac{dx}{dt} = A(x) + h(t) \quad \left(\text{resp. } \frac{d^2x}{dt^2} = A(x) + h(t)\right), \tag{7.2.7}$$

where $h(t)$ is a vector-valued function of the argument t.

If $h(t) \in X$ for each t, then at each t we can represent $h(t)$ as the sum of a series with respect to the basis $\{e_n\}$ of the space Y,

$$h(t) = \sum_{n=1}^{\infty} h_n(t) e_n. \tag{7.2.8}$$

Substituting (7.2.8) in (7.2.7) and applying the basic functionals f_i, $i = 1, \ldots$, to both sides of the resulting formula, we get the infinite system of differential equations

$$\frac{d}{dt}\varphi_i = \lambda_i \varphi_i + h_i(t) \quad \left(\text{resp. } \frac{d^2}{dt^2}\varphi_i = \lambda_i \varphi_i + h_i(t)\right) \tag{7.2.9}$$

with the initial condition

$$\varphi_i(0) = f_i(x_0) \quad (\text{or } \varphi_i(0) = f_i(x_0),\ \varphi_i'(0) = f_i(x_1)),$$

where $\varphi_i(t) = f_i(x(t))$, $i = 1, 2, \ldots$

Further on, we represent the solution of equation (7.2.9) in the form of a series and investigate its convergence properties in the same way as in the homogeneous case.

In many particular problems $h(t)$ does not belong to X but to Y. Then the whole reasoning can be repeated without any changes. In our considerations we shall often neglect the question whether the formulae derived are in fact solutions in the classical sense and we shall confine ourselves to generalized solutions defined by series (7.2.5).

More detailed considerations devoted to this subject can be found in advanced textbooks on partial differential equations.

7.3. TEMPERATURE DISTRIBUTION IN A ROD WITH ZERO BOUNDARY CONDITIONS

Consider a homogeneous rod of the length S. We shall assume that the rod is insulated along its length, i.e. any heat exchange with the surroundings is possible only at the end-points of the rod. Let x denote a space variable and let t denote time. Then the temperature distribution in the rod, $Q(t, x)$, satisfies the heat diffusion equation

$$\frac{\partial Q}{\partial t} = a\frac{\partial^2 Q}{\partial x^2}, \quad 0 \leqslant x \leqslant S, \tag{7.3.1}$$

where a denotes a coefficient which depends on the linear density and the heat capacity of the rod (cf. Tikhonov, Samarski, 1966).

Without loss of generality we can assume $a = 1$. Indeed, substituting $x' = \sqrt{a}x$ in equation (7.3.1), we obtain the equation

$$\frac{\partial Q}{\partial t} = \frac{\partial^2 Q}{\partial x^2}, \quad 0 \leqslant x' \leqslant \sqrt{a}S. \tag{7.3.1'}$$

Hence, in the sequel we shall assume $a = 1$. To obtain the corresponding formulae for $a \neq 1$ we only have to substitute $\sqrt{a}S$ for S.

We assume that the surrounding temperature is constant and equals zero. Then, at the end-points of the rod, the following boundary condition is satisfied

$$\frac{\partial Q}{\partial x} = \alpha Q, \tag{7.3.2}$$

where the heat transfer coefficient α can take different values, e.g., $\alpha = \alpha_0$ for $x = 0$ and α_S for $x = S$.

It follows from a physical interpretation that $\alpha_0 \geqslant 0$ and $\alpha_S \leqslant 0$.

If the heat transfer coefficient α is equal to 0 (the end-points of the rod are insulated), then equation (7.3.2) takes the form

$$\frac{\partial Q}{\partial x} = 0. \tag{7.3.3}$$

If the heat exchange with the surroundings is total, which corresponds to $|\alpha| = +\infty$, then the boundary condition is of the form

$$Q(t, 0) = 0 \quad (\text{or } Q(t, S) = 0), \tag{7.3.4}$$

instead of (7.3.2).

7. Systems with distributed parameters

Now our aim is to solve equation (7.3.1) with the initial temperature distribution

$$Q(0, x) = Q_0(x) \tag{7.3.5}$$

and with the boundary condition (7.3.2) (in particular, with the boundary condition (7.3.3) or (7.3.4) at one or both end-points of the rod).

To solve equation (7.3.1) by the method described in the previous section we represent equation (7.3.1) as a differential equation in a Banach space.

First of all, a space X on which the operator A will be defined should be specified. This can easily be done in the case where the boundary conditions (7.3.4) are imposed at the end-points of the rod. Namely, as the space X we can choose the space $C_0[0, S]$ of all continuous functions F defined on the interval $[0, S]$ which takes the value 0 at the end-points of the interval, with the standard norm $||F|| = \sup_{0 \leqslant x \leqslant S} |F(x)|$.

The space $C_0[0, S]$ is a Banach space. The problem is more complicated if at one of the end-points, e.g. at the point 0, condition (7.3.3) is imposed and at the second end-point condition (7.3.4) is imposed. Then the space X can be chosen as the space of all continuous functions $f(x)$ defined on the interval $[0, S]$ which vanish at the point S and have the right derivative at zero equal to zero. Clearly, this is a linear space. By taking the sup norm we obtain a normed space which is not complete. To obtain a Banach space we introduce a norm by the formula

$$f = \max\left(\sup_{0 \leqslant x \leqslant S} |f(x)|, \sup_{0 \leqslant x \leqslant S} \left|\frac{f(x)-f(0)}{x}\right|\right). \tag{7.3.6}$$

We shall prove that the space X with the norm given by formula (7.3.6) is complete. Take a fundamental sequence $\{f_n\}$ of functions belonging to X. This sequence is also fundamental in the sup norm. Hence, it is uniformly convergent to a certain function $f(x)$. In order to prove that X is complete it is enough to show that $f(x)$ has the right derivative at 0 equal to 0 and that $\{f_n\}$ tends to $f(x)$ in the norm (7.3.6).

Let ε be an arbitrary positive number. Since $\{f_n\}$ is fundamental in the norm (7.3.6), there exists an n such that, for all m, $m > n$,

$$||f_n - f_m|| < \varepsilon. \tag{7.3.7}$$

7.3. Temperature distribution. Zero boundary conditions

This implies that

$$\left| \frac{(f_n(x)-f_m(x))-(f_n(0)-f_m(0))}{x} \right|$$
$$= \left| \frac{(f_n(x)-f_n(0))-(f_m(x)-f_m(0))}{x} \right| < \varepsilon \qquad (7.3.8)$$

for all x.

Since $f_m(x)$ has the right derivative at 0 equal to 0, there exists a $\delta > 0$ such that

$$\left| \frac{f_m(x)-f_m(0)}{x} \right| < \varepsilon \qquad (7.3.9)$$

for all x, $0 < x < \delta$.

Passing to the limit in (7.3.8) with n tending to infinity, we obtain by (7.3.9)

$$\left| \frac{f(x)-f(0)}{x} \right| < 2\varepsilon \qquad (7.3.10)$$

for all x, $0 < x < \delta$.

Since ε is arbitrary, formula (7.3.10) implies that the function $f(x)$ has the right derivative at 0 equal to 0. Hence, $f \in X$.

It follows from (7.3.7) that

$$|f_n(x)-f_m(x)| < \varepsilon. \qquad (7.3.11)$$

Passing to the limit in (7.3.8) and (7.3.11) with n tending to infinity, we obtain

$$\|f_m - f\| < \varepsilon, \qquad (7.3.12)$$

which implies the completeness of the space X.

If at both end-points of the rod condition (7.3.3) is satisfied, then the space X can be chosen as the space of all continuous functions with one-sided derivatives which vanish at the end-points of the rod. In this space we introduce a norm by the formula

$$\|f\| = \max\left(\sup_{0 \leqslant x \leqslant S} |f(x)|, \sup_{0 < x < S} \left| \frac{f(x)-f(0)}{x} \right|, \sup_{0 < x < S} \left| \frac{f(x)-f(S)}{x-S} \right| \right)$$
$$(7.3.13)$$

It can be shown that the space X is a Banach space.

If at both end-points of the rod condition (7.3.2) is satisfied, the space X can be chosen as the space of all continuous functions defined on the interval $[0, S]$, which have one-sided derivative at the end-points and satisfy condition (7.3.2). A norm in this space is defined by the formula

$$\|f\| = \max\left(\sup_{0\leqslant x\leqslant S}|f(x)|,\ \sup_{0< x< S}\left|\frac{f(x)-f(0)}{x}-f'(0)\right|,\right.$$
$$\left.\sup_{0< x< S}\left|\frac{f(x)-f(S)}{x-S}-f'(S)\right|\right). \tag{7.3.14}$$

By similar arguments to those used before, X is complete. In all those cases the operator A is the second derivative operator. The domain of the operator A is the space of all twice differentiable functions which belong to X together with their second derivatives.

Although formally in all the cases listed above A does not change (A is the second derivative operator), in fact in each case A is different since its domain changes. Thus, in different spaces X the eigenvalues and the eigenvectors of A are different.

EXAMPLE 7.3.1. Suppose that at both end-points of the rod the boundary condition (7.3.4) is satisfied. The space X is the space of all continuous functions which vanish at the end-points of the interval. To find the eigenvalues and the eigenvectors of the operator A we have to solve the differential equation

$$\frac{d^2f}{dx^2} = \lambda x \tag{7.3.15}$$

with the boundary conditions (7.3.4)

$$f(0) = f(S) = 0,$$

and to determine the λ's for which there exists a nonzero solution. The general form of the solution for $\lambda > 0$ can be written as

$$f(x) = ae^{\sqrt{\lambda}x} + be^{-\sqrt{\lambda}x} \tag{7.3.16}$$

and for $\lambda < 0$ as

$$f(x) = a\cos\sqrt{|\lambda|}\,x + b\sin\sqrt{|\lambda|}\,x. \tag{7.3.17}$$

7.3. Temperature distribution. Zero boundary conditions

Substituting in equation (7.3.16) $x = 0$ and $x = S$, we obtain the system of equations

$$a+b = 0, \quad ae^{\sqrt{\lambda}S}+be^{-\sqrt{\lambda}S} = 0. \tag{7.3.18}$$

For all λ this system has no nonzero solutions. Substituting $x = 0$ and $x = S$ in (7.3.17), we obtain

$$a = 0, \quad b\sin\sqrt{|\lambda|}\,S = 0. \tag{7.3.19}$$

A nonzero solution of this system exists when $\sin\sqrt{|\lambda|}S = 0$, i.e., when $\lambda = \lambda_n = -\left(\dfrac{n\pi}{S}\right)^2$, $n = 1, 2, \ldots$ These λ's are the eigenvalues of A. A single eigenvector e_n corresponds to an eigenvalue λ_n, namely

$$e_n = \sin\frac{n x \pi}{S}.$$

For $\lambda = 0$ the solution is $at+b$. Under the boundary conditions (7.3.4) this solution is equal to zero.

EXAMPLE 7.3.2. Suppose that at both end-points of the rod equation (7.3.3) is satisfied, i.e.,

$$f'(0) = f'(S) = 0.$$

Substituting solutions (7.3.16) and (7.3.17) in equation (7.3.3) we find that in this case there is no positive eigenvalue and there exist only negative eigenvalues λ_n,

$$\lambda_n = -\left(\frac{n\pi}{S}\right)^2, \quad n = 0, 1, \ldots$$

They are the same eigenvalues as in Example 7.3.1, but the eigenvectors are different, namely

$$e_n = \cos\frac{n x \pi}{S}, \quad n = 0, 1, \ldots$$

EXAMPLE 7.3.3. Suppose that at one end-point of the rod condition (7.3.4) is satisfied and at the second end-point of the rod condition (7.3.3) is satisfied. Substituting the solution of equation (7.3.15) (formulae

(7.3.16) and (7.3.17)) in equation (7.3.3) and in equation (7.3.4), we obtain the following system of equations for $\lambda \geqslant 0$:

$$a+b = 0, \quad a\sqrt{\lambda}e^{\sqrt{\lambda}S}-b\sqrt{\lambda}e^{-\sqrt{\lambda}S} = 0.$$

This system has a unique solution and that solution is equal to 0 for each positive λ. This implies that there is no positive eigenvalue in this case. If λ is negative, we obtain the system

$$a = 0, \quad b\sqrt{|\lambda|}\cos\sqrt{|\lambda|}S = 0.$$

This system determines the eigenvalues λ_n,

$$\lambda_n = -\left(\frac{\pi}{S}\left(n-\frac{1}{2}\right)\right)^2, \quad n = 1, 2, \ldots,$$

and the corresponding eigenvectors e_n,

$$e_n = \sin\frac{x\pi}{S}\left(n-\frac{1}{2}\right).$$

EXAMPLE 7.3.4. Suppose that at one end-point of the rod, say $x = 0$, condition (7.3.4) is satisfied and at the other end-point condition (7.3.2) is satisfied with $\alpha_S < 0$.

By substituting formulae (7.3.16) and (7.3.17) in equation (7.3.4) and in equation (7.3.2) we obtain for $\lambda > 0$ the following system of equations:

$$a+b = 0, \quad a(\sqrt{\lambda}-\alpha_S)e^{\sqrt{\lambda}S}-b(\sqrt{\lambda}+\alpha_S)e^{-\sqrt{\lambda}S} = 0. \quad (7.3.20)$$

The determinant of the matrix of this system is equal to

$$(\sqrt{\lambda}-\alpha_S)e^{\sqrt{\lambda}S}+(\sqrt{\lambda}+\alpha_S)e^{-\sqrt{\lambda}S}. \quad (7.3.21)$$

Since $\alpha_S < 0$, the absolute value of the first element in the sum (7.3.21) is less than the absolute value of the second element. Hence, the determinant (7.3.21) differs from zero. This implies that there is no positive eigenvalue in this case. For $\lambda < 0$ we obtain the system

$$a = 0, \quad b\sqrt{|\lambda|}\cos\sqrt{|\lambda|}S = \alpha_S b\sin\sqrt{|\lambda|}S.$$

This gives the eigenvalues λ_n,

$$\lambda_n = -\mu_n^2,$$

7.3. Temperature distribution. Zero boundary conditions

where μ_n are positive solutions of the equation

$$\mu \cot \mu S = \alpha_S,$$

and the corresponding eigenvectors e_n,

$$e_n = \sin \mu_n x.$$

For $\lambda = 0$ we obtain the solution $at+b$. Together with the boundary condition this solution is found to be identically equal to zero, i.e., $\lambda = 0$ is not an eigenvalue of the operator A in this case.

EXAMPLE 7.3.5. Assume that at one end-point of the rod, say $x = 0$, the boundary condition (7.3.3) is satisfied and at the other end-point condition (7.3.2) is satisfied with $\alpha_S < 0$.

Substituting formulae (7.3.16) and (7.3.17) in equation (7.3.3) and in equation (7.3.2) we obtain for $\lambda > 0$

$$a\sqrt{\lambda} - b\sqrt{\lambda} = 0,$$
$$a(\sqrt{\lambda} - \alpha_S)e^{\sqrt{\lambda}S} - b(\alpha_S + \sqrt{\lambda})e^{-\sqrt{\lambda}S} = 0. \qquad (7.3.22)$$

The absolute value of the determinant of the matrix of system (7.3.22) is given by the formula

$$(\sqrt{\lambda} - \alpha_S)e^{\sqrt{\lambda}S} - (\sqrt{\lambda} + \alpha_S)e^{-\sqrt{\lambda}S}.$$

Since $\alpha_S < 0$, this determinant is nonzero. Hence for $\lambda > 0$ the only solution of the system is zero. This means that there is no positive eigenvalue. For $\lambda < 0$ we obtain the system

$$b = 0, \quad -a\sqrt{|\lambda|} \sin \sqrt{|\lambda|} S = \alpha_S a \cos S.$$

This gives the eigenvalues λ_n,

$$\lambda_n = -\mu_n^2,$$

where μ_n are positive solutions of the equation

$$\mu \tan \mu S = \alpha_S \qquad (7.3.23)$$

(recall that $\alpha_S < 0$), and the eigenvectors e_n,

$$e_n = \cos \mu_n x.$$

The solution which corresponds to $\lambda = 0$ and satisfies the boundary condition is identically equal to zero.

EXAMPLE 7.3.6. Suppose that at both end-points of the rod condition (7.3.2) is satisfied for $\alpha_0 > 0$ and $\alpha_S < 0$. Consider first $\lambda > 0$. Substituting formula (7.3.16) in equation (7.3.2) we obtain the system

$$a(\sqrt{\lambda}-\alpha_0)-b(\sqrt{\lambda}+\alpha_0) = 0,$$
$$a(\sqrt{\lambda}-\alpha_S)e^{\sqrt{\lambda}S}-b(\sqrt{\lambda}+\alpha_S)e^{-\sqrt{\lambda}S} = 0. \quad (7.3.24)$$

The absolute value of the determinant of system (7.3.24) is

$$(\sqrt{\lambda}+\alpha_0)(\sqrt{\lambda}-\alpha_S)e^{\sqrt{\lambda}S}-(\sqrt{\lambda}-\alpha_0)(\sqrt{\lambda}+\alpha_S)e^{-\sqrt{\lambda}S}.$$

Since $\alpha_0 > 0$ and $\alpha_s < 0$, the determinant is nonzero and, consequently, there is no positive eigenvalue.

For $\lambda < 0$ we obtain the system of equations

$$a\alpha_0 - b\sqrt{|\lambda|} = 0,$$
$$a(\sqrt{|\lambda|}\sin\sqrt{|\lambda|}S + \alpha_S\cos\sqrt{|\lambda|}S) -$$
$$-b(\sqrt{|\lambda|}\cos\sqrt{|\lambda|}S - \alpha_S\sin\sqrt{|\lambda|}S) = 0.$$

If λ is an eigenvalue, then the determinant of this system is equal to 0,

$$\sqrt{|\lambda|}(\sqrt{|\lambda|}\sin\sqrt{|\lambda|}S + \alpha_S\cos\sqrt{|\lambda|}S) -$$
$$-\alpha_0(\sqrt{|\lambda|}\cos\sqrt{|\lambda|}S - \alpha_S\sin\sqrt{|\lambda|}S) = 0.$$

Dividing both sides of this equation by $\cos\sqrt{\lambda}S$ and putting $\mu = \sqrt{|\lambda|}$ we obtain

$$(\mu^2 + \alpha_0\alpha_S)\tan\mu S = (\alpha_0 - \alpha_S)\mu. \quad (7.3.25)$$

The eigenvalues λ_n are then $\lambda_n = -\mu_n^2$, where μ_n are positive solutions of equation (7.3.25). The corresponding eigenvectors e_n are the functions

$$e_n = \frac{\alpha_0}{\mu_n}\sin\mu_n x + \cos\mu_n x.$$

In Examples 7.3.1, 7.3.2, 7.3.3 the eigenvalues λ_n are given by explicit formulae. In Examples 7.3.4, 7.3.5, 7.3.6 the eigenvalues λ_n are expressed as functions of μ_n, where μ_n are solutions of certain equations. Now we analyse the asymptotic behaviour of those μ_n. Observe first that the equation

$$\mu\tan\mu S = |\alpha_S|$$

7.3. Temperature distribution. Zero boundary conditions

is equivalent to the equation

$$\frac{\mu}{|\alpha_S|} = \cot \mu S.$$

The solutions of this equation can be interpreted geometrically as the points lying on the intersection of the curve

$$y = \cot \mu S$$

and the straight line $y = x/|\alpha_S|$ (see Fig. 7.3.1). This implies that

$$\lim_{n \to \infty} (\mu_n S - n\pi) = 0. \tag{7.3.26}$$

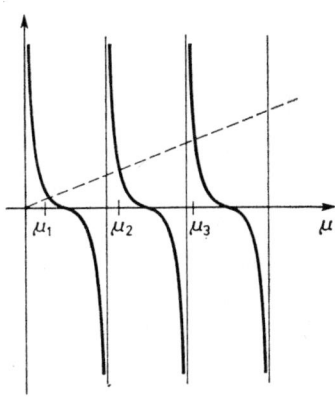
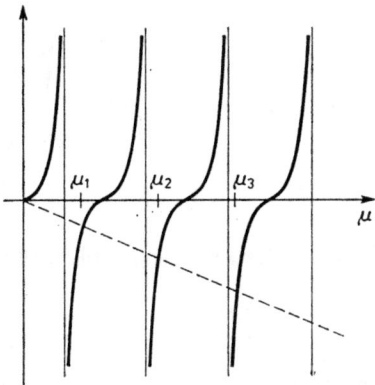

FIG. 7.3.1.   FIG. 7.3.2.

Similar arguments applied to the equation

$$\mu \cot \mu S = \alpha_S$$

lead us to the formula

$$\lim_{n \to \infty} \left(\mu_n S - \left(n - \frac{1}{2} \right) \pi \right) = 0 \tag{7.3.27}$$

(see Fig. 7.3.2).

For equation (7.3.25) formula (7.3.26) can also be derived from the equation

$$\cot \mu S = \frac{\mu}{\alpha_0 - \alpha_S} - \frac{\alpha_S}{\mu(\alpha_0 - \alpha_S)}, \tag{7.3.28}$$

which is equivalent to equation (7.3.25). Note that $\alpha_0 - \alpha_S > 0$ and $\alpha_S < 0$. Hence, if $\mu \to 0$, then the right-hand side of equation (7.3.28) tends to $-\infty$ (see Fig. 7.3.3).

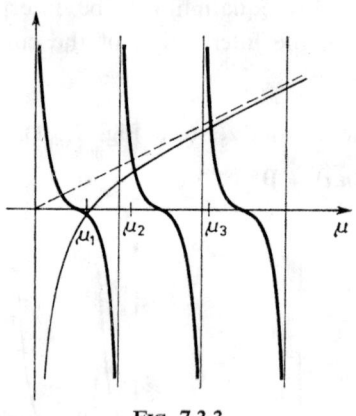

Fig. 7.3.3.

Consider now the problem of solving the equation

$$\frac{dx}{dt} = Ax$$

in the space X, where X is one of the spaces introduced in Examples 7.3.1, 7.3.2, 7.3.3, 7.3.4, 7.3.5 and 7.3.6.

In order to make use of the results of the previous section we have to determine a space Y which contains X as a linear subset and $\{e_n\}$ as a basis in Y.

For all the spaces X which appear in the above examples we choose the space $L^2[0, S]$ as the space Y. The sequences $\{e_n\}$, where either

$$e_n(x) = \sin\frac{n\pi x}{S}, \quad n = 1, 2, \ldots \quad \text{(cf. Example 7.3.1)},$$

or

$$e_n(x) = \cos\frac{n\pi x}{S}, \quad n = 0, 1, \ldots \quad \text{(cf. Example 7.3.2)},$$

forms an orthogonal basis in the space $L^2[0, S]$.

7.3. Temperature distribution. Zero boundary conditions

As can readily be verified, the sequence $\{e_n\}$,

$$e_n(x) = \sin\frac{\pi x}{S}\left(n - \frac{1}{2}\right), \quad n = 1, 2, \ldots \quad \text{(cf. Example 7.3.3)}$$

is also an orthogonal basis in the space $L^2[0, S]$.

To prove that the sequences $\{e_n\}$ from the above examples form orthonormal bases in the space $L^2[0, S]$ we need the following simple fact.

Let $f(x)$ and $g(x)$ be twice differentiable functions which belong to one of the spaces X mentioned in Examples 7.3.4, 7.3.5 or 7.3.6. Then

$$\int_0^S f''(x)g(x)\,dx = f'(x)g(x)|_0^S - \int_0^S f'(x)g'(x)\,dx$$

$$= f'(x)g(x)|_0^S - f(x)g'(x)|_0^S + \int_0^S f(x)g''(x)\,dx.$$

We shall prove that for the boundary conditions (7.3.2), (7.3.3) and (7.3.4) we have

$$f'(x)g(x)|_0^S - f(x)g'(x)|_0^S = 0.$$

Indeed, for the boundary condition (7.3.2) (and clearly for condition (7.3.3)) we have

$$f'(x)g(x)|_0^S - f(x)g'(x)|_0^S$$
$$= f'(S)g(S) - f'(0)g(0) - (f(S)g'(S) - f(0)g'(0))$$
$$= \alpha_S f(S)g(S) - \alpha_0 f(0)g(0) - f(S)\alpha_S g(S) + f(0)\alpha_0 g(0) = 0.$$

By a similar development we obtain the same result for the boundary condition (7.3.4). Finally we get

$$\int_0^S f''(x)g(x)\,dx = \int_0^S f(x)g''(x)\,dx.$$

By considering the space X as a subset of the space $L^2[0, S]$ and by taking the operator A equal to the second derivative operator defined on a subset of the space $L^2[0, S]$ we can rewrite the last formula in the form

$$(Af, g) = (f, Ag).$$

Operators A which are defined on subsets of Hilbert spaces and have this property are called *symmetric*.

If A is a symmetric operator and e_n, e_m are its eigenvectors which correspond to the different eigenvalues λ_n, λ_m, then the vectors e_n, e_m are orthogonal. Indeed,

$$\lambda_n(e_n, e_m) = (Ae_n, e_m) = (e_n, Ae_m) = \lambda_m(e_n, e_m),$$

and $(e_n, e_m) = 0$ since $\lambda_n \neq \lambda_m$.

By the above arguments, the eigenvectors $e_n(x)$ from Examples 7.3.4, 7.3.5 and 7.3.6 form orthogonal sequences and thereby the sequences $\{e_n\}$ form basic sequences. But still we do not know whether these sequences form bases in the space $L^2[0, S]$. To show this we shall apply Theorem 7.1.7.

We begin our analysis by considering the space X which appears in Example 7.3.5.

LEMMA 7.3.1. *If $\{\mu_n\}$ is an increasing sequence of positive solutions of the equation $\mu\tan\mu S = |\alpha_S|$, then there exists a constant $C > 0$ such that*

$$n\pi < \mu_n S < n\pi + \frac{C}{n}. \qquad (7.3.29)$$

Proof. The first inequality is obvious (cf. Fig. 7.3.1). We shall prove the second inequality. Clearly, the n-th μ_n lies in the interval $[n\pi, (n+\tfrac{1}{2})\pi]$. In this interval

$$\tan\mu S > (\mu S - n\pi).$$

Setting a_n as a solution of the quadratic equation

$$\mu(\mu S - n\pi) = |\alpha_S|, \qquad (7.3.30)$$

which lies in the interval $[n\pi, (n+\tfrac{1}{2})\pi]$, we find that μ_n is not greater than a_n. The value of a_n can easily be calculated, namely,

$$a_n = \frac{n\pi + \sqrt{(n\pi)^2 + 4|\alpha_S|}}{2S}.$$

7.3. Temperature distribution. Zero boundary conditions

After subtracting $n\pi$ from both sides of the latter equality we obtain

$$\mu_n S - n\pi \leqslant a_n S - n\pi = \frac{n\pi + \sqrt{(n\pi)^2 + 4|\alpha_S|}}{2} - n\pi$$

$$= \frac{\sqrt{(n\pi)^2 + 4|\alpha_S|} - n\pi}{2} = \frac{1}{2} \frac{4|\alpha_S|}{\sqrt{(n\pi)^2 + 4|\alpha_S|} + n\pi}$$

$$< \frac{|\alpha_S|}{n\pi}. \blacksquare$$

LEMMA 7.3.2. *If $\{\mu_n\}$ is an increasing sequence of positive solutions of the equation $\mu \cot \mu S = \alpha_S$, then there exists a positive number C such that*

$$\left(n - \frac{1}{2}\right)\pi < \mu_n S < \left(n - \frac{1}{2}\right)\pi + \frac{C}{n}. \tag{7.3.31}$$

Proof. The proof proceeds in the same way as the proof of Lemma 7.3.1 and is motivated by the analysis of Fig. 7.3.2. ∎

LEMMA 7.3.3. *If $\{\mu_n\}$ is an increasing sequence of positive solutions of equation (7.3.25),*

$$(\mu^2 + \alpha_S) \tan \mu S = (\alpha_0 - \alpha_S) \mu$$

then there exists a positive constant C such that

$$n\pi < \mu_n S < n\pi + \frac{C}{n}. \tag{7.3.32}$$

Proof. Equation (7.3.25) is equivalent to equation (7.3.28),

$$\cot \mu S = \frac{\mu}{\alpha_0 - \alpha_S} + \frac{\alpha_S}{\alpha_0 - \alpha_S} \cdot \frac{1}{\mu}.$$

Since $\alpha_0 - \alpha_S > 0$ and $\alpha_S < 0$, the right-hand side of equation (7.3.28) is an increasing function and there exists an asymptote to this function. Hence, for μ large enough, the right-hand side of equation (7.3.28) can be estimated by a linear function. Thus, inequalities (7.3.32) follow from Lemma 7.3.1. ∎

From Lemmas 7.3.1, 7.3.2, 7.3.3, we obtain

LEMMA 7.3.4. *Let $\{\mu_n\}$ be an increasing sequence of positive solutions of equation* (7.3.23),

$$\mu \tan \mu S = \alpha_S.$$

Then there exists a $\tilde{C} > 0$ such that

$$\left\| \cos \mu_n x - \cos \frac{n\pi x}{S} \right\| \leq \frac{\tilde{C}}{n},$$

where $\| \ \|$ is the norm in the Hilbert space $L^2[0, S]$.

Proof. We have

$$\cos \mu_n x - \cos \frac{n\pi x}{S} = 2 \cos \frac{1}{2}\left(\mu_n + n\frac{\pi}{S}\right) x \cdot \sin \frac{1}{2}\left(\mu_n - n\frac{\pi}{S}\right) x.$$

Hence, by Lemma 7.3.1,

$$\left| \cos \mu_n x - \cos \frac{n\pi x}{S} \right| \leq 2 \cdot \left| \sin \frac{1}{2}\left(\mu_n - \frac{n}{S}\pi\right) x \right|$$

$$\leq 2 \cdot \frac{1}{2}\left| \mu_n - \frac{n}{S}\pi \right| \cdot S = |\mu_n S - n\pi| < \frac{C}{n}.$$

This immediately implies the lemma. ∎

LEMMA 7.3.5. *Let $\{\mu_n\}$ be an increasing sequence of positive solutions of the equation*

$$\mu \cot \mu S = \alpha_S.$$

Then there exists a constant $\tilde{C} > 0$ such that

$$\left\| \sin \mu_n x - \sin \frac{1}{S}\left(n - \frac{1}{2}\right)\pi x \right\| < \frac{\tilde{C}}{n},$$

where $\| \ \|$ is the norm in the Hilbert space $L^2[0, S]$.

Proof. The proof proceeds in the same way as the proof of Lemma 7.3.4, the only modification being that the difference of the sin functions is to be estimated and Lemma 7.3.2 is to be applied instead of Lemma 7.3.1. ∎

7.3. Temperature distribution. Zero boundary conditions

LEMMA 7.3.6. *Let $\{\mu_n\}$ be an increasing sequence of positive solutions of the equation*

$$(\mu^2 + \alpha_S)\tan\mu S = (\alpha_0 - \alpha_S)\mu.$$

There exists a constant $\tilde{C} > 0$ such that

$$\left\|e_n - \cos\frac{n\pi x}{S}\right\| < \frac{\tilde{C}}{n},$$

where

$$e_n = \frac{\alpha_0}{\mu_n}\sin\mu_n x + \cos\mu_n x,$$

and $\|\ \|$ is the norm in the Hilbert space $L^2[0, S]$.

Proof. The definition of e_n and the fact that $n\pi - \mu_n$ tends to zero as $n \to \infty$ imply that there exists a constant C_1 such that

$$\|e_n - \cos\mu_n x\| = \left\|\frac{\alpha_0}{\mu_n}\sin\mu_n x\right\| < \frac{C_1}{n}. \qquad (7.3.33)$$

On the other hand, by Lemma 7.3.3 and by arguments similar to those used in Lemma 7.3.4, we infer that there exists a constant $C_2 > 0$ such that

$$\left\|\cos\mu_n x - \cos\frac{n\pi x}{S}\right\| < \frac{C_2}{n}. \qquad (7.3.34)$$

Formulae (7.3.33) and (7.3.34) immediately imply the assertion of the lemma. ∎

Lemmas 7.3.4, 7.3.5, 7.3.6 and Theorem 7.1.7 allow to formulate the following theorem.

THEOREM 7.3.7. *In Examples 7.3.4, 7.3.5, 7.3.6 the eigenvectors $\{e_n\}$ form bases which are equivalent to some orthonormal bases.*

Now we can apply the method developed in Section 7.2. We require $Q_0(x)$ to belong to the corresponding space X described in Examples 7.3.1–7.3.6. Let the space Y be defined as $L^2[0, S]$. Clearly, X is a linear subset of Y and, as we have shown, the eigenvectors of the operator A form a basis in the space Y. This basis is orthogonal and equivalent to an orthonormal basis (note that we choose e_n so as to have $0 < \inf\|e_n\| \leq \sup\|e_n\| < +\infty$).

Since X is a linear subset of Y, Q_0 can be regarded as an element of the space Y, and Q_0 can be represented as

$$Q_0 = \sum_{n=1}^{\infty} c_n e_n. \tag{7.3.35}$$

However, we should be aware of the fact that this series need not be summable in the space X.

Since the basis $\{e_n\}$ is orthogonal, we have

$$c_n = \left(\frac{1}{\|e_n\|}\right)^2 (Q_0, e_n). \tag{7.3.36}$$

Now we can form the series

$$Q(t, x) = \sum_{n=1}^{\infty} c_n e^{\lambda_n t} e_n(x). \tag{7.3.37}$$

The formal expression which gives the derivative of this series with respect to t is

$$\frac{\partial}{\partial t} Q(t, x) = \sum_{n=1}^{\infty} c_n \lambda_n e^{\lambda_n t} e_n(x). \tag{7.3.38}$$

Differentiating the series (7.3.37) twice with respect to x, we obtain

$$\frac{\partial^2}{\partial x^2} Q(t, x) = \sum_{n=1}^{\infty} c_n \lambda_n^2 e^{\lambda_n t} e_n(x), \tag{7.3.39}$$

since $\{e_n\}$ are eigenvectors of the second derivative operator A.

Thus, the series (7.3.38) and (7.3.39) are the same. To prove that $Q(t, x)$ satisfies equation (7.3.1) it is enough to show the convergence of the series (7.3.37) and (7.3.38).

Note that, by the asymptotic formulae derived above, there exists the limit

$$C = \lim_{n \to \infty} \frac{\lambda_n}{-n^2} \neq 0 \tag{7.3.40}$$

and $C > 0$. It follows from (7.3.40) that the series (7.3.37) and (7.3.38) converge in Y for $t \geq 0$ and, moreover, converge in X for $t > 0$.

Hence, $Q(t, x)$ given by formula (7.3.37) solves equation (7.3.1). Moreover, $Q(t, x)$ satisfies the initial condition

$$\lim_{t \to 0} Q(t, x) = Q_0(x), \tag{7.3.41}$$

where the convergence in formula (7.3.41) is understood in the sense of the space $L^2[0, S]$. A somewhat stronger convergence result can also be shown, but we shall not consider this problem here.

7.4. Nonhomogeneous heat diffusion equation with zero boundary conditions

In the previous section we considered equation (7.3.1),

$$\frac{\partial Q}{\partial t} = \frac{\partial^2 Q}{\partial x^2}, \quad 0 \leq x \leq S, \quad 0 \leq t \leq T, \quad Q(0, x) = Q_0(x),$$

with zero boundary conditions. Different Banach spaces X correspond to different boundary conditions. Those Banach spaces X can be regarded as subspaces of the space $L^2[0, S]$ and, moreover, the eigenvectors $\{e_n\}$ of the operator A, which belong to X, form an orthogonal basis in the space $L^2[0, S]$.

Applying the method described in Section 7.2, we found a generalized solution of equation (7.3.1) and this generalized solution also solved equation (7.3.1) in the classical sense.

In engineering applications the following nonhomogeneous equation often appears

$$\frac{\partial Q}{\partial t} = \frac{\partial^2 Q}{\partial x^2} + H(t, x), \tag{7.4.1}$$

where the function $H(t, x)$ is called the *heat source density*. This equation models a situation where the rod is noninsulated and the ingoing or outgoing heat density at each point x and at each moment t is assumed to be known (cf. Tikhonov and Samarski, 1966, Section 3.1).

As regards the case of a homogeneous equation we assume that the initial temperature distribution is known, i.e.,

$$Q(0, x) = Q_0(x), \tag{7.4.2}$$

where $Q_0(x) \in L^2[0, S]$ and zero boundary conditions of the type (7.3.2) and (7.3.3) or (7.3.4) are imposed at the end-points of the rod.

Now we apply the method of Section 7.2. Take $Y = L^2[0, S]$. Assume that $H(t, x)$ belongs to the space M_{L^2} of all measurable functions $h(t, x)$ such that

$$\|h\| = \operatorname*{ess\,sup}_{0 \leqslant t \leqslant T} \left(\int_0^S |h(t, x)|^2 dx \right)^{1/2} < +\infty. \tag{7.4.3}$$

Clearly the space M_{L^2} is a normed space. It can be interpreted as the space of all measurable functions defined on the interval $[0, T]$, which are essentially bounded and take values in the space $L^2[0, S]$. Completeness of the space $L^2[0, S]$ implies completeness of the space M_{L^2}. The proof of this fact is—mutatis mutandis—exactly the same as the corresponding proof for the space $M[0, T]$ of scalar-valued functions.

Together with the space M_{L^2} we consider the space L_{L^2} of all measurable functions $g(t, x)$ such that

$$\|g\| = \int_0^T \left(\int_0^S |g(t, x)|^2 dx \right)^{1/2} < +\infty. \tag{7.4.4}$$

Let $h(t, x) \in M_{L^2}$; then the expression

$$F(g) = \int_0^T \int_0^S h(t, x) g(t, x) dx dt \tag{7.4.5}$$

defines a continuous linear functional on the space L_{L^2}. By simple calculations we obtain the norm of that functional,

$$\|F\| = \|h\|, \tag{7.4.6}$$

where $\|h\|$ is the norm of h in the space M_{L^2}. The proof of the fact that every continuous linear functional F defined on the space L_{L^2} is of form (7.4.5) is more complicated.

To do this, consider an orthogonal basis $\{e_n\}$ in the space $L^2[0, S]$. Without loss of generality we can assume that there exist constants m_1, m_2, both positive, such that $m_1 \leqslant \|e_n\| \leqslant m_2$. Let $f_n = e_n/\|e_n\|^2$ denote the basic functionals defined by the basis $\{e_n\}$. Note that the norms of the functionals f_n are uniformly bounded, i.e.,

$$\|f_n\| \leqslant M = \frac{1}{m_1}, \quad n = 1, 2, \ldots$$

7.4. Nonhomogeneous heat diffusion equation

Recall that for each n the space $L_{(n)}$ which consists of elements of the form $g(t)e_n$, $g \in L^1$, is isomorphic to L^1. Hence $F(g(t)e_n)$ is of the form

$$F(g(t)e_n) = \int_0^T h_n(t)g(t)\,dt, \qquad (7.4.7)$$

where $h_n \in M[0, T]$. Formula (7.4.7) implies that the functional F is of the form

$$F(g) = \int_0^T \sum_{n=1}^\infty h_n(t)g_n(t)\,dt, \qquad (7.4.8)$$

where

$$g_n(t) = \int_0^S f_n(x)g(t, x)\,dx. \qquad (7.4.9)$$

Since the basis $\{e_n\}$ is equivalent to an orthonormal basis, $g = g(t, x)$ belongs to the space L_{L^2} if and only if

$$\int_0^T \sum_{n=1}^\infty |g_n(t)|^2\,dt < +\infty. \qquad (7.4.10)$$

By (7.4.10) and by the fact that the integral (7.4.8) is bounded for each g, we have

$$\sup_t \sum_{n=1}^\infty |h_n(t)|^2 < +\infty. \qquad (7.4.11)$$

This implies that the function

$$h(t, x) = \sum_{n=1}^\infty h_n(t)f_n(x) \qquad (7.4.12)$$

belongs to the space M_{L^2} and defines a functional F by formula (7.4.5).

Now we want to solve equation (7.4.1) by the method described in Section 7.2. To begin with, we represent $Q(t, x)$ in the form

$$Q(t, x) = \sum_{n=1}^\infty q_n(t)e_n(x), \qquad (7.4.13)$$

where the coefficients $q_n(t)$ are given by the formula

$$q_n(t) = \int_0^S Q(t, x) f_n(x) \, dx. \qquad (7.4.14)$$

Recall that e_n is the eigenvector of the second derivative operator, which corresponds to the eigenvalue λ_n. Substituting $Q(t, x)$ in the form (7.4.13) in equation (7.4.1), we obtain the equation

$$\sum_{n=1}^{\infty} \left(\frac{d}{dt} q_n - \lambda_n q_n - H_n \right) e_n = 0, \qquad (7.4.15)$$

where

$$H_n(t) = \int_0^S H(t, x) f_n(x) \, dx. \qquad (7.4.16)$$

Equation (7.4.15) is equivalent to an infinite system of equations

$$\frac{d}{dt} q_n = \lambda_n q_n + H_n, \quad n = 1, 2, \ldots \qquad (7.4.17)$$

The initial condition (7.4.2), applied to the series (7.4.13) determines an initial condition which involves the functions q_n. Namely, we have

$$q_n(0) = q_{n,0}, \qquad (7.4.18)$$

where

$$q_{n,0} = \int_0^S Q_0(x) e_n(x) \, dx, \qquad (7.4.19)$$

are coefficients of the representation of Q_0 with respect to the basis $\{e_n\}$.

From (7.4.17) and (7.4.18) we obtain the formula

$$q_n(t) = e^{\lambda_n t} q_{n,0} + \int_0^t e^{\lambda_n(t-\tau)} H_n(\tau) \, d\tau. \qquad (7.4.20)$$

Substituting (7.4.20) in (7.4.13) we obtain

$$Q(t, x) = \sum_{n=1}^{\infty} e^{\lambda_n t} q_{n,0} e_n(x) + \sum_{n=1}^{\infty} e_n(x) \cdot \int_0^t e^{\lambda_n(t-\tau)} H_n(\tau) \, d\tau. \qquad (7.4.21)$$

7.4. Nonhomogeneous heat diffusion equation

As was shown in the previous section, the first series appearing in formula (7.4.21) is convergent for $t > 0$ and its sum is a differentiable function with respect to t and a twice differentiable function with respect to x.

To estimate the terms of the second series observe that, by the definition of the space M_{L^2},

$$|H_n(t)| \leqslant M\|H\|, \qquad (7.4.22)$$

where M is a uniform bound of the sequence of basic functionals f_n. Since $\lambda_n < 0$, we obtain

$$\left|\int_0^t e^{\lambda_n(t-\tau)} H_n(\tau)\,d\tau\right| \leqslant M\|H\|\int_0^t e^{\lambda_n(t-\tau)}\,d\tau = \frac{M}{|\lambda_n|}\|H\|. \qquad (7.4.23)$$

In all the examples considered in Section 7.3 we have

$$\lim_{n\to\infty} \frac{|\lambda_n|}{n^2} > 0. \qquad (7.4.24)$$

Hence, the second series appearing in formula (7.4.21) is uniformly convergent. Thus, its limit and therefore the function $Q(t, x)$ are continuous.

Unfortunately, we do not know whether the function $Q(t, x)$ is differentiable since the series of derivatives with respect to t,

$$\sum_{n=1}^\infty \lambda_n e_n(x) \int_0^t e^{\lambda_n(t-\tau)} H_n(\tau)\,d\tau + e_n(x) H_n(t) \qquad (7.4.25)$$

need not be convergent.

Now the question arises in what sense the function $Q(t, x)$ solves equation (7.4.1), since, as we have seen, the function $Q(t, x)$ is not a classical solution of equation (7.4.1).

We shall not discuss this question here. The reader interested in the problem should refer to the book of Lions, 1968, where it is discussed in detail.

The same concerns the question in what sense the boundary conditions of types (7.3.2) and (7.3.4) are satisfied (the sense of condition (7.3.3) is clear and the function $Q(t, x)$ satisfies this condition).

As was shown by Lions (1968), the solution $Q(t, x)$ should be regarded as a solution in the sense of distributions or as a generalized Sobolev solution.

For our considerations, however, it suffices to know that $Q(t, x)$ is a generalized solution in the sense defined in Section 7.2.

We have proved that the series

$$Q_1(t, x) = \sum_{n=1}^{\infty} \int_0^t e^{\lambda_n(t-\tau)} H_n(\tau) d\tau \cdot e_n(x) \qquad (7.4.26)$$

defines a continuous linear operator A which maps the space M_{L^2} into the space $C([0, T] \times [0, S])$ of all continuous functions defined on the rectangle $0 \leq t \leq T$, $0 \leq x \leq S$, with the standard norm,

$$\|Q_1(t, x)\| = \sup_{\substack{0 \leq t \leq T \\ 0 \leq x \leq S}} |Q_1(t, x)|. \qquad (7.4.27)$$

Recall that M_{L^2} is the conjugate space of the space L_{L^2}. Hence, in the space M_{L^2} we have the weak-$*$-topology generated by the topology of the space L_{L^2}. We shall show that the operator A transforms sequences convergent in the weak-$*$-topology into norm convergent sequences.

To see this, let a sequence $\{H^m\}$ converge in the weak-$*$-topology to $H(t, x)$; then the sequence $\{H^m\}$ is bounded and, by formula (7.4.22), the double sequence

$$H_n^m(t) = \int_0^S H^m(t, x) f_n(x) dx$$

is uniformly bounded.

Note that the weak convergence of the sequence $\{H^m\}$ to $H(t, x)$ implies that for each n the sequence $\{H_n^m(t)\}$ weakly converges to $H_n(t)$. Hence, by inequality (7.4.23) and by formula (7.4.24) we obtain

$$A(H^m) \to A(H). \qquad (7.4.28)$$

7.5. HOMOGENEOUS HEAT DIFFUSION EQUATION WITH NONHOMOGENEOUS BOUNDARY CONDITIONS

Consider the homogeneous heat diffusion equation (7.3.1):

$$\frac{\partial Q}{\partial t} = \frac{\partial^2 Q}{\partial x^2}, \quad Q(0, x) = Q_0(x),$$

with the coefficient a equal to 1.

7.5. Nonhomogeneous boundary conditions

Suppose now that we want to find solutions of this equation which satisfy not conditions (7.3.2), (7.3.3) and (7.3.4) (as in Section 7.3) but the nonhomogeneous boundary conditions

$$\left.\frac{\partial Q}{\partial x}\right|_{(t,0)} = \alpha_0 (Q(t, 0) - \theta_1(t)),$$

$$\left.\frac{\partial Q}{\partial x}\right|_{(t,S)} = \alpha_S (Q(t, S) - \theta_2(t))$$

(7.5.1)

or

$$\left.\frac{\partial Q}{\partial x}\right|_{(t,0)} = \nu_1(t), \quad \left.\frac{\partial Q}{\partial x}\right|_{(t,S)} = \nu_2(t) \qquad (7.5.2)$$

(if we know the quantity of heat ingoing or outgoing from a given end-point of the rod) or

$$Q(t, 0) = \mu_1(t), \quad Q(t, S) = \mu_2(t) \qquad (7.5.3)$$

(if we know the total heat exchange at a given rod end-point and at a given surrounding temperature).

In classical textbooks the case of nonhomogeneous boundary conditions is reduced to the case of homogeneous boundary conditions for a nonhomogeneous equation, i.e., the problem is reduced to that considered in the previous section. This reduction is made by means of the substitution

$$Q(t, x) = v(t, x) + U(t, x), \qquad (7.5.4)$$

where the function $U(t, x)$ satisfies the corresponding boundary conditions. Table 7.5.1 (p. 368) contains examples of the function $U(t, x)$ which can be used for the corresponding pairs of boundary conditions.

Under the assumption that the corresponding derivatives exist we can substitute formula (7.5.4) in equation (7.3.1). This yields the equation

$$\frac{\partial v}{\partial t} = \frac{\partial^2 v}{\partial x^2} + H(t, x) \qquad (7.5.5)$$

where

$$H(t, x) = \frac{\partial^2 U}{\partial x^2} - \frac{\partial U}{\partial t}. \qquad (7.5.6)$$

Further considerations are similar to those presented in Section 5.4.

Table 7.5.1

Boundary condition for $x=0$ \ Boundary condition for $x=S$	$Q = \mu_2$	$\dfrac{dQ}{dx} = v_2$	$\dfrac{dQ}{dx} = \alpha_S(Q-\theta_2)$
$Q = \mu_1$	$U(t,x) = \mu_1 + \dfrac{x}{S}(\mu_2 - \mu_1)$	$U(t,x) = \mu_1 + xv_2$	$U(t,x) = \mu_1 + \dfrac{\alpha_S(\mu_1 - \theta_2)}{1 - S\alpha_S} x$
$\dfrac{dQ}{dx_1} = v_1$	$U(t,x) = (x-S)v_1 + \mu_2$	$U(t,x) = xv_1 + \dfrac{x^2}{2S}(v_2 - v_1)$	$U(t,x) = v_1 \cdot x + \left(\dfrac{v_1}{\alpha_S} - v_1 S + \theta_2\right)$
$\dfrac{dQ}{dx} = \alpha_0(Q-\theta_1)$	$U(t,x) = \mu_2 + (x-S)\dfrac{\alpha_0(\mu_2-\theta_1)}{1+S\alpha_0}$	$U(t,x) = v_2(x-S) + \left(\dfrac{v_2}{\alpha_0} + v_1 S - \theta_1\right)$	$U(t,x) = \alpha_0 \dfrac{2\alpha_0 \alpha_S \theta_1 + (\alpha_0 + \alpha_S)\theta_2}{\alpha_0 - \alpha_S - \alpha_0 \alpha_S} +$ $+ x \dfrac{\alpha_0 \alpha_S \theta_1 + \alpha_S \theta_2 + \alpha_0 \theta_1}{\alpha_0 - \alpha_S - \alpha_0 \alpha_S}$, if $\alpha_0 - \alpha_S - \alpha_0 \alpha_S \neq 0$.

7.5. Nonhomogeneous boundary conditions

Observe that the functions $U(t, x)$ from Table 7.5.1 are polynomials with respect to x and thus $U(t, x)$ are differentiable with respect to x. On the other hand, the function $U(t, x)$ is differentiable with respect to t if and only if the functions μ, ν, θ, which appear in the definition of the function $U(t, x)$, are differentiable with respect to t. Unfortunately, we cannot assume the differentiability of μ, ν, θ, since in practice these functions are frequently nondifferentiable and even not continuous.

Then, to be able to apply the Fourier method we must consider, instead of equation (7.3.1) with the initial condition $Q(0, x) = Q_0(x)$, the equivalent integro-differential equation

$$Q(t, x) - Q_0(x) = \int_0^t \frac{\partial^2 Q}{\partial x^2} d\tau. \qquad (7.5.7)$$

Substituting now formula (7.5.4) in (7.5.7), we obtain the integro-differential equation with respect to $v(t, x)$,

$$v(t, x) - Q_0(x) = \int_0^t \frac{\partial^2 v}{\partial x^2} d\tau - U(t, x) + \int_0^t \frac{\partial^2 U}{\partial x^2} d\tau. \qquad (7.5.8)$$

Note that $\partial^2 U/\partial x^2$ is identically equal to zero in most of the functions in Table 7.5.1.

Clearly, the function $v(t, x)$ satisfies homogeneous boundary conditions. Hence, we can apply the method developed in Section 7.3. We start by representing the function $v(t, x)$ with respect to the basis $\{e_n\}$ in the space Y. Thus, similarly to the series (7.4.13), we obtain

$$v(t, x) = \sum_{n=1}^{\infty} v_n(t) e_n(t). \qquad (7.5.9)$$

To calculate the functions $v_n(t)$ from formula (7.5.8) we should represent the function

$$U(t, x) - \int_0^t \frac{\partial^2 U}{\partial x^2} d\tau$$

with respect to the basis $\{e_n\}$. Assume first that the function $U(t, x)$ is linear with respect to x. This means that

$$U(t, x) = u_1(t) + x u_2(t), \qquad (7.5.10)$$

where the functions $u_1(t)$, $u_2(t)$ are linear combinations of functions appearing in the boundary conditions. Those functions, in turn, can be easily determined for given boundary conditions from Table 7.5.1.

If formula (7.5.10) holds, the second derivative of the function $U(t, x)$ with respect to x is identically equal to zero. Hence, in this case, equation (7.5.8) reduces to the simpler equation

$$v(t, x) - Q_0(x) = \int_0^t \frac{\partial^2 v}{\partial x^2} d\tau - U(t, x). \tag{7.5.11}$$

Now, consider the representation of the function $f(x) = 1$ and the function $f(x) = x$ with respect to the basis $\{e_n\}$. Denote the coefficients of representation of the function $f(x) = 1$ by $\{c_n^0\}$ and the coefficients of representation of the function $f(x) = x$ by $\{c_n^1\}$.

Recall that in our considerations the functions $e_n(x)$ are of the form $\sin \sqrt{\lambda_n} x$ (or $\cos \sqrt{\lambda_n} x$ or linear combinations of both). Hence, by the already proved estimations of λ_n, we obtain

$$C_i = \sup n|c_n^i| < +\infty, \quad n = 1, 2, \ldots, \quad i = 0, 1. \tag{7.5.12}$$

Substituting (7.5.9) and (7.5.10) to equation (7.5.11), we obtain

$$\sum_{n=1}^{\infty} \left[v_n(t) - q_{n,0} - \int_0^t \lambda_n v_n(\tau) d\tau + c_n^0 u_1(t) + c_n^1 u_2(t) \right] e_n(x) = 0,$$

$$\tag{7.5.13}$$

where $q_{n,0}$ are coefficients of the representation of the function $Q_0(x)$ with respect to the basis $\{e_n\}$.

Equation (7.5.13) induces an infinite system of linear Volterra integral equations of the second kind, namely

$$v_n(t) = \lambda_n \int_0^t v_n(\tau) d\tau + q_{n,0} - u_1(t) c_n^0 - u_2(t) c_n^1, \tag{7.5.14}$$

$n = 1, 2, \ldots$ Solving this system, we obtain

$$v_n(t) = e^{\lambda_n t} q_{n,0} - [c_n^0 u_1(t) + c_n^1 u_2(t)] -$$
$$- \lambda_n \int_0^t e^{\lambda_n(t-\tau)} [c_n^0 u_1(\tau) + c_n^1 u_2(\tau)] d\tau. \tag{7.5.15}$$

7.5. Nonhomogeneous boundary conditions

Hence, we can write the solution $v(t, x)$ of equation (7.5.11) in the form of the series

$$v(t, x) = \sum_{n=1}^{\infty} v_n(t) e_n(x) = Q^0(t, x) - U(t, x) + Q_2(t, x), \quad (7.5.16)$$

where

$$Q^0(t, x) = \sum_{n=1}^{\infty} e^{\lambda_n t} q_{n,0} e_n(x) \quad (7.5.17)$$

solves equation (7.3.1) with zero boundary conditions ($\mu = \nu = \theta = 0$) and

$$Q_2(t, x) = \sum_{n=0}^{\infty} |\lambda_n| \int_0^t e^{\lambda_n(t-\tau)} [c_n^0 u_1(\tau) + c_n^1 u_2(\tau)] d\tau \cdot e_n(x). \quad (7.5.18)$$

Formulae (7.5.16) and (7.5.4) imply

$$Q(t, x) = Q^0(t, x) + Q_2(t, x), \quad (7.5.19)$$

where Q_2 corresponds to zero initial conditions and nonzero boundary conditions.

Suppose now that the boundary functions which appear in conditions (7.5.1)–(7.5.3) belong to the space $M[0, T]$. Under this hypothesis we shall prove that the function $Q_2(t, x)$ belongs to the space M_{L^2}. Indeed, if the functions μ, ν, θ which appear in the boundary conditions belong to the space $M[0, T]$, then the functions $u_1(t)$ and $u_2(t)$ belong to the space $M[0, T]$ as linear combinations of the functions μ, ν, θ (those combinations can be found in Table 7.5.1).

Now we estimate the integral $\int_0^S (Q_2(t, x))^2 dx$. We have

$$\int_0^S [Q_2(t, x)]^2 dx \leq K \sum_{n=1}^{\infty} (|\lambda_n| \int_0^t e^{\lambda_n(t-\tau)} [c_n^0 u_1(\tau) + c_n^1 u_2(\tau)] d\tau)^2$$

$$\leq K \sum_{n=1}^{\infty} (|c_n^0| \cdot ||u_1|| + |c_n^1| \cdot ||u_2||)^2 (|\lambda_n| \int_0^t e^{\lambda_n(t-\tau)} d\tau)^2$$

$$\leq 2K \sum_{n=1}^{\infty} [|c_n^0|^2 \cdot ||u_1||^2 + |c_n^1|^2 \cdot ||u_2||^2]$$

$$\leq 2KL \max(C_0^2, C_1^2) [||u_1||^2 + ||u_2||^2], \quad (7.5.20)$$

where the constants C_0, C_1 are given by the formula (7.5.12) and $L = \sum_{n=1}^{\infty} (1/n^2)$.

Since the estimation (7.5.20) holds for each t, the function $Q_2(t, x)$ belongs to M_{L^2}. Basing ourselves on (7.5.20), we prove also that the function $Q_2(t, x)$ regarded as a function of the argument t which takes values in the space $L^2[0, S]$ is continuous.

Indeed, take an arbitrary $\varepsilon > 0$. It follows from (7.5.20) that there exists an N such that the norm of the function $\tilde{Q}_N$,

$$\tilde{Q}_N = \sum_{n=N}^{\infty} |\lambda_n| \int_0^t e^{\lambda_n(t-\tau)}[c_n^0 u_1(\tau) + c_n^1 u_2(\tau)] d\tau \cdot e_n(x), \quad (7.5.21)$$

in the space M_{L^2} is less than $\varepsilon/4$.

On the other hand, the function

$$\tilde{Q}^N = Q_2(t, x) - \tilde{Q}_N(t, x)$$
$$= \sum_{n=1}^{N-1} \lambda_n \int_0^t e^{\lambda_n(t-\tau)}[c_n^0 u_1(\tau) + c_n^1 u_2(\tau)] d\tau \cdot e_n(x)$$

is continuous with respect to t and x. Since $\tilde{Q}^N$ is defined on a bounded closed set, it is uniformly continuous. This implies that there exists a $\delta > 0$ such that

$$|\tilde{Q}^N(t, x) - \tilde{Q}^N(t', x)| \leqslant \frac{\varepsilon}{2\sqrt{S}} \quad (7.5.22)$$

for all t, t', $|t-t'| < \delta$, and all x. Hence,

$$\|\tilde{Q}^N(t, \cdot) - \tilde{Q}^N(t', \cdot)\|_{L^2} < \frac{1}{2}\varepsilon,$$

where $\|\ \|$ denotes the norm of the space $L^2[0, S]$. By the triangle inequality,

$$\|Q_2(t, \cdot) - Q_2(t', \cdot)\|_{L^2} \leqslant \|Q_2(t, \cdot) - \tilde{Q}^N(t, \cdot)\|_{L^2} +$$
$$+ \|\tilde{Q}^N(t, \cdot) - \tilde{Q}^N(t', \cdot)\|_{L^2} + \|\tilde{Q}^N(t', \cdot) - Q_2(t', \cdot)\|_{L^2}$$
$$< \frac{1}{4}\varepsilon + \frac{1}{2}\varepsilon + \frac{1}{4}\varepsilon = \varepsilon, \quad (7.5.23)$$

7.5. Nonhomogeneous boundary conditions

This means that the function $Q_2(t, x)$ is continuous as a function of the argument t which takes values in the space $L^2[0, S]$.

Denote by C_{L^2} the space of all continuous functions $G(t)$ of the argument t which take values in $L^2[0, S]$ with the norm

$$\|G\| = \sup \|G(t)\|_{L^2}. \tag{7.5.24}$$

This is a subspace of the space M_{L^2}. It can be proved that C_{L^2} is complete. The following theorem summarizes our results.

THEOREM 7.5.1. *The operator which assigns the function $Q_2(t, x)$ to the functions u_1, u_2 by formula (7.5.18) is a continuous linear operator which maps the space $M[0, T] \times M[0, T]$ into the space C_{L^2}.*

Since u_1 and u_2 are linear combinations of functions appearing in boundary conditions, Theorem 7.5.1 implies that the operator which assigns to boundary conditions and to an initial condition the generalized solution is a continuous linear operator which maps the space $M[0, T] \times M[0, T]$ into the space C_{L^2}.

The space $M[0, T] \times M[0, T]$ can be regarded as the conjugate space of the space $L[0, T] \times L[0, T]$. Thus, the weak-$*$-topology is defined in $M[0, T] \times M[0, T]$.

Consider the function $\tilde{Q}^N(t, x)$,

$$\tilde{Q}^N(t, x) = \sum_{n=1}^{N-1} |\lambda_n| e^{\lambda_n t} e_n(x) \int_0^t e^{|\lambda_n|\tau} [c_n^0 u_1(\tau) + c_n^1 u_2(\tau)] d\tau. \tag{7.5.25}$$

Since $e^{\lambda_n \tau} \in L[0, T]$, the representation (7.5.25) entails the continuity (sequential continuity) of the finite-dimensional operator A_N, which assigns to functions u_1 and u_2 the function $\tilde{Q}^N$. This operator is defined on the space $M[0, T] \times M[0, T]$ endowed with the weak-$*$-topology and takes values in the space C_{L^2} endowed with the norm topology. Hence, by Theorem 7.4.9, we obtain

THEOREM 7.5.2. *The operator defined in Theorem 7.5.1 is a sequentially continuous operator which maps the space $M[0, T] \times M[0, T]$ endowed* with the weak-∗-topology into the space C_{L^2}, endowed with the norm topology.

So far we have assumed the linearity of the function $U(t, x)$ which is determined by boundary conditions. Unfortunately, this assumption is in applications often too restrictive. For instance, if the rod which we investigate here is insulated at both end-points, the following quadratic function appears

$$U(t, x) = u_1(t) + xu_2(t) + \tfrac{1}{2}x^2 u_3(t), \tag{7.5.26}$$

(the coefficient 1/2 is introduced to simplify calculations). For such a function $U(t, x)$ we have

$$-U(t, x) + \int_0^t \frac{\partial^2 U}{\partial x^2} d\tau$$

$$= -u_1(t) + \int_0^t u_2(\tau) d\tau - xu_2(t) - \frac{x^2}{2} u_3(t).$$

Write

$$\tilde{u}(t) = u_1(t) - \int_0^t u_3(\tau) d\tau. \tag{7.5.27}$$

The operator A which assigns the function u to functions u_1 and u_3 is a continuous linear operator and maps the space $M[0, T] \times M[0, T]$ into the space $M[0, T]$. Moreover, this operator is continuous in the weak-∗-topologies.

Using the notation of (7.5.27) we can rewrite equation (7.5.8) in the form

$$v(t, x) - Q_0(x) = \int_0^t \frac{\partial^2 v}{\partial x^2} d\tau - \tilde{u}(t) - xu_2(t) - \frac{x^2}{2} u_3(t). \tag{7.5.28}$$

To be able to apply our method again we represent the function $\frac{x^2}{2}$ with respect to the basis $\{e_n\}$,

$$\frac{x^2}{2} = \sum_{n=1}^{\infty} c_n^2 e_n(x). \tag{7.5.29}$$

7.5. Nonhomogeneous boundary conditions

Simple calculations give an estimation which is similar to (7.5.12). Namely,

$$\sup_n n|c_n^2| = C_2. \tag{7.5.30}$$

Now we can apply to equation (7.5.28) the same technique as that applied to equation (7.5.11), i.e., we represent both sides of the equation with respect to the basis $\{e_n\}$, we solve the equation for coefficients and obtain a representation of a generalized solution in the form

$$v(t, x) = Q^0(t, x) - U(t, x) + Q_2(t, x), \tag{7.5.31}$$

where

$$Q_2(t, x) = \sum_{n=1}^{\infty} |\lambda_n| \int_0^t e^{\lambda_n(t-\tau)} [c_n^0 \tilde{u}(\tau) + c_n^1 u_2(\tau) + c_n^2 u_3(\tau)] d\tau \cdot e_n(x). \tag{7.5.32}$$

Hence, by (7.5.4), we obtain from (7.5.31)

$$Q(t, x) = Q^0(t, x) + Q_2(t, x). \tag{7.5.33}$$

Let us observe that by the same arguments as in Theorem 7.5.1 and in Theorem 7.5.2, it can be shown that the operator A which assigns the function $Q_2(t, x)$ to functions $\tilde{u}$, u_2, u_3, is a continuous linear operator. This operator maps the space $M[0, T] \times M[0, T] \times M[0, T]$ into the space C_{L^2}. Moreover, the operator A is sequentially compact when regarded as a mapping of the space $M[0, T] \times M[0, T] \times M[0, T]$ equipped with the weak-*-topology induced by the topology of the space $L[0, T] \times L[0, T] \times L[0, T]$ into the space C_{L^2} equipped with the norm topology.

Since A is continuous and u_1, u_2, u_3 are linear combinations of two of the functions μ, ν, θ, the operator which assigns to the zero initial conditions and to given boundary conditions the corresponding generalized solution is a continuous linear operator. This operator maps $M[0, T] \times M[0, T]$ into C_{L^2}. Moreover, this operator is sequentially continuous in the weak-*-topology of the space $M[0, T] \times M[0, T]$ and the norm topology of the space C_{L^2}.

So far we have considered either a nonhomogeneous equation with zero boundary conditions or a homogeneous equation with nonzero boundary conditions. It may happen in practice that a nonhomogeneous equation of type (7.4.1) appears with nonzero boundary conditions; then the generalized solution is the sum of three functions,

$$Q(t, x) = Q^0(t, x) + Q_1(t, x) + Q_2(t, x), \qquad (7.5.34)$$

where $Q^0(t, x)$ solves the homogeneous equation with zero boundary conditions and nonhomogeneous initial conditions, and is given by formula (7.3.37), $Q_1(t, x)$ solves the nonhomogeneous equation (7.4.1) with zero initial condition and zero boundary condition, and is given by formula (7.4.26), $Q_2(t, x)$ solves the homogeneous equation with zero initial condition and nonzero boundary conditions, and is given either by formula (7.5.18) or formula (7.5.32) depending on $U(t, x)$ which is either linear or quadratic with respect to x.

7.6. Rod heating control

The rod heating process can be controlled in two ways: a rod can be heated (or cooled) either at the end-points (at one chosen end-point or at both) or along all its length. The heating patterns are given by the time and spatial distribution of ingoing heat (we should rather use the notion of heat density). Heating along the length of the rod can be realized by passing through it an electric current of a known intensity provided that the exact resistance distribution of the rod is given. In view of the fact that the temperature distribution depends on initial conditions, the input space X should be taken as

$$X = L^2[0, S] \times M_{L^2} \times M[0, T] \times M[0, T], \qquad (7.6.1)$$

where the space $L^2[0, S]$ is the space of the initial conditions, the space M_{L^2} corresponds to the nonhomogeneous part of equation (7.4.1) and $M[0, T] \times M[0, T]$ is the space of the boundary conditions, since at each end-point of the rod the controls are independent.

The operator A assigns to the initial value and to the homogeneous part of the equation and to the boundary values the generalized solution $Q(t, x)$ given by formula (7.5.34).

It was shown in the previous section that the operator A is a continuous linear operator and maps the space X into the space $\square = C_{L^2}$.

7.6. Rod heating control

Moreover, it was shown that if the space X is endowed with the weak-$*$-topology generated by the topology of the space

$$X_- = L^2[0, S] \times L_{L^2} \times L[0, T] \times L[0, T], \tag{7.6.2}$$

then A is a sequentially continuous operator which maps the space X equipped with the weak-$*$-topology into the space $\square$ equipped with the norm topology.

The output operator B can be defined in many different ways. For applications the output B is important which assigns to the space and time temperature distribution the space temperature distribution at the moment T, $B(Q(t, x)) = Q(T, x)$. Then $Y = L^2[0, S]$.

Thus, we have obtained a linear system

$$(X \xrightarrow{A} \square \xrightarrow{B} Y). \tag{7.6.3}$$

The first question is whether the system is controllable, i.e., whether $Y = BA(X)$. The answer is negative. The coefficients q_n^T of the representation of the function $Q(T, x)$ with respect to the basis $\{e_n(x)\}$ tend to zero as fast as $1/n$. This is a consequence of the definitions of Q^0, Q_1, Q_2 and the respective estimations appearing in the proofs of continuity. More precisely,

$$\sup_n |q_n^T| \cdot n < +\infty.$$

Hence, the element $\sum_{n=1}^{\infty} \frac{1}{n^2} e_{2^n}$ does not belong to $BA(X)$ and $BA(X) \neq Y$. On the other hand, $BA(X)$ is dense in Y, and thus $BA(X)$ is not a closed subspace.

Let us consider a point of controllability y_0, i.e., $y_0 \in BA(X)$ and let us introduce a norm in the following way. Denote by E an arbitrary four-dimensional space whose elements are quadruples $(x_1, \ldots, x_4)$. Each element of the space X can be represented as a system of four functions $(Q_0, H, \omega_1, \omega_2)$, where the choice of ω equal to μ, or ν, or θ depends on a boundary condition.

In the space X we define a norm as

$$\|(Q_0, H, \omega_1, \omega_2)\|^E = \left\| (\|Q_0\|_{L^2}, \|H\|_{M_{L^2}}, \|\omega_1\|_M, \|\omega_2\|_M) \right\|_E, \tag{7.6.4}$$

where $\| \ \|_E$ denotes a norm in the space E.

Recall that the space E is reflexive, since E is a finite-dimensional space. Hence, E is the conjugate space of E^*, where E^* denotes the conjugate space of E.

Now, in the space X_- defined by formula (7.6.2) we introduce a norm, namely

$$\|(f_1, f_2, f_3, f_4)\|^{E^*} = \left\|(\|f_1\|_{L^2}, \|f_2\|_{L_{L^2}}, \|f_3\|_L, \|f_4\|_L)\right\|_{E^*}.$$

By simple estimations we can show that the norm of continuous linear functionals defined on the space X_- (i.e., the norm of elements of the space X) equipped with the norm $\| \|^{E^*}$ is equal to the norm $\| \|^E$ given by the formula (7.6.4). Since the space X_- is separable, each ball in the space X endowed with the norm $\| \|^E$ is sequentially compact in the weak-$*$-topology which is determined on X by the topology of the space X_-. Hence, in view of Example 5.3.1, the minimum-norm problem has a solution and the following theorem holds.

THEOREM 7.6.1. *For every $y_0 \in BA(X)$ there exists an element $x_0 \in X$ such that*

$$\|x_0\|^E = \inf\{\|x\|^E : BA(x) = y_0\}. \tag{7.6.5}$$

Another important norm in the space X is the following. Let us consider the space $M[0, T] \times M[0, T]$ as the space $M_E[0, T]$ related to a two-dimensional Banach space E (cf. Section 6.3). Let F be a three-dimensional Banach space; in the space X we introduce the following norm

$$\|(Q_0, H, \omega_1, \omega_2)\|^F = \left\|(\|Q_0\|_{L^2}, \|H\|_{M_{L^2}}, \|(\omega_1, \omega_2)\|_{M_E})\right\|_F, \tag{7.6.6}$$

where $\| \|_F$ denotes a norm in the space F.

As before, we define a norm in X_- as

$$\|(f_1, f_2, f_3, f_4)\|^{F^*} = \left\|(\|Q_0\|_{L^2}, \|H\|_{L_{L^2}}, \|(\omega_1, \omega_2)\|_{L_{E^*}})\right\|_{F^*}. \tag{7.6.7}$$

The norm of continuous linear functionals defined on X_- with the norm (7.6.7) is equal to the norm of elements of X with the norm given by formula (7.6.6). Obviously, the minimum-norm problem corresponding to the norm (7.6.7) has a solution.

7.6. Rod heating control

In applications we are often interested not in the whole X but in a certain linear manifold V,

$$V = \{x \in X: P(x) = x_0\},$$

where P is a certain continuous projection operator acting in X. Thus, instead of the system (7.6.3) we consider the system

$$(X_1 \underset{A}{\rightarrow} \square \underset{B}{\rightarrow} Y), \tag{7.6.8}$$

where $X_1 = \{x \in X: P(x) = 0\}$. Moreover, instead of the minimum-norm problem for the equation $BA(x) = y_0$ we solve the minimum-norm problem for the equation $BA(x) = y_0 - BA(x_0)$. We encounter such situations when some elements of our system (e.g., the initial condition, the boundary condition or the nonhomogeneous part of the equation) are fixed; then the system is controlled by non-fixed elements.

If we assume that the space X_1 is closed in the weak-$*$-topology (as happens in applications), then the minimum-norm problem for system (7.6.8) has a solution, i.e., Theorem 7.6.1 holds in the space X_1.

Consider now the case where the initial conditions are fixed, the equation is homogeneous and the system is controlled by the boundary value control only. Assume that $U(t, x)$ is linear with respect to x,

$$U(t, x) = u_1(t) + x u_2(t).$$

In this case, by (7.5.18) and by the definition of B, the operator BA is of the form

$$BA(\omega_1, \omega_2) = \sum_{n=1}^{\infty} |\lambda_n| \int_0^T e^{\lambda_n(T-\tau)} [c_n^0(a_1\omega_1 + a_2\omega_2) + \\ + c_n^1(b_1\omega_1 + b_2\omega_2)] d\tau \cdot e_n(x),$$

where ω_1 and ω_2 are functions appearing in the boundary conditions. The operator BA maps the space $M[0, T] \times M[0, T]$ into the space $L^2[0, S]$.

Since the basis $\{e_n\}$ is equivalent to an orthonormal basis, the equation

$$BA(x) = y_0 \tag{7.6.9}$$

is equivalent to the infinite system of equations

$$|\lambda_n| \int_0^T e^{\lambda_n(T-\tau)}[(c_n^0 a_1 + c_n^1 b_1)\omega_1(\tau) + (c_n^0 a_2 + c_n^1 b_2)\omega_2(\tau)]d\tau = y_n,$$

(7.6.10)

$n = 1, 2, \ldots$, where $\{y_n\}$ are coefficients of representation of y_0 with respect to the basis $\{e_n\}$. Since $\{e_n\}$ is equivalent to an orthonormal basis, the sequence $\{y_n\} \in l^2$ and the system of equations (7.6.10) can be regarded as a continuous linear operator which maps the space $M[0, T] \times M[0, T]$ into the space l^2.

By Theorem 7.6.1, the minimum-norm problem in the space $X_1 = M[0, T] \times M[0, T]$ has a solution.

$BA(X_1)$ is dense in the output space Y. Hence, by Theorem 5.2.8, we cannot formulate the maximum principle. However, a certain substitute of this principle can be applied to investigate the problem below, which is close to the minimum-norm problem.

Let r be a positive number. Find ω_1^0, ω_2^0 such that

$$\|(\omega_1^0, \omega_2^0)\| = \inf\{\|(\omega_1, \omega_2)\|: \|BA(\omega_1, \omega_2) - y_0\| \leq r\},$$

(7.6.11)

where $\|\ \|$ is an arbitrary norm in the space $M[0, T] \times M[0, T]$ such that balls generated by that norm are compact in the weak-∗-topology.

Theorem 5.2.12 states that there exists a functional G defined by a sequence $\{g_n\} \in l^2$ and such that

$$G(BA(\omega_1, \omega_2)) \leq G(BA(\omega_1^0, \omega_2^0)) = G(y_0) - r \cdot \|G\|, \quad (7.6.12)$$

for all (ω_1, ω_2) such that

$$\|(\omega_1, \omega_2)\| \leq R, \tag{7.6.13}$$

where

$$R = \|(\omega_1^0, \omega_2^0)\|. \tag{7.6.14}$$

It follows from (7.6.12) and (7.6.13) that the element (ω_1^0, ω_2^0) maximizes the functional $G(BA(\omega_1, \omega_2))$ on the set of elements (ω_1, ω_2) satisfying (7.6.13).

Consider now the space $M[0, T] \times M[0, T]$ as the space $M_E[0, T]$ where E is a two-dimensional Banach space. $G(BA(\omega_1, \omega_2))$ can be rewritten in the form

$$G(BA(\omega_1, \omega_2)) = \sum_{n=1}^{\infty} g_n \lambda_n \int_0^T e^{\lambda_n(T-\tau)}[(c_n^0 a_1 + c_n^1 b_1)\omega_1 +$$
$$+ (c_n^0 a_2 + c_n^1 b_2)\omega_2]d\tau$$
$$= \int_0^T [G_1(\tau)\omega_1(\tau) + G_2(\tau)\omega_2(\tau)]d\tau, \quad (7.6.15)$$

where

$$G_i(\tau) = \sum_{n=1}^{\infty} g_n \lambda_n (c_n^0 a_i + c_n^1 b_i) e^{\lambda_n(T-\tau)}, \quad i = 1, 2, \quad (7.6.16)$$

are analytic functions of τ, $0 < \tau < T$.

As in Section 6.3, we obtain for almost all τ

$$(\omega_1^0(\tau), \omega_2^0(\tau)) \in K_R \cap H_\tau, \quad (7.6.17)$$

where

$$K_R = \{(v_1, v_2) \in E: \|(v_1, v_2)\|_E \leqslant R\} \quad (7.6.18)$$

and

$$H_\tau = \left\{(v_1, v_2) \in E: \frac{R \cdot [G_1(\tau)v_1 + G_2(\tau)v_2]}{\|(G_1(\tau), G_2(\tau))\|_{E^*}} = 1\right\}. \quad (7.6.19)$$

This means, geometrically, that H_τ is a supporting hyperplane of the set K_R. If $K_R \cap H_\tau$ consists of a single point p_τ for almost all τ, then, by (7.6.17), the minimum-norm control is determined uniquely. Clearly, p_τ is an extreme point of the set K_R.

The set K_R is a bounded convex closed two-dimensional set. Its boundary consists of extremal points and a certain, at most countable, number of segments. If the set $K_R \cap H_\tau$ for $\tau \in D$, where D is a set of positive measure, is not a single point set, then there exists a set D_0 of positive measure such that for $\tau \in D_0$ the set $K_R \cap H_\tau$ is a boundary segment W_0 of the set K_R. Furthermore, since the functions G_1, G_2 are analytic, $K_R \cap H_\tau = W_0$ for all τ. This implies that the functions $G_1(\tau)$, $G_2(\tau)$ are colinear. In view of the form of the functions $G_1(\tau)$, $G_2(\tau)$, by the strong linear independence of the functions $e^{\lambda_n \tau}$, we obtain

$$\frac{c_n^0 a_1 + c_n^1 b_1}{c_n^0 a_2 + c_n^1 b_2} = C, \quad C \text{ — constant}, \quad n = 1, 2, \ldots \quad (7.6.20)$$

From (7.6.20) we get

$$\frac{a_1}{a_2} = \frac{b_1}{b_2} = C. \tag{7.6.21}$$

Note that this case differs from those listed in Table 7.5.1. This is not surprising since in Table 7.5.1 $U(t, x)$ is chosen so as to reduce to zero the boundary functions ω_1, ω_2. Thus, the matrix

$$\begin{bmatrix} a_1 & b_1 \\ a_2 & b_2 \end{bmatrix}$$

is of rank 2. This contradicts (7.6.20). Hence, we have

THEOREM 7.6.2. *Consider a boundary value control of a rod heating system with a fixed initial condition and a fixed nonhomogeneous part of the heat diffusion equation. Assume that the boundary value control belongs to $M_E[0, T]$. Moreover, assume that the system can be reduced to a homogeneous system by using a function $U(t, x)$ which is linear with respect to x. Then the minimum-norm control satisfying (7.6.17) is of the form*:

(ω_1^0, ω_2^0) *is an extreme point of K_R, where K_R is the ball of radius*

$$R = \inf \{||(\omega_1, \omega_2)||: ||BA[(\omega_1, \omega_2)] - y_0|| \leq r\}. \tag{7.6.22}$$

Theorem 7.6.2 is of the bang-bang principle type for the boundary value control of the rod heating system. In applications we often encounter situations where one boundary value is under control and another one is fixed. Then, in this case, we have to investigate the operator BA given by formula (7.6.15), where $G_2 = 0$. Since the function G_1 is assumed to be analytic, the optimal control $\omega^0(\tau)$ is of the form $|\omega^0(\tau)| = R$ almost everywhere.

Theorem 7.6.2 is formulated for $U(t, x)$ which is linear with respect to x. However, it often happens that $U(t, x)$ cannot be represented as a linear function of x. We shall illustrate this general case by considering one special form of $U(t, x)$ given in Table 7.5.1, namely, $U(t, x) = xu_2(t) + \frac{1}{2}x^2 u_3(t)$ (note that there is no constant term in $U(t, x)$).

By formula (7.5.32) and by the definition of the operator B^T, we can derive the general form of the operator BA. As before, we investigate the existence of minimum-norm solutions corresponding to the operator

7.6. Rod heating control

BA, which maps $M_E[0, T]$ into the space l^2. As (7.6.10), BA can be represented in the form

$$|\lambda_n| \int_0^T e^{\lambda_n(T-\tau)} \left[c_n^0 \int_0^\tau u_3(s)\,ds - c_n^1 u_2(\tau) - c_n^2 u_3(\tau) \right] d\tau = y_n,$$

(7.6.23)

for $n = 1, 2, \ldots$ Integrating by parts, we obtain

$$|\lambda_n| \int_0^T e^{\lambda_n(T-\tau)} \left(\int_0^\tau u_3(s)\,ds \right) d\tau = \int_0^T u_3(\tau)\,d\tau - \int_0^T e^{\lambda_n(T-\tau)} u_3(\tau)\,d\tau.$$

(7.6.24)

By (7.6.24), formula (7.6.23) can be rewritten in the form

$$y_n = -\int_0^T \{ (|\lambda_n| e^{\lambda_n(T-\tau)} c_n^1 b_1 + [|\lambda_n| e^{\lambda_n(T-\tau)} c_n^2 -$$
$$- (1 - e^{\lambda_n(T-\tau)}) c_n^0] d_1) \omega_1 + (|\lambda_n| e^{\lambda_n(T-\tau)} c_n^1 b_2 +$$
$$+ [|\lambda_n| e^{\lambda_n(T-\tau)} c_n^2 - (1 - e^{\lambda_n(T-\tau)}) c_n^0] d_2) \omega_2 \} d\tau.$$

(7.6.25)

Just as in the previous case, to derive formula (7.6.15) we choose a functional G satisfying (7.6.12); in this case, however, the functions G_1 and G_2 are implied by (7.6.25) and they are obviously different from the functions given by (7.6.16). In this case the functions $G_1(\tau)$ and $G_2(\tau)$ are also analytic. Further argumentation proceeds in the same way as in the case where $U(t, x)$ is linear with respect to x, although the corresponding formulae are more complicated. Hence, Theorem 7.6.2 is valid also for $U(t, x) = xu_2(t) + \frac{1}{2}x^2 u_3(t)$.

Now we consider the following minimum-time problem: under a given initial condition, $Q(0, x) = Q_0(x)$ we have to drive our system to a given set Y_0 (in particular, to a given y_0) in a minimal time by controlling either the boundary values and/or the nonhomogeneous part of the equation. Let us note that $Q(t, x)$ is a continuous function in C_{L^2}. Write $Y(t) = Y_0 - Q(t, \cdot)$. This is a linearly continuous multifunction with values in $L^2[0, S]$.

Suppose now that our control (H, ω_1, ω_2) belongs to a certain set U which is compact in the weak-$*$-topology in the space $M_{L^2} \times M[0, T] \times$

$\times M[0, T]$. By previous arguments, $A(U)$ is compact in the norm topology of the space C_{L^2}. A family of operators B_t given by the formula

$$B_t Q(\,\cdot\,, x) = Q(t, x)$$

maps the space C_{L^2} into the space $L^2[0, S]$ and has the property that $B_t y$ is a continuous function of t for each $y \in C_{L^2}$. Hence, by Theorem 5.4.11, $B_t A(U)$ is a linearly upper semicontinuous multifunction and, by Theorem 5.4.4, there exists a solution of the minimum-time problem.

7.7. Observability of temperature distribution in a rod

Suppose we are given a homogeneous rod of length S. Without loss of generality we can assume that $S = 1$ because this assumption simplifies calculations. Moreover, we assume that the coefficient a in the heat diffusion equation is equal to 1, i.e., the temperature distribution in the rod satisfies the equation

$$\frac{\partial Q}{\partial t} = \frac{\partial^2 Q}{\partial x^2}. \tag{7.7.1}$$

We also assume zero boundary conditions of types (7.3.2)–(7.3.4). The initial conditions are not assumed to be known.

Our problem is to decide whether it is possible to determine the temperature distribution $Q(T, x)$ at the moment T by measuring the temperature $Q(t, \vartheta)$ at different moments t, $t \in [0, T]$, and at a fixed point ϑ of the rod.

We deal with the linear system

$$(X \underset{A}{\rightarrow} \square \underset{B}{\rightarrow} Y), \tag{7.7.2}$$

where the input operator A, described in Section 7.3, assigns to the initial conditions the solution of equation (7.7.1), $X = L^2[0, 1]$, $\square = \overline{A(X)} \subset C_{L^2}$. The output operator B is of form

$$B_\vartheta Q(\,\cdot\,,\,\cdot\,) = Q(\,\cdot\,, \vartheta)$$

and maps the space $\square$ into $Y = L^p[0, T]$, $2 \leqslant p < +\infty$. It can be shown that B is continuous but we shall not use this fact in further considerations.

7.7. Observability of the temperature in a rod

We define E to be either $C[0, 1]$ or $L^p[0, 1]$, $1 \leqslant p < +\infty$. Let $\tilde{F}$ be an operator defined on $\square$ by the formula

$$\tilde{F}(Q(\cdot\,,\,\cdot)) = Q(T,\,\cdot) \tag{7.7.3}$$

and let $F = \tilde{F}A$. The operator F is well-defined and continuous. To see this, observe that in Examples 7.3.1–7.3.6 $e_n(x)$ are uniformly bounded functions.

Let

$$M = \sup_{0 \leqslant x \leqslant 1} \sup_n |e_n(x)|.$$

By formula (7.3.37),

$$\|Q(T, x)\| \leqslant M \sum_{n=1}^{\infty} |c_n| e^{\lambda_n T} \leqslant MK \Big(\sum_{n=1}^{\infty} |c_n|^2\Big)^{1/2} \leqslant MK \|Q_0\|,$$

where

$$K = \Big(\sum_{n=1}^{\infty} e^{2\lambda_n T}\Big)^{1/2}.$$

The problem of observability is to determine a continuous linear operator $\Phi_{\vartheta, T}$ such that the diagram

$$\begin{array}{c} X \xrightarrow{A} \square \xrightarrow{B} Y_0 \\ {}_{F} \searrow \;\; \downarrow {\tilde{F}} \;\; \swarrow {\Phi_{\vartheta, T}} \\ E \end{array} \tag{7.7.4}$$

is commutative, $Y_0 = \overline{BA(X)} \subset Y$.

All the results of this section rely on the following theorem of the approximation theory.

THEOREM 7.7.1 (Luxemburg, Korevaar, 1971). *Suppose we are given a sequence $\{\tilde{\lambda}_n\}$ of complex numbers which have the following properties*:

(1) $\sum_{n=1}^{\infty} \dfrac{1}{|\tilde{\lambda}_n|} < +\infty,$

(2) *there exists a positive number ϱ such that*

$$|\tilde{\lambda}_n - \tilde{\lambda}_m| \geqslant \varrho |n - m|,$$

(3) *there exists a positive number δ such that*

$$\operatorname{Re} \tilde{\lambda}_n > \delta |\tilde{\lambda}_n|$$

for all n sufficiently large. Then the elements of the sequence $\{\exp(-\tilde{\lambda}_n t)\}$ *are strongly linearly independent in the space E (i.e. either in the space* $C[0, T]$ *or in the space* $L^p[0, T]$, $1 \leq p < +\infty$). *Moreover, if*

$$d_n = \inf\{\|\exp(-\tilde{\lambda}_n t) - z\|: z \in \lin\{\exp(-\tilde{\lambda}_1 t), \ldots$$
$$\ldots, \exp(-\tilde{\lambda}_{n-1} t), \exp(-\tilde{\lambda}_{n+1} t), \ldots\}\},$$

then for every positive ε *and all n sufficiently large*

$$\frac{1}{d_n} \leq \exp(\varepsilon \operatorname{Re} \tilde{\lambda}_n) \leq \exp(\varepsilon \delta |\tilde{\lambda}_n|). \tag{7.7.5}$$

The proof of Theorem 7.7.1 requires a thorough knowledge of the theory of analytic functions and it is far beyond the scope of this book.

From Theorem 7.7.1 immediately follows

THEOREM 7.7.2 (Dolecki, 1973). *Let* $\{\varphi_n(x)\}$ *be an arbitrary sequence of functions which belong to E. Let a sequence* $\{\tilde{\lambda}_n\}$ *satisfy the assumptions of Theorem 7.7.1. If there exists a sequence of positive numbers* $\{B_n\}$ *such that*

$$\|\varphi_n\| \leq B_n |\varphi_n(\vartheta)| \tag{7.7.6}$$

and for some positive ε

$$\sum_{n=1}^{\infty} B_n \exp(-(T-\varepsilon) \operatorname{Re} \tilde{\lambda}_n) < +\infty, \tag{7.7.7}$$

then the operator $\Phi_{\vartheta, T}$ *which assigns the function*

$$u(T, x) = \sum_{n=1}^{\infty} \frac{\tilde{c}_n}{\varphi_n(\vartheta)} \varphi_n(x) e^{-\tilde{\lambda}_n T} \tag{7.7.8}$$

to the function

$$u(t, \vartheta) = \sum_{n=1}^{\infty} \tilde{c}_n \exp(-\tilde{\lambda}_n t) \tag{7.7.9}$$

$(u(t, \vartheta) \in \tilde{Y} = \lin_\infty \exp(-\tilde{\lambda}_n t)$, *where, as before,* $\lin_\infty\{e_n\}$ *denotes the set consisting of all elements x which can be represented as the sum of a series* $\sum_{n=1}^{\infty} t_n e_n$) *is a continuous linear operator which maps* $\tilde{Y}$ *into E.*

7.7. Observability of the temperature in a rod

Proof. We have

$$\|u(T,x)\|_E \leq \sum_{n=1}^{\infty} \frac{|\tilde{c}_n|}{|\varphi_n(\vartheta)|} \|\varphi_n(x)\| e^{-\tilde{\lambda}_n T} \leq \sum_{n=1}^{\infty} |\tilde{c}_n| \cdot B_n \cdot e^{-\tilde{\lambda}_n T}. \tag{7.7.10}$$

By the definition of d_n,

$$|\tilde{c}_n| \leq \frac{1}{d_n} \|u(t,\vartheta)\|_{\tilde{Y}}, \tag{7.7.11}$$

and hence

$$\|u(T,x)\|_E \leq \|u(t,\vartheta)\|_{\tilde{Y}} \sum_{n=1}^{\infty} \frac{1}{d_n} B_n e^{-\tilde{\lambda}_n T}. \tag{7.7.12}$$

By taking $\tfrac{1}{2}\varepsilon$ instead of ε in (7.7.5) and by (7.7.7), we conclude that the series

$$\sum_{n=1}^{\infty} \frac{1}{d_n} B_n e^{-\tilde{\lambda}_n T} \tag{7.7.13}$$

converges. This completes the proof. ∎

Let $\{\lambda_n\} = \{-\mu_n^2\}$ be a sequence of eigenvalues of the second derivative operator in the case where the boundary conditions are of the forms (7.3.2)–(7.3.4). Then the sequence $\{\tilde{\lambda}_n\} = \{-\lambda_n\} = \{\mu_n^2\}$ satisfies the assumptions of Theorem 7.7.1.

Observe that, by formula (7.3.37), the generalized solution $Q(t,x)$ is of the form

$$Q(t,x) = \sum_{n=1}^{\infty} c_n e^{-\mu_n^2 t} e_n(x), \tag{7.7.14}$$

where $-\mu_n^2 = \lambda_n$ and c_n are coefficients of the representation of the function $Q_0(x)$ which appears in an initial condition (cf. (7.3.35)).

By putting $\tilde{c}_n = c_n e_n(\vartheta)$, the function $u(t,\vartheta)$ given by (7.7.9) can be represented as

$$u(t,\vartheta) = \sum_{n=1}^{\infty} c_n \exp(-\mu_n^2 t) e_n(\vartheta), \tag{7.7.15}$$

and the function $u(T, x)$ given by (7.7.8) can be represented as

$$u(T, x) = \sum_{n=1}^{\infty} c_n \exp(-\mu_n^2 T) e_n(x). \tag{7.7.16}$$

Summarizing, let $\tilde{\lambda}_n = -\lambda_n = \mu_n^2$, where λ_n are eigenvalues of the second derivative operator defined on the corresponding spaces induced by boundary conditions. If there exists a sequence of positive numbers B_n and a positive number ε such that

$$\|e_n(x)\| \leq B_n e_n(\vartheta) \tag{7.7.17}$$

and

$$\sum_{n=1}^{\infty} B_n \exp\left(-(T-\varepsilon)\mu_n^2\right) < +\infty, \tag{7.7.18}$$

then there exists a continuous linear operator $\Phi_{\vartheta, T}$ such that the diagram (7.7.4) is commutative.

The main result of this section is the following:

THEOREM 7.7.3 (Dolecki, 1973). *For almost all $\vartheta \in [0, 1]$ there exists an operator $\Phi_{\vartheta, T}$ such that the diagram (7.7.4) is commutative.*

The following lemmas are applied in the proof of this theorem.

LEMMA 7.7.4. *Let v be an arbitrary number greater than 1. Let a and b be real numbers such that $0 \leq a \leq 1$ and $0 \leq b \leq \tfrac{1}{4}$. Then the Lebesgue measure of the set*

$$H_{v,a,b} = \{\vartheta \colon 0 \leq \vartheta \leq 1, \ |v\vartheta - a| \leq b\}, \tag{7.7.19}$$

where $|c| = \inf\{|c-n|\colon n = 0, \pm 1, \pm 2, \ldots\}$, is not greater than $8b$.

Proof. Consider the set of real numbers. Determine those real ϑ for which $|v\vartheta - a| \leq b$. Simple calculations show that this inequality holds only for ϑ belonging to the intervals

$$\left[\frac{a+n}{v} - \frac{b}{v}, \ \frac{a+n}{v} + \frac{b}{v}\right], \quad n = 0, \pm 1, \pm 2, \ldots \tag{7.7.20}$$

The intervals (7.7.20) are of the length $2b/v$ and the intervals lying between subsequent intervals (7.7.20) are of the length $(1-2b)/v$.

7.7. Observability of the temperature in a rod

Suppose that a certain number, say k, of the intervals (7.7.20) have nonempty intersections with the interval $[0, 1]$. Since, by assumption, $v > 1$ and $b \leqslant \frac{1}{4}$, we obtain $k \geqslant 2$. Hence, the interval $[0, 1]$ contains $k-1$ intervals whose elements do not satisfy the inequality $|v\vartheta - a| \leqslant b$. This implies

$$|H_{v,a,b}| \leqslant \frac{2kb}{v} \tag{7.7.21}$$

and

$$|[0, 1]\setminus H_{v,a,b}| \geqslant (k-1)\frac{1-2b}{v}, \tag{7.7.21'}$$

where $|B|$ denotes the Lebesgue measure of the set B.

Since $k \geqslant 2$, formulae (7.7.21) and (7.7.21') imply

$$\frac{|H_{v,a,b}|}{|[0, 1]\setminus H_{v,a,b}|} \leqslant \frac{2kb/v}{(k-1)(1-2b)/v} \leqslant 4\frac{b}{1-2b} \leqslant 8b. \tag{7.7.22}$$

This immediately implies the assertion of the lemma. ∎

LEMMA 7.7.5. *Let $\{\mu_q\}$ be an arbitrary sequence of numbers such that $\liminf_{q \to \infty} \mu_q > 1$. Let $\{a_q\}$ and $\{b_q\}$ be arbitrary sequences of real numbers such that $b_q > 0$, $q = 1, 2, \ldots$, and the series $\sum_{q=1}^{\infty} b_q$ converges. Then the set H of all numbers ϑ, $\vartheta \in [0, 1]$, such that there exists an infinite number of a_q satisfying the inequality*

$$|\mu_q \vartheta - a_q| \leqslant b_q, \tag{7.7.23}$$

where, as before,

$$|r| = \inf\{|r-n|: n = 0, \pm 1, \pm 2, \ldots\}, \tag{7.7.24}$$

has the Lebesgue measure zero, $|H| = 0$.

Proof. From the definition,

$$H \subset H_Q = \bigcup_{q=Q}^{\infty} H_{\mu_q, a_q, b_q}. \tag{7.7.25}$$

By Lemma 7.7.4, for all Q sufficiently large,

$$|H_Q| \leq \sum_{q=Q}^{\infty} |H_{\mu_q, a_q, b_q}| \leq \sum_{q=Q}^{\infty} 8b_q. \tag{7.7.26}$$

Since the series $\sum_{q=1}^{\infty} b_q$ converges, formulae (7.7.25) and (7.7.26) imply

$$0 \leq |H| \leq \lim_{Q \to \infty} |H_Q| = 0. \quad \blacksquare \tag{7.7.27}$$

LEMMA 7.7.6. *Let $\{\mu_q\}$ be an arbitrary sequence of positive numbers such that*

$$\liminf_{q \to \infty} \frac{\mu_q}{q} > 0. \tag{7.7.28}$$

Let

$$B_{q,\vartheta} = \frac{1}{|\cos \pi \mu_q \vartheta|} \quad \left(B_{q,\vartheta} = \frac{1}{|\sin \pi \mu_q \vartheta|} \right).$$

Then $B_{q,\vartheta} < +\infty$ for almost all ϑ and

$$\|\cos \pi \mu_q x\| \leq B_{q,\vartheta} |\cos \pi \mu_q \vartheta| \tag{7.7.29}$$
$$(\|\sin \pi \mu_q x\| \leq B_{q,\vartheta} |\sin \pi \mu_q \vartheta|).$$

Moreover,

$$\sum_{q=1}^{\infty} B_{q,\vartheta} e^{-\mu_q^2 T} < +\infty \tag{7.7.30}$$

for all positive T.

Proof. Formula (7.7.29) immediately follows since

$$\|\cos \pi \mu_q x\| \leq \|1\| \leq B_{q,\vartheta} |\cos \pi \mu_q \vartheta|$$
$$(\|\sin \pi \mu_q x\| \leq \|1\| \leq B_{q,\vartheta} |\sin \pi \mu_q \vartheta|).$$

To prove formula (7.7.30), put $a_q = 1/2$ ($a_q = 0$) and $b_q = \mu_q^2 e^{-\mu_q^2 T}$. Clearly, the series $\sum_{q=1}^{\infty} b_q$ converges.

7.7. Observability of the temperature in a rod

By Lemma 7.7.5, the inequality

$$\left|\mu_q \vartheta - \frac{1}{2}\right| \geq b_q \quad (|\mu_q \vartheta| \geq b_q) \tag{7.7.31}$$

holds for almost all ϑ and almost all q (i.e., for all but a finite number of q's, where that finite number may depend on ϑ).

Formula (7.7.31) implies

$$\frac{1}{\mu_q^2} \geq \frac{e^{-\mu_q^2 T}}{\left|\mu_q \vartheta - \frac{1}{2}\right|} \geq \frac{e^{-\mu_q^2 T}}{\pi|\cos \pi \mu_q \vartheta|} \tag{7.7.32}$$

$$\left(\frac{1}{\mu_q^2} \geq \frac{e^{-\mu_q^2 T}}{|\mu_q \vartheta|} \geq \frac{e^{-\mu_q^2 T}}{\pi|\sin \pi \mu_q \vartheta|}\right).$$

By the definition of $B_{q,\vartheta}$ and by (7.7.28) we obtain (7.7.30). ∎

Proof of Theorem 7.7.3. By Theorem 7.7.2 and by Lemma 7.7.6, the assertion of Theorem 7.7.3 follows in the case where the boundary conditions are of forms (7.3.3) and (7.3.4) and at most at one end-point of the rod the boundary conditions are of form (7.3.2) with a satisfying $0 < a < +\infty$, i.e., in the case where the eigenvectors of A are of the form

$$e_n(x) = \cos \mu_n x \quad (\text{or } \sin \mu_n x).$$

When at both end-points of the rod the boundary conditions are of form (7.3.2), i.e., when the eigenvectors are of the form

$$e_n(x) = \cos \mu_n x + \frac{1}{\mu_n} \sin \mu_n x \tag{7.7.33}$$

the proof is more complicated. Letting d_n be a number such that

$$\frac{1}{\mu_n} = \tan d_n, \tag{7.7.34}$$

we obtain

$$\left|\frac{1}{e_n(\vartheta)}\right| = \left|\frac{\cos d_n}{\sin(\mu_n \vartheta + d_n)}\right| \leq \frac{1}{|\sin(\mu_n \vartheta + d_n)|} \leq \frac{\pi}{|\mu_n \vartheta + d_n|}. \tag{7.7.35}$$

Putting $a_q = -d_q$, by Lemma 7.7.5 and by arguments similar to those in the proof of Lemma 7.7.6, we obtain the assertion of the theorem. ∎

It is easy to show that there exist ϑ such that the corresponding $\Phi_{\vartheta,T}$ does not exist. For instance, if $e_n(\vartheta) = 0$ for at least one n, the operator $\Phi_{\vartheta,T}$ does not exist. The set

$$\theta = \{\vartheta : \text{there exists an } n \text{ such that } e_n(\vartheta) = 0\}$$

is dense in the interval $[0, 1]$.

For $\mu_n = 0, 1, 2, \ldots$ or $\mu_n = c, 1+c, 2+c, \ldots$ the following results hold.

THEOREM 7.7.7 (Dolecki, 1973). *Let* $\mu_q = q$ *(or* $q+c$*),* $e_n(x) = {\cos \atop \sin} qx$. *Then the set* θ_0 *of all those* ϑ *such that* $e_n(\vartheta) \neq 0$ *and the corresponding operator* $\Phi_{\vartheta,T}$ *does not exist for any* T *is dense in the interval* $[0, 1]$.

THEOREM 7.7.8 (Dolecki, 1973). *Let* $\{\mu_n\}$ *and* $\{e_n(x)\}$ *be as in Theorem 7.7.7. Let* T *be a fixed positive number. Then the set* θ_1 *of all those numbers* ϑ *of the interval* $[0, 1]$ *for which there exists no operator* $\Phi_{\vartheta,T}$ *ensuring the commutativity of the diagram* (7.7.4), *but there exists for every positive* ε *an operator* $\Phi_{\vartheta,T+\varepsilon}$ *ensuring the commutativity of the diagram* (7.7.4), *is dense in the interval* $[0, 1]$.

We omit here the proofs of these theorems. They exploit the theory of diophantine inequalities and are beyond the scope of this book.

So far we have considered measurement of temperature (by thermometers or thermocouples) at a fixed point. Now we are interested in the following problem. We move the measuring instrument along the rod with a constant speed b. Is it possible to determine the temperature distribution in the rod at the moment T? More precisely, let

$$Q(t, \vartheta + bt) = \sum_{n=1}^{\infty} c_n \exp(-\mu_n^2 t) e_n(\vartheta + bt), \quad 0 \leqslant \vartheta < 1,\ b > 0 \tag{7.7.36}$$

and let

$$Q(T, x) = \sum_{n=1}^{\infty} c_n \exp(-\mu_n^2 T) e_n(x). \tag{7.7.37}$$

7.7. Observability of the temperature in a rod

Assign the function (7.7.37) to the function (7.7.36). Denote this correspondence by $\Phi_{\vartheta,b}$.

THEOREM 7.7.9 (Dolecki, 1973). *Let T be a positive number such that $\vartheta+bt < 1$. The correspondence $\Phi_{\vartheta,b}$ is a continuous linear operator which maps the space Y_0 of functions of the type (7.7.36) with the standard norm of the space $C[0, T]$ or $L^p[0, T]$, $1 \leqslant p < +\infty$, into the space E, which is one of the spaces $C[0, 1]$, $L^p[0, 1]$, $1 \leqslant p < +\infty$.*

Proof. We exploit the particular form of eigenvectors $e_n(x)$ which, according to the boundary conditions, are of the type

$$e_n(x) = \cos \mu_n x,$$

$$e_n(x) = \sin \mu_n x,$$

$$e_n(x) = \frac{1}{\mu_n} \sin \mu_n x + \cos \mu_n x = \alpha_n \cos(\mu_n x + d_n),$$

where $\alpha_n \to 1$ and $d_n \to 0$.

Consider the case where $e_n(x) = \cos \mu_n x$. Then

$$Q(t, \vartheta+bt) = \sum_{n=1}^{\infty} c_n e^{-\mu_n^2 t} \frac{e^{in(\vartheta+bt)} + e^{-in(\vartheta+bt)}}{2}$$

$$= \sum_{n=1}^{\infty} \frac{c_n}{2} e^{in\vartheta} e^{-t(\mu_n^2 - inb)} + \sum_{n=1}^{\infty} \frac{c_n}{2} e^{-in\vartheta} e^{-t(\mu_n^2 + inb)}.$$

(7.7.38)

Letting

$$\tilde{\lambda}_{2n} = \mu_n^2 - inb, \quad \tilde{\lambda}_{2n+1} = \mu_n^2 - inb,$$

we obtain the sequence $\{\tilde{\lambda}_n\}$, which satisfies the assumptions of the Luxemburg–Korevaar theorem (Theorem 7.7.1). Hence, by formula (7.7.5),

$$|c_n| \leqslant \frac{2}{d_n} |e^{in\vartheta}| \, \|Q(t, \vartheta+bt)\|_{Y_0} \leqslant 2e^{(\varepsilon|\tilde{\lambda}_n|)} \|Q(t, \vartheta+bt)\|_{Y_0}$$

(7.7.39)

for every $\varepsilon > 0$ and for n sufficiently large.

If $T > \varepsilon$, then by (7.7.39),

$$\|Q(T,x)\|_E \leq \sum_{n=1}^{\infty} |c_n| e^{-|\tilde{\lambda}_n|T} \|e_n(x)\|_E$$

$$\leq 2K \sum_{n=1}^{\infty} e^{-|\tilde{\lambda}_n|(T-\varepsilon)} \|Q(t, \vartheta+bt)\|_{Y_0}, \qquad (7.7.40)$$

where

$$2K = \sup_n |c_n d_n| (\|Q(t, \vartheta+bt)\|_{Y_0})^{-1},$$

by (7.7.38), is finite. In view of the fact that the series

$$\sum_{n=1}^{\infty} e^{-|\tilde{\lambda}_n|(T-\varepsilon)}$$

is convergent, (7.7.40) implies the continuity of the operator $\Phi^{\vartheta,b}$. For

$$e_n(x) = \sin \mu_n x$$

or

$$e_n(x) = \frac{\alpha_0}{\mu_n} \sin \mu_n x + \cos \mu_n x$$

the proof proceeds in the same way. ∎

7.8. Other problems related to heat diffusion in one-dimensional objects

In previous sections we considered two cases; either a rod is insulated along all its length or the quantity of ingoing heat is known. However, in applications we often deal with rod cooling processes. Assuming the surrounding temperature to be equal to zero, we get the following temperature distribution equation

$$\frac{\partial Q}{\partial t} = a \frac{\partial^2 Q}{\partial x^2} - \sigma Q, \quad 0 \leq x \leq S, \quad 0 \leq t \leq T, \qquad (7.8.1)$$

where the heat transfer parameter $\sigma \geq 0$ depends on the particular rod chosen.

7.8. Other problems related to heat diffusion

If the surrounding temperature is not equal to zero, then we have to consider the distribution of the surrounding temperature at a given moment t and at a given point x, $H(t, x)$. The equation then takes the form

$$\frac{\partial Q}{\partial t} = a \frac{\partial^2 Q}{\partial x^2} - \sigma Q + \sigma H. \tag{7.8.2}$$

Clearly, for equations (7.8.1) and (7.8.2) one has to specify appropriate boundary values (7.3.2)–(7.3.4) and initial values.

The method of solving the above equations consists in determining the space X induced by the boundary values (cf. Examples 7.3.1–7.3.6) and investigating the properties of the eigenvalues and eigenvectors of the operator

$$A(f) = \frac{d^2 f}{dx^2} - \sigma f.$$

This leads to the same equations as for the second derivative operator, namely, to the equations

$$\frac{d^2 f}{dx^2} = (\lambda + \sigma) f. \tag{7.8.3}$$

This implies that the operators $\frac{d^2}{dx^2} - \sigma$ and $\frac{d^2}{dx^2}$ have the same eigenvectors, and the eigenvalues of the first operator are obtained by subtracting σ from the eigenvalues of the second operator. Clearly, the fact that λ_n are negative implies that $\lambda_n - \sigma$ are negative numbers.

Since the eigenvectors of both operators coincide, we can easily get generalized solutions of equations (7.8.1) and (7.8.2) and, moreover, all estimations previously derived are valid also for them. This implies in turn that all qualitative conclusions, such as the existence of solutions of the minimum-norm problem and of the minimum-time problem and the Dolecki theorem (Theorem 7.7.3, stating that the set of those points ϑ for which the temperature observations at ϑ during the time interval $[0, T]$ determine the temperature distribution at T for the whole segment $[0, S]$ is of measure S), are also valid for these equations.

Another interesting example is the case where only part of the rod s insulated (e.g., the segment $[0, s]$ is insulated and the segment $[s, S]$ is not). Then the heat diffusion equation has the form

$$\frac{\partial Q}{\partial t} = \begin{cases} a_1 \dfrac{\partial^2 Q}{\partial x^2} & \text{for } 0 \leqslant x < s, \\ a_2 \dfrac{\partial^2 Q}{\partial x^2} - \sigma Q & \text{for } s < x \leqslant S. \end{cases} \quad (7.8.4)$$

This equation is a model of the following practical problem. A steel landing platform whose cross-section is depicted in Fig. 7.8.1, is mounted

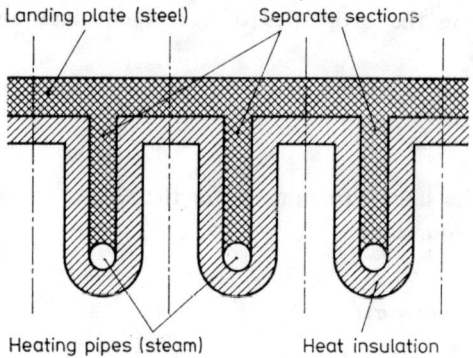

FIG. 7.8.1.

on a research vessel. The platform is insulated from below and heated up by steam circulating in pipes. The aim of this heating system is to melt the ice which covers the platform and to protect the platform from icing. We assume that the steam flow in all sections is the same and the insulation is perfect. From the surface of the platform the heat is dissipated proportionally to the difference between the temperature of the surface and the temperature of the surroundings. The coefficient σ depends on several factors, such as wind speed for example. We assume however that σ is constant. We can formulate several problems related to this system, for instance how to control the heating process so as to make the platform usable in the shortest possible time, how much time it takes, how to protect the platform from icing at the lowest expedi-

7.8. Other problems related to heat diffusion

ture of heat, whether it is possible to determine the temperature distribution in the platform by a single-point measurement, etc.

To answer these questions we have to solve an appropriate heat diffusion equation. Because of our assumptions we can confine ourselves to one section of the heating system. Furthermore, assuming that one section is uniformly heated and cooled lengthwise we reduce our system to a one-dimensional system of type T (cf. Figure 7.8.2).

Our assumptions imply that the temperature distribution in the horizontal part of our system is symmetric. Therefore, the system can be identified with a single rod system which is described by equation (7.8.4). This rod system may have (e.g., because of the nonhomogeneity of the construction of the platform) different coefficients a_1 and a_2 in the heat diffusion equation (cf. Fig. 7.8.3).

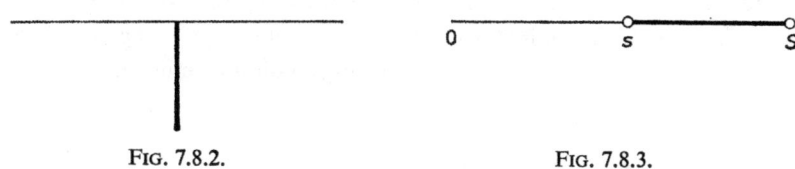

FIG. 7.8.2.   FIG. 7.8.3.

The following boundary conditions must be satisfied at the point s

$$\lim_{x \to s-0} u(t, x) = \lim_{x \to s+0} u(t, x) \qquad (7.8.5)$$

and

$$\lim_{x \to s-0} \frac{\partial u}{\partial x} = k \lim_{x \to s+0} \frac{\partial u}{\partial x}, \qquad (7.8.6)$$

where k depends on both parts of the rod. By a simple substitution of variables the coefficient k can be reduced to 1. In the sequel we shall assume $k = 1$.

We shall now consider two cases. Namely, either the quantity of ingoing heat (the quantity and parameters of steam) is known or the quantity of ingoing heat is null. To analyse those cases we apply the method described in Section 7.2. The operator A of the right-hand side of the heat diffusion equation is defined by the formula

$$A(f) = \begin{cases} a_1 f''(x) & \text{for } 0 \leqslant x \leqslant s, \\ a_2 f''(x) - \sigma f & \text{for } s \leqslant x \leqslant S, \end{cases} \qquad (7.8.7)$$

on the space of twice differentiable functions over $0 \leqslant x \leqslant S$, $x \neq s$, with boundary conditions

$$\frac{\partial f}{\partial x}\bigg|_{x=0} = \frac{\partial f}{\partial x}\bigg|_{x=S} = 0. \tag{7.8.8}$$

We determine the eigenvalues of the operator A and the corresponding eigenvectors.

Assume that λ is an eigenvector of the operator A. Then, the corresponding eigenvector e_λ has, by (7.8.7), the form

$$e_\lambda(x) = \begin{cases} c_1 \sin\sqrt{\dfrac{\lambda}{a_1}}\,x + d_1 \cos\sqrt{\dfrac{\lambda}{a_1}}\,x & \text{for } 0 \leqslant x \leqslant s, \\ c_2 \sin\sqrt{\dfrac{\lambda-\sigma}{a_2}}\,x + d_2 \cos\sqrt{\dfrac{\lambda-\sigma}{a_2}}\,x & \text{for } s \leqslant x \leqslant S, \end{cases}$$

where the constants c_1, d_1, c_2, d_2 are unknown. By the boundary conditions (7.8.5) and (7.8.6), by the differentiability of e_λ and by the boundary conditions (7.8.8), we find that the following conditions must be satisfied:

$$c_1 = 0,$$

$$d_1\left(-\sqrt{\frac{\lambda}{a_1}}\sin\sqrt{\frac{\lambda}{a_1}}\,s\right)$$
$$= c_2 \sqrt{\frac{\lambda-\sigma}{a_2}}\cos\sqrt{\frac{\lambda-\sigma}{a_2}}\,s - d_2 \sqrt{\frac{\lambda-\sigma}{a_2}}\sin\sqrt{\frac{\lambda-\sigma}{a_2}}\,s,$$

$$d_1 \cos\sqrt{\frac{\lambda}{a_1}}\,s = c_2 \sin\sqrt{\frac{\lambda-\sigma}{a_2}}\,s + d_2 \cos\sqrt{\frac{\lambda-\sigma}{a_2}}\,s,$$

$$0 = c_2 \sqrt{\frac{\lambda-\sigma}{a_2}}\cos\sqrt{\frac{\lambda-\sigma}{a_2}}\,S - d_2 \sqrt{\frac{\lambda-\sigma}{a_2}}\sin\sqrt{\frac{\lambda-\sigma}{a_2}}\,S.$$

This system has nonzero solutions c_1, d_1, c_2, d_2 if and only if its determinant

$$\begin{vmatrix} \sqrt{\dfrac{\lambda}{a_1}}\sin\sqrt{\dfrac{\lambda}{a_1}}\,s & \sqrt{\dfrac{\lambda-\sigma}{a_2}}\cos\sqrt{\dfrac{\lambda-\sigma}{a_2}}\,s & -\sqrt{\dfrac{\lambda-\sigma}{a_2}}\sin\sqrt{\dfrac{\lambda-\sigma}{a_2}}\,s \\ -\cos\sqrt{\dfrac{\lambda}{a_1}}\,s & \sin\sqrt{\dfrac{\lambda-\sigma}{a_2}}\,s & \cos\sqrt{\dfrac{\lambda-\sigma}{a_2}}\,s \\ 0 & \sqrt{\dfrac{\lambda-\sigma}{a_2}}\cos\sqrt{\dfrac{\lambda-\sigma}{a_2}}\,S & -\sqrt{\dfrac{\lambda-\sigma}{a_2}}\sin\sqrt{\dfrac{\lambda-\sigma}{a_2}}\,S \end{vmatrix}$$

7.8. Other problems related to heat diffusion

is equal to zero. Expanding this determinant by its first column, we obtain the following equation

$$-\sqrt{\frac{\lambda}{a_1}}\sin\sqrt{\frac{\lambda}{a_1}}s\sqrt{\frac{\lambda-\sigma}{a_2}}\cos\sqrt{\frac{\lambda-\sigma}{a_2}}(S-s)+$$
$$+\cos\sqrt{\frac{\lambda}{a_1}}s\left(\sqrt{\frac{\lambda-\sigma}{a_2}}\right)^2\sin\sqrt{\frac{\lambda-\sigma}{a_2}}(s-S) = 0.$$

Then

$$\tan\sqrt{\frac{\lambda-\sigma}{a_2}}(s-S) = \sqrt{\frac{\lambda}{\lambda-\sigma}}\sqrt{\frac{a_2}{a_1}}\tan\sqrt{\frac{\lambda}{a_1}}S. \tag{7.8.9}$$

It is difficult to obtain an estimate characterizing the asymptotic behaviour of solutions of equation (7.8.9). Therefore, we confine ourselves to the case $a_1 = a_2 = 1$.

Now equation (7.8.9) takes the form

$$\tan\sqrt{\lambda-\sigma}(s-S) = \sqrt{\frac{\lambda}{\lambda-\sigma}}\tan\sqrt{\lambda}S. \tag{7.8.10}$$

But

$$\lim_{\lambda\to\infty}\sqrt{\frac{\lambda}{\lambda-\sigma}} = 1, \tag{7.8.11}$$

whence, the asymptotic behaviour of the solutions of (7.8.10) is analogous to that of the solutions of the equation

$$\tan\sqrt{\lambda-\sigma}(s-S) = \tan\sqrt{\lambda}s. \tag{7.8.12}$$

By rearragements of the terms of (7.8.12) we obtain

$$\tan\sqrt{\lambda-\sigma}(s-S) - \tan\sqrt{\lambda}s = 0, \tag{7.8.13}$$

whence

$$\sin(\sqrt{\lambda-\sigma}(s-S) - \sqrt{\lambda}s) = 0 \tag{7.8.14}$$

and finally

$$\sqrt{\lambda-\sigma}(s-S) - \sqrt{\lambda}s = n\pi.$$

By further rearragements and by squaring both sides of the resulting equation we obtain a quadratic equation with respect to $\sqrt{\lambda}$:

$$(\lambda-\sigma)(s-S) = (n\pi)^2 - 2n\pi s\sqrt{\lambda} + \lambda s.$$

By estimations similar to those given in Section 7.3 we find that there are constants c, C such that

$$|\sqrt{\bar{\lambda}_n} - cn| \leq C\left(\frac{1}{n}\right). \tag{7.8.15}$$

Let us observe that the vectors which correspond to two different eigenvalues are orthogonal (Section 7.3). Therefore, the method previously developed can be applied in this case. In particular, the form of the eigenvectors implies the following theorem.

THEOREM 7.8.1 (Dolecki, 1973). *Let us assume that all assumptions of this section hold. The set of ϑ such that there exists an operator $\Phi_{\vartheta,T}$ for which the diagram (7.7.4) is commutative is of a measure equal to the length of the interval* $[0, S]$ *(i.e., the set of points which are not observable is of measure zero).*

Consider now the problem of heat diffusion in a torus. If there is no heat exchange with the surroundings, the heat diffusion equation for a torus is analogous to that for a rod, namely

$$\frac{\partial Q}{\partial t} = a\frac{\partial^2 Q}{\partial x^2}. \tag{7.8.16}$$

For this equation we have to specify the initial condition

$$Q(0, x) = Q_0(x). \tag{7.8.17}$$

There is no boundary condition in this case, but instead of that the solutions must be periodical with respect to x. We assume that the perimeter of the torus is 2π and the coefficient a in (7.8.16) is equal to 1.

Applying the methods of Section 7.2, we have to determine the eigenvalues of the second derivative operator in the space of continuous periodical functions of period 2π.

By simple calculations we obtain the eigenvalues $\lambda_n = -n^2$, $n = 0, 1, 2, \ldots$

The eigenvalue $\lambda_0 = 0$ corresponds to the eigenvector $e_0 = 1$. The eigenvalues $\lambda_n = -n^2$ correspond to two linearly independent eigenvectors $\sin nx$, $\cos nx$.

7.8. Other problems related to heat diffusion

By arguments similar to those of Section 7.3 we obtain a generalized solution of the heat diffusion equation in the form

$$Q(t, x) = \sum_{n=1}^{\infty} (q_{0,n}^c \cos nx + q_{0,n}^s \sin nx) e^{-n^2 t}, \qquad (7.8.18)$$

where

$$q_{0,n}^c = \int_0^{2\pi} Q_0(x) \cos nx \, dx, \quad n = 0, 1, 2, \ldots,$$
$$q_{0,n}^s = \int_0^{2\pi} Q_0(x) \sin nx \, dx, \quad n = 1, 2, \ldots \qquad (7.8.19)$$

If the temperature distribution is controlled by an external heat inflow, as described by the function $H(t, x)$ (which introduces non-homogeneity to the heat diffusion equation), we obtain the equation

$$\frac{\partial Q}{\partial t} = \frac{\partial^2 Q}{\partial x^2} + H(t, x), \qquad (7.8.20)$$

and the generalized solution of (7.8.20) has the form

$$Q(t, x) = \sum_{n=0}^{\infty} (q_{0,n}^c \cos nx + q_{0,n}^s \sin nx) e^{-n^2 t} +$$
$$+ \sum_{n=0}^{\infty} e^{-n^2 t} \left[\int_0^t e^{-n^2(t-\tau)} h_n^c(\tau) d\tau \cos nx + \right.$$
$$\left. + \int_0^t e^{-n^2(t-\tau)} h_n^s(\tau) d\tau \sin nx \right], \qquad (7.8.21)$$

where

$$h_n^c(t) = \int_0^{2\pi} H(t, x) \cos nx \, dx, \quad h_n^s(t) = \int_0^{2\pi} H(t, x) \sin nx \, dx,$$
$$n = 0, 1, 2, \ldots$$

Formula (7.8.21) implies that the correspondence which assigns to the boundary condition (7.8.17) and to the function $H(t, x)$ the generalized solution (7.8.21) constitutes a linear continuous operator which maps the space $X = L^2[0, 2] \times M_{L^2}$ into the space $\square = C_{L^2}$. This operator is sequentially continuous when X is endowed with the weak-∗-

topology and □ is endowed with the norm topology. Therefore, all minimum-time problems have solutions (compare Section 7.6).

The same results are obtained when the temperature distribution is controlled by varying the surrounding temperature, i.e., when the heat diffusion equation has the form (7.8.2).

Another interesting problem is the temperature observability in a torus. Measuring the temperature at one point of a torus, even during an infinitely long period of time, we cannot determine the temperature distribution in the whole torus. For each point ϑ one can easily choose numbers $q_{0,n}^c$, $q_{0,n}^s$, not all equal to zero, such that

$$q_{0,n}^c \cos n\vartheta + q_{0,n}^s \sin n\vartheta = 0, \quad n = 0, 1, \ldots \quad (7.8.22)$$

Therefore, there exist nonzero temperature distributions in a torus such that at the point ϑ the temperature is constant and equal to zero.

Thus, it is important to investigate multipoint observations. The exact formulation of this problem is as follows.

Given a sequence of numbers $\{\tilde{\lambda}_n\}$ satisfying the assumptions of Theorem 7.7.1, to each number $\tilde{\lambda}_n$ there corresponds a system of functions $\varphi_{n,1}(x), \ldots, \varphi_{n,k}(x)$, where k is a fixed number.

Write

$$u(t, x) = \sum_{n=1}^{\infty} \sum_{i=1}^{k} a_{n,i} \varphi_{n,i}(x) e^{-\tilde{\lambda}_n t}. \quad (7.8.23)$$

With fixed $\vartheta_1, \ldots, \vartheta_k$ we obtain a vector function $u(t, \vartheta_1), \ldots, u(t, \vartheta_k)$ which is an element of the space $X \times X \times \ldots \times X$, where X is either $C[0, T]$ or $L^p[0, T]$, $1 \leq p < +\infty$. On the other hand, fixing T in (7.8.23), we obtain a function $u(T, x) \in Y$, where Y is either $C[0, 2\pi]$ or $L^p[0, 2\pi]$, $1 \leq p < +\infty$.

If the correspondence between $u(t, x)$ and the vector function $u(t, \vartheta_1), \ldots, u(t, \vartheta_k)$ is one-to-one, then we obtain an operator $\Phi_{\vartheta_1, \ldots, \vartheta_k, T}$ which maps $X \times X \times \ldots \times X$ into Y.

THEOREM 7.8.2 (Dolecki, 1973). *If there exists a sequence of positive numbers $\{B_n\}$ such that for each function $f_n \in \text{lin}\{\varphi_{n,1}, \ldots, \varphi_{n,k}\}$*

$$\|f_n\| \leq B_n \sup_{1 \leq j \leq k} |f_n(\vartheta_j)| \quad (7.8.24)$$

and condition (7.7.7) *is satisfied, then the operator $\Phi_{\vartheta_1, \ldots, \vartheta_k, T}$ is continuous.*

The proof of this theorem is analogous to the proof of Theorem 7.7.2.

In the case of a torus, because of the particular form of the eigenvectors, the functions $f_n \in \lin\{\varphi_{n,1}, \varphi_{n,2}\}$ have the form $a\sin(nx-a)$. Thus, B_n can easily be estimated by

$$B_n \leqslant \sup_a \frac{1}{\max[|\sin(n\vartheta_1 - a)|, |\sin(n\vartheta_2 - a)|]}$$

$$= \frac{1}{\inf_a \max[|\sin(n\vartheta_1 - a)|, |\sin(n\vartheta_2 - a)|]}$$

$$= \frac{1}{|\sin \pi^{-1}|n\vartheta_1 - n\vartheta_2\|} \leqslant \frac{\pi}{n\vartheta_1 - n\vartheta_2}. \qquad (7.8.25)$$

Let us fix ϑ_2. As in Theorem 7.7.3, we can show that for almost all ϑ_1 formula (7.7.7) is valid. This implies

THEOREM 7.8.3 (Dolecki, 1973). *The set of points $(\vartheta_1, \vartheta_2)$ for which there exists an operator $\Phi_{\vartheta_1, \vartheta_2, T}$ such that the diagram (7.7.4) is commutative, is of the product measure equal to the measure of the product of intervals* $[0, 2\pi] \times [0, 2\pi]$.

7.9. ROD VIBRATION CONTROL

We are given a rod of length S with a constant elasticity coefficient. The equation of longitudinal vibrations of the rod is

$$\frac{\partial^2 V}{\partial t^2} = a\frac{\partial^2 V}{\partial x^2} + F(t, x), \quad 0 \leqslant x \leqslant S, \, 0 \leqslant t \leqslant T, \qquad (7.9.1)$$

where a denotes a constant which describes the physical properties of the rod and $F(t, x)$ denotes the tension of the rod multiplied by the corresponding coefficient (see Tikhonov and Samarski, 1966, Chapter II, Section 1).

A similar equation describes the transverse vibrations of the rod.

Without loss of generality we can assume that the coefficient a in (7.9.1) is equal to one.

The function $V(t, x)$ describes the deviations from the equilibrium of the point x at the moment t.

Equation (7.9.1) alone does not constitute a full description of rod vibrations. We have to specify also the initial conditions

$$V(0, x) = V_0(x) \quad \text{and} \quad \frac{\partial V}{\partial t}(0, x) = V_1(x) \qquad (7.9.2)$$

and the boundary conditions. Various boundary conditions may be specified. The examples are:

$$V(t, 0) = \mu(t) \quad (V(t, S) = \mu(t)), \qquad (7.9.3)$$

which corresponds to the case where the end-points of the rod are fixed rigidly to points which move and at the moment t their deviation from the equilibrium is equal to $\mu(t)$;

$$\frac{\partial V}{\partial x} = v(t), \qquad (7.9.4)$$

where at a given end-point of the rod and at the moment t a force $v(t)$ is acting;

$$\frac{\partial V}{\partial x} = \alpha_0 (V(t, 0) - \theta(t)) \quad \left(\frac{\partial V}{\partial x} = \alpha_S (V(t, S) - \theta(t)) \right) \qquad (7.9.5)$$

which corresponds to what is called *elastic suspension* with a deviation of the suspension point from the equilibrium.

The general solution of equation (7.8.1) can be represented as a sum

$$V(t, x) = V^0(t, x) + V^1(t, x) + V^2(t, x), \qquad (7.9.6)$$

where $V^0(t, x)$ corresponds to nonzero initial conditions and to the homogeneous equation (i.e., $F = 0$) and to zero boundary conditions, $V^1(t, x)$ corresponds to zero initial conditions, to a nonzero force $F(t, x)$ and to zero boundary conditions and, finally, $V^2(t, x)$ corresponds to zero initial conditions, to the homogeneous equation and to nonhomogeneous boundary conditions.

We determine $V^0(t, x)$ first. To do this we have to solve the homogeneous equation

$$\frac{\partial^2 V}{\partial t^2} = \frac{\partial^2 V}{\partial x^2} \qquad (7.9.7)$$

with the initial condition (7.9.2) and zero boundary conditions of type (7.9.3)–(7.9.5), where these conditions may be different for each end-point.

7.9. Rod vibration control

Equation (7.9.7) can be interpreted as a differential equation in a Banach space X

$$\frac{d^2}{dt^2} V(t, \cdot) = A(V(t, \cdot)), \tag{7.9.8}$$

where A is the second derivative operator and the space X should be specified as a space of functions which are continuous with respect to x.

For each type of boundary conditions the space X is determined as in Section 7.3. Moreover, since we obtain the same spaces as in Section 7.3 and since we deal with the same operator we obtain also the same eigenvalues and eigenvectors. In the same way we choose the space Y as the space $L^2[0, S]$ and we conclude that the eigenvectors $\{e_n(x)\}$ constitute a basis in the space Y which is orthogonal and equivalent to an orthonormal basis.

From now on we shall consider equation (7.9.8) in the space Y, taking into account that λ_n are the eigenvalues of the operator A. Let us recall that all λ_n are negative. To simplify the notation we assume

$$\mu_n = \sqrt{|\lambda_n|}.$$

We represent $V_0(x)$ and $V_1(x)$ with respect to the basis $\{e_n(x)\}$:

$$V_0(x) = \sum_{n=1}^{\infty} v_{0,n} e_n(x), \quad V_1(x) = \sum_{n=1}^{\infty} v_{1,n} e_n(x). \tag{7.9.9}$$

The basis $\{e_n(x)\}$ is equivalent to an orthonormal basis. Therefore the sequences $\{v_{0,n}\}$, $\{v_{1,n}\}$ belong to l^2.

Let $\{f_n(x)\}$ denote the system of basic functionals corresponding to $\{e_n(x)\}$,

$$f_n(x) = \frac{e_n(x)}{\|e_n\|^2}.$$

Applying the functionals $f_n(x)$ to both sides of equation (7.9.8) and observing that $A(e_n) = \lambda_n e_n$ we obtain

$$\frac{d^2}{dt^2} v_n(t) = \lambda_n v_n(t), \quad n = 1, 2, \ldots, \tag{7.9.10}$$

where

$$v_n(t) = f_n(V(t, \cdot)) = \int_0^S V(t, x) f_n(x) dx. \tag{7.9.11}$$

The initial conditions (7.9.9) determine the initial conditions for v_n, namely

$$v_n(0) = v_{0,n}, \quad \frac{d}{dt} v_n(0) = v_{1,n}. \tag{7.9.12}$$

The following functions solve equations (7.9.10) with the initial conditions (7.9.12)

$$v_n(t) = v_{0,n} \cos \mu_n t + \frac{v_{1,n}}{\mu_n} \sin \mu_n t, \tag{7.9.13}$$

where, as before,

$$\mu_n = \sqrt{|\lambda_n|}. \tag{7.9.14}$$

According to the method developed in Section 7.2, we claim that the required solution has the form

$$V(t, x) = \sum_{n=1}^{\infty} \left(v_{0,n} \cos \mu_n t + v_{1,n} \frac{1}{\mu_n} \sin \mu_n t \right) e_n(x). \tag{7.9.15}$$

By the fact that the coefficients $v_{0,n}$, $v_{1,n}$ constitute sequences belonging to l^2 and by the asymptotic behaviour of μ_n analysed in Section 7.3, the operator A, which assigns to the pair of functions V_0, V_1 the function $V(t, x)$ given by (7.9.15), is a linear continuous operator which maps the space $L^2[0, S] \times L^2[0, S]$ into the space $L^2([0, T] \times [0, S])$.

Let us assume that $V_0(x)$ belongs to the space $C^1[0, S]$ of continuously differentiable functions with a norm defined as

$$\|V_0\| = \sup_{0 \leq x \leq S} |V_0(x)| + \sup_{0 \leq x \leq S} \left| \frac{d}{dx} V_0(x) \right|. \tag{7.9.16}$$

Multiplying $V_0(x)$ by $\cos ax$ and integrating by parts, we obtain the following estimation

$$\left| \int_0^S V_0(x) \cos ax \, dx \right|$$

$$= \left| -\frac{1}{a} \sin ax V_0(x) \Big|_0^S + \frac{1}{a} \int_0^S \sin ax V_0'(x) \, dx \right|$$

$$\leq \frac{2+S}{a} \|V_0\|. \tag{7.9.17}$$

7.9. Rod vibration control

Similarly, multiplying $V_0(x)$ by $\sin ax$ instead of $\cos ax$, we obtain

$$\left|\int_0^S V_0(x)\sin ax\,dx\right| \leqslant \frac{2+S}{a}\|V_0\|. \tag{7.9.18}$$

It immediately follows from (7.9.17) and (7.9.18) from the form of $\{e_n(x)\}$ and from the asymptotic behaviour of μ_n that for the function $V_0(x) \in C^1[0, S]$ there exists a positive constant K such that

$$|v_{0,n}| < \frac{K}{n}\|V_0\|. \tag{7.9.19}$$

If $V_1 \in L^2[0, S]$, then there exists a $K_1 > 0$ such that

$$|v_{1,n}| \leqslant \|V_1\|K_1. \tag{7.9.20}$$

From (7.9.19) and (7.9.20) we deduce, by arguments similar to that applied in the proof of (7.5.20), that the operator A which assigns to the pair of functions V_0, V_1 the function $V(t, x)$ given by (7.9.15) is a linear continuous operator which maps the space $C^1[0, S] \times L^2[0, S]$ into the space $C_{L^2}[0, S]$.

The main disadvantage of this approach is the circumstance that a ball in the space $C^1[0, S]$ is not compact in the weak topology and the weak-∗-topology cannot be introduced in this space. Therefore, we consider here also other spaces of absolutely continuous functions (cf. Section 2.6).

Recall that a function $f(x) \in C[0, S]$ is called *absolutely continuous*, if for each $\varepsilon > 0$ there exists a $\delta > 0$ such that for an arbitrary sequence of disjoint intervals $I_i = (a_i, b_i)$ such that

$$\sum_{i=1}^\infty |b_i - a_i| < \delta, \tag{7.9.21}$$

we have

$$\sum_{i=1}^\infty |f(b_i) - f(a_i)| < \varepsilon. \tag{7.9.22}$$

A fundamental theorem on absolutely continuous functions is the Radon–Nikodym theorem which states that every function absolutely continuous has almost everywhere the derivative $f'(x)$ and this derivative is measurable and integrable and

$$f(x) = f(0) + \int_0^x f'(y)\,dy. \tag{7.9.23}$$

Another assertion which we need is that the product of two absolutely continuous functions is an absolutely continuous function and for the product of two absolutely continuous functions the formula for integration by parts is valid.

These statements will not be proved in this book. The reader who is interested in this topic should refer to a textbook of calculus, e.g. Sikorski (1969).

By $DM[0, S]$ $(DL^2[0, S])$ we denote the space of absolutely continuous functions $V(x)$ such that $V'(x) \in M[0, S]$ $(V'(x) \in L^2[0, S])$. In this space a norm can be introduced by formula (7.9.16) with sup replaced by ess sup

$$\left(\|V\| = \left(\int_0^S |V'(x)|^2 dx + V^2(0) + V^2(S) \right)^{1/2} \right). \tag{7.9.24}$$

The space $DM[0, S]$ is isomorphic to the space $M[0, S]$ and therefore the weak-$*$-topology can be introduced in it. The space $DL^2[0, S]$ is a Hilbert space, thus, it is reflexive.

In the space $DM[0, S]$ formulae (7.9.17) and (7.9.18) are valid and therefore also (7.9.19) and (7.9.20) are valid. This implies that the operator A which assigns to the pair of functions $V_0(x), V_1(x)$ the function $V(t, x)$ given by (7.9.15), is a continuous linear operator mapping the space $DM[0, S] \times L^2[0, S]$ into the space C_{L^2}.

This operator is also a sequentially continuous mapping of the space $DM[0, S] \times L^2[0, S]$ endowed with the weak-$*$-topology into C_{L^2} with the norm topology. The proof of this fact follows from formulae (7.9.19) and (7.9.20) and Theorem 4.4.9. Namely, by these formulae we can approximate the operator A by finite dimensional operators A_n of the form

$$A_n(V_0, V_1) = \sum_{i=1}^n \left(v_{0,i} \cos \mu_i t + \frac{1}{\mu_i} v_{1,i} \sin \mu_i(t) \right) e_i(x), \tag{7.9.25}$$

7.9. Rod vibration control

where

$$v_{0,i} = \int_0^S V_0(x)f_i(x)\,dx, \quad v_{1,i} = \int_0^S V_1(x)f_i(x)\,dx \quad (7.9.26)$$

are linear functionals determined by V_0 and V_1, continuous in the weak-$*$-topology.

Formulae analogous to (7.9.17) and (7.9.18) are also valid in the space $DL^2[0, S]$. Hence, the operator A can also be regarded as a continuous linear operator which maps the space $DL^2[0, S] \times L^2[0, S]$ into the space C_{L^2}.

So far we have been concerned with the function $V_0(x)$ but it can be shown that the operator A may be regarded as a continuous linear operator which maps the space $DM[0, S] \times M[0, S]$ $(DL^2[0, S] \times M[0, S])$ into the space C_{L^2} and, moreover, it may be regarded as a sequentially continuous linear operator which maps the space $DM[0, S] \times M[0, S]$ $(DL^2[0, S] \times M[0, S])$ endowed with the weak-$*$-topology into the space C_{L^2} endowed with the norm topology.

We now consider the nonhomogeneous equation

$$\frac{\partial^2 V}{\partial t^2} = \frac{\partial^2 V}{\partial x^2} + H(t, x) \quad (7.9.27)$$

with zero initial conditions

$$V(0, x) = 0 = \frac{\partial V}{\partial t}(0, x)$$

and zero boundary conditions. We assume that $H(t, x)$ belongs to the space M_{L^2} (or $L_{L^2}[0, T]$).

The basic functionals f_n (cf. formula (7.9.10)) applied to equation (7.9.27) give an infinite system of equations, namely

$$\frac{d^2}{dt^2} v_n(t) = \lambda_n v_n(t) + h_n(t), \quad n = 1, 2, \ldots, \quad (7.9.28)$$

where $v_n(t)$'s are determined by (7.9.11) and

$$h_n(t) = f_n(H) = \int_0^S H(t, x) f_n(x)\,dx.$$

By zero initial conditions,
$$v_n(0) = v'_n(0) = 0. \tag{7.9.29}$$

Solutions of (7.9.28) with initial conditions (7.9.29) are given by the following functions
$$v_n(t) = \frac{1}{\mu_n} \int_0^t \sin\mu_n(t-\tau)h_n(\tau)d\tau, \tag{7.9.30}$$

where, as before, $\mu_n = \sqrt{|\lambda_n|}$.

Thus, the generalized solution $V(t, x)$ of equation (7.9.27) with zero initial conditions is of the form
$$V(t, x) = \sum_{n=1}^{\infty} \frac{1}{\mu_n} \int_0^t \sin\mu_n(t-\tau)h_n(\tau)d\tau \cdot e_n(x). \tag{7.9.31}$$

By asymptotic estimations for μ_n, there exists a constant K such that
$$|v_n(t)| \leqslant \frac{K}{n}\|H\|_{M_{L^2}} \quad \left(\sum_{n=1}^{\infty} |v_n(t)|^2 \leqslant K\|H\|_{L^2}\right). \tag{7.9.32}$$

Thus, the operator A which assigns to the function $H(t, x)$ the function $V(t, x)$ given by (7.9.31), is a continuous operator mapping the space M_{L^2} into the space C_{L^2} (the space $L^2[0, T]$ into $L^2([0, T] \times [0, S])$). Moreover, formula (7.9.32) implies that the operator A can be approximated in the operator norm topology by finite dimensional operators
$$A_n(H) = \sum_{i=1}^{n} \frac{1}{\mu_i} \int_0^t \sin\mu_i(t-\tau)h_i(\tau)d\tau \cdot e_i(x), \tag{7.9.33}$$

where the functionals
$$\int_0^t \sin\mu_i(t-\tau)h_i(\tau)d\tau \cdot e_i(x) \tag{7.9.34}$$

are continuous in the weak-$*$-topology. By Theorem 4.4.9, A is a sequentially continuous operator which maps the space M_{L^2} endowed with

7.9. Rod vibration control

the weak-∗-topology into the space C_{L^2} endowed with the norm topology.

We now consider the homogeneous equation with zero initial conditions. As has been done in Section 7.5, at each end-point of the rod one can specify conditions of different type. We assume that the boundary conditions are given by measurable and bounded functions ω_1, ω_2 such that the vector-valued function $(\omega_1(\tau), \omega_2(\tau))$ belongs to $L^2[0, S]$.

The idea which we develop is to reduce the homogeneous equation with nonzero boundary conditions to a nonhomogeneous equation with zero boundary conditions (as in Section 7.5). This is achieved by the substitution

$$V(t, x) = v(t, x) + U(t, x), \qquad (7.9.35)$$

where $U(t, x)$ is a certain function determined by Table 7.5.1. However, $U(t, x)$ is nondifferentiable and noncontinuous with respect to t. Therefore, the idea is to replace the equation

$$\frac{\partial^2 V}{\partial t^2} = \frac{\partial^2 V}{\partial x^2} \qquad (7.9.36)$$

with zero conditions

$$V(0, x) = 0 = \frac{\partial V}{\partial t}(0, x) \qquad (7.9.37)$$

by the equivalent integral equation

$$V(t, x) = \int_0^t \left(\int_0^s \frac{\partial^2 V}{\partial x^2}(\tau, x) \, d\tau \right) ds. \qquad (7.9.38)$$

Substituting (7.9.35) we obtain

$$v(t, x) = \int_0^t \left(\int_0^s \frac{\partial^2 v}{\partial x^2}(\tau, x) \, d\tau \right) ds +$$

$$+ \int_0^t \left(\int_0^s \frac{\partial^2 U}{\partial x^2}(\tau, x) \, ds \right) d\tau - U(t, x). \qquad (7.9.39)$$

Assume now that $U(t, x)$ is linear with respect to x, i.e., it has the form

$$U(t, x) = u_1(t) + x u_2(t). \tag{7.9.40}$$

The second derivative with respect to x vanishes and the equation has the form

$$v(t, x) = \int_0^t \left(\int_0^s \frac{\partial^2 v}{\partial x^2}(\tau, x) \, ds \right) d\tau - U(t, x). \tag{7.9.41}$$

The basic functionals f_n applied to equation (7.9.41) give rise to the following infinite system of integral equations

$$v_n(t) = \lambda_n \int_0^t \left(\int_0^s v_n(\tau) \, ds \right) d\tau - c_n^0 u_1(t) - c_n^1 u_2(t), \tag{7.9.42}$$

where

$$v_n(t) = \int_0^s v(t, x) f_n(x) \, dx,$$

and c_n^0 (c_n^1) are coefficients of the representation of 1 (of the function x) with respect to the basis $\{e_n\}$.

Equations (7.9.42) have solutions

$$v_n(t) = \frac{1}{\mu_n} \int_0^t \sin \mu_n (t-\tau) [c_n^0 u_1(\tau) + c_n^1 u_2(\tau)] d\tau -$$
$$- [c_n^0 u_1(t) + c_n^1 u_2(t)]. \tag{7.9.43}$$

We shall show now that there exists a constant K such that

$$\sum_{n=1}^\infty |v_n(t)|^2 \leq K \sqrt{||u_1||^2 + ||u_2||^2}, \tag{7.9.44}$$

where $||u_1||$, $||u_2||$ denote the norms of the functions $u_1(t)$ and $u_2(t)$ in the space $L^2[0, S]$.

By the triangle inequality and by the fact that

$$\sup_n |\mu_n| c_n^i < +\infty, \quad i = 0, 1, \tag{7.9.45}$$

7.9. Rod vibration control

it is enough to show that there exists a constant K_1 such that

$$\sum_{n=1}^{\infty}\left(\int_0^t \sin\mu_n(t-\tau)u(\tau)d\tau\right)^2 \leqslant K_1\|u\|^2, \tag{7.9.46}$$

where $\|u\|$ denotes the norm in $L^2[0, S]$. But the sequence $\{\mu_n\}$ is such that the sequence $\{\sin\mu_n t\}$ is a basic sequence for each interval $[0, T]$, provided T is large enough. Moreover, this basic sequence is equivalent to a certain orthonormal basic sequence. This immediately implies (7.9.46).

Furthermore, formula (7.9.44) implies that the operator A which assigns to the functions v_1 and v_2 (and therefore to ω_1 and ω_2) the function

$$v(t, x) = \sum_{n=1}^{\infty} v_n(t)e_n(x), \tag{7.9.47}$$

is a continuous linear operator mapping the space $L^2[0, S] \times L^2[0, S]$ into the space $L^2([0, T] \times [0, S])$.

Thus, the operator A which assigns to the functions ω_1, ω_2 the function

$$V(t, x) = v(t, x) + U(t, x)$$
$$= \sum_{n=1}^{\infty} \frac{1}{\mu_n} \int_0^t \sin\mu_n(t-\tau)[c_n^0 u_1(\tau) + c_n^1 u_2(\tau)]d\tau, \tag{7.9.48}$$

is also a continuous linear operator mapping the space $L^2[0, S] \times L^2[0, S]$ into the space $L^2([0, T] \times [0, S])$.

If $U(t, x)$ is a quadratic function with respect to x, we proceed as in Section 7.5. Write

$$U(t, x) = u_1(t) + xu_2(t) + \tfrac{1}{2}x^2 u_3(t). \tag{7.9.49}$$

After all necessary estimations we find that the generalized solution is

$$V(t, x) = \sum_{n=1}^{\infty} \frac{1}{\mu_n} \int_0^t \sin\mu_n(t-\tau)[c_n^0 \tilde{u}(\tau) + c_n^1 u_2(\tau) + c_n^2 u_3(\tau)]d\tau,$$

$$\tag{7.9.50}$$

where (cf. formula (7.5.27))

$$\tilde{u}(t) = u_1(t) - \int_0^t u_3(\tau) d\tau, \qquad (7.9.51)$$

and c_n^2 are the coefficients of the expansion of the function $x^2/2$ with respect to the basis $\{e_n(x)\}$. Similarly, we infer that the operator which assigns to the functions ω_1, ω_2 (these functions define boundary conditions) the function $V(t, x)$ given by (7.9.50) is a continuous linear operator mapping the space $L^2[0, S] \times L^2[0, S]$ into the space $L^2([0, T] \times [0, S])$.

Because of the assumption that ω_1, ω_2 belong to the space $M[0, T]$, we cannot prove by the methods presented in this chapter that $V(t, x) \in C_{L^2}$. As we shall soon see, this is a serious drawback of these methods. However, $V(t, x) \in C_{L^2}$, as can be shown by the general theory (see Lions, 1968; Lions and Magenes, 1968).

The generalized solution of the wave equation (7.9.1) with nonhomogeneous boundary conditions and nonzero initial conditions has the form as in (7.9.6):

$$V(t, x) = V^0(t, x) + V^1(t, x) + V^2(t, x),$$

where V^0 is the solution of the homogeneous equation with zero boundary conditions and nonzero initial conditions, V^1 is the solution of the nonhomogeneous equation with zero boundary and initial conditions, and, finally, V^2 is the solution of the homogeneous equation with zero initial conditions and nonzero boundary conditions.

The operator A which assigns to functions $(V_0(x), V_1(x), H(t, x), \omega_1(t), \omega_2(t))$, where $\omega_1(t), \omega_2(t)$ are functions which define boundary conditions, the function $V(t, x)$ given by (7.9.6) can—as it follows from our considerations—be regarded as a continuous linear operator which maps the input space $X = L^2[0, S] \times L^2[0, S] \times L^2([0, T] \times [0, S]) \times L^2[0, T] \times L^2[0, T]$ into the trajectory space $\square = L^2([0, T] \times [0, S])$.

Thus, we have defined the input and trajectory spaces and the operator A which maps X into $\square$. To have a complete description of the system investigated we have to define the output space and the operator B. They can be introduced in many different ways. For the heat diffusion equation we have defined the operator B by the formula $B[V(t, x)]$

7.9. Rod vibration control

$= V(T, x)$. This operator is well defined on the space C_{L^2} but on the space $L^2([0, T] \times [0, S])$ (and on the space M_{L^2}) the above definition is meaningless since we cannot determine the values of functions at points.

Therefore, we define an operator B_h by the formula

$$B_h(V) = \int_{T-h}^{T} \frac{1}{h} V(t, x) \, dx. \tag{7.9.52}$$

Clearly, the operator B_h is a continuous linear operator which maps the space $\square = L^2([0, T] \times [0, S])$ into the output space $Y = L^2[0, S]$. Thus, we have determined a linear system

$$(X \xrightarrow{A} \square \xrightarrow{B_h} Y). \tag{7.9.53}$$

We consider now the minimum-time problem for an arbitrary convex functional $F(x)$, i.e., we seek an $x_0 \in X$ such that $B_h A(x_0) = y_0$ and

$$F(x_0) = \inf \{F(x) \colon B_h A(x) = y_0\}. \tag{7.9.54}$$

By Theorem 5.3.4, x_0 exists under the assumption that for each r the set $\{x \colon F(x) \leq r\}$ is bounded.

EXAMPLE 7.9.1. Let U be a convex closed set in X which contains 0. Let

$$F(x) = \inf \left\{ t > 0 \colon \frac{x}{t} \in U \right\}, \tag{7.9.55}$$

and, by definition, $F(x) = +\infty$, if x/t does not belong to U for any t. From our previous considerations we conclude that if $y_0 \in cBA(U)$ for a certain c, then there exists an x_0 such that the minimum (7.9.54) is attained at x_0. The functional $F(x)$ can be interpreted as a seminorm in a certain Banach space. Thus, the minimum-norm problem has a solution.

Consider now the minimum-time problem. We assume that the initial conditions $V_0(x)$ and $V_1(x)$ satisfy $V_0(x) \in DL^2[0, S]$, $V_1(x) \in L^2[0, S]$. Then we consider a reduced linear system

$$(X_1 \xrightarrow{A} \square \xrightarrow{B_h} Y), \tag{7.9.56}$$

where $X_1 = L^2([0, T] \times [0, S]) \times L^2[0, T] \times L^2[0, T]$. We now define a family of operators $B_{t,h}$,

$$B_{t,h}(V) = \int_{t-h}^{t} \frac{1}{h} V(\tau, x) \, d\tau. \tag{7.9.57}$$

The operators $B_{t,h}^*$, conjugate to the operators $B_{t,h}$, can be regarded as continuous linear operators which assign to each function $F \in L^2[0, S]$ the function

$$B_{t,h}^* F = \frac{1}{h} F(x) \chi_{[t-h, t]} \in L^2([0, S] \times [0, T]).$$

Observe that, for each F, $B_{t,h}^* F$ is a continuous function of the variable t. Hence, by Theorem 5.4.7, $B_{t,h}(U)$ is a linearly upper semicontinuous multifunction in the weak topology.

Let $Y(t) = Y_0 - V^0(t, x)$, where Y_0 is a fixed subset of the space Y and $V^0(t, x)$ is a solution which corresponds to the homogeneous equation with zero boundary conditions and initial conditions determined by V_0 and V_1. Hence, $V^0(t, x) \in C_{L^2}$ and the multifunction $Y(t)$ is linearly continuous.

As before,

$$T = \inf\{t: B_{t,h}A(U) \cap Y(t) \neq \emptyset\}.$$

Then, by Theorem 5.4.4,

$$B_{T,h}A(U) \cap Y(T) \neq \emptyset.$$

This means that in this case the minimum-time problem has a solution.

We can also consider the minimum-norm and minimum-time problems when system with zero boundary conditions is controlled by the nonhomogeneous part of the wave equation. If the control belongs to M_{L^2} then $V^1(t, x) \in C_{L^2}$. In this case the operators B and B_t can be defined as in Section 7.6, namely

$$B_t(V) = V(t, \cdot), \quad B = B_T; \tag{7.9.58}$$

then we can easily prove the existence of solutions for the minimum-norm and the minimum-time problems.

As we have mentioned before (by applying methods which are not outlined in this book), it can be proved (see Lions, 1968; Lions–Magenes,

7.9. Rod vibration control

1968) that when initial conditions are zero and $H(t, x) = 0$, to non-homogeneous boundary conditions ω_1, ω_2 corresponds a solution $V(t, x) \in C_{L^2}$. With this result we can assume that

$$B(V(\cdot, \cdot)) = V(T, \cdot).$$

The boundary value control system takes the form

$$L^2[0, T] \times L^2[0, T] \underset{A}{\to} C_{L^2} \underset{B}{\to} L^2[0, S].$$

By the results of Russell (1967), there exists a T_0 such that for $T > T_0$

$$BA(L^2[0, T] \times L^2[0, T]) = L^2[0, S],$$

i.e., the system is controllable and for $T < T_0$

$$BA(L^2[0, T] \times L^2[0, T]) \neq L^2[0, S],$$

i.e., the system is not controllable.

This result is significantly different from that obtained for boundary value control systems for the temperature distributions in a rod. In that case, as we have shown in Section 7.6, the system is not controllable for any T.

Chapter 8

Differential and integral equations in Banach spaces

8.1. DERIVATIVES AND INTEGRALS OF FUNCTIONS WITH VALUES IN BANACH SPACES

Let $(X, \| \ \|)$ be a Banach space. Let $[a, b]$, $-\infty < a < b < +\infty$, be a closed interval. By $C_X[a, b]$ we denote the space of all continuous functions $x(t)$ defined in the interval $[a, b]$ with values in the space X. Observe that

$$\||x(\cdot)\|| = \sup_{a \leqslant t \leqslant b} \|x(t)\| \qquad (8.1.1)$$

is finite. It is easy to verify that $\|| \ \||$ is a norm in the space $C_X[a, b]$. The spaces $C_{n,\tilde{\varrho}}[a, b]$ for the metric $\tilde{\varrho}$ induced by a norm in n-dimensional spaces are particular cases of the spaces $C_X[a, b]$ (compare Example 1.1.15).

The space $C_X[a, b]$ is complete. The proof is on the same lines as the considerations in Example 1.5.8.

Let $x(t) \in C_X[a, b]$. We recall that by the derivative of $x(t)$ at the point t we mean the limit

$$x'(t) = \lim_{h \to 0} \frac{x(t+h) - x(t)}{h}. \qquad (7.0.4)$$

Now we shall define an integral of $x(t)$ as an analogy of a Riemann integral. We shall do this as follows. By $\{\Delta_n\}$ we denote a sequence of subdivisions of the interval $[a, b]$,

$$\Delta_n = \{t_i^n\}, \quad i = 0, 1, \ldots, m_n,$$

$$a = t_0^n < t_1^n < \ldots < t_{m_n}^n = b.$$

We say that the sequence $\{\Delta_n\}$ is *normal* if

$$\lim_{n \to \infty} \sup_{1 \leqslant i \leqslant m_n} |t_i^n - t_{i-1}^n| = 0. \qquad (8.1.2)$$

8.1. Differentiation and integration in Banach spaces

We say that a *function $x(t)$ defined on the closed interval $[a, b]$ has the Bochner–Riemann integral* if for each normal sequence of subdivisions $\{\Delta_n\}$ and for an arbitrary choice of θ_i^n, $t_{i-1}^n < \theta_i^n < t_i^n$ the sequence of Riemann sums

$$S^n = \sum_{i=1}^{m_n} (t_i^n - t_{i-1}^n) x(\theta_i^n)$$

is convergent. In the same way as in the classical calculus we can show that this limit does not depend on the choice of subdivisions and that it exists for all continuous functions. This limit we shall denote by $\int_a^b x(t) dt$ and we shall call it the *Bochner–Riemann integral*.

There is also an analogy of the Lebegue integral, the so called Bochner–Lebesgue integral. We shall define it as follows.

A function $x(t) = x_i$ on E_i, where $\{E_i\}$ is a finite system of measurable disjoint sets, is called a *simple function*.

A function $x(t)$ defined on the closed interval $[a, b]$ is called *strongly measurable* if there is a sequence of simple functions $\{x_n(t)\}$ such that

$$\lim_{n \to \infty} \|x(t) - x_n(t)\| = 0, \qquad (8.1.3)$$

for almost all t (almost everywhere).

For a simple function $x(t)$ we can trivially define the Bochner–Riemann integral as

$$\int_a^b x(t) dt = \sum_{i=1}^n |E_i| x_i,$$

where $|E_i|$, $i = 1, 2, \ldots, n$ denote the Lebesgue measures of the sets $E_1, \ldots, E_n$.

We say that a strongly measurable function $x(t)$ is *Bochner–Lebesgue integrable* if there is a sequence of simple functions $\{x_n(t)\}$ such that

$$\lim_{n \to \infty} \int_a^b \|x(t) - x_n(t)\| dt = 0. \qquad (8.1.4)$$

The limit

$$\lim_{n \to \infty} \int_a^b x_n(t) dt = \int_a^b x(t) dt \qquad (8.1.5)$$

we shall call the *Bochner–Lebesgue integral of the function* $x(t)$. In the same way as in the classical measure theory we can show that a strongly measurable function $x(t)$ has the Bochner–Lebesgue integral if and only if the positive valued function $\|x(t)\|$ is integrable.

If a function $x(t)$ is Bochner–Riemann integrable, then it is also Bochner–Lebesgue integrable and the integrals are equal.

In the same way as in the classical calculus and in measure theory we can prove all arithmetical rules for Bochner–Riemann and Bochner–Lebesgue integrals, i.e.

$$\int_a^b x(t)\,dt = \int_a^c x(t)\,dt + \int_c^b x(t)\,dt \quad \text{for} \quad a \leqslant c \leqslant b$$

and

$$\int_a^b (Ax(t) + By(t))\,dt = A\int_a^b x(t)\,dt + B\int_a^b y(t)\,dt$$

for all scalars A, B.

If $x(t)$ is continuous

$$\frac{d}{dt}\int_a^t x(s)\,ds = x(t).$$

In the same way as in the finite dimensional case we can define the spaces $L_X^1[a, b]$, $L_X^2[a, b]$, $-\infty \leqslant a < b \leqslant +\infty$ as a completion of the space $C_X^0[a, b]$ of continuous functions with finite support (i.e. consisting of such functions $x(t)$ that sup $\{|t|: x(t) \neq 0\} < +\infty$) in the metrics induced by the norms

$$\|x\|_1 = \int_a^b \|x(t)\|\,dt$$

in the first case and

$$\|x\|_2 = \left(\int_a^b \|x(t)\|^2\,dt\right)^{1/2}$$

in the second case.

8.2. Continuous differential equations

By $L_X^\infty[a, b]$, $-\infty \leqslant a < b \leqslant +\infty$, we shall denote the space of all essentially bounded functions with the norm

$$||x||_\infty = \underset{a \leqslant t \leqslant b}{\text{ess sup}}\, ||x(t)||$$

$$= \inf_{|E|=0}\, \sup_{t \in [a,b] \setminus E}\, ||x(t)||.$$

8.2. VOLTERRA INTEGRAL EQUATIONS AND DIFFERENTIAL EQUATIONS WITH CONTINUOUS RIGHT-HAND SIDES

Let $(X, ||\ ||)$ be a Banach space. Let $K(t, s, x)$ be a continuous function defined on the product $[a, b]^2 \times X$, $-\infty < a < b < +\infty$, with values in the space X.

We consider a Volterra integral operator

$$K(x) = \int_a^t K(t, s, x(s))ds. \tag{8.2.1}$$

We shall assume that the function $K(t, s, x)$ satisfies the Lipschitz condition with respect to x uniformly with respect to t, i.e.

$$||K(t, s, x) - K(t, s, y)|| \leqslant L(s)||x-y||, \tag{8.2.2}$$

and we shall assume that the Lipschitz constant $L(s)$ is a locally integrable function of s.

In the space $C_X[a, b]$ we can introduce a *Bielecki norm* in the same way as in Example 1.6.1

$$||x||_{B, p} = \sup_{a \leqslant t \leqslant b} e^{-p \int_a^t L(s)\,ds} ||x(t)||. \tag{8.2.3}$$

By similar consideration to those in Example 1.6.1 we can show that the norm $||\ ||_{B, p}$ is equivalent to the norm $|||\ |||$ and that the operator K satisfies the Lipschitz condition with the Lipschitz constant $1/p$ with respect to the Bielecki norm $||\ ||_{B, p}$, i.e.

$$||K(x) - K(y)||_{B, p} \leqslant \frac{1}{p}||x-y||_{B, p}. \tag{8.2.4}$$

Taking $p > 1$, we find that the operator K is a contraction in the Bielecki norm. Thus by the Banach theorem (Theorem 1.6.1) the equation

$$x(t) = \int_a^t K(t, s, x(s))ds + y(t) \qquad (8.2.5)$$

has a unique solution x, and this solution depends in a continuous way on y.

Now we shall apply this result to the following differential equation

$$\frac{dx}{dt} = f(t, x) \qquad (8.2.6)$$

with the initial condition

$$x(a) = x_0. \qquad (8.2.7)$$

We shall assume that $f(t, x)$ is continuous with respect to t and that it satisfies the Lipschitz condition with respect to x

$$\|f(t, x) - f(t, y)\| \leqslant L(t)\|x - y\| \qquad (8.2.8)$$

where the function $L(t)$ is locally integrable. We request that equation (8.2.6) should be satisfied almost everywhere.

The differential equation (8.2.6) together with the initial condition (8.2.7) is equivalent to the integral equation

$$x(t) = x_0 + \int_a^t f(s, x(s))ds \qquad (8.2.9)$$

which is a particular case of equation (8.2.5).

Thus by the previous considerations we infer that the solution of equation (8.2.9) and thus the solution of equation (8.2.6) with the initial condition (8.2.7) are unique and depend in a continuous way on x_0.

Here we shall make an important observation, which we shall use later. It is not essential that $K(t, s, x)$ ($f(s, x)$) should be continuous with respect to s. It is enough to assume that for fixed t, x it is a measurable locally integrable function of s and that the Lipschitz condition (8.2.2) (respectively (8.2.8)) holds. In this case, even in the absence of continuity conditions, the integral operator K maps $C_X[a, b]$ into itself, and, without changes in the proof, we can show that it satisfies the Lipschitz condition with the same constant.

8.2. Continuous differential equations

Using the classical method of reducing equations of higher order to a system of equations of order one, we can show that the equation

$$\frac{d^n x}{dt^n} = f(t, x, x', \ldots, x^{(n-1)}) \tag{8.2.10}$$

with the initial conditions

$$x^{(i)}(a) = x_i, \quad i = 0, 1, \ldots, n-1, \tag{8.2.11}$$

has a unique solution $x(t)$ depending in a continuous way on $x_0, \ldots, x_{n-1}$, provided that $f(t, y_1, \ldots, y_n)$ is measurable and locally integrable with respect to t and satisfies the Lipschitz condition with respect to $y_1, \ldots, y_n$, i.e.

$$\|f(t, y_1, \ldots, y_n) - f(t, z_1, \ldots, z_n)\| \leq L(t) \max_{1 \leq i \leq n} \|y_i - z_i\|$$

where $L(t)$ is locally integrable.

Now we shall show the existence of solutions of differential-difference equations in Banach spaces.

Let $(E, \|\ \|)$ be a Banach space. We consider a differential-difference equation

$$\frac{dx}{dt} = f(t, x_t), \quad 0 \leq t \leq T \tag{8.2.12}$$

with the initial condition

$$x(t) = x_0(t), \quad -h \leq t \leq 0, \tag{8.2.13}$$

where $x(\cdot)$ is an element of the space $C_E[-h, T]$, $x(t)$ is its value at the point t, $x_t(s)$ is an element of the space $C_E[-h, 0]$ defined as

$$x_t(s) = x(t+s), \quad 0 \leq t \leq T, \tag{8.2.14}$$

and f is a continuous function defined on the product $[0, T] \times C_E[-h, 0]$ satisfying the Lipschitz condition with respect to the second variable, i.e., there is a constant $L > 0$ such that

$$\|f(t, y) - f(t, z)\| < L \|y - z\|_{C_E[-h, 0]}, \tag{8.2.15}$$

where as before

$$\|y(\cdot)\|_{C_E[-h, 0]} = \sup_{-h \leq t \leq 0} \|y(t)\|. \tag{8.2.16}$$

Using the classical "step by step" method, we shall show that equation (8.2.12) with the initial condition (8.2.13) has a unique solution depending in a continuous way on x_0.

Of course equation (8.2.12) with the initial condition (8.2.13) is equivalent to the integral equation

$$x(t) = \int_0^t f(s, x_s)\,ds + x_0(0), \quad 0 \leqslant t \leqslant T \qquad (8.2.17)$$

with the initial condition

$$x(t) = x_0(t), \quad -h \leqslant t \leqslant 0. \qquad (8.2.13)$$

Now we can represent $x(t)$ as a sum of two functions

$$x(t) = \bar{x}_0(t) + y(t),$$

where

$$\bar{x}_0(t) = \begin{cases} x_0(t) & \text{for } t \leqslant 0, \\ x_0(0) & \text{for } t \geqslant 0, \end{cases}$$

and

$$y(t) = \begin{cases} 0 & \text{for } t \leqslant 0, \\ x(t) - x_0(0) & \text{for } t \geqslant 0. \end{cases}$$

Of course

$$\frac{dx}{dt} = \frac{dy}{dt} \quad \text{for } t > 0.$$

Thus we can rewrite equation (8.2.17) as an equation with respect to $y(t)$, considering x_0 as a fixed function

$$y(t) = \int_0^t g(s, y_s)\,ds, \quad 0 \leqslant t \leqslant T, \qquad (8.2.18)$$

where

$$g(s, y_s) = f(s, x_{0,s} + y_s).$$

The initial condition will take the form

$$y(t) = 0 \quad \text{for } -h \leqslant t \leqslant 0. \qquad (8.2.19)$$

8.2. Continuous differential equations

Since the function f satisfies the Lipschitz condition with the constant L, the function g also satisfies the Lipschitz condition with the same constant L.

Let $T \leqslant h$. Then, having y_T, we know y_s for all $0 < s < T$. Let

$$K(y) = \int_0^t g(s, y_s)\,ds. \tag{8.2.20}$$

The operator K can be treated as an operator mapping $C_E[-h, 0]$ into itself and assigning $(Ky)_T$ to each y_T. The operator K satisfies the Lipschitz condition. Indeed, the function g satisfies the Lipschitz condition with a constant L with respect to the second variable, and thus for two arbitrary elements $x, y \in C_E[-h, 0]$,

$$\|(Ky)(t) - (Kz)(t)\| \leqslant tL\|y_s - z_s\|_{C_E[-h, 0]}$$
$$= tL \sup_{-h \leqslant \tau \leqslant 0} \|y(\tau) - z(\tau)\|. \tag{8.2.21}$$

Thus

$$\|(Ky)_T - (Kz)_T\|_{C_E[-h, 0]} \leqslant TL\|y_T - z_T\|_{C_E[-h, 0]}. \tag{8.2.22}$$

Suppose that $T \leqslant h$ and $TL < 1$. Then by (8.2.22) the operator K is a contraction. This implies that equation (8.2.18) with the initial condition (8.2.19) has a unique solution.

Of course, starting from different initial functions x_0, we have different operators K_{x_0}. However, it is not difficult to show a uniform continuous dependence of K_{x_0} on x_0. Thus the solution of equation (8.2.17) with the initial condition (8.2.13) depends in a continuous way on x_0 (see Theorem 1.6.2).

Replacing interval $[-h, 0]$ by interval $[-h+T, T]$, we can prove the existence of a solution of equation (8.2.17) with the initial condition (8.2.13) and its continuous dependence on the initial function in the interval $[0, 2T]$. By induction, we can prove this step by step for an arbitrary interval $[0, nT]$.

Further on in this section we shall consider a linear differential equation.

Suppose we are given a linear homogeneous equation

$$\frac{dx}{dt} = A(t)x(t), \quad a \leqslant t \leqslant b. \tag{8.2.23}$$

We assume that for each t the operator $A(t)$ is linear and continuous. Moreover, we assume that $A(t)$, as a function defined in the interval $[a, b]$ with values in the space $B(X)$ of linear continuous operators mapping a Banach space $(X, \|\ \|)$ into itself, is strongly measurable and locally integrable.

Simultaneously with equation (8.2.23) we shall consider the equation

$$\frac{dU}{dt} = A(t)U(t), \quad a \leqslant t \leqslant b, \tag{8.2.24}$$

where $U(t)$ is a continuous function defined in the interval $[a, b]$ with values in the space $B(X)$.

We shall say that $x(t)$ $(U(t))$ is a solution of equation (8.2.23) (resp. (8.2.24)) if it is continuous and satisfies this equation almost everywhere.

It is easy to observe that, if $U(t)$ is a solution of equation (8.2.24) corresponding to the initial condition

$$U(a) = I, \tag{8.2.25}$$

then $U(t)x_0$ is a solution of equation (8.2.23) corresponding to the initial condition

$$x(a) = x_0. \tag{8.2.26}$$

Now we shall show the existence of a solution of equation (8.2.24) with the initial condition (8.2.25).

Of course, equation (8.2.24) with the initial condition (8.2.25) is equivalent to the integral equation

$$U(t) = \int_a^t A(t)U(t)dt + I, \quad a \leqslant t \leqslant b. \tag{8.2.27}$$

Equation (8.2.27) can be regarded as a linear operator equation

$$U = K(U) + I, \tag{8.2.28}$$

where the operator

$$K(U) = \int_a^t A(t)U(t)dt$$

is a continuous linear operator mapping the space $C_{B(X)}[a, b]$ into itself.

8.2. Continuous differential equations

In the space $C_{B(X)}[a, b]$ we can introduce the Bielecki norm in the same way as in Section 1.6

$$\|V(\cdot)\|_{B,p} = \sup_{a \leqslant t \leqslant b} e^{-p\int_a^t \|A(s)\| ds} \|V(t)\|. \tag{8.2.29}$$

Repeating the considerations of Section 1.6 (see also the beginning of this section), we can show that the operator K is a contraction in the space $(C_{B(X)}[a, b], \|\ \|_{B,p})$, provided that $p > 1$.

Thus equation (8.2.28) with the initial condition (8.2.25) has a unique solution $U(t)$.

Now we shall show that for each t the operator $U(t)$ is invertible. We shall repeat here step by step the considerations of Section 6.1. Introducing a new variable $s = b+a-t$ and putting it in equation (8.2.23), we obtain

$$\frac{d\hat{x}}{ds} = -\hat{A}(s)\hat{x}(s), \tag{8.2.30}$$

where $\hat{x}(s) = x(b+a-s)$, $\hat{A}(s) = A(b+a-s)$. Since equation (8.2.30) satisfies the hypotheses of the Bielecki theorem, it has a solution for an arbitrary initial condition

$$\hat{x}(a) = x_0.$$

Returning to the variable t, we find in this way that equation (8.2.23) has a unique solution for an arbitrary final condition $x(b) = x_0$. Thus $U(b)$ is a one-to-one mapping onto, and hence $U(b)$ is invertible. Replacing b by an arbitrary $\hat{t}$, $a \leqslant \hat{t} \leqslant b$, we find that $U(\hat{t})$ is invertible.

In general, there is no explicit formula for $U(t)$. However, in the case where equation (8.2.23) is autonomous, i.e., $A(t) = A$ does not depend on time,

$$U(t) = e^{(t-a)A} = \sum_{i=0}^{\infty} \frac{1}{i!}(t-a)^i A^i. \tag{8.2.31}$$

Having $U(t)$ we can write explicitly a formula for a solution of the nonhomogeneous equation

$$\frac{dx}{dt} = A(t)x + f(t), \tag{8.2.32}$$

where the function $f(t)$ is strongly measurable and locally integrable, with the initial condition

$$x(a) = x_0. \tag{8.2.33}$$

The solution is

$$x(t) = U(t)x_0 + \int_a^t U(t) U^{-1}(s) f(s) \, ds. \tag{8.2.34}$$

Indeed,

$$\frac{dx}{dt} = \frac{dU}{dt} x_0 + \frac{dU}{dt} \int_a^t U^{-1}(s) f(s) \, ds + U(t) U^{-1}(t) f(t)$$

$$= A(t) U(t) x_0 + A(t) U(t) \int_a^t U^{-1}(s) f(s) \, ds + f(t)$$

$$= A(t) U(t) x_0 + A(t) \left(x(t) - U(t) x_0 \right) + f(t)$$

$$= A(t) x(t) + f(t). \tag{8.2.35}$$

Since $U(a) = I$, we have $x(a) = x_0$, i.e., initial condition (8.2.23) is satisfied.

So far we have considered only finite intervals $[a, b]$. Observe that the Bielecki norm $\| \, \|_{B,p}$ can also be considered in the interval $[a, +\infty)$ and also in this interval the operator K is a contraction provided $p > 1$. Thus also in this case we are able to show the existence of a solution $U(t)$ of the homogeneous equation and to show that formula (8.2.35) gives the solution of the nonhomogeneous equation.

8.3. AUTONOMOUS DIFFERENTIAL EQUATION WITH A DISCONTINUOUS RIGHT-HAND SIDE AND SEMIGROUPS OF OPERATORS

In the previous section we considered the linear differential equation

$$\frac{dx}{dt} = A(t)x(t) + f(t), \quad a \leqslant t \leqslant b, \tag{8.2.23}$$

8.3. Discontinuous differential equations

with the initial condition

$$x(a) = x_0 \tag{8.2.33}$$

under the hypothesis that for each t the operators $A(t)$ are continuous.

Unfortunately, in many engineering examples (for example, for systems described by partial differential equations) this assumption is too restrictive.

For that reason we shall consider the system (8.2.32) without this restrictive assumption and an essential role will be played by notions corresponding to the fundamental matrix of solutions, i.e., the notions of evolution operators and a semigroup of operators.

We shall start with an autonomous homogeneous equation

$$\frac{\mathrm{d}x}{\mathrm{d}t} = Ax, \quad 0 \leqslant t < +\infty, \tag{8.3.1}$$

with the initial condition

$$x(0) = x_0. \tag{8.3.2}$$

The operator A does not depend on time and it is a linear operator which is defined on a domain D_A dense in X. We do not assume that A is continuous.

Equation (8.3.1) is considered in the interval $[0, +\infty)$. By a solution of equation (8.3.1) we shall understand a function $x(t)$ defined in the interval $[0, +\infty)$ with values in $D_A \subset X$ and such that for all t, $\frac{\mathrm{d}x}{\mathrm{d}t} = Ax(t)$. The requirement that $x(t)$ should admit values in D_A automatically implies that x_0 in the initial condition (8.3.2) should also belong to D_A.

An essential role in the construction of solutions of equation (8.3.1) is played by semigroups of operators. We say that a family of continuous linear operators $U(t)$, $0 \leqslant t < +\infty$, constitutes a *semigroup* if

$$U(t+s) = U(t)U(s), \quad 0 < s, t < +\infty, \tag{8.3.3}$$

We say that a semigroup $U(t)$ is *continuous* if $U(t)x$ is a continuous function for all $x \in X$. A continuous semigroup $U(t)$ is called a *semigroup of the class C_0* if for all $x \in X$

$$\lim_{t \to 0+0} U(t)(x) = x. \tag{8.3.4}$$

In this case we can extend a semigroup to the closed interval $[0, +\infty)$, putting $U(0) = I$.

The reader can find the theory of semigroups of linear operator in several books. The first was a book by Hille (1948), which played an essential role in the development of the theory (see also Hille–Phillips, 1957; Yoshida, 1965). Here we shall restrict ourselves to the fundamental facts.

PROPOSITION 8.3.1. *Let $U(t)$ be a semigroup of linear operators of the class C_0. Then there are constants $M > 0$ and a such that*

$$\|U(t)\| \leqslant Me^{at}.$$

Proof. Since $U(t)x$ is continuous for all $x \in X$ with respect to t, $0 \leqslant t \leqslant 1$, we have, by the Banach–Steinhaus theorem (Theorem 3.1.1),

$$M = \sup_{0 \leqslant t \leqslant 1} \|U(t)\| < +\infty.$$

Let

$$a = \ln \|U(1)\|.$$

Then, by (8.3.3)

$$\|U(t)\| = \|U([t])U(t-[t])\|$$
$$\leqslant \|U(1)\|^t \|U(t-[t])\| \leqslant Me^{a[t]} \leqslant Me^{at},$$

where, as usual, $[t]$ denotes the greatest integer not greater than t. ∎

Let $U(t)$ be a semigroup of the class C_0. Let

$$Ax = \lim_{t \to 0+0} \frac{U(t)x - x}{t}. \qquad (8.3.5)$$

The operator A is a linear operator. In general it need not be continuous. The domain D_A of the operator A is the set of those x for which the limit on the right-hand side of formula (8.3.5) exists. The operator A is called the *infinitesimal generator* of the semigroup $U(t)$.

Now we shall give examples of semigroups and their infinitesimal generators.

8.3. Discontinuous differential equations

EXAMPLE 8.3.1. Let X be a Banach space. Let A be a continuous linear operator mapping X into itself. Then

$$U(t) = e^{tA} = \sum_{i=0}^{\infty} \frac{t^i A^i}{i!} \qquad (8.3.6)$$

is a semigroup of operators. The proof is on the same lines as the classical proof that $e^{x+y} = e^x e^y$ for scalar x, y. Also in the same way as in the scalar case we can show that

$$\lim_{t \to 0} \frac{e^{tA}x - x}{t} = Ax, \qquad (8.3.7)$$

i.e., A is the infinitesimal generator of the semigroup e^{tA}.

EXAMPLE 8.3.2. Let X be a Banach space over either real or complex numbers with a basis $\{e_n\}$. We shall assume that the basis $\{e_n\}$ is unconditional, i.e., that for any bounded sequence $\{a_n\}$ and each $x = \sum_{n=1}^{\infty} t_n e_n$, the series $\sum_{n=1}^{\infty} a_n t_n e_n$ is convergent.

Suppose that A is a diagonal operator in X, i.e., that A is of the following form:

$$A\left(\sum_{n=1}^{\infty} t_n e_n\right) = \sum_{n=1}^{\infty} \lambda_n t_n e_n, \qquad (8.3.8)$$

where $\{\lambda_n\}$ is a sequence of numbers which may be unbounded. We shall assume

$$\sup_n \lambda_n < +\infty \qquad (8.3.9)$$

in the real case and

$$\sup_n \operatorname{Re} \lambda_n < +\infty \qquad (8.3.9')$$

in the complex case.

Under the assumption of (8.3.9) (or (8.3.9′)) the operator

$$U(t)\left(\sum_{n=1}^{\infty} t_n e_n\right) = \sum_{n=1}^{\infty} e^{\lambda_n t} t_n e_n \qquad (8.3.10)$$

is, for all $t \geq 0$, a well determined linear continuous operator. It is easy to check that the family of operators $U(t)$, $0 \leq t < +\infty$, constitutes a semigroup of operators of the class C_0.

By a simple calculation we find that the infinitesimal generator of the semigroup $U(t)$ is just the operator A.

The example given above plays a fundamental role in the Fourier method, presented in Chapter 7.

EXAMPLE 8.3.3. Let $X = C[-\infty, +\infty]$ be the space of uniformly continuous bounded functions defined on the whole real line with the norm
$$\|x\| = \sup_t |x(t)|.$$
Let
$$U(t)(x)|_s = x(t+s). \tag{8.3.11}$$
Then $U(t)$ is a semigroup of linear operators of the class C_0. The infinitesimal generator of the semigroup $U(t)$ is the operator
$$A(x) = \frac{d}{ds} x. \tag{8.3.12}$$

EXAMPLE 8.3.3'. Replacing the space $C[-\infty, +\infty]$ by the space $C_0(-\infty, +\infty)$ of continuous functions vanishing at $\pm \infty$ or the space $C[0, +\infty)$ of bounded uniformly continuous functions defined in the interval $[0, +\infty)$ with the norm $\|\ \|$, we find that the operators $U(t)$ given by formula (8.3.11) constitutes a semigroup of linear operators of the class C_0. The infinitesimal generator is given by formula (8.3.12). Of course, we ought to remember that the operators $U(t)$, A being formally identical, are in fact different, since they act in different spaces, have different domains, different ranges and so on.

Formula (8.3.11) also determines a semigroup of linear operators in the spaces $L^p(-\infty, +\infty)$ and $L^p[0, +\infty)$, $p = 1, 2$. The infinitesimal generator of this semigroup will also be given by formula (8.3.12).

EXAMPLE 8.3.4. Let $X = C[-\infty, +\infty]$ be the space of bounded uniformly continuous functions defined on the whole real line with the norm $\|x\| = \sup_t |x(t)|$.

8.3. Discontinuous differential equations

Let
$$N_t(u) = (2\pi t)^{-1/2} e^{-u^2/2t}, \quad -\infty < u < +\infty, \; 0 < t < +\infty.$$

We put
$$U(t)(x)|_s = \int_{-\infty}^{+\infty} N_t(s-u) x(u) \, du \quad \text{for} \quad t > 0,$$

and
$$U(0) = I.$$

The operators $U(t)$, $0 \leq t < +\infty$, form a semigroups of linear operators of the class C_0. We shall not present the proof here, since it is based on the theory of Fourier transforms. The reader can find it in Yoshida (1965), Chapter IX, Section 2. The infinitesimal generator of the group $U(t)$ is $\dfrac{1}{2} \dfrac{d^2}{ds^2}$.

Now we shall consider the relation between solutions of equation (8.3.1) and semigroups of linear operators.

PROPOSITION 8.3.2. *Let X be a Banach space. Let $U(t)$ be a semigroup of linear operators of the class C_0 acting in X. Let A be an infinitesimal generator of the semigroup $U(t)$.*
Then the linear equation

$$\frac{dx}{dt} = Ax, \quad 0 \leq t < +\infty, \tag{8.3.1}$$

with the initial condition

$$x(0) = x_0, \quad x_0 \in D_A \tag{8.3.2}$$

has the solution

$$x(t) = U(t)x_0. \tag{8.3.13}$$

Proof. We shall begin by showing that $x(t) \in D_A$ for all $t > 0$, provided $x_0 \in D_A$. Indeed,

$$AU(t)x_0 = \lim_{h \to 0+0} \frac{U(h) - I}{h} U(t) x_0$$

$$= \lim_{h \to 0+0} \frac{U(t+h) - U(t)}{h} x_0$$

$$= U(t) \lim_{h \to 0+0} \frac{U(h) - I}{h} x_0 = U(t) A x_0. \tag{8.3.14}$$

Thus $x(t) = U(t) x_0 \in D_A$ for all $t > 0$.

Now take an arbitrary linear continuous functional $f_0 \in X^*$. We have

$$f_0(U(t) x_0 - x_0) = \int_0^t \frac{d^+ f_0(U(s) x_0)}{ds} ds, \tag{8.3.15}$$

where

$$\frac{d^+ g}{dt} = \lim_{h \to 0+0} \frac{g(t+h) - g(t)}{h}.$$

denotes the right derivative of the function g. Observe that, by the existence of the limit in formula (8.3.14) and the linearity and continuity of f_0, $\dfrac{d^+ f_0(U(s) x_0)}{ds}$ exists.

Thus, by the linearity and continuity of f_0 and by (8.3.15), we obtain

$$f_0(U(t) x_0 - x_0) = f_0 \left(\int_0^t U(s) A x_0 \, ds \right).$$

The arbitrariness of f_0 implies

$$U(t) x_0 - x_0 = \int_0^t U(s) A x_0 \, ds.$$

The function $U(s) A x_0$ is continuous with respect to s. Hence

$$\frac{d}{dt} x(t) \Big|_t = \frac{d}{dt} U(t) x_0 \Big|_t$$

$$= \lim_{h \to 0+0} \frac{1}{h} \int_t^{t+h} U(s) A x_0 \, ds$$

$$= U(t) A x_0 = A U(t) x_0$$

by formula (8.3.14). ∎

8.3. Discontinuous differential equations

Since for fixed t, $U(t)$ is a continuous linear operator, the value of the solution $x(t)$ at each point t depends in a continuous way on the initial value x_0. Of course $x(t) = U(t)x$ if $x \notin D_A$ cannot be regarded as a solution of equation (8.3.1) defined as above. For all $x \in X$ we shall call $U(t)x$ a *mild solution* of equation (8.3.1).

Suppose that equation (8.3.1) with the initial condition (8.3.2) has a unique solution $x(t, x_0)$ which for each fixed t depends on x_0 in a continuous way. Thus we can define an operator $U(t)$ in the following way
$$U(t)x_0 = x(t, x_0).$$
Since equation (8.3.1) is linear, the operator $U(t)$ is linear. Since $x(t, x_0)$ depends on x_0 in a continuous way for all t, the operator $U(t)$ is continuous for all t. The operator $U(t)$ is defined on D_A. Under our assumption the domain D_A is dense in X. Thus $U(t)$ can be extended to the whole space X in a unique way.

Equation (8.3.1) is autonomous, whence
$$x(t+s, x_0) = x(t, x(s, x_0)), \tag{8.3.16}$$
which implies that $U(t)$ satisfies the semigroup property
$$U(t+s) = U(t)U(s), \quad 0 \leqslant t, s \leqslant +\infty. \tag{8.3.3}$$
The continuity of solutions of the differential equation (8.3.1) implies that $U(t)$ is a semigroup of the class C_0.

These observations permit us to give another important example of a semigroup of linear operators.

EXAMPLE 8.3.5. Let E be a Banach space. We shall consider an autonomous differential-difference equation
$$\frac{dx}{dt} = Ax_t, \quad 0 \leqslant t \leqslant T, \tag{8.3.17}$$
where
$$x_t(s) = x(t+s), \quad -h \leqslant s \leqslant 0, \tag{8.3.18}$$
and A is a continuous linear operator mapping the space $C_E[-h, 0]$ into E, with the initial condition
$$x(t) = x_0(t), \quad -h \leqslant t \leqslant 0. \tag{8.3.19}$$

By the considerations of the previous section there is a unique solution of equation (8.3.17) with the initial condition (8.3.19). Moreover, this solution depends in a continuous way on the initial function $x_0(t)$. Thus equation (8.3.17) induces a semigroup $U(t)$ of linear operators of class C_0 mapping the space $C_E[-h, 0]$ into itself.

Now we shall look for the infinitesimal generator of this semigroup.

Let x_0 be an arbitrary element, $x_0 \in C_E[-h, 0]$. By the definition of the solution of equation (8.3.17)

$$U(t)(x_0)|_s = \begin{cases} x_0(t+s) & \text{for } t+s < 0, \\ \int_0^t Ax_s \, ds + x_0(0) & \text{for } t+s \geq 0. \end{cases} \qquad (8.3.20)$$

Thus

$$\lim_{t \to 0+0} \frac{1}{t}(U(t)-I)(x_0) = \begin{cases} \dfrac{d^+(x_0)}{dt}\bigg|_s & \text{for } s < 0, \\ Ax_0 & \text{for } s = 0, \end{cases} \qquad (8.3.21)$$

where $\dfrac{d^+}{dt}$ denotes the right derivative.

The domain of the infinitesimal generator is the set of all continuously differentiable functions $y(s)$ such that $y(0) = Ay$.

EXAMPLE 8.3.6. Let Ω be a domain in an n-dimensional Euclidean space. We assume that the boundary Γ of the domain Ω is smooth. We consider an elliptic operator A on Ω, i.e.,

$$Au(x) = \sum_{i,j=1}^{\infty} a_{i,j} \frac{\partial^2 u}{\partial x_i \partial x_j}, \qquad (8.3.22)$$

where the matrix $[a_{i,j}]$ is positively defined.

Let $Q(t, x)$ be a scalar-valued function defined on the product $\Omega \times [0, T]$. We consider a parabolic equation

$$\frac{\partial Q(t, x)}{\partial t} = AQ(t, x) \qquad (8.3.23)$$

8.3. Discontinuous differential equations

with an initial condition

$$Q(0, x) = Q_0(x) \tag{8.3.24}$$

and with a homogeneous boundary condition

$$Q(t, x) = 0 \quad \text{for} \quad x \in \Gamma \quad \text{and} \quad 0 \leqslant t \leqslant T. \tag{8.3.25}$$

It is possible to show, and the reader can find this in any book devoted to the theory of parabolic equations, that there is a continuous solution of equation (8.3.23) with initial condition (8.3.24) satisfying the boundary condition (8.3.25).

The solution $Q(t, x)$ induces a semigroup of linear operators acting on the space $C(\Omega|\Gamma)$ of continuous functions on Ω vanishing on Γ by the formula

$$U(t)Q_0|_x = Q(t, x).$$

EXAMPLE 8.3.7. Let Ω, Γ, A be as in Example 8.3.6. We consider a hyperbolic differential equation

$$\frac{\partial^2 Q(t, x)}{\partial^2 t} = AQ(t, x), \quad x \in \Omega, \quad 0 \leqslant t \leqslant T \tag{8.3.26}$$

with the initial conditions

$$Q(0, x) = Q_0(x), \quad \left.\frac{\partial Q}{\partial t}\right|_{t=0} = Q_1(x) \tag{8.3.27}$$

and with homogeneous boundary conditions

$$Q(t, x) = 0, \quad \frac{\partial Q}{\partial n}(t, x) = 0 \quad \text{for } x \in \Gamma \text{ and } 0 \leqslant t \leqslant T, \tag{8.3.28}$$

where $\dfrac{\partial Q}{\partial n}$ denotes the normal derivative of Q at the boundary Γ.

The problem of finding a solution can be put in the framework of the theory of differential equations in Banach spaces. Namely, we consider a vector function $V(t, x) = \big(Q(t, x), R(t, x)\big)$ and rewrite the hyperbolic equation (8.3.26) in the form

$$\frac{\partial R(t, x)}{\partial t} = AQ(t, x),$$

$$\frac{\partial Q(t, x)}{\partial t} = R(t, x), \tag{8.3.29}$$

i.e., in the form

$$\frac{\partial V}{\partial t} = \tilde{A}V,$$

where

$$\tilde{A}\begin{pmatrix} Q \\ R \end{pmatrix} = \begin{pmatrix} 0 & I \\ A & 0 \end{pmatrix}\begin{pmatrix} Q \\ R \end{pmatrix}. \tag{8.3.30}$$

We shall assume that equation (8.3.26) with the initial condition (8.3.27) and the boundary condition (8.3.28) has a differentiable solution vanishing together with its derivative on the boundary. The condition for this can be found in any book devoted to the theory of hyperbolic equations. The solution $Q(t, x)$ induces a semigroup of linear operators of the class C_0 in the space of two-dimensional vector functions continuous on Ω and vanishing on Γ in the following way

$$U(t)\begin{pmatrix} Q_0(x) \\ Q_1(x) \end{pmatrix} = \begin{pmatrix} Q(t, x) \\ \dfrac{\partial Q(t, x)}{\partial t} \end{pmatrix}. \tag{8.3.31}$$

So far only homogeneous equations have been considered. Now we shall consider autonomous non-homogeneous equation

$$\frac{dx}{dt} = Ax + f(t), \tag{8.3.32}$$

where A is an infinitesimal generator of a semigroup $U(t)$ of linear operators of the class C_0 and $f(t)$ is a measurable locally integrable function with the initial condition

$$x(0) = x_0. \tag{8.3.33}$$

8.3. Discontinuous differential equations

PROPOSITION 8.3.3. *Suppose that $x_0 \in D_A$ and $f(t)$ is continuous with values in D_A. Then*

$$x(t) = U(t)x_0 + \int_0^t U(t-s)f(s)\,ds \qquad (8.3.34)$$

is the solution of equation (8.3.32) *corresponding to the initial condition* (8.3.33).

Proof. Recall that $U(t)D_A \subset D_A$. Thus

$$\frac{dx}{dt} = \frac{dU(t)}{dt}x_0 + f(t) + \int_0^t \frac{d}{dt}(U(t-s)f(s))\,ds$$

$$= AU(t)x_0 + f(t) + \int_0^t AU(t-s)f(s)\,ds$$

$$= A\left(U(t)x_0 + \int_0^t U(t-s)f(s)\,ds\right) + f(t) = Ax(t) + f(t). \blacksquare$$

Of course, if either $x_0 \notin D_A$ or $f(t) \notin D_A$, then

$$x(t) = U(t)x_0 + \int_0^t U(t-s)f(s)\,ds$$

may not belong to D_A.

In this case we shall call $x(t)$ given by (8.3.34) a *mild solution* of equation (8.3.32) corresponding to the initial condition (8.3.33).

Using Proposition 8.3.3, we can solve by semigroup methods a partial differential equation with a nonhomogeneous right-hand side and homogeneous boundary conditions, provided that we know a semigroup corresponding to the homogeneous problem with homogeneous boundary conditions (see Examples 8.3.6 and 8.3.7). Moreover, using a simple method, we can reduce a problem with nonhomogeneous boundary conditions to a problem with homogeneous boundary conditions. We shall show how to do this only in the case of a parabolic equation (i.e., that given in Example 8.3.6), but the method is general.

We now consider the problem

$$\frac{\partial Q}{\partial t} = AQ(t, x) + f(t, x), \quad x \in \Omega, \quad 0 \leqslant t \leqslant T,$$

with an initial condition

$$Q(0, x) = Q_0(x)$$

and boundary conditions

$$Q(t, x) = Q_1(t, x) \quad \text{for} \quad x \in \Gamma, \quad 0 \leqslant t \leqslant T.$$

Let $g(t, x)$ be an arbitrary function such that, for each t, $g(t, x)$ belongs to the domain of A and, moreover, $Ag(t, x)$ regarded as a function of t with values in $C(\Omega)$ is continuous.

We now introduce a new function $\hat{Q}(t, x)$ as the difference

$$\hat{Q}(t, x) = Q(t, x) - g(t, x).$$

First of all we observe that $\hat{Q}(t, x)$ satisfies homogeneous boundary conditions (i.e., $\hat{Q}(t, x) = 0$ for $x \in \Gamma$). Moreover,

$$\frac{\partial \hat{Q}}{\partial t} = A\hat{Q}(t, x) + f(t, x) + Ag(t, x)$$

and

$$\hat{Q}(0, x) = Q_0(x) + g(0, x).$$

Thus, knowing the semigroup $U(t)$ of a parabolic equation, we can explicitly write the solution

$$\hat{Q}(t, x) = U(t)(Q_0(x) - g(0, x)) +$$
$$+ \int_0^t U(t-s)(f(s, x) + Ag(s, x)) ds$$

and find $Q(t, x)$. Of course it is a solution, provided that $Q_0 \in D_A$, $f(t, x) \in D_A$, $Ag(t, x) \in D_A$. Otherwise it may not be a solution. In that case we shall call it a *mild solution*.

This method was used in Section 7.5.

8.4. PROPERTIES OF THE INFINITESIMAL GENERATOR

Let $U(t)$ be a semigroup of linear operators of the class C_0 acting in a Banach space X. Let A denote the infinitesimal generator of the semigroup $U(t)$.

8.4. Properties of infinitesimal generator

THEOREM 8.4.1. *The domain D_A of the infinitesimal generator A is dense in the space X.*

Proof. Let E denote the class of differentiable functions $k(t)$ with finite support (i.e., such that $\sup\{|t|: k(t) \neq 0\} < +\infty$). By E_U we denote the set of elements of X which can be represented in the form

$$y = \int_0^\infty k(s) U(s) x \, ds \qquad (8.4.1)$$

where x is an arbitrary element of X.

We shall show that $y \in D_A$. Indeed,

$$\frac{1}{h}(U(h)y - y) = \frac{1}{h} \int_0^\infty k(s)(U(s+h) - U(s)) x \, ds$$

$$= \int_h^\infty \frac{1}{h}(k(s-h) - k(s)) U(s) x \, ds - \frac{1}{h} \int_0^h k(s) U(s) x \, ds. \qquad (8.4.2)$$

When h tends to 0, the right-hand side of equality (8.4.2) tends to

$$- \int_0^\infty k'(s) U(s) x \, ds + k(0) x,$$

which implies that $E_U \subset D_A$.

To complete the proof it is enough to show that E_U is dense in the whole space X. Suppose that this is not true. Then there is a continuous linear functional $f \neq 0$ such that

$$f\left(\int_0^\infty k(s) U(s) x \, ds\right) = \int_0^\infty k(s) f(U(s) x) \, ds = 0 \qquad (8.4.3)$$

for all $x \in X$ and all differentiable functions $k(t)$ with finite support.

Since (8.4.3) holds for all differentiable functions with finite support, we obtain

$$f(U(s) x) = 0$$

for all s, $0 \leqslant s < +\infty$, and all $x \in X$. Putting $s = 0$, we have $f(x) = 0$ for all $x \in X$. Thus $f = 0$ and we obtain a contradiction. ∎

THEOREM 8.4.2. *The infinitesimal generator A is a closed operator.*

Proof. Suppose that $x_n \in D_A$, $x_n \to x_0$, $Ax_n \to y$. Then

$$\frac{1}{h}\int_0^h U(s)Ax_n \, ds = \frac{1}{h}\int_0^h \frac{dU(s)x_n}{ds} \, ds = \frac{1}{h}(U(h)-I)x_n. \quad (8.4.4)$$

As $n \to \infty$ we obtain

$$\frac{1}{h}\int_0^h U(s)y \, ds = \frac{1}{h}(U(h)-I)x_0. \quad (8.4.5)$$

Then as $h \to 0$ we obtain $y = Ax_0$, which completes the proof. ∎

The first two theorems of this section just as all the results of the preceding two sections are valid both for the Banach spaces over real numbers and for Banach spaces over complex numbers. The same applies to the next two theorems of this section. However, the formulation for the real and the complex case are slightly different. Thus we shall first prove them for the real case.

THEOREM 8.4.3. *Let $(X, \|\ \|)$ be a Banach space over real numbers. Let $U(t)$ be a semigroup of linear operators of the class C_0 acting in X. Let A denote the infinitesimal generator of the semigroup $U(t)$. Then the operator $A - \lambda I$ is invertible for sufficiently large λ.*

Proof. By Proposition 8.3.1 there are constants $M > 0$ and a such that

$$\|U(t)\| \leq Me^{at}. \quad (8.4.6)$$

Let

$$R(\lambda)x = -\int_0^\infty e^{-\lambda t} U(t)x \, dt. \quad (8.4.7)$$

The integral on the right-hand side of formula (8.4.7) is well determined, provided $\lambda > a$.

8.4. Properties of infinitesimal generator

Take an arbitrary $x_0 \in D_A$. Then

$$R(\lambda)Ax_0 = -\int_0^\infty e^{-\lambda t} U(t) Ax_0 \, dt = -\int_0^\infty e^{-\lambda t} \frac{dU(t)x_0}{dt} dt$$

$$= -e^{-\lambda t} U(t) x_0 \Big|_0^\infty - \lambda \int_0^\infty e^{-\lambda t} U(t) x_0 \, dt = x_0 + \lambda R(\lambda) x_0. \quad (8.4.8)$$

Hence

$$R(\lambda)(A - \lambda I)x_0 = x_0. \quad (8.4.9)$$

On the other hand, for all $x \in X$

$$AR(\lambda)x = -\lim_{h \to 0+0} \frac{U(h)-I}{h} \int_0^\infty e^{-\lambda t} U(t) x \, dt$$

$$= -\lim_{h \to 0+0} \frac{1}{h} \left(\int_0^\infty e^{-\lambda t} U(t+h) x \, dt - \int_0^\infty e^{-\lambda t} U(t) x \, dt \right)$$

$$= -\lim_{h \to 0+0} \frac{1}{h} \left(\int_h^\infty (e^{-\lambda(t-h)} - e^{-\lambda t}) U(t) x \, dt + \int_0^h e^{-\lambda t} U(t) x \, dt \right)$$

$$= -\lambda \int_0^\infty e^{-\lambda t} U(t) x \, dt + x, \quad (8.4.10)$$

i.e.,

$$(A - \lambda I) R(\lambda) x = x. \quad (8.4.11)$$

By (8.4.9) and (8.4.11) we find that $A - \lambda I$ is an invertible operator mapping D_A onto X and that its inverse $(A - \lambda I)^{-1}$ is equal to $R(\lambda)$,

$$(A - \lambda I)^{-1} = R(\lambda). \quad\blacksquare \quad (8.4.12)$$

As a consequence of (8.4.12) and the definition of $R(\lambda)$ (formula (8.4.7)) we obtain

THEOREM 8.4.4. *Let $U(t)$ be a semigroup of linear operators of the class C_0 acting in a Banach space over real numbers $(X, \| \ \|)$. Let constants M and a be such that (8.4.6) holds.
Then*

$$\|(A-\lambda I)^{-m}\| \leqslant \frac{M}{(\lambda-a)^m}, \quad m = 1, 2, \ldots, \tag{8.4.13}$$

provided that $\lambda > a$.

Proof. Let x be an arbitrary element of X. Then

$$\|(A-\lambda I)^{-m} x\| = \|R^m(\lambda) x\|$$

$$= \left\| \int_0^\infty \ldots \int_0^\infty e^{-\lambda s_1} U(s_1) \ldots e^{-\lambda s_m} U(s_m) x \, ds_1 \ldots ds_m \right\|$$

$$= \left\| \int_0^\infty \ldots \int_0^\infty e^{-\lambda(s_1+\ldots+s_m)} U(s_1+\ldots+s_m) x \, ds_1 \ldots ds_m \right\|$$

$$\leqslant \|x\| \int_0^\infty \ldots \int_0^\infty M e^{-(\lambda-a)(s_1+\ldots+s_m)} \, ds_1 \ldots ds_m$$

$$= \frac{\|x\| M}{(\lambda-a)^m} \cdot \blacksquare \tag{8.4.14}$$

Replacing in all calculations λ by $\operatorname{Re}\lambda$ we obtain theorems similar to Theorems 8.4.3 and 8.4.4 for the spaces over complex numbers.

THEOREM 8.4.3'. *Let X be a Banach space over complex numbers. Let $U(t)$ be a semigroup of linear operators of the class C_0 acting in X. Let A denote the infinitesimal generator of the semigroup $U(t)$. Then the operator $A-\lambda I$ is invertible, provided that the real part of λ, $\operatorname{Re} \lambda$, is sufficiently large.*

THEOREM 8.4.4'. *Let $X, U(t), A$ be as in Theorem 8.4.3'. Let M and a be constants such that (8.4.6) holds. Then*

$$\|(A-\lambda I)^{-m}\| \leqslant \frac{M}{(\operatorname{Re} \lambda - a)^m}, \quad m = 1, 2, \ldots \tag{8.3.13'}$$

8.5. INFINITESIMAL GENERATOR OF BOUNDED SEMIGROUPS

Let $(X, \| \ \|)$ be a Banach space either over the real numbers or over the complex numbers. Let $U_0(t)$ be a semigroup of linear operator of class C_0 acting in the space X. We say that the semigroup $U_0(t)$ is *bounded* if there is an $M > 0$ such that

$$\|U_0(t)\| \leqslant M, \quad 0 \leqslant t < +\infty. \tag{8.5.1}$$

Let $U(t)$ be an arbitrary semigroup of linear operators of class C_0 acting in X. By Proposition 8.3.1 there are $M > 0$ and a such that

$$\|U(t)\| \leqslant Me^{at}, \quad 0 < t < +\infty. \tag{8.4.6}$$

Let

$$U_0(t) = e^{-at}U(t). \tag{8.5.2}$$

It is easy to verify that $U_0(t)$ is a bounded semigroup. We shall denote by A the infinitesimal generator of the semigroup $U(t)$ and by A_0 the infinitesimal generator of the semigroup $U_0(t)$. Observe that for arbitrary $x \in D_{A_0}$ we have

$$A_0 x = \lim_{h \to 0+0} \frac{U_0(h) - I}{h} x = \lim_{h \to 0+0} \frac{e^{-ah}U(h) - I}{h} x$$

$$= \lim_{h \to 0+0} \left(e^{-ah} \frac{U(h) - I}{h} x + \frac{e^{-ah} - 1}{h} x \right) = Ax - ax. \tag{8.5.3}$$

By formula (8.5.3) we obtain

PROPOSITION 8.5.1. *The domain of the infinitesimal generators A, A_0 of the semigroups $U(t)$ and $U_0(t)$ satisfying (8.5.2) are the same and, moreover,*

$$A_0 = A - aI. \tag{8.5.3'}$$

Proposition 8.5.1 permits us to reduce the considerations of arbitrary semigroups of class C_0 to the consideration of bounded semigroups.

Of course, if $U_0(t)$ is a bounded semigroup, by Theorems 8.4.4 and 8.4.4' we have

$$\|(A_0 - \lambda I)^{-m}\| \leqslant \frac{M}{\lambda^m}, \quad m = 1, 2, \ldots \tag{8.5.4}$$

in the real case and

$$\|(A_0-\lambda I)^{-m}\| \leqslant \frac{M}{(\operatorname{Re}\lambda)^m}, \quad m = 1, 2, \ldots \qquad (8.5.4')$$

in the complex case.

THEOREM 8.5.2. *Let $(X, \|\ \|)$ be a Banach space over real numbers (or complex numbers). Let A be a closed operator acting in X, and such that the domain D_A of the operator A is dense in X. Suppose that the operators $A - nI$ are invertible for $n = 1, 2, \ldots$ Then the operator A is an infinitesimal generator of a bounded semigroup $U(t)$ if and only if there is a constant $M_1 > 0$ such that*

$$\|J_n^m\| \leqslant M_1, \quad m, n = 1, 2, \ldots, \qquad (8.5.5)$$

where

$$J_n = \left(I - \frac{A}{n}\right)^{-1}. \qquad (8.5.6)$$

Proof. Necessity. Observe that

$$J_n = n(nI - A)^{-1}. \qquad (8.5.7)$$

Thus by (8.5.4) (or (8.5.4'))

$$\|J_n^m\| \leqslant Mn^m \frac{1}{n^m} = M. \qquad (8.5.8)$$

Sufficiency. We shall show that

$$\lim_{n \to \infty} J_n x = x \qquad (8.5.9)$$

for all $x \in X$.

Indeed, by the definition of J_n, for $x \in D_A$

$$x = J_n\left(I - \frac{A}{n}\right)x \qquad (8.5.10)$$

and

$$x = J_n x - \frac{J_n A}{n} x.$$

8.5. Infinitesimal generator of bounded semigroups

Therefore

$$(J_n - I)x = \frac{J_n}{n}(Ax). \tag{8.5.11}$$

By (8.5.5) the right-hand side of formula (8.5.11) tends to 0. Thus (8.5.9) holds for all $x \in D_A$. Since D_A is dense in X and the operators J_n are simultaneously bounded by (8.5.5), formula (8.5.6) holds for all $x \in X$.

Let

$$U_n(t) = e^{tAJ_n} = e^{tn(J_n - I)} = e^{-nt}e^{ntJ_n}. \tag{8.5.12}$$

The operators $U_n(t)$ are well determined because for all n the operators J_n are bounded (see Example 8.3.1).

In view of (8.5.5)

$$\|e^{ntJ_n}\| \leq \sum_{m=0}^{\infty} \frac{(nt)^m}{m!} \|J_n^m\| \leq M_1 e^{nt} \tag{8.5.13}$$

and

$$\|U_n(t)\| \leq M_1$$

for all n and all t. This means that the semigroups $U_n(t)$ are bounded for $n = 1, 2, \ldots$

Observe that the operators J_n commute. This implies that J_k commute with $U_n(t)$, $k, n = 1, 2, \ldots$ By the definition of $U_n(t)$ (see (8.5.12))

$$\frac{d}{dt} U_n(t)x = AJ_n U_n(t)x = U_n(t) AJ_n x. \tag{8.5.14}$$

Thus

$$\|U_n(t)x - U_k(t)x\| = \|U_k(t-s) U_n(s)\big|_0^t x\|$$

$$= \left\| \int_0^t \left(\frac{d}{ds} U_k(t-s) U_n(s) \right) ds \right\|$$

$$= \left\| \int_0^t U_k(t-s) U_n(s) (AJ_n - AJ_k) x \, ds \right\| \leq tM_1^2 \|(J_n - J_k) Ax\|.$$

$$\tag{8.5.15}$$

and by (8.5.9) the right-hand side of formula (8.5.15) converges to 0 for all t and all $x \in D_A$ as n, k tend to infinity. This implies that $\{U_n(t)x\}$ is a Cauchy sequence for all t and $x \in D_A$. The uniform boundedness of $\{U_n(t)\}$ and the density of D_A imply that $\{U_n(t)x\}$ is a Cauchy sequence for all t and all $x \in X$, and thus it has a limit $U(t)$. By the Banach–Steinhaus theorem (Corollary 3.1.2), for each t, $U(t)$ is a continuous linear operator. We shall show that $U(t)$ is a semigroup of linear operators. Indeed, $U_n(t)$ are semigroups for all n, whence

$$\|U(t+s)x - U(t)U(s)x\|$$
$$\leqslant \|U(t+s)x - U_n(t+s)x\| + \|U_n(t+s)x - U_n(t)U_n(s)x\| +$$
$$+ \|U_n(t)U_n(s)x - U_n(t)U(s)x\| + \|U_n(t)U(s)x - U(t)U(s)x\|$$
$$\leqslant \|U(t+s)x - U_n(t+s)x\| + M_1\|U_n(s)x - U(s)x\| +$$
$$+ \|(U_n(t) - U(t))U(s)x\|. \qquad (8.5.16)$$

The right-hand side of formula (8.5.16) tends to 0 as n tends to infinity, which implies the semigroup property of the family $U(t)$. The semigroup $U(t)$ is of class C_0 since $U(t)x$ is the limit of the family of uniformly continuous functions $\{U_n(t)x\}$ for all $x \in X$.

Now we shall show that A is an infinitesimal generator of the semigroup $U(t)$. Let $x \in D_A$. Then

$$\lim_{n \to \infty} U_n(t)AJ_n x = U(t)Ax \qquad (8.5.17)$$

and this limit is uniform in each finite interval of t, because

$$\|U(t)Ax - U_n(t)AJ_n x\|$$
$$\leqslant \|U(t)Ax - U_n(t)Ax\| + \|U_n(t)Ax - U_n(t)AJ_n x\|$$
$$\leqslant \|(U(t) - U_n(t))\|Ax + M_1\|Ax - J_n Ax\| \to 0 \qquad (8.5.18)$$

by the definition of $U(t)$ and (8.5.9). Thus by (8.5.14)

$$U(t)x - x = \lim_{n \to \infty} (U_n(t)x - x) = \lim_{n \to \infty} \int_0^t U_n(s)AJ_n x\,ds$$
$$= \int_0^t \left(\lim_{n \to \infty} U_n(s)AJ_n x\right)ds = \int_0^t U(s)Ax\,ds \qquad (8.5.19)$$

and

$$\lim_{h\to 0+0} \frac{1}{h}(U(h)x-x) = \lim_{h\to 0+0} \frac{1}{h}\int_0^h U(s)Ax\,ds = Ax. \quad (8.5.20)$$

Therefore the infinitesimal generator of U is identical with the operator A on the domain D_A. Since A is closed and the domain D_A is dense in X, the operator A is the infinitesimal generator of the semigroup $U(t)$.

To complete the proof we shall show that the operator A uniquely determines the semigroup. Indeed, suppose that the operator A is also the infinitesimal generator of another semigroup of linear operators $\hat{U}(t)$. Then, of course, A and J_n commute with $\hat{U}(t)$ and

$$\|U_n(t)x - \hat{U}(t)x\| = \left\|\int_0^t \left(\frac{d}{ds}\hat{U}(t-s)U_n(s)x\right)ds\right\|$$

$$= \left\|\int_0^t (\hat{U}(t-s)U_n(s)(A-AJ_n)x)\,ds\right\| \to 0 \quad (8.5.21)$$

for all $x \in D_A$. Hence

$$U(t)x = \lim_{n\to\infty} U_n(t)x = \hat{U}(t)x \quad (8.5.22)$$

for $x \in D_A$. Since the set D_A is dense in X, (8.5.22) holds for all $x \in X$. ∎

8.6. Nonautonomous differential equations in Banach spaces

We shall consider a differential homogeneous equation

$$\frac{dx}{dt} = A(t)x, \quad a \leq t \leq b, \quad (8.6.1)$$

where the operators $A(t)$ are closed linear operators such that their domains $D_{A(t)}$ are dense in X, with an initial condition

$$x(a) = x_0. \quad (8.6.2)$$

By an *evolution operator* we understand a family of linear continuous operators $\{U(t, s)\}$, $a \leq s \leq t \leq b$, such that

$$U(t, t) = I, \qquad (8.6.3)$$

$$U(v, t)U(t, s) = U(v, s) \quad \text{for} \quad a \leq s \leq t \leq v \leq b, \qquad (8.6.4)$$

and $U(t, s)x$ is continuous with respect to t for all s and x.

Suppose that equation (8.6.1) has a unique solution for each initial condition

$$x(s) = x_0, \quad a \leq s \leq b, \quad x_0 \in D_{A(s)}. \qquad (8.6.5)$$

We shall denote this solution by $U(t, s)x_0$. We shall assume that the solution depends in a continuous way on the initial data, i.e., that $U(t, s)x_0$ is a continuous linear operator acting in X. Directly from the definition of $U(t, s)$ we have that

$$U(v, t)(U(t, s)x) = U(v, s)x$$

(i.e., (8.6.4) holds) and

$$U(s, s)x = x$$

(i.e., (8.6.3) holds). The continuity of $U(t, s)x$ with respect to t follows from the hypothesis that $U(t, s)x_0$ is a solution of equation (8.6.1). The density of $D_{A(t)}$ implies that for each t and s $U(t, s)$ can be uniquely extended to the whole space X. We recall that, by definition,

$$\frac{\partial}{\partial t} U(t, s)x_0 = A(t)U(t, s)x_0$$

for all $x_0 \in D_{A(s)}$.

In this way we have shown how the existence of a solution of equation (8.6.1) induces an evolution operator. Unfortunately, such a general formulation of the conditions ensuring the existence of a solution of equation (8.6.1) is much more difficult than in the case which we considered before. The reader interested in the topic can find those conditions in S. G. Krein (1967).

Now we shall give examples of evolution operators on the basis of our previous considerations.

8.6. Nonautonomous differential equations in Banach spaces

EXAMPLE 8.6.1. Let $A(t)$, $a \leqslant t \leqslant b$, be a family of linear continuous operators. Let $U(t)$ be the solution of an equation

$$\frac{dU}{dt} = A(t)U(t), \quad a \leqslant t \leqslant b, \tag{8.2.24}$$

corresponding to an initial value

$$U(a) = I. \tag{8.2.25}$$

The operators $U(t, s) = U(t)U^{-1}(s)$ form an evolution operator.

EXAMPLE 8.6.2. Let $U(t)$ be a semigroup of linear operators of the class C_0 acting in a Banach space X. Then $U(t, s) = U(t-s)$, $0 \leqslant s \leqslant t < +\infty$, is an evolution operator.

In the same way as in the previous sections, having an evolution operator corresponding to equation (8.6.1), we can find a solution of a nonhomogeneous linear equation

$$\frac{dx}{dt} = A(t)x + f(t), \quad a \leqslant s \leqslant t \leqslant b \tag{8.6.6}$$

corresponding to an initial condition

$$x(s) = x_0. \tag{8.6.7}$$

Namely, we have

PROPOSITION 8.6.1. *Let $U(t, s)$ be an evolution operator corresponding to equation (8.6.1). Let $x_0 \in D_{A(s)}$. If $f(t)$ is a continuous function such that $f(t) \in D_{A(t)}$ for all $a \leqslant s \leqslant t \leqslant b$, then*

$$x(t) = U(t, s)x_0 + \int_s^t U(t, u)f(u)du, \quad a \leqslant s \leqslant t \leqslant b, \tag{8.6.8}$$

is the solution of equation (8.6.6) corresponding to the initial condition (8.6.7).

Proof. By the definition of the solution of equation (8.6.1), $U(t, s)D_{A(s)} \subset D_{A(t)}$. Thus, differentiating (8.6.8), we have

$$\frac{dx}{dt} = \frac{\partial}{\partial t}U(t, s)x_0 + f(t) + \int_s^t \frac{\partial}{\partial t}U(t, u)f(u)du$$

$$= A(t)U(t,s)x_0 + f(t) + A(t)\left(\int_s^t U(t,u)f(u)\,du\right)$$

$$= A(t)\left(U(t,s)x_0 + \int_s^t U(t,u)f(u)\,du\right) + f(t) = A(t)x(t) + f(t). \blacksquare$$

The conditions of the continuity of f and the conditions that $x_0 \in D_{A(s)}$ and $f(t) \in D_{A(t)}$ are very restrictive. Thus we define as before the *mild solution* of equation (8.6.6) corresponding to the initial condition (8.6.7) by formula (8.6.8).

8.7. Controllability and approximative controllability of linear systems described by differential equations in Banach spaces

Let $(E, \|\,\|_E)$, $(U, \|\,\|_U)$ be two Banach spaces over real numbers. We consider a linear system described by a linear differential equation

$$\frac{dx}{dt} = A(t)x + S(t)u, \quad a \leqslant t \leqslant b < +\infty, \tag{8.7.1}$$

where $x(t) \in E$, $u(t) \in U$ for all t, with an initial condition

$$x(a) = x_0. \tag{8.7.2}$$

We assume that $A(t)$ is a family of closed operators satisfying the conditions which ensure the existence of an evolution operator $\{U(t,s)\}$, $a \leqslant s \leqslant t \leqslant b$, corresponding to the homogeneous differential equation

$$\frac{dx}{dt} = A(t)x. \tag{8.6.1}$$

About $u(t)$ and $S(t)$ we shall assume that they are measurable functions with values in U and, respectively, in $B(U \to E)$ and such that $S(t)u(t)$ is an integrable function with values in E. Those assumptions are satisfied for example if

(1) $u(\cdot) \in L_U^\infty[a,b]$, $S(\cdot) \in L_{B(U \to E)}^1[a,b]$,

(2) $u(\cdot) \in L_U^2[a,b]$, $S(\cdot) \in L_{B(U \to E)}^2[a,b]$,

(3) $u(\cdot) \in L_U^1[a,b]$, $S(\cdot) \in L_{B(U \to E)}^\infty[a,b]$.

8.7. Controllability

Observe that the above assumptions ensure the existence of the mild solution

$$x(t) = U(t, a)x_0 + \int_a^t U(t, s)S(s)u(s)ds. \tag{8.7.3}$$

Functions $u(\cdot)$ will be called *control*, the space of the functions $u(\cdot)$ will be called the *space of controls* and denoted by $\hat{U}$; thus $\hat{U}$ is equal to $L_U^\infty[a, b]$ in the first case, to $L_U^2[a, b]$ in the second case and to $L_U^1[a, b]$ in the third case.

Equation (8.7.1) with the initial condition (8.7.2) determines a linear system in the following way. As the space of inputs X we shall take the product of the space of initial data by the space of controls, $X = E \times \hat{U}$. As the space of trajectories we shall take the space $C_E[a, b]$, as the space of outputs we shall take the space E. The input operator A is determined by formula (8.7.3). The output operator B is of the form $Bx = B_T x = x(T)$, where T is a fixed point of the interval $[a, b]$. Taking the fixed initial condition $x(a) = 0$ we obtain a reduced linear system with the space of input $\hat{U}$, with the same space of trajectories and outputs, with the same output operator and with the input operator A_0 given by the formula

$$A_0 u = \int_a^t U(t, s)S(s)u(s)ds. \tag{8.7.4}$$

We say that the system described by the differential equation (8.7.1) is *controllable from 0 at the time* T if, for an arbitrary $x_1 \in E$, we can find a control $u(\cdot) \in \hat{U}$ such that the mild solution $x(\cdot)$ corresponding to the control $u(\cdot)$ and to the initial condition $x(a) = 0$ satisfies the final condition

$$x(T) = x_1. \tag{8.7.5}$$

In other words the system is controllable if and only if $BA_0 \hat{U} = E$.

Suppose we are given a system controllable from 0 at the time T. Then we can find a control $u(\cdot) \in \hat{U}$ such that

$$BA_0 u = x_1 - BA_0 x_0. \tag{8.7.6}$$

It is easy to observe that the solution $x(\cdot)$ corresponding to $u(\cdot)$ satisfies the following initial and final conditions

$$x(a) = x_0, \quad x(T) = x_1. \tag{8.7.7}$$

Since x_0 and x_1 have been arbitrary elements of E, we have shown

PROPOSITION 8.7.1. *Suppose we are given a system controllable from* 0. *Then, for arbitrary* $x_0, x_1 \in E$, *there is a control* $u(\cdot)$ *such that the mild solution corresponding to the control* $u(\cdot)$ *and to the initial condition* $x(a) = x_0$ *satisfies the final condition* $x(T) = x_1$.

Systems with the property described in Proposition 8.7.1 will be called *controllable at the time T*.

We say that the system described by equation (8.7.1) is *approximatively controllable from* 0 *at the time T* if, for arbitrary $x_1 \in E$ and $\varepsilon > 0$, there is a control $u(\cdot)$ such that the mild solution $x(\cdot)$ of equation (8.7.1) corresponding to the control $u(\cdot)$ and to the initial condition $x(a) = 0$, satisfies the final condition $\|x_1 - x(T)\|_E < \varepsilon$, in other words

$$\overline{BA_0(\hat{U})} = E. \tag{8.7.8}$$

As a consequence of the Hahn–Banach theorem (Theorem 2.5.3) we obtain

PROPOSITION 8.7.2. *The system described by equation* (8.7.1.) *is not approximatively controllable from* 0 *at the time T if and only if there is a continuous linear functional* $f \neq 0$, $f \in E^*$, *such that*

$$f(BA_0 u) = 0 \quad \text{for all} \quad u \in \hat{U}, \tag{8.7.9}$$

in other words

$$A_0^* B^* f = 0. \tag{8.7.10}$$

We say that the system described by equation (8.7.1) is *approximatively controllable at the time T* if for arbitrary $x_0, x_1 \in E$, $\varepsilon > 0$, there is a control $u(\cdot)$ such that the mild solution $x(\cdot)$ corresponding to the control $u(\cdot)$ and to the initial condition $x(a) = x_0$ satisfies the final condition

$$\|x(T) - x_1\|_E < \varepsilon. \tag{8.7.11}$$

Reasoning as in the case of Proposition 8.7.1, we can obtain

PROPOSITION 8.7.3. *If the system described by equation* (8.7.1) *is approximatively controllable from* 0 *at the time T, then it is approximatively controllable at the time T.*

THEOREM 8.7.4. *Let the family $A(t)$ consist of continuous operators. Let $a \leqslant T \leqslant T_1 \leqslant b$. Then*

(a) *if the system described by equation* (8.7.1) *is controllable at the time T, then it is also controllable at the time T_1;*

(b) *if the system described by equation* (8.7.1) *is approximatively controllable at the time T, then it is also approximatively controllable at the time T_1.*

Proof. The proof is based on the fact that, if $A(t)$ are continuous operators then the corresponding evolution operator $U(t, s)$ is of the form $U(t, s) = U(t)U^{-1}(s)$, where $U(t)$ is a continuous function with values in the space $C_{B(E)}[a, b]$, and is such that $U^{-1}(t)$ exists for all t and is a continuous linear operator (see Section 8.2).

Now we shall prove (a). Let x_0, x_1 be two arbitrary elements of E. Since the system is controllable at the time T, there is a control u such that the corresponding mild solution satisfies the following initial and final conditions

$$x(a) = x_0, \quad x(T) = U(T)U^{-1}(T_1)x_1. \qquad (8.7.12)$$

Now we shall define a new control

$$\hat{u}(t) = \begin{cases} u(t) & \text{for} \quad a \leqslant t \leqslant T, \\ 0 & \text{for} \quad T < t \leqslant T_1. \end{cases} \qquad (8.7.13)$$

Thus the mild solution corresponding to $\hat{u}$ and the initial condition $x(a) = x_0$ satisfies the following equality

$$x(T_1) = U(T_1)U^{-1}(a)x_0 + \int_a^{T_1} U(T_1)U^{-1}(s)S(s)\hat{u}(s)\,ds$$

$$= U(T_1)U^{-1}(T)\Big(U(T)U^{-1}(a)x_0 + \int_a^T U(T)U^{-1}(s)S(s)u(s)\,ds\Big).$$

By the choice of u, the term in parenthesis is equal to $U(T)U^{-1}(T_1)x_1$, hence the right-hand side of the equality is equal to x_1.

(b) The proof is quite similar to the proof of (a). We take an arbitrary $\varepsilon > 0$. Since the system is approximatively controllable at the time T, there are $y \in E$ and a control u such that

$$\|y - U(t)U^{-1}(T_1)x_1\|_E < \frac{\varepsilon}{\|U(T_1)\|_{B(E)}\|U^{-1}(T)\|_{B(E)}}$$

and the mild solution $x(t)$ of equation (8.7.1) corresponding to the control u and the initial condition $x(a) = x_0$ satisfies the equality

$$x(T) = y.$$

Determining $\hat{u}$ as in (8.7.13), we find that the mild solution $x(t)$ of equation (8.7.1) corresponding to the control $\hat{u}$ and to the initial condition $x(a) = x_0$ satisfies the equality

$$x(T_1) = U(T_1)U^{-1}(T)y.$$

Thus

$$\|x_1 - x(T_1)\|_E$$
$$\leqslant \|U(T_1)U^{-1}(T)U(T)U^{-1}(T_1)x_1 - U(T_1)U^{-1}(T)y\|_E$$
$$\leqslant \|U(T_1)\|_{B(E)}\|U^{-1}(T)\|_{B(E)}\|U(T)U^{-1}(T_1)x_1 - y\|_E < \varepsilon.$$

Without the continuity of $A(t)$ Theorem 8.7.4 does not always hold, as can be seen from the following examples.

EXAMPLE 8.7.1. Let $U = E = C$. Let the space of controls $\hat{U}$ be equal to $L_U^\infty[a, b]$. We consider the equation

$$\frac{dx}{dt} = A(t)x + u, \quad -1 \leqslant t \leqslant 1, \tag{8.7.14}$$

where

$$A(t) = \begin{cases} 0 & \text{for } -1 \leqslant t \leqslant 0, \\ Ax & \text{for } 0 \leqslant t \leqslant 1, \end{cases}$$

and A is the following closed noncontinuous operator

$$A(\{x_1, \ldots, x_n, \ldots\}) = \{-x_1, -2x_2, \ldots, -nx_n, \ldots\}.$$

8.7. Controllability

Observe that for $t > 0$ the operator A generates the semigroup $U(t)$ of linear operators of class C_0,
$$U(t)(\{x_1, \ldots, x_n, \ldots\}) = \{e^{-t}x_1, \ldots, e^{-nt}x_n, \ldots\}.$$

The evolution operator $\{U(t, s)\}$ corresponding to equation (8.7.14) is the following:
$$U(t, s) = \begin{cases} I & \text{for} \quad -1 \leq s \leq t \leq 0, \\ U(t) & \text{for} \quad -1 \leq s \leq 0 \leq t \leq 1, \\ U(t-s) & \text{for} \quad 0 \leq s \leq t \leq 1. \end{cases}$$

The system described by equation (8.7.14) is controllable at the time T for $T < 0$. Indeed, the control $u_0(t) = \dfrac{x_1 - x_0}{T+1}$ has the property that the mild solution $x(t)$ corresponding to u_0 and to the initial condition $x(-1) = x_0$ satisfies the final condition $x(T) = x_1$.

For $T > 0$, the system described by equation (8.7.14) is not controllable from 0, since for $t > 0$ the operator $U(t)$ is compact and the operator BA_0,

$$BA_0 u = \int_0^T U(T, s)u(s)\,ds$$
$$= \int_{-1}^0 U(T, s)u(s)\,ds + \int_0^T U(T, s)u(s)\,ds$$
$$= U(T)\left(\int_{-1}^0 u(s)\,ds\right) + \int_0^T U(T-s)u(s)\,ds$$

is compact as a sum of two compact operators. Then the range of BA_0 cannot be the whole space E, and the system is not controllable at the time T.

Observe that a similar effect is obtained by taking as the space of control $\hat{U}$ the space $L_U^1[a, b]$ or the space $L_U^2[a, b]$.

The construction of a similar example for approximative controllability is more complicated. The example presented below will be based on the so-called *pointwise degeneracy* of differential-difference systems (see Weiss, 1967; Popov, 1972; Zverkin, 1973).

We consider a differential-difference equation

$$\frac{dx}{dt} = Ax(t) + A_1 x(t-h), \quad 0 \leq t \leq T, \tag{8.7.15}$$

with an initial condition

$$x(t) = x_0(t), \quad -h \leq t \leq 0, \tag{8.7.16}$$

where $x(t)$ admits values in an n-dimensional real space R^n.

We say that equation (8.7.15) is *pointwise degenerated*, if there are a vector $r \in R^n$ and a time T_0 such that the value of every solution of equation (8.7.15) independent of the initial condition (8.7.16) is orthogonal to r for $t \geq T_0$,

$$(r, x(t)) = 0 \quad \text{for} \quad t \geq T_0. \tag{8.7.17}$$

The first example of this type was given by Weiss (1967). However, we shall present here the example of Popov (1972), since it is the first example with constant coefficients.

EXAMPLE 8.7.2 (Popov, 1972). We consider differential-difference equation in R^3

$$\frac{d}{dt} x(t) = 2y(t), \tag{8.7.18.i}$$

$$\frac{d}{dt} y(t) = -z(t) + x(t-1), \tag{8.7.18.ii}$$

$$\frac{d}{dt} z(t) = 2y(t-1), \tag{8.7.18.iii}$$

in the interval $[0, T]$ with an initial condition

$$(x(t), y(t), z(t)) = (x_0(t), y_0(t), z_0(t)), \quad -1 \leq t \leq 0. \tag{8.7.19}$$

Let $t \geq 1$. Replacing in (8.7.18.i) t by $(t-1)$ and then subtracting (8.7.18.iii) from (8.7.18.i), we get

$$\frac{d}{dt} x(t-1) - \frac{d}{dt} z(t) = 0. \tag{8.7.20}$$

8.7. Controllability

Therefore $x(t-1) - z(t)$ is a constant. We shall denote this constant by $-c_0$. By (8.7.18.ii)

$$y(t) = c_0 t + c_1$$

and by (8.7.18.i)

$$x(t) = c_0 t^2 + 2c_1 t + c_2.$$

Having $x(t)$ and $y(t)$ by (8.7.18.ii), we have

$$z(t) = -c_0 + c_0(t-1)^2 + 2c_1(t-1) + c_2.$$

Therefore we can obtain by simple calculation

$$x(t) - 2y(t) - z(t) = 0 \quad \text{for} \quad t \geq 1.$$

Thus, taking $T_0 = 1$ and $r = (1, -2, -1)$, we find that system (8.7.18) is pointwise degenerated.

Having Example 8.7.2, we can easily construct an example showing that the approximative controllability in a shorter interval does not imply the approximative controllability in a longer one.

EXAMPLE 8.7.3. Let $\hat{U} = C_3[0, 1]$ be the space of three-dimensional continuous vector-valued functions.

We consider a system described by the following differential-difference equation

$$\frac{dx}{dt} = \begin{cases} y(t) + 2u_1(t) + u_2(t) & \text{for } 0 \leq t \leq 1, \\ 2y(t) & \text{for } 1 < t \leq T, \end{cases}$$

$$\frac{dy}{dt} = \begin{cases} x(t) + 2u_1(t) - u_3(t) & \text{for } 0 \leq t \leq 1, \\ -z(t) + x(t-1) & \text{for } 1 < t \leq T, \end{cases}$$

$$\frac{dz}{dt} = \begin{cases} z(t) + u_1(t) - u_3(t) & \text{for } 0 \leq t \leq 1, \\ 2y(t-1) & \text{for } 1 < t \leq T. \end{cases}$$

As the space of trajectories we consider the space $C_3[0, T]$. As the space of outputs we consider the space R^3. The output operator is given by $B(x(\cdot), y(\cdot), z(\cdot)) = (x(T), y(T), z(T))$.

In the interval [0, 1], the system is described by ordinary differential equations. And by the criterion of Kalman (Theorem 6.2.1) it is controllable for all T, $0 < T \leqslant 1$.

For $T \geqslant 2$ the system is not approximatively controllable by Example 8.7.2 and Proposition 8.7.2.

In the case where a system is autonomous, i.e., where A and S do not depend on time, we have

PROPOSITION 8.7.5. *Let a system be described by an autonomous differential equation in a Banach space*

$$\frac{dx}{dt} = Ax + Su, \quad 0 \leqslant t < +\infty, \tag{8.7.21}$$

where A is an infinitesimal generator of a semigroup of linear operators $U(t)$ of class C_0.

(a) *If the system described by equation (8.7.21) is controllable at the time T, then it is controllable at the time T_1 for all $T_1 \geqslant T$.*

(b) *If the system described by equation (8.7.21.) is approximatively controllable at the time T, then it is also approximatively controllable at the time T_1 for every $T_1 \geqslant T$.*

Proof. (a) Let x_0 and x_1 be arbitrary elements of E. By our assumption there is a control $u(\cdot)$ such that the mild solution $x(t)$ corresponding to the control $u(\cdot)$ and to the initial condition

$$x(0) = U(T_1 - T)x_0$$

satisfies the final condition

$$x(T) = x_1.$$

Let

$$\hat{u}(t) = \begin{cases} 0 & \text{for } 0 \leqslant t \leqslant T_1 - T, \\ u(t - T_1 + T) & \text{for } T_1 - T < t \leqslant T_1. \end{cases}$$

Then the mild solution $\hat{x}(t)$ corresponding to the control $\hat{u}$ and the initial condition

$$\hat{x}(0) = x_0$$

satisfies

$$\hat{x}(T_1) = U(T_1)x_0 + \int_0^{T_1} U(T_1-s)S\hat{u}(s)\,ds$$

$$= U(T)(U(T_1-T)x_0) + \int_{T_1-T}^{T_1} U(T_1-s)S\hat{u}(s)\,ds$$

$$= U(T)(U(T_1-T)x_0) + \int_0^{T} U(T_1-s-T_1+T)S\hat{u}(s+T_1-T)\,ds$$

$$= U(T)(U(T_1-T)x_0) + \int_0^{T} U(T-s)Su(s)\,ds = x_1$$

by the choice of the control u.

(b) The proof follows the same line, the only difference being that for each $\varepsilon > 0$ we choose a control u in such a way that the mild solution $x(t)$ corresponding to the control u and the initial condition

$$x(0) = U(T_1-T)x_0$$

satisfies the final condition

$$\|x(T)-x_1\|_E < \varepsilon. \quad \blacksquare$$

We say that a system described by the differential equation (8.7.1) is *null-controllable* (or *controllable to* 0) *at the time* T if for arbitrary $x_0 \in E$ there is a control u such that the mild solution corresponding to the control u and to the initial condition

$$x(a) = x_0$$

satisfies the final condition

$$x(T) = 0.$$

PROPOSITION 8.7.6. *If a linear system described by equation* (8.7.1) *is null-controllable at the time* T, *then it is null-controllable at the time* T_1 *for each* $T_1 > T$.

Proof. Let x_0 be an arbitrary element of E. Let $u(\cdot)$ denote a control such that the mild solution corresponding to the control u and to the initial condition

$$x(a) = x_0$$

satisfies the final condition

$$x(T) = 0.$$

We put

$$\hat{u}(t) = \begin{cases} u(t) & \text{for} \quad a \leqslant t \leqslant T, \\ 0 & \text{for} \quad T < t \leqslant T_1. \end{cases}$$

It is easy to verify that the mild solution corresponding to the control $\hat{u}$ and to the initial condition $x(a) = x_0$ satisfies the required final condition $x(T_1) = 0$. ∎

As a consequence of Example 8.7.1 and Proposition 8.7.6 we find that there are null-controllable systems which are not controllable.

We shall say that the system described by equation (8.7.1) is *approximatively null-controllable* (or *approximatively controllable to* 0) *at the time T* if for any initial state $x_0 \in E$ and arbitrary $\varepsilon > 0$ there is a control u such that the mild solution $x(t)$ corresponding to the control u and to the initial condition $x(a) = x_0$ satisfies the final condition

$$\|x(T)\| < \varepsilon.$$

PROPOSITION 8.7.7. *If a linear system described by equation (8.7.1) is approximatively null-controllable at the time T, then it is also approximatively null-controllable at the time T_1 for each $T_1 > T$.*

Proof. Let x_0 be an arbitrary element of E and let ε be an arbitrary positive number. Let $u(\cdot)$ be a control such that the mild solution corresponding to the control u and to the initial condition $x(a) = x_0$ satisfies the final condition

$$\|x(T)\|_E < \frac{\varepsilon}{\|U(T_1, T)\|_{B(E)}}.$$

Let
$$\hat{u}(t) = \begin{cases} u(t) & \text{for } a \leqslant t \leqslant T, \\ 0 & \text{for } T < t \leqslant T_1. \end{cases}$$

Then the mild solution $\hat{x}(t)$ corresponding to the control $\hat{u}(\cdot)$ and to the initial condition $\hat{x}(a) = x_0$ satisfies the final condition

$$\|\hat{x}(T_1)\|_E < \varepsilon.$$

Example 8.7.3 and Proposition 8.7.7 imply that there are approximatively null-controllable systems which are not approximatively controllable.

8.8. LONG-TERM CONTROLLABILITY AND LONG-TERM APPROXIMATIVE CONTROLLABILITY

Let a linear system be described by the differential equation in a Banach space $(E, \|\ \|_E)$,

$$\frac{dx}{dt} = A(t)x + S(t)u, \quad a \leqslant t \leqslant +\infty, \tag{8.7.1}$$

be given. We say that the system is *long-term controllable* if for arbitrary $x_0, x_1 \in E$ there is a time T and a control u such that the mild solution $x(t)$ of equation (8.7.1) corresponding to the control u and to the initial condition $x(a) = x_0$ satisfies the final condition

$$x(T) = x_1.$$

By Corollary 5.4.19 we trivially obtain

THEOREM 8.8.1. *Suppose that a system described by equation* (8.7.1) *is long-term controllable. Suppose that the ranges of operators* $B_T A_0$, *where*

$$B_T A_0 u = \int_a^T U(T, s) S(s) u(s) \, ds, \tag{8.8.1}$$

forms an increasing family of sets. Then there is a universal time T_u *such that the system is controllable at the time* T_u.

By considerations similar to those used in the proofs of Theorems 8.7.4 and 8.7.5 and by Theorem 8.8.1 we obtain

COROLLARY 8.8.2. *Let either the family $A(t)$ consist of bounded operators or the system be autonomous. If the system is long-term controllable, then there is a universal time T_u such that the system is controllable at the time T_u.*

We say that the system described by equation (8.7.1) is *long-term approximatively controllable* if for arbitrary $x_0, x_1 \in E$ there is a time T such that for arbitrary $\varepsilon > 0$, there is a control u such that the mild solution $x(t)$ of equation (8.7.1) corresponding to the control u and to the initial condition

$$x(a) = x_0$$

satisfies the final condition

$$\|x_1 - x(T)\|_E < \varepsilon.$$

We shall say that the system described by equation (8.7.1) is *long-term controllable from* 0 if for arbitrary $x_1 \in E$ there are a time T and a control u, such that the mild solution of equation (8.7.1) corresponding to the control u and to the initial condition

$$x(a) = 0$$

satisfies the final condition

$$x(T) = x_1.$$

We shall say that the system described by equation (8.7.1) is *long-term approximatively controllable from* 0 if, for an arbitrary $x_1 \in E$, there is a time T such that for arbitrary $\varepsilon > 0$ there is a control u such that the mild solution of equation (8.7.1) corresponding to the control u and to the initial condition

$$x(a) = 0$$

satisfies the final condition

$$\|x_1 - x(T)\|_E < \varepsilon.$$

In the same way as in Proposition 8.7.1 and 8.7.3 we can show

PROPOSITION 8.8.3. *If the system described by equation (8.7.1) is long-term controllable (long-term approximatively controllable) from 0, then it is long-term controllable (respectively, long-term approximatively controllable).*

By Theorem 5.4.22 we trivially obtain

THEOREM 8.8.4 (Dolecki, 1974). *Let the system described by equation (8.7.1) be long-term approximatively controllable. If the ranges of operators $B_T A_0$ (see (8.8.1) form a nondecreasing family of sets, then there is a universal time T_u, such that the system is approximatively controllable at the time T_u.*

By Theorem 8.7.4 and Proposition 8.7.5 and by Theorem 8.8.4 we trivially obtain

COROLLARY 8.8.5 (Dolecki, 1974). *Let the system described by equation (8.7.1) be long-term approximatively controllable. If either $A(t)$ are continuous or the system is autonomous, then there is a universal time T_u, such that the system is approximatively controllable at the time T_u.*

In the same way as in Propositions 8.7.1 and 8.7.3 we can show that if a system is long-term controllable from 0 (long-term approximatively controllable from 0), then it is long-term controllable (long-term approximatively controllable).

In the notion of long-term approximative controllability the time T depends on x_0, x_1 but it is not dependent on ε. We can introduce a weaker notion. We shall say that the system described by differential equation (8.7.1) is *weakly long-term approximatively controllable (weakly long-term approximatively controllable from* 0) if for arbitrary $x_0, x_1 \in E$ and $\varepsilon > 0$ (respectively, $x_1 \in E$, $\varepsilon > 0$) there are a time T and a control u such that the mild solution $x(t)$ of equation (8.7.1) corresponding to the control u and the initial condition $x(a) = x_0$ (resp. $x(a) = 0$) satisfies the final condition

$$||x_1 - x(T)||_E < \varepsilon.$$

There are systems which are weakly long-term approximatively controllable and which are not long-term approximatively controllable. This follows from

EXAMPLE 8.8.1. Let $E = L^2[0, +\infty)$. Let $A = 0$, i.e., let the corresponding semigroup be equal to $U(t) = I$. Let $U = E$ and let

$$S(t)u|_s = \begin{cases} u(s) & \text{for } 0 \leq s \leq t, \\ 0 & \text{for } t < s < +\infty. \end{cases}$$

For an arbitrary control u, the mild solution $x(t)$ has property that, for a fixed t, $x(t)$ as an element of the space $L^2[0, +\infty)$ has the support in the interval $[0, t]$, Thus, the range of $B_t A_0$ cannot be dense in E for any t, which implies that the system is not long-term approximatively controllable.

Now we shall show that the system is weakly long-term controllable. Let $x_1(\cdot)$ be an arbitrary element belonging to the space $E = L^2[0, +\infty)$. Let ε be an arbitrary positive number. Then there is an $x_2(\cdot)$ with a finite support such that

$$\|x_1 - x_2\|_E < \varepsilon.$$

Let $[0, t_0]$ denote the closed interval containing the support of x_2. Let $T > t_0$. We put

$$u(t)|_s = \begin{cases} 0 & \text{for } 0 \leq t < t_0, \\ \dfrac{x_2(s)}{T - t_0} & \text{for } t_0 \leq t \leq T. \end{cases}$$

Thus

$$S(t)u(t) = \begin{cases} 0 & \text{for } 0 \leq t < t_0, \\ \dfrac{x_2}{T - t_0} & \text{for } t_0 \leq t \leq T \end{cases}$$

and the mild solution $x(t)$ corresponding to the control u and the initial state $x(0) = 0$ satisfies the final condition $x(T) = x_2$, i.e.,

$$\|x_1 - x(T)\| < \varepsilon,$$

and the system is long-term controllable.

Dolecki (1974), constructed an autonomous system which is not long-term approximatively controllable, but which is weakly long-term approximatively controllable.

We say that the system described by equation (8.7.1) is *long-term null-controllable* (or *long-term controllable to* 0) if for each $x_0 \in E$ there are a time T and a control u such that the mild solution $x(t)$ corresponding to the control u and to the initial condition $x(a) = x_0$ satisfies the final condition $x(T) = 0$.

By Theorem 5.4.17 we obtain

THEOREM 8.8.5 (Dolecki, 1974). *If the system described by equation (8.7.1) is long-term null-controllable, then there is a universal time T_u such that the system is null-controllable at the time T_u.*

We say that the system described by equation (8.7.1) is *long-term approximatively null-controllable* (*long-term approximatively controllable to* 0) if for an arbitrary $x_0 \in E$ there is a time T such that for each $\varepsilon > 0$ there is a control u such that the mild solution $x(t)$ corresponding to the control u and to the initial state $x(T) = x_0$ satisfies the final condition

$$\|x(T)\|_E < \varepsilon.$$

As a consequence of Theorem 5.4.21 we obtain

THEOREM 8.8.6 (Dolecki, 1974). *If the system described by equation (8.7.1) is long-term approximatively null-controllable, then there is a universal time T_u, such that the system is approximatively null-controllable at the time T_u.*

8.9. LINEAR SYSTEMS WITH ANALYTIC SEMIGROUPS

Let $(X, \|\ \|)$ be a Banach space over complex numbers. Let D be a domain on the complex plane. Let $x(z)$ be a function defined on D with values in X. We say that the function $x(z)$ is *analytic* if for each $z_0 \in D$ there is an $r > 0$ such that for z, $|z - z_0| < r$, the function $x(z)$ can be represented as a sum of a power series,

$$x(z) = \sum_{i=0}^{\infty} (z-z_0)^i x_i, \tag{8.9.1}$$

where x_i are fixed elements of the space X. As in the theory of analytic functions, we can show that for $|z-z_0| < r$ the series (8.9.1) converges absolutely, i.e.,

$$\sum_{i=0}^{\infty} |z-z_0|^i \|x_i\| < +\infty.$$

Let $(X, \|\ \|)$ be a Banach space over real numbers. Let (a, b) be an open interval. We say that a function $x(t)$, $a < t < b$, is *analytic*, if, for an arbitrary t_0, $a < t_0 < b$, there is an $r > 0$ such that for $|t-t_0| < r$, the function $x(t)$ can be represented in the form of a power series

$$x(t) = \sum_{i=0}^{\infty} (t-t_0)^i x_i, \qquad (8.9.2)$$

where x_i are elements of X.

The theory of analytic functions of real variables with values in Banach spaces is similar to the theory of analytic functions of real variables with values in finite-dimensional spaces. However, we shall not present here the theory of analytic functions. For our purposes the following proposition will be sufficient.

PROPOSITION 8.9.1. *Let $x(t)$ be an analytic function of real variable with values in a Banach space X defined in the interval (a, b). If $x(t) = 0$ in a certain subinterval (a_1, b_1), $a \leqslant a_1 \leqslant b_1 \leqslant b$, then it is equal to 0 in the whole interval (a, b).*

Proof. Let f be an arbitrary linear continuous functional defined on X, $f \in X^*$. Thus, for t, $|t-t_0| < r$, by (8.9.2),

$$f(x(t)) = f\left(\sum_{i=0}^{\infty} (t-t_0)^i x_i\right) = \sum_{i=0}^{\infty} (t-t_0)^i f(x_i)$$

and $f(x(t))$ is a scalar-valued analytic function.

Observe that $f(x(t))$ is equal to zero for $a_1 < t < b_1$, and thus by the theory of analytic functions $f(x(t)) = 0$ for $a < t < b$. The arbitrariness of f implies that $x(t) = 0$, $a < t < b$. ∎

8.9. Linear systems with analytic semigroups

We say that an operator function $A(t)$, $a < t < b$, where $A(t) \in B(X \to Y)$ is *analytic* if it is analytic as a function with values in the space $B(X \to Y)$.

It is easy to prove that, if $A(t)$ is an analytic operator function with values in $B(X \to Y)$, $B(t)$ is an analytic operator function with values in $B(Y \to Z)$, and both functions are defined in the same interval (a, b), then their superposition $B(t)A(t)$ is an analytic operator function with values in the space $B(X \to Z)$ defined in the interval (a, b).

PROPOSITION 8.9.2. *Let $A(t)$, $a < t < b$, be an analytic operator function with values in $B(X \to Y)$. Then the conjugate function $A^*(t)$ is an analytic operator function with values in the space $B(Y^* \to X^*)$.*

Proof. Let t_0 be an arbitrary point of the interval (a, b). Since $A(t)$ is analytic, there is an $r > 0$ such that for all t, $|t-t_0| < r$,

$$A(t) = \sum_{i=0}^{\infty} (t-t_0)^i A_i, \tag{8.9.3}$$

where $A_i \in B(X \to Y)$, and the series (8.9.3) is absolutely convergent.

Since $||A_i|| = ||A_i^*||$,

$$A^*(t) = \sum_{i=0}^{\infty} (t-t_0)^i A_i^* \tag{8.9.4}$$

where the series (8.9.4) is also absolutely convergent. ∎

THEOREM 8.9.3 (Fattorini, 1967). *Let*

$$\frac{dx}{dt} = Ax + Su, \quad 0 < t < +\infty,$$

be an autonomous system.

Suppose that A is an infinitesimal generator of an analytic semigroup $U(t)$ (i.e., $U(t)$ is analytic as an operator function with values in $B(E)$). If the system is approximatively controllable at a time T, then it is also approximatively controllable for any time T_1, $0 < T_1 < T$.

Proof. Suppose that the system is not approximatively controllable at the time T_1. Thus it is also not approximatively controllable from 0 at the time T_1 and by Proposition 8.7.2 there is a linear continuous functional f, $f \neq 0$, $f \in E^*$, such that

$$0 = f\left(\int_0^{T_1} U(T_1-s)Su(s)ds\right) = \int_0^{T_1} f(U(T_1-s)Su(s))ds$$

$$= \int_0^{T_1} (S^*U^*(T_1-s)f)(u(s))ds = 0. \qquad (8.9.5)$$

Since (8.9.5) holds for all continuous functions $u(t)$, we can show, by similar considerations to those relating to in the fundamental lemma of the calculus of variations, that

$$S^*U^*(T_1-s)f = 0 \quad \text{for all} \quad 0 < s < T_1. \qquad (8.9.6)$$

By the analyticity of the semigroup $U(t)$ and by Proposition 8.9.1 we have

$$S^*U^*(s)f = 0 \quad \text{for} \quad 0 < s < +\infty, \qquad (8.9.7)$$

and for all T, $0 < T$ we have

$$f\left(\int_0^T U(T-s)Su(s)\right)ds = \int_0^T (S^*U^*(T-s)f)u(s)ds = 0$$

for all possible controls u. Thus by Proposition 8.7.2 the system is not approximatively controllable at the time T. ∎

REMARK 8.9.4. Observe that the proof proceeds without any changes if S is not a constant operator but $S(s)$ is an analytic operator function with values in the space $B(U \to E)$.

Now we shall give examples of analytic semigroups.

EXAMPLE 8.9.1. Let A be a continuous linear operator. Then

$$e^{tA} = \sum_{i=0}^{\infty} \frac{t^i}{i!} A^i$$

is an analytic semigroups of operators.

8.9. Linear systems with analytic semigroups

EXAMPLE 8.9.2. The semigroup describing the solutions of the one-dimensional heat equation (see Section 7.3) is an analytic semigroup.

PROPOSITION 8.9.5. *Let A, S be continuous linear operators. Then the system is approximatively controllable for certain T if and only if*

$$\overline{\operatorname{lin} \bigcup_{n=0}^{\infty} A^n S(U)} = E. \tag{8.9.8}$$

Proof. By (8.9.7) the system is approximatively controllable at a time T if and only if

$$S^* U^*(s) f = 0 \quad \text{for} \quad 0 < s < T \tag{8.9.9}$$

implies that

$$f = 0.$$

Since A^* is bounded, (8.9.9) means that

$$S^* e^{tA^*} f = 0, \quad 0 < t < T. \tag{8.9.10}$$

The function e^{tA} is analytic on the whole real line. Thus (8.9.10) is equivalent to

$$S^* e^{tA^*} f = 0, \quad -\infty < t < +\infty. \tag{8.9.11}$$

Since e^{tA} is analytic, it holds if and only if all the derivatives at 0 are equal to 0, i.e.,

$$0 = \frac{d^k}{dt^k} S^* e^{tA^*} \bigg|_{t=0} = S^*(A^*)^k f, \quad k = 0, 1, 2, \ldots \tag{8.9.12}$$

Observe that (8.9.12) means simply that

$$f \in \bigcap_{k=0}^{m} \ker S^*(A^*)^k,$$

and we can reformulate the conditions for approximative controllability, saying that the system is approximatively controllable if and only if

$$\bigcap_{k=0}^{\infty} \ker S^*(A^*)^k = 0. \tag{8.9.13}$$

The last step of the proof is to show that (8.9.13) is equivalent to (8.9.8). Indeed, by the definition,

$$f \in \bigcap_{k=0}^{\infty} \ker S^*(A^*)^k$$

if and only if

$$f(A^k Su) = 0, \quad k = 1, 2, \ldots \tag{8.9.14}$$

for arbitrary u.

The fact that (8.9.14) implies $f = 0$ is equivalent to (8.9.8). ∎

Proposition 8.9.5 was proved without the assumption that A is continuous for Hilbert spaces by Fuhrmann (1972). However, his proof required a deeper knowledge of the theory of linear operators in Hilbert spaces.

8.10. OPTIMIZATION PROBLEMS FOR SYSTEMS DESCRIBED BY LINEAR DIFFERENTIAL EQUATIONS IN BANACH SPACES

Let a system be described by a linear differential equation

$$\frac{dx}{dt} = A(t)x + S(t)u, \quad a \leqslant t \leqslant b. \tag{8.7.1}$$

We consider the following initial and final conditions:

$$x(a) = x_0, \quad x(b) = x_1. \tag{8.10.1}$$

Let $F_0(x, u)$ be a continuous convex functional defined on the product $\square \times \hat{U}$ of the space of trajectories $\square$ by the space of controls $\hat{U}$. For example, it can be a functional of integral form

$$F_0(x(\cdot), u(\cdot)) = \int_a^b f(t, x(t), u(t)) dt, \tag{8.10.2}$$

where $f(t, x, u)$ is a convex continuous function defined on the product $[a, b] \times E \times U$. It can also be a norm in the space E or U, and so on. Observe that such a functional $F_0(x, u)$ induces a continuous linear functional on the space of inputs $\hat{U} \times E$ by formula

$$F(u) = F_0(A(u, x_0), u),$$

8.10. Optimization problems

where A is the input operator assigning to the control u and to the initial values x_0, the trajectory x (cf. Section 5.1).

Suppose that $U(t, s)$ is the evolution operator of equation (8.7.1). Then the mild solution corresponding to the control u and to the initial condition $x(a) = x_0$ is of the form

$$x(t) = U(t, a)x_0 + \int_a^t U(t, s)S(s)u(s)\,ds. \tag{8.7.3}$$

Thus the final condition (8.10.1) implies that

$$BA_0(u) = \int_a^b U(b, s)S(s)u(s)\,ds = x_1 - U(b, a)x_0. \tag{8.10.3}$$

If the system described by equation (8.7.1) is controllable, then by Corollary 5.2.7 the maximum principle of Pontryagin holds. If the system is not controllable but is approximatively controllable, then by Theorem 5.2.8 there are points y_0 for which the maximum principle of Pontryagin does not hold.

In sequel we shall assume that the maximum principle of Pontryagin holds at the point $y_0 = x_1 - U(b, a)x_0$. This means that there is a continuous linear functional $\varphi_0 \in E^*$ such that

$$\inf \{F(u)\colon BA_0(u) = y_0\}$$
$$= \inf \{F(u)\colon \varphi_0(BA_0(u)) = \varphi_0(y_0)\}, \tag{8.10.4}$$

where

$$BA_0(u) = \int_a^b U(b, s)S(s)u(s)\,ds.$$

Now we shall calculate $A_0^* B^* \varphi_0$.

$$(A_0^* B^* \varphi_0)(u) = \varphi_0\left(\int_a^b U(b, s)S(s)u(s)\,ds\right)$$
$$= \int_a^b \psi(s)S(s)u(s)\,ds, \tag{8.10.5}$$

where

$$\psi(s) = U^*(b, s)\varphi_0. \tag{8.10.6}$$

8. Differential and integral equations in Banach spaces

If either the operators $A(t)$ are continuous for all t or $A(t) = A$ is a constant operator A being an infinitesimal generator of a semigroup of linear operators $U(t)$ of the class C_0, then $\psi(s)$ is a solution of the conjugate equation

$$\frac{d\psi}{ds} = -A^*(s)\psi(s). \tag{8.10.7}$$

Indeed, in the first case $U(b, s) = U(b)U^{-1}(s)$, where $U(s)$ is a solution of the operator equation

$$\frac{dU(s)}{ds} = A(s)U(s). \tag{8.10.8}$$

Repeating without change the calculations of Section 6.3 (relating there to finite-dimensional spaces), we obtain

$$\frac{d}{ds}U(b, s) = U(b)\frac{d}{ds}U^{-1}(s) = -U(b)U^{-1}(s)\frac{dU}{ds}U^{-1}(s)$$
$$= -U(b)U^{-1}(s)A(s)U(s)U^{-1}(s) = -U(b, s)A(s).$$

Thus

$$\frac{d}{ds}U^*(b, s) = -A^*(s)U^*(b, s), \tag{8.10.9}$$

which implies (8.10.6).

In the second case, by differentiation,

$$\frac{d}{ds}U^*(b-s) = -A^*U^*(b-s) \tag{8.10.10}$$

and (8.10.9) holds.

Formula (8.10.7) permits us to say something more about the form of the optimal solution in the case of minimum norm-problem.

THEOREM 8.10.1. *Let the space of control $\hat{U}$ be of the form $L_U^\infty[a, b]$. Let $F(u)$ be the standard norm in $\hat{U}$,*

$$F(u) = \|u\|_{\hat{U}} = \underset{a \leqslant t \leqslant b}{\mathrm{ess\ sup}}\ \|u(t)\|_U. \tag{8.10.11}$$

8.10. Optimization problems

Suppose that at the point $y_0 = x_1 - U(b, a)x_0$ *the maximum principle of Pontryagin holds. Then the optimal control* $u_0(t)$ *satisfies the following condition: There is* $\psi(s) = U^*(b, s)\varphi_0$ *such that*

$$\psi(s)S(s)u_0(s) = \sup\{\psi(s)S(s)u : \|u\|_U \leq \|u_0\|_U\} \qquad (8.10.12)$$

for almost all s.

Proof. Take $\varphi_0 \in E$ such that the maximum principle of Pontryagin holds. Let

$$r = \inf\{\|u\|_{\hat{U}} : BA_0(u) = y_0\}$$
$$= \inf\{\|u\|_{\hat{U}} : \varphi_0(BA_0 u) = \varphi_0(y_0)\}.$$

Then $A_0^* B^* \varphi_0$ is a supporting functional of the ball of radius r of the space $\hat{U} = L_U^\infty[a, b]$ supporting this ball at the point y_0 (cf. Section 5.2). This trivially implies (8.10.12). ∎

COROLLARY 8.10.2. *If the unit ball in the space U is strictly convex, and all the assumptions of Theorem 8.10.1 hold, then the optimal control $u_0(t)$ is uniquely determined.*

Let A, S be constant continuous linear operators. We say that the unit ball in the space U *is in general position* with respect to equation (8.7.1) if for each vector $h \neq 0$ parallel to a face of the unit ball (i.e., such that there is a u_h, $\|u_h\|_U = 1$ such that

$$\|u_h - h\|_U = \|u_h + h\|_U = 1)$$

we have

$$\overline{\operatorname{lin}\{Sh, ASh, \ldots, AS^n h, \ldots\}} = E. \qquad (8.10.13)$$

COROLLARY 8.10.3. *Suppose that A, S are continuous linear operators and that all the assumptions of Theorem 8.10.1 hold. If the unit ball in U has a countable number of faces and it is in general position with respect to equation (8.7.1), then the optimal control $u_0(t)$ is uniquely determined.*

Proof. Suppose that there are two different optimal controls $u_0(s) \neq u_1(s)$. Thus $u_0(s) - u_1(s) \neq 0$ on a set D of positive measure. On the other hand, by (8.10.12) for a given s, $u_0(s)$ and $u_1(s)$ belong to the same face of the ball of radius, which implies that $h(s) = u_1(s) - u_0(s)$ is parallel to the face A of the unit ball. Since the unit ball has only a countable number of faces, there is a face A_0 such that for all vectors h parallel to A_0

$$\varphi_0(Be^{sS}h) = 0$$

for a certain $\varphi_0 \neq 0$, and on a certain set D_0 of positive measure. By a similar argument as in the proof of Proposition 8.9.5 we can show that (8.10.13) implies that either $\varphi_0 = 0$ or $h = 0$. Thus we obtain a contradiction. ∎

8.11. Duality between observability and controllability of systems governed by autonomous linear differential equations in Banach spaces

Let E be a reflexive Banach space. Let $U(t)$ be a semigroup of linear operators of class C_0. By D we shall denote the infinitesimal generator of the semigroup $U(t)$. Let $\square = L_E^p[0, T]$, $1 < p < +\infty$. The semigroup $U(t)$ determines the operator A mapping E into $\square$ by the formula

$$Ax_0 = U(\cdot)x_0.$$

Let V be another reflexive Banach space and let B_0 be a linear operator mapping E into V. We do not assume that B_0 is continuous, but we assume that the domain D_0 of the operator B_0 is dense in the space E and it is invariant under the semigroup $U(t)$, $U(t)D_0 \subset D_0$. The operator B_0 induces an output operator B defined on the set

$$D_B = \{u(\cdot) \in L_U^p[0, T]: u(t) \in D_0 \text{ almost everywhere}\}$$

in the following way

$$Bu = B_0 u(t). \tag{8.11.1}$$

Observe that the superposition $C = BA$ maps D_0 into Y and we obtain a system

$$E = X \supset D_C \xrightarrow{C} Y = L_V^p[0, T] \tag{5.8.1}$$

where the domain D_C of the operator C is equal to D_0.

8.11. Duality

It may occur that the operator C can be extended in a continuous way onto the whole space E, even if B cannot be extended in a continuous way onto the space of trajectories $\Box$.

Now we shall observe that the conjugate operator C^* mapping $Y^* = L^q_{V^*}[0, T]$, $q = \dfrac{p}{p-1}$, into E is of the form

$$(C^*h)(x_0) = \left(\int_0^T U^*(t) B_0^* h(t)\,dt\right)(x_0)$$

$$= \left(\int_0^T U^*(T-t) B_0^* f(t)\,dt\right)(x_0), \qquad (8.11.2)$$

where

$$f(t) = h(T-t). \qquad (8.11.3)$$

The domain D_{C^*} of the conjugate operator C^* consists of such $h(\cdot) \in L^q_{V^*}[0, T]$ that

$$\left(\int_0^T U^*(t) B_0^* h(t)\,dt\right)(x)$$

defines a continuous linear functional on E.

In this framework C^* may be regarded as an operator assigning to the control f and the initial state $x(0) = 0$ the final state

$$y_f(T) = \int_0^T U^*(T-s) B_0^* f(s)\,ds$$

of the control system

$$\frac{dy}{dt} = -D^* y + B_0^* f, \quad 0 \leq t \leq T. \qquad (8.11.4)$$

In this way we obtain a duality between the problem of the controllability of the system (8.11.4) and the observability of the initial state x_0 of the system

$$\frac{dx}{dt} = Dx,$$
$$y = B_0 x. \qquad (8.11.5)$$

As a consequence of the Dolecki, Russell theorems (Theorems 5.8.4 and 5.8.5) we find that the operator C induced by the system (8.11.5) is estimating if and only if the system described by equation (8.11.4) is controllable.

We say that the system described by equation (8.11.5) admits an *initial state observation* if the operator C induced by the system is estimating.

THEOREM 8.11.1 (Dolecki and Russell, 1977). *Suppose that B_0 is a continuous linear operator and D is an infinitesimal generator of a semigroup of linear operators $U(t)$ of class C_0. Suppose that the system admits an initial state observation for a certain $T > 0$, i.e., there is a $K > 0$ such that, for all $x_0 \in E$*

$$K \|B_0 U(\cdot) x_0\|_{L^p_E[0, T]} \geq \|x_0\|_E. \tag{8.11.6}$$

Suppose that for all $t > 0$, the range of $U(t)$ is dense in E. Then $U(t)$ can be extended to a group of continuous linear operators $U(t)$, $-\infty < t < +\infty$.

Proof. The semigroup $U(t)$ is of class C_0. Then by the Banach–Steinhaus theorem (Theorem 3.1.1)

$$M(s) = \sup_{0 \leq t \leq s} \|U(t)\|_{B(E)} < +\infty. \tag{8.11.7}$$

Of course, $M(s)$ is a non-decreasing function of s.

At the beginning we shall show that there is an $s_0 > 0$, such that the operator $U(s_0)$ is invertible. Suppose that it is not true. This means that the operators $U(s)$ are not invertible for any s. Thus for all $c > 0$ and $s > 0$ there is an $x_0 = x_0(c, s) \in E$ such that

$$\|U(s) x_0\|_E \leq e^{-cs} \|x_0\|_E. \tag{8.11.8}$$

By (8.11.8), for $t, s \leq t \leq T$,

$$\|B_0 U(t) x_0\|_V = \|B_0 U(t-s) U(s) x_0\|_V$$
$$\leq \|B_0\|_{B(E \to V)} M(T) e^{-cs} \|x_0\|_E, \tag{8.11.9}$$

and for $0 \leq t \leq s$

$$\|B_0 U(t) x_0\|_V \leq \|B_0\|_{B(E \to V)} M(s) \|x_0\|_E. \tag{8.11.10}$$

8.11. Duality

Hence

$$\left(\|B_0 U(\cdot)x_0\|_{L^p_V[0,T]}\right)^p = \int_0^T \|B_0 U(t)x_0\|_V^p dt$$

$$= \int_0^s \|B_0 U(t)x_0\|_V^p dt + \int_s^T \|B_0 U(t)x_0\|_V^p dt \qquad (8.11.11)$$

$$= \left(\|B_0\|^p_{B(E\to V)} s(M(s))^p + \|B_0\|^p_{B(E\to V)} T(M(T))^p T \cdot e^{-pcs}\right)\|x_0\|_E.$$

Taking $c \to \infty$ and $s = c^{-1/2}$, we observe that the expression in parenthesis tends to 0, which leads to a contradiction with formula (8.11.6).

Thus there is an $s_0 > 0$ such that the operator $U(s_0)$ is invertible. This means that there is a $c > 0$ such that

$$\|U(s_0)x\|_E \geq e^{-cs_0}\|x\|_E \qquad (8.11.12)$$

for all $x \in E$.

Let t be an arbitrary positive number. We shall represent t in the form

$$t = ks_0 + s,$$

where k is a positive integer and s, $0 < s < s_0$.

Since

$$\|U(s_0)x\|_E = \|U(s_0-s)U(s)x\|_E$$
$$\leq M(s_0)\|U(s)\|_E, \qquad (8.11.13)$$

we have

$$\|U(t)x\|_E = \|U^k(s_0)U(s)x\|_E$$

$$\geq e^{-kcs_0}\|U(s)x\|_E \geq \frac{e^{-kcs_0}}{M(s_0)}\|U(s_0)x\|_E$$

$$\geq \frac{e^{-(k+1)cs_0}}{M(s_0)}\|x\|_E \geq \frac{e^{-cs_0}}{M(s_0)} e^{-ct}\|x\|_E, \qquad (8.11.14)$$

and $U(t)$ is invertible for all $t > 0$.

We have assumed that the range of $U(t)$ is dense in E. Thus $U^{-1}(t)$ has a unique extension onto the whole space E.

We define

$$U(-t) = U^{-1}(t), \quad 0 < t, \qquad (8.11.15)$$

and it is easy to verify that $U(t)$, $-\infty < t < +\infty$, obtained in this way form a group. We shall show that $U(t)$ is strongly continuous. Let t be an arbitrary negative number. Let $T > |t|$. Then

$$U(t+h)x - U(t)x = U(t+T+h)U(-T)x - U(t+T)U(-T)x.$$
(8.11.16)

If h tends to 0, $h > 0$, then the right-hand side of equality (8.11.16) tends to 0, because $U(t)$ is strongly continuous for positive t. This finishes the proof. ∎

Passing to the conjugate systems and using Theorem 5.8.1 we obtain the following

THEOREM 8.11.2. *Suppose we are given the following system*:

$$\frac{dy}{dt} = -D^*y + B_0^*f, \quad 0 \leq t \leq T,$$

$$y(0) = 0,$$
(8.11.17)

where D^* is the infinitesimal generator of a semigroup of linear operators $U^*(t)$ of the class C_0 and the operator B_0^* is continuous. If for a $T > 0$, the system (8.11.17) is controllable, then the semigroup $U^*(t)$ can be extended to a strongly continuous group $U^*(t)$, $-\infty < t < +\infty$.

In many practical problems we are interested in the final observation, i.e., we consider the operator $F = U(T)$ and we ask: Is the operator F estimable? (see Section 5.8). The final state observability problem consists in the determining a continuous linear operator G mapping E into itself and such that $F = GC$ on a domain C, where C is the operator induced by system (8.11.5).

The dual problem consists of finding for each $y_0 \in E$ a solution h of the operator equation

$$F^*y_0 = C^*h$$

(cf. Section 5.8), i.e., such an $f(t) = h(T-t)$ that

$$U^*(T)y_0 = -\int_0^T U^*(T-t)B_0^*f(t)dt,$$

which correspond to a solution of the conjugate system

$$\frac{dy}{dt} = -D^*y + B_0^* f, \quad 0 \leqslant t \leqslant T \tag{8.11.17}$$

with the initial condition $y(0) = y_0$ and the final condition $y(T) = 0$. Thus the problem of final observability of system (8.11.5) is dual to the problem of null-controllability of system (8.11.17).

8.12. STABILITY OF SOLUTIONS OF LINEAR DIFFERENTIAL EQUATIONS IN BANACH SPACES

Let

$$\frac{dx}{dt} = A(t)x, \quad 0 \leqslant t < +\infty, \tag{8.12.1}$$

be a homogeneous linear equation.

We assume that $U(t, s)$, $0 \leqslant s \leqslant t < +\infty$, is the corresponding evolution operator (see Section 8.6).

The problem of stability is the behaviour at infinity of the solutions of equation (8.12.1). In the case where the space is finite-dimensional, the theory of stability is well developed and there are several books devoted to the subject. For infinite-dimensional Banach spaces, there is a fundamental book of Daleckii and Krein (1970), devoted to the problem of stability.

Of course, we are not able to expound this vast theory here. We shall concentrate on a characterization of so-called *uniform exponential equistability*.

We shall say that an evolution operator $U(t, s)$ is *uniformly exponentially equistable* if there are an $M > 0$ and a $\beta > 0$ such that

$$\|U(t, s)\| \leqslant M e^{-\beta(t-s)}. \tag{8.12.2}$$

Let $N(a, u)$ be a continuous function of two variables defined for $a > 0$ and $u \geqslant 0$. Suppose that, for every fixed a, $N(a, u)$ is a nondecreasing function of u. Assume that for all $a > 0$, $N(a, u) = 0$ if and only if $u = 0$.

We say that an evolution operator $\{U(t,s)\}$ is *uniformly N-equistable* if for every $x \in E$, there is an $a(x)$ such that

$$\sup_{0 \le s < +\infty} \int_s^\infty N\big(a(x), \|U(t,s)x\|\big) dt < +\infty. \tag{8.12.3}$$

We say that an evolution operator $\{U(t,s)\}$ is *uniformly bounded* if

$$K_1 = \sup_{0 \le t-s \le 1} \|U(t,s)\| < +\infty. \tag{8.12.4}$$

The main theorem of this section is

THEOREM 8.12.1 (Rolewicz, 1986). *Let an evolution operator $\{U(t,s)\}$ be uniformly bounded. If $\{U(t,s)\}$ is uniformly N-equistable, then it is uniformly exponentially equistable.*

The proof of Theorem 8.12.1 is based on several lemmas and propositions.

LEMMA 8.12.2 (Daleckii and Krein, 1970). *If an evolution operator $\{U(t,s)\}$ is uniformly bounded, then for all $T > 0$*

$$K_T = \sup_{0 \le t-s \le T} \|U(t,s)\| < +\infty. \tag{8.12.5}$$

Proof. Let s be an arbitrary positive number. Then

$$\|U(s+T,s)\| \le \|U(s+T, s+T-1)\| \|U(s+T-1, s+T-2)\| \ldots$$
$$\ldots \|U(s+T-[T], s)\| \le K_1^{[T]+1} < +\infty,$$

where, as usual, $[T]$ denotes the greatest integer not greater than T. ∎

LEMMA 8.12.3 (Daleckii and Krein, 1970). *Let $\{U(t,s)\}$, $0 \le s \le t < +\infty$, be a uniformly bounded evolution operator. If there are a $T > 0$ and a q, $0 < q < 1$, such that for all $x \in E$ and t, $0 < t < +\infty$, there is a $\theta_{x,t}$, $0 < \theta_{x,t} \le T$, such that*

$$\|U(t+\theta_{x,t}, t)x\| \le q\|x\|, \tag{8.12.6}$$

then the evolution operator $\{U(t,s)\}$ is uniformly exponentially equistable.

8.12. Stability

Proof. Let s, t and $x \in E$ be fixed. Let $\hat{t}_1$ be such that
$$\|U(\hat{t}_1, s)\| \leq q\|x\|, \quad s \leq \hat{t}_1 \leq s+T.$$
Such a $\hat{t}_1$ exists by (8.12.6). Then by induction we construct a sequence $\{\hat{t}_i\}$ such that
$$\hat{t}_{i-1} < \hat{t}_i \leq \hat{t}_{i-1}+T$$
and
$$\|U(\hat{t}_i, \hat{t}_{i-1})U(\hat{t}_{i-1}, s)x\| \leq q\|U(\hat{t}_{i-1}, s)x\|. \tag{8.12.7}$$

Now we have two possibilities: either $\lim_{i \to \infty} \hat{t}_i = +\infty$ or $\{\hat{t}_i\}$ is a bounded sequence. In the first case we simply put $t_i = \hat{t}_i$. In the second case the construction of the sequence $\{t_i\}$ is more complicated. Namely, let m denote a fixed positive integer such that
$$q^{m-1}K_{T/2} < 1. \tag{8.12.8}$$
Let i_0 be the smallest positive integer such that
$$\hat{t}_{i_0+m}-\hat{t}_{i_0} < \frac{T}{2}.$$
We put
$$t_i = \hat{t}_i \quad \text{for} \quad i = 1, \ldots, i_0,$$
$$t_{i_0+1} = \hat{t}_{i_0+m}+\frac{T}{2}.$$
Of course
$$\|U(t_{i_0+1}, t_{i_0})U(t_{i_0}, s)x\| \leq K_{T/2}\|U(\hat{t}_{i_0+m}, t_{i_0}), U(t_{i_0}, s)x\|$$
$$\leq q^m K_{T/2}\|U(t_{i_0}, s)x\| \leq q\|U(t_{i_0}, s)x\|. \tag{8.12.9}$$
Repeating this construction, we can find a sequence $\{t_i\}$ such that
$$\|U(t_{i+1}, t_i)U(t_i, s)x\| \leq q\|U(t_i, s)x\|, \tag{8.12.10}$$
$$t_i < t_{i+1} \leq t_i+T, \tag{8.12.11}$$
$$\lim_{i \to \infty} t_i = +\infty. \tag{8.12.12}$$
By (8.12.10) we obtain by induction
$$\|U(t_i, s)x\| \leq q^i\|x\|. \tag{8.12.13}$$

Let the index j be such that
$$t_j < t \leqslant t_j + T.$$
By (8.12.11)
$$j > \frac{t-s}{T},$$
and by (8.12.13)
$$\|U(t,s)x\| \leqslant \|U(t,t_j)\| \|U(t_j,s)x\|$$
$$\leqslant K_T q^j \|x\| \leqslant K_T q^{\frac{t-s}{T}} \|x\| = K_T e^{-(t-s)\frac{1}{T}\log\frac{1}{q}} \|x\|. \blacksquare$$
(8.12.14)

Now we shall reformulate Lemma 8.12.3 in the following way, more useful for our purposes.

LEMMA 8.12.3′ (Daleckii and Krein, 1970). *Let $\{U(t,s)\}$ be a uniformly bounded evolution operator. If the evolution $\{U(t,s)\}$ is not uniformly exponentially equistable, then for all $T > 0$ and all $q < 1$, there are a t and an $x \in E$, such that*
$$\|U(t+s,t)x\| > q\|x\| \quad \text{for all } s, \quad 0 \leqslant s \leqslant T. \quad (8.12.15)$$

PROPOSITION 8.12.4 (Rolewicz, 1986). *Let $N(u)$ be a continuous nondecreasing function such that $N(0) = 0$ and $N(u) > 0$ for $u > 0$. Let $\{U(t,s)\}$ be a uniformly bounded evolution operator.*

If the evolution operator $\{U(t,s)\}$ is not uniformly exponentially equistable, then the set E_0 of such x that
$$\sup_s \int_s^\infty N(\|U(t,s)x\|) dt = +\infty \quad (8.12.16)$$

is dense in the space E.

Proof. Let x_0 be an arbitrary element of E. Let δ_0 be an arbitrary positive number. We want to show that there is an element $\hat{x}$ such that $\|x_0 - \hat{x}\| \leqslant 2\delta_0$ and
$$\sup_s \int_s^\infty N(\|U(t,s)\hat{x}\|) dt = +\infty. \quad (8.12.17)$$

8.12. Stability

If

$$\sup_s \int_s^\infty N(\|U(t,s)x_0\|)dt = +\infty, \qquad (8.12.18)$$

we simply put $\hat{x} = x_0$.

Suppose that (8.12.18) does not hold. We write

$$M_0 = \sup_s \int_s^\infty N(\|U(t,s)x_0\|)dt. \qquad (8.12.19)$$

The definition of M_0 implies that for all s the Lebesgue measure of the set $\{t: \|U(t,s)x_0\| > \varepsilon\}$ is less than $\dfrac{M_0}{N(\varepsilon)}$,

$$|\{t: \|U(t,s)x_0\| > \varepsilon\}| < \frac{M_0}{N(\varepsilon)}, \qquad (8.12.20)$$

where $|F|$ denotes the Lebesgue measure of the set F. Let q be an arbitrary positive number less than 1, $0 < q < 1$. By the continuity of the function N we can find a number $\varepsilon > 0$ such that

$$N(q\delta_0 - \varepsilon) > \tfrac{1}{2}N(q\delta_0). \qquad (8.12.21)$$

Let T_1 be an arbitrary number such that

$$T_1 > \frac{M_0}{N(\varepsilon)} + 2^2 \frac{1}{N(q\delta_0)}. \qquad (8.12.22)$$

By Lemma 8.12.3', there are an element $x \in E$, $\|x\| = \delta_0$ and a t_1 such that

$$\|U(t_1+s, t_1)x\| \geq q\delta_0 \quad \text{for} \quad 0 \leq s \leq T. \qquad (8.12.23)$$

We put $x_1 = x_0 + x$. Since

$$\|U(t_1+s, t_1)x_1\| \geq \|U(t_1+s, t_1)x\| - \|U(t_1+s, t_1)x_0\|, \qquad (8.12.24)$$

by (8.12.22) and (8.12.23) the set of those s, $0 \leq s \leq T_1$, for which

$$\|U(t_1+s, t_1)x_1\| \geq q\delta_0 - \varepsilon \qquad (8.12.25)$$

486   8. Differential and integral equations in Banach spaces

has a measure greater than

$$T_1 - \frac{M_0}{N(\varepsilon)} > 2^2 \frac{1}{N(q\delta_0)}. \tag{8.12.26}$$

Thus by (8.12.22)

$$\int_0^{T_1} N(\|U(t_1+s, t_1)x_1\|)ds \geq 2^2 \frac{1}{N(q\delta_0)} \frac{1}{2} N(q\delta_0) = 2. \tag{8.12.27}$$

Hence we have constructed an element x_1 such that

$$\|x_1 - x_0\| < \delta_0 \tag{8.12.28}$$

and we have found numbers t_1, T_1 such that

$$\int_0^{T_1} N(\|U(t_1+s, t_1)x_1\|)ds > 2. \tag{8.12.29}$$

Now we shall construct by induction a sequence of elements $\{x_0, x_1, \ldots, x_n, \ldots\}$ and sequences of positive numbers $\{t_1, t_2, \ldots, t_n, \ldots\}$, $\{T_1, T_2, \ldots, T_n, \ldots\}$, $\{\delta_0, \delta_1, \ldots, \delta_n, \ldots\}$ such that

$$\delta_{n+1} < \frac{1}{2}\delta_n, \quad n = 1, 2, \ldots, \tag{8.12.30}$$

$$\|x_{n+1} - x_n\| \leq \delta_n, \quad n = 0, 1, \ldots, \tag{8.12.31}$$

$$\int_0^{T_i} N(\|U(t_i+s, t_i)x_n\|)ds > 2^i, \quad i = 1, 2, \ldots, n. \tag{8.12.32}$$

The first element x_1 was constructed above. Suppose that $x_0, \ldots, x_n$, $\delta_0, \ldots, \delta_{n-1}$, $t_1, \ldots, t_n$, $T_1, \ldots, T_n$ such that (8.12.30)–(8.12.32) hold are found. By the continuity of the function N and by the uniform boundedness of the evolution operator we can find $\delta_n > 0$ such that for all $x \in E$, $\|x - x_n\| \leq \delta_n$, we have

$$\int_0^{T_i} N(\|U(t_i+s, t_i)x\|)ds > 2^i, \quad i = 1, \ldots, n. \tag{8.12.33}$$

8.12. Stability

Without loss of generality we may assume that

$$\delta_n \leq \frac{1}{2}\delta_{n-1} \tag{8.12.30}$$

holds.

Now we have two possibilities. Either

$$\sup_t \int_0^\infty N(\|U(t+s,t)x_n\|)ds = +\infty$$

or

$$M_n = \sup_t \int_0^\infty N(\|U(t+s,t)x_n\|)ds < +\infty.$$

In the first case we put $x_{n+1} = x_n$, $t_{n+1} = t_n$, $T_{n+1} = T_n$. In the second case we repeat the construction given for x_1. Namely, we take an arbitrary q, $0 < q < 1$, and we choose $\varepsilon > 0$ in such a way that

$$N(q\delta_n - \varepsilon) > \frac{1}{2} N(q\delta_n). \tag{8.12.34}$$

Let T_{n+1} be an arbitrary number such that

$$T_{n+1} > \frac{M_n}{N(\varepsilon)} + 2^{n+2}\frac{1}{N(q\delta_n)}. \tag{8.12.35}$$

By Lemma 8.12.3' there are an element x, $\|x\| = \delta_n$, and a number t_{n+1} such that

$$\|U(t_{n+1}+s, t_{n+1})x\| \geq q\delta_n, \quad \text{for} \quad 0 \leq s \leq T_{n+1}. \tag{8.12.36}$$

Let $x_{n+1} = x_n + x$. Of course,

$$\|x_{n+1} - x_n\| = \delta_n \leq \delta_n. \tag{8.12.31'}$$

By the choice of δ_n we have

$$\int_0^{T_i} N(\|U(t_i+s, t_i)x_{n+1}\|)ds > 2^i, \quad i = 1, 2, \ldots, n. \tag{8.12.37}$$

We shall show that (8.12.37) holds also for $i = n+1$. Indeed, by the definition of M_n, for all t the Lebesgue measure of the set $\{s: 0 \leq s \leq T_{n+1}, \|U(t+s, t)x_n\| > \varepsilon\}$ is smaller than $\dfrac{M_n}{N(\varepsilon)}$.

Since
$$\|U(t_{n+1}+s, t_{n+1})x_{n+1}\|$$
$$\geqslant \|U(t_{n+1}+s, t_{n+1})x\| - \|U(t_{n+1}+s, t_{n+1})x_n\|, \quad (8.12.38)$$

by the definition of T_{n+1} the set of such s, $0 \leqslant s \leqslant T_{n+1}$, that
$$\|U(t_{n+1}+s, t_{n+1})x_{n+1}\| \geqslant q\delta_n - \varepsilon \quad (8.12.39)$$

has the Lebesgue measure greater than
$$T_{n+1} - \frac{M_n}{N(\varepsilon)} > 2^{n+2}\frac{1}{N(q\delta_n)}. \quad (8.12.40)$$

Thus by (8.12.34)
$$\int_0^{T_{n+1}} N(\|U(t_{n+1}+s, t_{n+1})x_{n+1}\|)\,ds$$
$$> 2^{n+2}\frac{1}{N(q_n)}\frac{1}{2}N(q_n) = 2^{n+1}. \quad (8.12.41)$$

Let $\hat{x} = \lim\limits_{n\to\infty} x_n$. This limit exists by (8.12.30) and (8.12.31) and
$$\|x - \hat{x}\| \leqslant \sum_{i=0}^{\infty} \delta_i \leqslant 2\delta_0.$$

Moreover, by passing to the limit in (8.12.33) we find that there are sequences $\{t_n\}$ and $\{T_n\}$ such that
$$\int_0^{T_n} N(\|U(t_n+s, t_n)\hat{x}\|)\,ds \geqslant 2^n, \quad n = 1, 2, \ldots \quad (8.12.42)$$

This implies
$$\sup_s \int_s^{\infty} N(\|U(t, s)\hat{x}\|)\,dt = +\infty.$$

The arbitrariness of δ_0 implies the density of E_0. ∎

COROLLARY 8.12.5. *Let $\{U(t, s)\}$ be a uniformly bounded evolution operator. If for a certain $p > 0$*
$$\sup_s \int_s^{\infty} \|U(t, s)x\|^p\,dt < +\infty \quad (8.2.43)$$

8.12. Stability

for all $x \in E$, *then the evolution operator is uniformly exponentially equistable.*

Proof. Suppose that $\{U(t,s)\}$ is not exponentially equistable. Then by Proposition 8.12.4 the set E_0 is dense. This contradicts (8.12.43). ∎

Corollary 8.12.5 was obtained for $p \geq 1$ by Daleckii and Krein (1970).

Proof of Theorem 8.12.1. By the continuity of the function N and the uniform boundedness of the evolution operator $\{U(t,s)\}$ the sets

$$E_{s,T,a,M} = \left\{x : \int_s^{s+T} N(a, \|U(t,s)x\|)dt \leq M\right\}$$

are closed. Thus the sets

$$E_{a,M} = \bigcap_{s>0} \bigcap_{T>0} E_{s,T,a,M}$$
$$= \left\{x : \sup_s \int_s^\infty N(a, \|U(t,s)x\|)dt < +\infty\right\}$$

are also closed for all a and M. By our assumption

$$E = \bigcup_{a>0} \bigcup_{M>0} E_{a,M}. \tag{8.12.44}$$

Observe that

$$E_{a,M} \subset E_{a',M'} \quad \text{for} \quad a \geq a', \, M \leq M'. \tag{8.12.45}$$

Thus, using the Baire category method, we find by (8.12.44) and (8.12.45) that there are indices a_0, M_0 such that the set E_{a_0, M_0} contains a ball. Then the set E_0 defined in Proposition 8.12.4 is not dense and the evolution operator $\{U(t,s)\}$ is uniformly exponentially equistable. ∎

COROLLARY 8.12.6. *Let N be a nondecreasing continuous function such that $N(0) = 0$, $N(u) > 0$, for $u > 0$. Let $\{U(t,s)\}$ be a uniformly bounded evolution operator. If for each $x \in E$ there is an $a(x)$ such that*

$$\sup_s \int_0^\infty N(a(x)\|U(t,s)x\|) dt < +\infty,$$

then the evolution operator $\{U(t,s)\}$ is uniformly exponentially equistable.

COROLLARY 8.12.7. *Let $U(t)$ be a semigroup of linear operators of class C_0. Let N be as in Corollary 8.12.6. If for each $x \in E$, there is $a(x)$ such that*

$$\int_0^\infty N(a(x)\|U(t)x\|)\,dt < +\infty,$$

then the semigroup is exponentially equistable.

Proof. Observe that in this case the evolution operator $U(t, s)$ equals $U(t-s)$. Thus

$$\int_s^\infty N(a(x)\|U(t,s)x\|)\,dt = \int_s^\infty N(a(x)\|U(t-s)x\|)\,dt$$

$$= \int_0^\infty N(a(x)\|U(t)x\|)\,dt. \blacksquare$$

For N being, in addition, a convex function, Corollary 8.12.7 was proved by Zabczyk, 1974.

COROLLARY 8.12.8 (Przyłuski and Rolewicz, 1984, 1985, 1985b). *Let $\{U(t, s)\}$ be a uniformly bounded operator. If for each $x \in E$ there is a $p(x)$ such that*

$$\sup_s \int_s^\infty \|U(t,s)x\|^{p(x)}\,dt < +\infty,$$

then the evolution operator $\{U(t, s)\}$ is uniformly exponentially equistable.

Observe that if $N(u)$ is a convex function then for each uniformly exponentially equistable evolution operator $\{U(t, s)\}$

$$\sup_s \int_s^\infty N(\|U(t,s)\|)\,dt < +\infty.$$

Without the convexity of N the above may not hold, as follows from

EXAMPLE 8.12.1 (Rolewicz, 1985). Let
$$N(u) = \begin{cases} 0 & \text{for } u = 0, \\ \dfrac{1}{|\log u|} & \text{for } 0 < u \leq \dfrac{1}{e}, \\ u & \text{for } u > \dfrac{1}{e}. \end{cases}$$

The function $N(u)$ is a continuous increasing function. It is easy to verify that for $U(t) = e^{-at}I$

$$\int_0^\infty N(\|U(t)\|)dt = +\infty$$

for all $a > 0$.

Till now we have considered the problem of stability on the whole space X. The results proved before can be easily extended on the case when we are considering the problem of stability on a convex cone.

More precisely, let $(X, \|\ \|)$ be a Banach space. Let C be a convex cone with the vertex at 0. We shall not assume that the cone C is closed. We shall make a weaker assumption, namely that for each convergent series $\sum_{n=1}^\infty x_n$ of elements of C, the sum of the series also belong to C. The convex cones with this property will be called *positively closed*. Of course, each closed cone C is also positively closed.

But the open cones are also positively closed. To show it we shall introduce another notion. Let C be a convex cone in a linear space X. Let C_1 be a subcone of the cone C. We say that C_1 *has the ideal property* if for all $x \in C$, $y \in C_1$, the sum $x+y \in C_1$.

Let C be a convex cone in a Banach space. It is easy to observe that if C has the nonempty interior, Int C, then the cone Int C has the ideal property in the cone C.

PROPOSITION 8.12.9. *Let a convex cone C be positively closed. Let C_1 be a subcone of the cone C with the ideal property. Then the cone C_1 is also positively closed.*

The proof is trivial.

COROLLARY 8.12.10. *Each open cone C_0 is positively closed.*

Proof. Let $C = \overline{C}_0$. The closed cone C is, of course, positively closed. Since $C_0 = \text{Int } C$ has the ideal property, by Proposition 8.12.9, the cone C_0 is also positively closed. ∎

PROPOSITION 8.12.11. *In finite-dimensional spaces all convex cones are positively closed.*

Proof. Let C be a convex cone contained in R^n, $C \subset R^n$. Without loss of generality we may assume that $\text{Int } C \neq 0$, since otherwise we can replace R^n by the space of smaller dimension. Let $\sum_{n=1}^{\infty} x_n$ be a convergent series, such that $x_n \in C$. If the sum does not belong to $\text{Int } C = \text{Int } \overline{C}$, then all elements belong to an extremal subset $\overline{C}_1$ of the closed cone $\overline{C}$. Let $C_1 = C \cap \overline{C}_1$. Now we shall repeat our considerations for the cone C_1 of dimension smaller than n. Since the space is n-dimensional after finite number of steps (non greater than n) we obtain that the sum of the series belongs to a certain $C_k \subset C$. ∎

However, in infinite-dimensional spaces, there are convex cones which are not positively closed as follows from

EXAMPLE 8.12.2. Let $X = C[0, 1]$ with the standard sup norm. Let
$$C = \{x(\cdot) \in X \colon x(\cdot) \in C^1[0, 1], \, x(t) \geqslant 0, \text{ for } 0 \leqslant t \leqslant 1\}$$
be the convex cone consisting of nonnegative continuous and continuously differentiable functions.

Let
$$x_n(t) = \frac{1}{2^n} t^{2^n}.$$

It is easy to verify that $x_n \in C$ and that the series $\sum_{n=1}^{\infty} x_n$ is convergent. However the sum $\sum_{n=1}^{\infty} x_n(t)$ is not differentiable at the point 1, thus it does not belong to the cone C.

8.12. Stability

Let $(X, \|\ \|)$ be a Banach space. Let C be a convex cone in X. Let $\{U(t,s)\}$ be an evolution operator. We say that the evolution operator $\{U(t,s)\}$ *is preserving C* (or briefly *C-preserving*) if

$$U(t,s)C \subset C \quad \text{for} \quad 0 \leq s \leq t < +\infty.$$

Since $U(t,s)$ are continuous operators, then each C-preserving evolution operator is also $\overline{C}$-preserving.

We say that a C-preserving evolution operator $\{U(t,s)\}$ *is C-uniformly bounded* if there is a constant $K_1 \geq 0$ such that

$$\sup_{0 \leq t-s \leq 1} \|U(t,s)x\| \leq K_1 \|x\|$$

for all $x \in C$. We say that a C-preserving evolution operator $\{U(t,s)\}$ is *C-uniformly exponentially equistable* if there are two positive constants $M > 0$ and $\beta > 0$ such that

$$\|U(t,s)x\| \leq M e^{-\beta(t-s)} \|x\|$$

for all $x \in C$ and $0 \leq s \leq t < +\infty$.

PROPOSITION 8.12.4'. *Let $(X, \|\ \|)$ be a Banach space. Let a convex cone $C \subset X$ be positively closed. Let $\{U(t,s)\}$ be a C-preserving C-uniformly bounded evolution operator. Let $N(u)$ be a function satisfying the assumptions of Proposition 8.12.4. If the evolution operator $\{U(t,s)\}$ is not C-uniformly exponentially equistable, then the set*

$$C_0 = \left\{ x \in C \colon \sup_s \int_s^\infty N(\|U(t,s)x\|)\,dt = +\infty \right\}$$

is dense in the cone C.

Proof. The proof is on the same lines as the proof of Proposition 8.12.4, except that in construction only elements of the cone C are used. ∎

Having Proposition 8.12.4' we trivially obtain

THEOREM 8.12.1'. *Let $(X, \|\ \|)$ be a Banach space. Let C be a closed cone in X. Let $N(\alpha, u)$ be a function of two variables satisfying the assump-*

tions of Theorem 8.12.1. *Let* $\{U(t, s)\}$ *be a C-preserving C-uniformly bounded evolution operator. If for each* $x \in C$ *there is an* $\alpha(x)$ *such that* (8.12.3) *holds, then the evolution operator* $\{U(t, s)\}$ *is C-uniformly exponentially equistable.*

Appendix

Necessary and sufficient condition for weak duality

The main purpose of this appendix is to prove the following theorem:
THEOREM A.1 (Dolecki, 1977). *Suppose we are given a linear system*
$(X \underset{A}{\to} \square \underset{B}{\to} Y)$.
Let $F(x)$ be a continuous convex functional defined on the space X. The weak duality holds for all $y_0 \in BA(X)$, which means that

$$\inf \{F(x): BA(x) = y_0\} = \sup_{\varphi \in y^*} \inf \{F(x): \varphi(y_0)$$
$$= \varphi(BA(x))\} \quad (5.2.10')$$

for all $y_0 \in BA(X)$, if and only if

$$\bigcap_{\varepsilon>0} BA(\{x: F(x) \leqslant r+\varepsilon\}) = \bigcap_{\varepsilon>0} \overline{BA(\{x: F(x) \leqslant r+\varepsilon\})}. \quad (A.1)$$

The result was mentioned in Section 5.2. However, the proof is based on the multifunction approach to optimization and that is why I have decided not to put it in the main body of the book, but to devote to it an appendix. The multifunction approach to optimization is now a strongly developed mathematical theory. Of course, I am not able to present this theory here, restricting myself only to those results which are necessary for the proof of Theorem A.1. I ought also to mention that Theorem A.1 is a particular case of a general theorem, which will be presented later.

Let X be an arbitrary set. Let Φ be an arbitrary set of real-valued functions defined on X. For an arbitrary real-valued function f we define the *generalized Fenchel conjugate (dual)* f^* as a function defined on Φ by the following formula

$$f^*(\varphi) = \sup_{x \in X} (\varphi(x) - f(x)). \quad (A.2)$$

Observe that $x \in X$ induces a function on Φ given by the formula
$$x(\varphi) = \varphi(x). \quad (A.3)$$

Thus we can consider X as a space of real-valued functions defined on Φ. Therefore we can define the second dual function as the dual function to a dual function

$$f^{**}(x) = (f^*)^*(x) = \sup_{\varphi \in \Phi} (\varphi(x) - f(\varphi)). \tag{A.4}$$

The function $f^{**}(x)$ is called also *bidual*. An essential role is played by the following proposition, which is an abstract generalization of the Moreau–Fenchel theorem.

PROPOSITION A.2 (Kurcyusz, 1975; Balder, 1977; Dolecki, Kurcyusz, 1978). *Suppose that the set of functions Φ is invariant under addition, which means that, for each real a and an arbitrary $\varphi \in \Phi$, $a + \varphi \in \Phi$. Then*

$$f^{**}(x) = \sup_{\varphi \leq f} \{\varphi(x) : \varphi \in \Phi\}, \tag{A.5}$$

where $\varphi \leq f$ means that

$$\varphi(x) \leq f(x) \quad \text{for all} \quad x \in X.$$

Proof. Write

$$f^0(x) = \sup_{\varphi \leq f} \varphi(x). \tag{A.6}$$

Observe that $\varphi \leq f$ if and only if

$$f^*(\varphi) = \sup_{x \in X} (\varphi(x) - f(x)) \leq 0. \tag{A.7}$$

Thus

$$f^0(x) = \sup_{f^*(\varphi) \leq 0} \varphi(x). \tag{A.8}$$

It is easy to show that

$$\sup_{f^*(\varphi) \leq 0} \varphi(x) = \sup_{f^*(\varphi) = 0} \varphi(x). \tag{A.9}$$

In fact, suppose that $f^*(\varphi) = -a < 0$. Then

$$f^*(\varphi + a) = \sup_{x \in X} (\varphi(x) + a - f(x))$$

$$= a + \sup_{x \in X} (\varphi(x) - f(x)) = a + f^*(\varphi) = 0.$$

Thus, for $a > 0$, putting $\psi = \varphi + a$ we have

$$\sup_{f^*(\varphi) = -a} \varphi(x) = \sup_{f^*(\psi) = 0} (\psi(x) - a) = \sup_{f^*(\varphi) = 0} (\varphi(x) - a)$$

$$\leqslant \sup_{f^*(\varphi) = 0} \varphi(x) \qquad (A.10)$$

which trivially implies (A.9).

By (A.8) and (A.9) we get

$$f^0(x) = \sup_{f^*(\psi) = 0} \varphi(x). \qquad (A.11)$$

On the other hand,

$$f^{**}(x) = \sup_{\varphi \in \Phi} (\varphi(x) - f^*(\varphi)).$$

Observe that for arbitrary real a

$$\varphi(x) - f^*(\varphi) = (\varphi(x) + a) - (f^*(\varphi) + a). \qquad (A.12)$$

Since

$$f^*(\varphi) + a = \sup_{x \in X} (\varphi(x) + a - f(x)) = f^*(\varphi + a), \qquad (A.13)$$

we have

$$\varphi(x) - f^*(\varphi) = (\varphi + a)(x) - f^*(\varphi + a) \qquad (A.14)$$

for all real a.

Let $a = -f^*(\varphi)$. Then $f^*(\varphi + a) = 0$ and

$$f^{**}(x) = \sup_{\varphi \in \Phi} (\varphi(x) - f^*(\varphi)) = \sup_{f^*(\varphi + a) = 0} (\varphi + a)(x)$$

$$= \sup_{f^*(\varphi) = 0} \varphi(x) = f^0(x). \blacksquare$$

The function $f^0(x)$ is called the *Φ-convexification* of the function f.

Let X, Y be two sets. Let $\Gamma\colon Y \to 2^X$ be a *multifunction*, i.e., a point-set mapping assigning to each $y \in Y$ a subset of X.

By Γ^- we shall denote the inverse multifunction, $\Gamma^-\colon X\to 2^Y$, defined as
$$\Gamma^- x = \{y\colon x \in \Gamma y\}.$$
We write
$$\Gamma_A = \bigcup_{y\in A} \Gamma y$$

Let f be a function given on X. We shall consider the following family of optimization problems
$$f(x) \to \inf, \, x \in \Gamma y.^{(1)} \qquad (A.15)$$

Suppose that Φ is a set of real-valued functions defined on Y and invariant with respect to the addition of constants, i.e., that for each real a if $\varphi \in \Phi$ then $\varphi + a \in \Phi$. For fixed $y_0 \in Y$ we construct the *Lagrangian* of problem (A.15):
$$\begin{aligned} L(x, \varphi, y_0) &= f(x) - \sup_{y\in\Gamma^- x} (\varphi(y) + \varphi(y_0)) \\ &= f(x) + \inf_{y\in\Gamma^- x} (-\varphi(y) + \varphi(y_0)). \end{aligned}$$

We say that the *weak duality holds at* y_0 if
$$\inf_{x\in\Gamma y_0} f(x) = \sup_{\varphi\in\Phi} \inf_{x\in X} L(x, \varphi, y_0). \qquad (A.16)$$

Suppose we are given a linear system
$$(X \xrightarrow{A} \Box \xrightarrow{B} Y).$$

Suppose that Φ is the class of all continuous affine functionals defined on $\tilde{Y}$, i.e., $\varphi \in \Phi$ if φ is of the form $\varphi = f + a$ where f is a continuous linear functional defined on Y and a is a real number. It is easy to check that Φ is invariant under the addition of constants. Let $\Gamma y = \{x\colon BA(x) = y\}$. Taking into account formula (5.2.9), we find that in this case the formula of weak duality is just the weak duality of linear systems, i.e., formula (5.2.10).

Now we shall give necessary and sufficient conditions for weak duality.

---

[1] i.e., the family of problems "find $\inf_{x\in\Gamma y} f(x)$".

By $\overline{f\Gamma}(y)$ we shall denote

$$\overline{f\Gamma}(y) = \inf_{x \in \Gamma y} f(x).$$

The function $\overline{f\Gamma}$ is called the *marginal function* or the *primal functional* of problem (A.15)

THEOREM A.3 (Dolecki and Kurcyusz, 1978). *There is weak duality at y_0 if and only if*

$$\overline{f\Gamma}(y_0) = \overline{f\Gamma}^0(y_0), \tag{A.17}$$

where, as before, $\overline{f\Gamma}^0$ denotes the Φ-convexification of the marginal function $\overline{f\Gamma}$.

Proof.

$$\inf_{x \in X} L(x, \varphi, y_0) = \inf_{x \in X} \left(f(x) + \inf_{y \in \Gamma^- x} [-\varphi(y) + \varphi(y_0)] \right)$$

$$= \inf_{x \in \Gamma y} \left(f(x) - \varphi(y) + \varphi(y_0) \right)$$

$$= \inf_{y \in Y} \left[\inf_{x \in \Gamma y} (f(x) - \varphi(y)) + \varphi(y_0) \right] = -(\overline{f\Gamma})^*(\varphi) + \varphi(y_0), \tag{A.18}$$

where $(\overline{f\Gamma})^*$ denotes the conjugate function to the marginal function $\overline{f\Gamma}$.

Taking the supremum over both sides of (A.18), we obtain

$$\sup_{\varphi \in \Phi} \inf_{x \in X} L(x, \varphi, y_0) = \sup_{\varphi \in \Phi} [-(\overline{f\Gamma})^*(\varphi) + \varphi(y_0)]$$

$$= -\inf_{\varphi \in \Phi} [(\overline{f\Gamma})^*(\varphi) - \varphi(y_0)] = -\left(-((\overline{f\Gamma})^*)^*(y_0)\right)$$

$$= \overline{f\Gamma}^{**}(y_0),$$

and by Proposition A.2

$$\sup_{\varphi \in \Phi} \inf_{x \in X} L(x, \varphi, y_0) = (\overline{f\Gamma})^0(y_0).$$

Hence there is weak duality at y_0 if and only if (A.17) holds. ∎

Now we shall look for conditions ensuring (A.17).

PROPOSITION A.4. *Let $(Y, \|\ \|)$ be a normed space. Let Φ denote the set of all continuous affine functionals defined on Y. Let f be a convex function defined on y. The equality*
$$f(y_0) = f^o(y_0)$$
holds if and only if f is lower semicontinuous at y_0.

Proof. Necessity. Let ε be an arbitrary positive number. Since $f(y_0) = f^o(y_0) = \sup_{\varphi \leq f} \varphi(y_0)$, there is a φ_0 such that

$$\varphi_0(y_0) \leq f(y_0) \leq \varphi_0(y_0) + \frac{\varepsilon}{2}. \tag{A.19}$$

The function φ_0 is continuous. Thus there is a $\delta > 0$ such that for all y, $\|y - y_0\| < \delta$, we have

$$|\varphi_0(y) - \varphi_0(y_0)| < \frac{\varepsilon}{2}. \tag{A.20}$$

By (A.19) and (A.20) for y, $\|y - y_0\| < \delta$, we have

$$f(y) \geq \varphi_0(y) \geq \varphi_0(y_0) - \frac{\varepsilon}{2} \geq f(y_0) - \varepsilon.$$

The arbitrariness of ε implies the lower semicontinuity of f.

Sufficiency. Take the epigraph
$$\text{epi} f = \{(y, r) \in Y \times R : r \geq f(y)\}$$
of the function f. Since f is convex, the epigraph is a convex set. Take any $\varepsilon > 0$. Observe that the point $(y_0, f(y_0) - \varepsilon)$ does not belong to the closure of the epigraph, $(y_0, f(y_0) - \varepsilon) \notin \overline{\text{epi} f}$. Indeed, if $(y_n, r_n) \in \text{epi} f$, $y_n \to y_0$ and $r_n \to r_0$, then by the lower semicontinuity of the function f, $r_0 \geq f(y_0)$.

Thus, by the separation theorem (Theorem 2.5.7), there is a continuous linear functional F defined on $Y \times R$ separating the point $(y_0, f(y_0) - \varepsilon)$ and the set $\overline{\text{epi} f}$.

Of course, the functional F is of the form
$$F(y, r) = \varphi(y) + ar,$$
where $\varphi(y)$ is a continuous linear functional on Y. Without loss of generality we may assume that $a = 1$ since if F is separating then $\frac{1}{a} F$ is also separating.

Since F is separating $(y_0, f(y_0) - \varepsilon)$ from $\overline{\text{epi } f}$,
$$F(y_0, f(y_0) - \varepsilon) \leq \inf\{F(y, r): (y, r) \in \text{epi } f\},$$
we have
$$\varphi(y_0) + f(y_0) - \varepsilon \leq \varphi(y) + r$$
for all $r \geq f(y)$. Thus
$$\varphi(y_0) + f(y_0) - \varepsilon \leq \varphi(y) + f(y)$$
and
$$f(y) \geq f(y_0) + \varphi(y_0) - \varphi(y) - \varepsilon. \qquad (A.21)$$
Let $\psi(y)$ be an affine continuous functional
$$\psi(y) = -\varphi(y) + \varphi(y_0) + f(y_0) - \varepsilon.$$
By (A.21)
$$\psi(y) \leq f(y).$$
By the definition of $\psi(y)$, $\psi(y_0) = f(y_0) - \varepsilon$. Hence
$$\sup_{\substack{\psi \leq f \\ \psi \in \Phi}} \psi(y_0) = f(y_0). \quad \blacksquare$$

Now we shall return to the theory of linear systems, putting $f(x) = F(x)$ and $\Gamma y = \{x \in X : BA(x) = y\}$. Since the functional F is convex, the marginal function $\overline{f\Gamma}$ is also convex. Thus the problem of the weak duality is reduced to the problem of the lower semicontinuity of the marginal function $\overline{f\Gamma}$.

Let Y be a topological space. Observe that
$$\inf_{r \in \Gamma y} f(x) \geq a$$
if and only if
$$\Gamma y \subset \{x: f(x) \geq a\}.$$
Thus the marginal function $\overline{f\Gamma}$ is lower semicontinuous at y_0 if and only if for each $\varepsilon > 0$ there is a neighbourhood G of y_0 such that
$$\Gamma y \subset \{x: f(x) > \overline{f\Gamma}(y_0) - \varepsilon\} \text{ for all } y \in G. \qquad (A.22)$$

Writing
$$\Gamma_{\exp} A = \{y : A \subset \Gamma y\},$$
we can rewrite (A.22) in the following form:
$$G \subset \Gamma_{\exp} \{x : f(x) > \overline{f\Gamma}(y_0) - \varepsilon\}. \tag{A.23}$$

PROPOSITION A.5 (Dolecki and Rolewicz, 1978). *The marginal function is lower semicontinuous at y_0 if and only if*
$$y_0 \in \bigcup_{0 < \varepsilon} \Gamma_{\exp} \{x : f(x) > r + \varepsilon\} \tag{A.24}$$
entails
$$y_0 \in \operatorname{Int} \Gamma_{\exp} \{x : f(x) > r\}. \tag{A.25}$$

Proof. Suppose that the marginal function $\overline{f\Gamma}(y_0)$ is lower semicontinuous at y_0 and that (A.24) holds. Thus there is an $\varepsilon_0 > 0$ such that for all ε, $0 < \varepsilon < \varepsilon_0$,
$$y_0 \in \Gamma_{\exp} \{x : f(x) > r + \varepsilon\}. \tag{A.26}$$
Hence by (A.26)
$$\overline{f\Gamma}(y_0) \geq r + \varepsilon. \tag{A.27}$$
The lower semicontinuity of $\overline{f\Gamma}$ implies (see (A.23))
$$y_0 \in \operatorname{Int} \Gamma_{\exp} \{x : f(x) \geq \overline{f\Gamma}(y_0) - \varepsilon\} \subset \operatorname{Int} \Gamma_{\exp} \{x : f(x) > r\},$$
i.e., (A.25) holds.

Conversely, suppose that (A.24) implies (A.25). Let ε be an arbitrary positive number. Let $r = \overline{f\Gamma}(y_0) - \varepsilon$. If (A.24) holds, then
$$y_0 \in \bigcup_{0 < \eta} \Gamma_{\exp} \{x : f(x) > r + \eta\}$$
and
$$y_0 \in \operatorname{Int} \Gamma_{\exp} \{x : f(x) > r\} = \operatorname{Int} \Gamma_{\exp} \{x : f(x) > \overline{f\Gamma}(y_0) - \varepsilon\}.$$
Thus by (A.23) $\overline{f\Gamma}$ is lower semicontinuous at y_0. ∎

Observe that
$$\Gamma^- A = (\Gamma_{\exp} A^c)^c, \tag{A.28}$$
where B^c denotes the complement of a set B.

Let
$$B(f, s) = \{x \in X: f(x) \leq s\}.$$
By (A.25) and (A.28) we trivially obtain

COROLLARY A.6 (Kurcyusz, 1975; Dolecki, 1977). *The marginal function $\overline{f\Gamma}$ is lower semicontinuous at y_0 if and only if $y_0 \in \overline{\Gamma^- B(f, r)}$ implies*
$$y_0 \in \bigcap_{\varepsilon > 0} \Gamma^- B(f, r + \varepsilon).$$

COROLLARY A.7 (Kurcyusz, 1975; Dolecki, 1977, 1977b). *The marginal function $\overline{f\Gamma}$ is lower semicontinuous for all $y \in Y$ if for all r*
$$\bigcap_{\varepsilon > 0} \overline{\Gamma^- B(f, r + \varepsilon)} = \bigcap_{\eta > 0} \Gamma^- B(f, r + \eta). \tag{A.29}$$

Proof of Theorem A.1. We put $\Gamma y = \{x: BAx = y\}$. Thus $\Gamma^- x$ consists of one element BAx. In this case (A.29) is nothing else than formula (A.1). By Theorem A.3, Proposition A.4 and Corollary A.7, we trivially obtain Theorem A.1. ∎

Bibliography

Alaoglu, L.: (1940), 'Weak topologies of normed spaces', *Ann. of Math.* (2), **41**, 252–267.

Baire, R.: (1899), 'Sur les fonctions de variables réelles', *Ann. di Math.* **3**, 1–123.
Balakrishnan, V. K.: (1976), *'Applied Functional Analysis'*, Springer-Verlag, Berlin.
Balder, E. J.: (1977), 'An extension of stability–duality relations to nonconvex optimization problems', *SIAM J. Control and Optim.* **15**, 329–343.
Banach, S.: (1922), 'Sur les opérations dans les ensembles abstraits et leur applications aux équations intégrales', *Fund. Math.* **3**, 133–181.
Banach, S.: (1929), 'Sur les fonctionelles linéaires'. Part I—*Stud. Math.* **1**, 211–216, Part II—*Stud. Math.* **1**, 223–229.
Banach, S.: (1931), *'Operator theory'* (in Polish), Monografie Matematyczne 1, Warszawa. French translation enlarged and extended, Kasa im. Mianowskiego, Warszawa, 1932. Ukrainian translation enlarged and extended, Naukova Dumka, Kiev 1948.
Banach, S., Steinhaus, H.: (1927), 'Sur le principle de la condensations de singularités', *Fund. Math.* **9**, 50–61.
Bielecki, A.: (1956), 'Une remarque sur la méthode de Banach–Cacciopoli–Tichonov dans la théorie des équations différentielles ordinaires', *Bull. Acad. Pol. Sci.* **4**, 261–264.
Bielecki, A.: (1956b), 'Une remarque sur l'application de la méthode de Banach–Cacciopoli–Tichonov dans la théorie de l'équation $s = f(x, y, z, p, q)$', *Bull. Acad. Pol. Sci.* **4**, 265–268.
Bohnenblust, H. F., Sobczyk, A.: (1938), 'Extensions of functionals on complex linear spaces', *Bull. Amer. Math. Soc.* **44**, 91–93.
Butkovski, A. G.: (1965), *'Theory of Optimal Control of Systems with Distributed Parameters'* (in Russian), Nauka, Moscow (English translation, Elsevier, Amsterdam, New York, 1969).
Butkovski, A. G.: (1975), *'Methods of Controls of Systems with Distributed Parameters'* (in Russian), Nauka, Moscow.

Carathéodory, C.: (1921), 'Über den Variabilitätbereich der Fourier Konstanten von positiven harmonischen Funktionen', *Rend. Circ. Mat. Palermo* **32**, 193–217.
Castaing, Ch.: (1967), 'Sur les multifunctions mesurables', *Revue Inf. Oper. Res.* **1**, 91–126.

Castaing, Ch., Valadier, M.: (1977), '*Convex Analysis and Measurable Multifunctions*', Lecture Notes in Mathematics 580, Springer-Verlag, Berlin, Heidelberg.

Christoffel, E.: (1858), 'Über die Gaussische Quadratur und eine Verallgemeinerung derselben', *J. Reine und Angew. Math.* **55**.

Curtain, R. F., Pritchard, A. J.: (1977), '*Functional Analysis in Modern Applied Mathematics*', Academic Press, New York, London.

Daleckii, Yu. L., Krein, M. G.: (1970), '*Stability of Solutions of Differential Equations in Banach Spaces*' (in Russian), Nauka, Moscow.

Dolecki, S.: (1973), 'On observability for one dimensional heat equation', *Stud. Math.* **48**, 291–305.

Dolecki, S.: (1976), 'A classification of controllability concepts for infinite dimensional spaces', *Control and Cybernetic* **5**, 33–44.

Dolecki, S.: (1977), 'Bounded controlling sequences, lower stability and certain penalty procedure', *Appl. Math. and Optim.* **4**, 15–26.

Dolecki, S.: (1977b), 'Remarks on semicontinuity', *Bull. Acad. Pol. Sci.* **25**, 863–867.

Dolecki, S., Kurcyusz, S.: (1978), 'On Φ-convexity in etremal problems', *SIAM J. Control and Optim.* **16**, 277–300.

Dolecki, S., Rolewicz, S.: (1978), 'A characterization of semicontinuity preserving multifunctions', *J. Math. Anal. and Appl.* **65**, 26–31.

Dolecki, S., Russell, D. L.: (1977), 'A general theory of observation and control', *SIAM J. Control and Optim.* **15**, 185–220.

Douglas, R.: (1966), 'On majorization, factorization and range inclusion of operators in Hilbert spaces', *Proc. Amer. Math. Soc.* **17**, 413–415.

Dunford, N., Schwartz, J.: (1958), '*Linear Operators* I', Interscience Publishers, New York, London.

Dvoretzky, A., Wald, A., Wolfowitz, J.: (1951), 'Relations among certain ranges of vector measures', *Pacific J. Math.* **1**, 59–74.

Eberlein, W. F.: (1950), 'Weak compactness in Banach spaces I', *Proc. Nat. Acad. USA* **36**, 162–166.

Engelking, R.: (1968), '*Outline of General Topology*', PWN — North-Holland, Warszawa.

Fattorini, H. O.: (1966), 'Some remarks on complete controllability', *SIAM J. Control and Optim.* **4**, 686–697.

Fattorini, H. O.: (1967), 'On complete controllability of linear systems', *J. Diff. Eq.* **3**, 391–402.

Fattorini, H. O.: (1974), 'The time optimal control problem in Banach spaces', *Appl. Math. Optim.* **1**, 163–188.

Fattorini H. O.: (1983), *The Cauchy Problem*, Addison-Wesley.

Fedorov, V. V.: (1971), '*Theory of Optimal Experiment*' (in Russian), Nauka, Moscow. English edition, Academic Press, New-York, 1972.

Fedorov, V. V.: (1980), 'Convex design theory', *Math. Operationsforschung und Statistik, ser. Statistik* **11**, 403–413.

Fuhrmann, P. A.: (1972), 'On weak and strong reachability and controllability of infinite dimensional systems', *J. Optim. Theory and Appl.* **9**, 77–89.

Fuhrmann, P. A.: (1973), 'On observability and stability in infinite dimensional systems', *J. Optim. Theory and Appl.* **12**, 173–181.

Gauss, C. F.: (1814), '*Methodus nova integralium valores per approximationem inveniendi*' (in: *Werke*, vol. 3), Göttingische gelehrte Anzeigen 1814 September 26, 202–206.

Gohberg, I. C., Krein, M. G.: (1957), 'Basic statements on defect numbers, root numbers and indices of linear operators' (in Russian), *Usp. Mat. Nauk* **12** issue 2, 43–118.

Goldberg, S.: (1959), 'Linear operators and their conjugates', *Pacific J. Math.* **9**, 69–79.

Goldberg, S.: (1968), '*Unbounded Linear Operators*', McGraw-Hill, New York.

Goldstine, H. H.: (1938), 'Weakly compact Banach spaces', *Duke Math. J.* **4**, 125–131.

Goodner, D. R.: (1950), 'Projections in normed spaces', *Trans. Amer. Math. Soc.* **69**, 89–108.

Hahn, H.: (1927), 'Über lineare Gleichungssysteme in linearen Räumen', *J. Reine und Angew. Math.* **157**, 214–229.

Halmos, P.: (1948), 'The range of a vector measure', *Bull. Amer. Math. Soc.* **54**, 416–421.

Haverkamp, R.: (1978), 'Approximationeigenschaften differenzierter Interpolationspolynome', *J. Approx. Theory* **23**, 261–266.

Haverkamp, R.: (1982), 'Approximationsgüte der Ableitungen bei trigonometrischen Interpolationen', *Math. Zeitschr.* **179**, 59–67.

Hermes, H., La Salle, J. P.: (1969), '*Functional Analysis and Time Optimal Control*', Mathematics in Science and Engineering, Vol. 56, Academic Press, New York, London.

Hille, E.: (1948), '*Functional Analysis and Semigroups*', Amer, Math. Soc.

Hille, E., Phillips, R. S.: (1957), 'Functional analysis and semigroups', *Amer. Math. Soc. Coll. Publ.* **31**.

Isbell, J. R., Semadeni, Z.: (1963), 'Projection constant and spaces of continuous functions', *Trans. Amer. Math. Soc.* **107**, 38–48.

Kakutani, S.: (1936), 'Über die Metrisation der topologischen Gruppen', *Proc. Imp. Acad. Tokyo* **12**, 82–84.

Kalman, R. E.: (1961), 'On general theory of control systems', *Proc. I IFAC Congress*, Moscow, Butterworths & Co. Ltd., vol. 4, 2020–2030.

Kalman, R. E.: (1968), '*Lectures on Controllability*', CIME Seminar.

Kalman, R. E., Falb, P. L., Arbib, M. A.: (1969), *Topics in Mathematical Control Theory*, McGraw Hill, New York.

Karlin, S., Studden, W. J.: (1966), 'Optimal experimental design', *Ann. Math. Stat.* **37**, 783–815.

Kato, T.: (1958), 'Perturbation theory for nullity, deficiency and other quantities of linear operators', *J. Analyse Math.* **64**, 273–322.

Kato, T.: (1966), *'Perturbation Theory for Linear Operators'*, Springer-Verlag, Berlin.

Kelley, J. L.: (1952), 'Banach spaces with extension property', *Trans. Amer, Math. Soc.* **72**, 323–326.

Kiefer, J., Wolfowitz, J.: (1959), 'Optimum design in regression problems', *Ann. Math. Stat.* **30**, 271–294.

Kolmogorov, A. N.: (1934), 'Zur Normierbarkeit eines allgemeinen topologischen linearen Raumes', *Stud. Math.* **5**, 29–34.

Krassovskii, N. N.: (1968), *'Theory of Control of Movement. Linear Systems'* (in Russian), Nauka, Moscow.

Krassovskii, N. N.: (1976), 'On equations with incomplete information' (in Russian), *Prikl. Mat. Mekh.* **40**, 197–206.

Krassovskii, N. N.: (1976b), 'To a problem of equations with incomplete information' (in Russian), *Izv. AN USSR, Tekh. Kiber.* **2**, No 2, 3–7.

Krein, M. G., Milman, D. P.: (1940), 'On extreme points of regularly convex sets', *Stud. Math.* **9**, 133–138.

Krein, M. G., Milman, D. P., Rutman, L. A.: (1940), 'On one property of basis in Banach space' (in Russian), *Zap. Chark. Mat. T-wa.* **4**, 106–110.

Krein, S. G.: (1967), *'Linear Differential Equations in Banach Space'* (in Russian), Nauka, Moscow.

Kulikowski, R.: (1965), *'Optimal and Adaptative Processes in Large Scale Systems'* (in Polish), PWN, Warszawa.

Kulikowski, R.: (1970), *'Control in Large Scale Systems'* (in Polish), WNT, Warszawa.

Kuratowski, K.: (1948), *'Topologie I'*, II edition, Monografie Matematyczne 20, PWN, Warszawa–Wrocław.

Kuratowski, K.: (1952), *'Topologie II'*, II edition, Monografie Matematyczne 21, PWN, Warszawa — Wrocław.

Kuratowski, K.: (1961), *'Introduction to Calculus'*, PWN, Pergamon Press, Warszawa.

Kuratowski, K., Ryll-Nardzewski, Cz.: (1965), 'A general theorem on selectors', *Bull. Acad. Pol. Sci.* **13**, 397–403.

Kurcyusz, S.: (1975), 'Some remarks on generalized Lagrangians', *Proc. 7-th IFIP Conference*, Nice.

Kurzhanskii, A. B.: (1965), 'Calculation of optimal control in systems with incomplete information' (in Russian), *Diff. Urav.* **1**, 360–373.

Kurzhanskii, A. B.: (1975), 'On minimax control and estimation strategies under incomplete information', *Problems of Control and Inf. Theory*, **4**, 205–218.

Kurzhanskii, A. B.: (1977), *'Control and Observation in Conditions of Indeterminity'* (in Russian), Nauka, Moscow.

La Salle, J. P.: (1960), 'The time optimal control problem', *Contribution to the theory of nonlinear oscillations* **5**, 1–24.

Lee, E. B., Markus, L.: (1968), *'Foundations of Optimal Control Theory'*, John Wiley, New York.

Lyapunov, A. A.: (1940), 'On fully additive vector functions I' (in Russian), *Izv. AN USSR* **3**, 465–478.

Lyapunov, A. A.: (1948), 'On fully additive vector functions II' (in Russian), *Izv. AN USSR* **10**, 277–279.

Lindenstrauss, J.: (1966), 'A short proof of Liapunoff's convexity theorem', *J. Math. Mech.* **15**, 971–972.

Lindenstrauss, J., Tzafiri, L.: (1971), 'On complemented subspace problem', *Israel J. Math.* **9**, 379–390.

Lions, J. L.: (1968), *'Contrôle optimale des systèmes gouvernés par des équations aux dérivées partielles'*, Dunod, Paris.

Lions, J. L., Magenes, E.: (1968), *'Problèmes aux limites non homogènes et applications'*, Dunod, Paris.

Luxemburg, W. A. J., Korevaar, J.: (1971), 'Entire functions and Müntz–Szász type approximations', *Trans. Amer. Math. Soc.* **157**, 23–37.

Malanowski, K.: (1966), 'On the optimal time control in the case of non-uniqueness', *Stud. Math.* **27**, 147–167.

Malanowski, K., Rolewicz S.: (1965), 'Application of a method of supporting hyperplanes to a determining of time optimal control' (in Polish), *Archiwum Automatyki i Telemechaniki* **10**, 148–176.

Malanowski, K., Rolewicz, S.: (1965b), 'Determining of time optimal control of linear multidimensional systems by the method of supporting hyperplanes' (in Russian), *Zhur. Vych. Mat. i Teor. Fiz.* **5**, 242–251.

Mazur, S.: (1933), 'Über konvexe Mengen in linearen normierten Räumen', *Stud. Math.* **4**, 70–84.

Mazur, S.: (1936). *'On Convex Sets and Convex Functions in Linear Spaces'* (in Polish), Lwów.

Nachbin, L.: (1950), 'A theorem of Hahn–Banach type for linear transformations', *Trans. Amer. Math. Soc.* **68**, 28–46.

Olech, Cz.: (1965), 'A note concerning set-valued measurable functions', *Bull. Acad. Pol. Sci.* **13**, 317–321.

Olech, Cz.: (1966), ' Extremal solutions of control systems', *J. Diff. Eq.* **2**, 74–101.

Pallaschke, D.: (1976), 'Differentiations-und Integrations-formeln in $C_0[a,b]$', *Num. Math.* **26**, 201–210.

Pallaschke, D.: (1977), 'Konvergenz optimaler Differentiationsformeln', *Num. Math.* **27**, 421–426.

Phan Quoc Khanh: (1979), 'On measurements in observation and optimal observation for linear lumped parametric systems', *Control and Cybernetics* **8**, 147–159.

Phan Quoc Khanh: (1985), 'On Pontryagin maximum principle for linear systems with constraints', *Math. Operationsforschung u. Stat., ser. Optimization*, **16**, 63–69.

Pontryagin, L. S., Boltyanskii, W. G., Gamkrelidze, R. W., Mishchenko, E. F.: (1961), '*Mathematical Theory of Optimal Processes*' (in Russian), Moscow, English translation, Interscience Publishers. Inc., New York, 1962.

Popov, V. M.: (1972), 'Pointwise degeneracy of linear time-invariant delay-difference equations', *J. Diff. Eq.* **11**, 541–561.

Porter, W. A.: (1966), '*Modern Foundations of System Engineering*', New York, London.

Pritchard, A. J., Wirth, A.: (1978), 'Unbounded control and observation systems and their duality', *SIAM J. Control and Optim.* **16**, 535–545.

Przeworska-Rolewicz, D.: (1987) '*Algebraic Analysis*', PWN, Reidel, Warszawa, Dordrecht.

Przyłuski, K. M., Rolewicz, S.: (1984), 'On stability of linear time-varying infinite dimensional discrete-time systems', *Syst. and Control. Lett.* **5**, 307–315.

Przyłuski, K. M., Rolewicz, S.: (1985), 'Stability of linear continuous processes on a Banach space. The discrete-time approach', *Syst. and Control Lett.* (to appear).

Przyłuski, K. M., Rolewicz, S.: (1985b), '*On Stability of Linear Time-Varying Infinite Dimensional Systems*', Lecture Notes in Economics and Mathematical Systems, 240 Springer-Verlag, Berlin, 159–173.

Ptak, V.: (1958), 'A remark on approximation of continuous functions', *Czechoslovak Math. J.* **8** (83), 251–256.

Riesz, F.: (1909), 'Sur les opérations fonctionelles linéaires', *C.R. Acad. Sci. Paris* **149**, 974–977.

Radström, H.: (1952), 'An embedding theorem for spaces of convex sets', *Proc. Amer. Math. Soc.* **3**, 165–169.

Rolewicz, S.: (1968), 'On a problem of moments', *Stud. Math.* **30**, 183–191.

Rolewicz, S.: (1972), '*Metric Linear Spaces*', Monografie Matematyczne 56, PWN, Warszawa.

Rolewicz, S.: (1972b), 'On optimal observability of linear systems with infinite dimensional states', *Stud. Math.* **46**, 411–416.

Rolewicz, S.: (1973), 'On observability of linear systems', *Berichte der Gesellschaft für Mathematik und Datenverarbeitung, Bonn.* **77**, 139–141.

Rolewicz, S.: (1973b), 'On a method of moments', in: '*Proc. II Conf. of Funct. Analysis*', *Bordeaux*, Publication Dept. Mathematiques de Lyon, 10–4, 27–29.

Rolewicz, S.: (1973c), 'On minimum time control problem', *IRIA Seminar Reports Analyse et Contrôle des Systèmes*, 253–259.

Rolewicz, S.: (1975), 'On minimum time control problem and continuous families of convex sets', *Stud. Math.* **56**, 39–45.
Rolewicz, S.: (1976), 'On universal time for controllability of time depending linear control systems', *Stud. Math.* **49**, 27–32.
Rolewicz, S.: (1976b), 'Linear systems in Banach spaces', in: *'Calculus of Variation and Control Theory'*, ed. D. L. Russell, Academic Press, New York, London, 245–256.
Rolewicz, S.: (1976c), 'On general theory of linear systems', *Beiträge zur Analysis* **8**, 119–127.
Rolewicz, S.: (1978), 'On convex sets containing only points of support', *Comm. Math. Ser. Spec.* **1**, 279–281.
Rolewicz, S.: (1979), 'On Pontryagin Maximum Principle for systems with non one-point target sets and systems with additional constraints', *Math. Operationsforschung und Statistik, ser. Optimization*, **10**, 97–100.
Rolewicz, S.: (1979b), 'On uppersemicontinuity of families of optimal control and families of supporting functionals', *Beiträge zur Analysis* **14**, 79–84.
Rolewicz, S.: (1985), 'Remarks on uniform N-stability', *Methods of Oper. Research*, **49**, 571–575.
Rolewicz, S.: (1985b), *'Metric Linear Spaces'*, PWN, Reidel, Warszawa, Dordrecht.
Rolewicz, S.: (1986), 'On uniform N-stability', *J. Math. Anal.* **115**, 434–441.
Rota, G. C.: (1958), 'Extension theory of differential operators' *Comm. Pure Appl. Math.* **11**, 83–90.
Russell, D. L.: (1967), 'Non-harmonic Fourier series in the control theory of distributed parameters systems', *J. Math. Anal. Appl.* **18**, 542–560.

Schauder, J.: (1927), 'Zur Theorie stetiger Abbildungen in Funktionenräumen', *Math. Zeitschr.* **26**, 47–65.
Sikorski, R.: (1969), *'Advanced Calculus. Functions of Several Variables'*, Monografie Matematyczne **51**, PWN, Warszawa.
Singer, I.: (1957), 'Asupra L-problemei teoriei momentelor în spatii Banach', *Bull. Sci. Acad. Rep. Pop. Romîne, Sec. Math. Fiz.* **9**, 19–28.
Singer, I.: (1970), *'Bases in Banach Spaces I'*, Springer-Verlag, Berlin.
Singer, I.: (1973), 'On a problem of moments of S. Rolewicz', *Stud. Math.* **48**, 95–98.
Singer, I.: (1979), 'On the Pontryagin Maximum Principle for constant-time linear control systems in Banach spaces', *J. Optim. Theory and Appl.* **27**, 315–321.
Singer, I.: (1980), 'A characterization of constant time linear control systems satisfying the Pontryagin Maximum Principle', *J. Optim. Theory and Appl.* **32**, 379–384.
Singer, I.: (1980b), 'Duality theorem for linear systems and convex systems', *J. Math. Anal. Appl.* **76**, 339–368.
Singer, I.: (1981), 'Duality theorems for perturbed convex optimization', *J. Math. Anal and Appl.* **81**, 437–452.

Singer, I.: (1981b), 'Optimization and best approximation', *Proc. Conf. Nonlinear Analysis* (Berlin—1979), 273–284.

Šmulian, V.: (1940), 'Über lineare topologische Räume', *Mat. Sb.* **7** (49), 425–448.

Tikhonov, A. N., Samarskii, A. A.: (1966), *'Equations of Mathematical Physics'* (in Russian), 3rd ed., Gos. Izd. Tech.-Teor. Lit, Moscow.

Weiss, L.: (1967) 'On the controllability of delay differential systems', *SIAM, J. Control and Optim.* **5**, 575–587.

Wojtaszczyk, P.: (1973), 'A theorem on convex sets related to the abstract Pontryagin Maximum Principle', *Bull. Acad. Pol. Sci.* **21**, 93–95.

Yoshida, K.: (1948), 'On the differentiability and the representation of one parametric semigroups of linear operators', *J. Math. Soć. Japan* **1**, 15–21.

Yoshida, K.: (1965), *'Functional Analysis'*, Springer-Verlag, Berlin, Heidelberg.

Zabczyk, J.: (1974), 'Remarks on the control of discrete-time distributed parameter systems', *SIAM J. Control and Optim.* **12**, 721–735.

Zabczyk, J.: (1976), 'Remarks on algebraic Riccati equations on Hilbert spaces', *Appl. Math. and Optim.* **2**, 251–258.

Zverkin, A. M.: (1973) 'The pointwise completeness of systems with key' (in Russian), *Diff. Urav.* **9**, 430–436.

Author index

Alaoglu, L. 153, 504
Arbib, M. A. 231, 506

Baire, R. 28, 504
Balder E. J. 496, 504
Balakrishnan, V. K. XIII, 504
Banach, S. 92, 93, 115, 116, 118, 120, 126, 131, 504
Bielecki, A. 25, 504
Bohnenblust, H. F. 94, 504
Boltyanskii, W. G. 275, 509
Borel, E. 50
Butkovski, A. G. XIII, 504

Cantor, G. 50
Carathéodory, C. 42, 237, 504
Castaing, Ch. 291, 504
Christoffel, E. 324, 504
Cohen, P. 93
Curtain, R. XIII, 505

Daleckii, Yu. L. 482, 484, 489, 505
Dolecki, S. 175, 176, 177, 189, 225, 254, 256, 257, 258, 386, 388, 392, 393, 400, 402, 403, 465, 466, 467, 478, 495, 496, 499, 502, 503, 505
Douglas, R. 129, 505
Dunford, N. 44, 159, 505
Dvoretzky, A. 283, 505

Eberlein, W. F. 159, 505
Engelking, R. 143, 146, 505

Falb, P. L. 231, 506
Fattorini, H. O. 469, 505
Fedorov, V. V. 505
Fuhrman, P. A. 472, 505, 506

Gamkrelidze, R. W. 275, 509
Gauss, C. F. 325, 506
Gohberg, I. C. 169, 506
Goldberg, S. 136, 506
Goldstine, H. H. 157, 506
Goodner, D. R. 243, 506

Hahn, H. 92, 93, 506
Halmos, P. 283, 506
Haverkamp, R. 327, 506
Hermes, H. XIII, 506
Hille, E. 430, 506

Isbell, J. R. 246, 506

Kakutani, S. 506
Kalman, R. E. 76, 231, 268, 506
Karlin, S. 507
Kato, T. 136, 507
Kelley, J. L. 243, 507
Kiefer, J. 507
Kolmogorov, A. N. 71, 72, 507
Korevaar, J. 385, 508
Krassowski, N. N. 231, 232, 249, 507
Krein, M. G. 162, 169, 336, 482, 489, 505, 506, 507
Krein, S. G. 430, 484, 507
Kulikowski, R. XIII, 507

Author index

Kuratowski, K. 20, 289, 507
Kupka, J. 289
Kurcyusz, S. 189, 195, 496, 499, 503, 505, 507
Kurzhanskii, A. B. 249, 507

La Salle, J. XIII, 278, 506, 508
Lebesgue, H. 36, 50
Lee, E. B. 508
Lindelöf, E. 48
Lindenstrauss, J. 131, 283, 508
Lions, J. XIII, 231, 329, 365, 414, 416, 508
Luxemburg, W. A. J. 385, 508
Lyapunov, A. A. 283, 508

Magenes, E. 329, 414, 416, 508
Malanowski, K. 298, 508
Markus, L. 508
Mazur, S. 96, 148, 508
Milman, D. P. 162, 336, 507
Mishchenko, E. F. 275, 509

Nachbin, L. 243, 508, 509
Nikodym, O. 104

Olech, Cz. 298, 508

Pallaschke, D. 321, 327, 508
Phan Quoc Khanh 201, 321, 509
Phillips, R. S. 430, 506
Pontryagin, L. S. 275, 509
Popov, V. M. 457, 458, 509
Porter, W. A. XIII, 509
Pritchard, A. J. XIII, 505, 509
Przeworska-Rolewicz, D. 125, 509
Przyłuski, K. M. 490, 509
Ptak, V. 308, 509

Radon, J. 104
Radström, H. 171, 172, 509
Riesz, F. 53, 111, 509
Rolewicz, S. 76, 159, 201, 223, 224, 227, 482, 484, 490, 491, 502, 505, 508, 509, 510
Rota, G. C. 136, 510
Russell, D. L. 254, 256, 257, 258, 417, 418, 505, 510
Rutman, L. A. 336, 507
Ryll-Nardzewski, Cz. 289, 507

Samarski, A. A. 345, 361, 403, 511
Schauder, J. 330, 510
Schwartz, J. 44, 159, 505
Semadeni, Z. 246, 506
Siemaszko, Cz. 176
Sikorski, R. 42, 510
Singer, I. 175, 188, 197, 199, 200, 201, 308, 339, 510
Sobczyk, A. 94, 504
Steinhaus, H. 115, 116, 504
Studden, W. J. 506
Šmulian, V. 159, 510

Tikhonov, A. N. 146, 345, 361, 403, 511
Tzafiri, L. 131, 508

Valadier, M. 504

Wald, A. 283, 505
Weiss, L. 457, 458, 511
Wirth, A. 509
Wojtaszczyk, P. 132, 190, 191, 511
Wolfowitz, J. 283, 505, 507

Yoshida, K. 430, 433, 511

Zabczyk, J. 490, 511
Zverkin, A. M. 457, 511

Subject index

absolutely continuous function 103
absolutely convex vector measure 282
affine hull 78
algebraic basis 58
— boundary 190
— — point 190
— interior 190
— — point 190
almost all elements of a sequence 11
analytic function 467, 468
— operator function 469
approximatively controllable system 224
— — to zero system 224
— null-controllable system 462
aureola 175
axiom of choice 93

ball 1
— closed 1
Banach space 66
bang-bang control 278
— principle 278
bases equivalent 335
basic functionals 332
— sequence 334
basis 58
     algebraic 58
— of neighbourhoods 139
— orthogonal 335
— orthonormal 335
— Schauder 329
bidual function 496
Bielecki metric 25
— norm 421
— space 25

Bochner–Lebesgue integrable function 419
— integral 420
Bochner–Riemann integral 418
B-space 66
B^*-space 65
boundary algebraic 190
— of a set 12
bounded semigroup 445
— set 71

canonical injection 96, 156
Cauchy inequality 30, 66
Cauchy sequence 17
Cartesian product of metric spaces 9
C-estimable operator 255
characteristic function 36
Christoffel coefficients 324
closed ball 1
— convex hull 161
— operator 122
— set 11, 140
closed-valued multifunction 170, 287
closure of a set 12
combination
— convex 60
— linear 57
compact space 52, 143
compact-valued multifunction 170, 287
complete metric space 17
completion 30
cone 199
— positively closed 491
— with an ideal property 491
conjugate equation 271

515

— operator 133
— space 80
continuous function 13, 140
   — linear functional 79
   — — operator 79
   — mapping 13
   — semigroup 429
contraction 24
contractive mapping 24
control 166, 453
   — bang-bang 278
controllable operator 255
   — system 168, 255
   — — (time depending) 223
   — to zero system 461
   — to zero system (time depending) 223
convergent sequence 9
   — with respect to an F-norm sequence 64
convex combination 60
   — functional 179
   — hull 61, 161
   — operator 200
   — set 60
   — system 200
coset 61
countably additive field 34
covering 48
C-preserving evolution operator 493
C-uniformly bounded evolution operator 493
C-uniformly exponentially equistable evolution operator 493

dense subset 12, 140
diagonal representation of an operator 342
difference 56
dimension of a linear manifold 78
   — — space 57

discrete metric 3
   — space 3
distance 1
domain of an operator 122
dual function 495
   — general position 316
   — operator 133
   — space 80, 133
duality principle 234
   — strong 190
   — weak 186

elastic suspension 404
element
   — neutral 57
   — optimal 201
elements
   — linearly dependent 57
   — — independent 57
   — orthogonal 66
epigraph 180
ε-net 46
equation pointwise degenerated 458
equivalence class 18, 37, 61
equivalent bases 335
   — metrics 14
   — F-norms 64
   — norms 69
   — semimetric 16
estimating operator 255
evolution operator 450
   — — C-preserving 493
   — — C-uniformly bounded 493
   — — C-uniformly exponentially equistable 493
   — — uniformly bounded 482
   — — uniformly exponentially equistable 481
   — — uniformly N-equistable 482
extension norm-preserving 91
   — of a functional 91

Subject index

extremal subset 159
extreme point 159

face 61
Φ-convexification 497
fixed point 24
F-controllable system 257
F-norm 63
— homogeneous 65
— stronger than 64
— weaker than 64
F-norms equivalent 64
F-space 66
F^*-space 64
function analytic 467, 468
— bidual 496
— Bochner–Lebesgue integrable 419
— characteristic 36
— continuous 13, 140
— dual 495
— generalized Fechel conjugate 495
— integrable 35
— lower semicontinuous on the right 228
— marginal 499
— of bounded variation 108
— sequentially continuous 142
— strongly measurable 419
functional continuous linear 79
— linear 78
— observable 232
— optimally observable 234
— primal 499
— quantified 307
functionals basic 332
fundamental sequence 17

Γ-compact set
Γ-topology
generalized Fenchel conjugate function 495
graph of an operator 122

Hausdorff space 140
heat source density 361
Hilbert space 66
homeomorphic spaces 13, 142
homeomorphism 13, 141
Hölder inequality 44
hull affine 78
hyperplane 78
— supporting 83
— tangent 84

image continuous family of linear operators 170
— lower semicontinuous family of linear operators 170
— upper semicontinuous family of linear operator 170
inequality Cauchy 30
— Hölder 44
— Minkowski 44
— Schwarz 30
— triangle 1
infinitesimal generator 430
initial state observation 478
inner product 65
input
— operator 165
— optimal 201
— space 165
integrable function 35
integral 35
interior 12, 140
— algebraic 190
— point 12
invariant
— metric 63
— semimetric 73
— topology 146
inverse image by multifunction 286
infinite dimension 57
infinite-dimensional space 57

isometric spaces 15
isometry 15

Kuratowski–Zorn lemma 93

Lagrangian 498
Lebesgue measure 43
limit of a sequence 10
linear
   — combination 57
   — complex space 56
   — functional 78
   — manifold 78, 180
   — metric space 62
   — operator 78
   — real space 56
   — semimetric space 73
   — space 55
   — subset 59
   — subspace 59
   — system 165
   — topological space 146
linearly
   — continuous multifunction 170
   — continuous multifunction at t_0 170
   — dependent elements 57
   — independent elements 57
   — independent set (system) 58
   — lower semicontinuous multifunction 170
   — lower semicontinuous multifunction at t_0 169
   — upper semicontinuous multifunction 170
   — upper semicontinuous multifunction at t_0 169
locally convex space 146
long-term
   — approximative controllable (from zero) system 464
   — controllable from 0 system 464
   — controllable system 463
   — controllable to 0 system 468
   — null-controllable system 467
lower semicontinuous on the right functions 228

manifold linear 78
mapping
   — continuous 13
   — contractive 24
marginal function 499
matrix of observation 300
matrix-valued function of bounded variation 309
maximal element 93
maximum principle 273
measurable function 34
measurable multifunction 286
measurable set 34
measure 34
   — Lebesgue 43
   — σ-finite 39
   — vector 282
method of observation 232, 254
metric 1
   — Bielecki 25
   — discrete 3
   — invariant 63
   — space 1
   — stronger than 14
   — weaker than 14
matrix-valued function with bounded variation 309
metrics equivalent 14
mild solution 435, 439, 440, 452
minimum-norm problem 182
Minkowski inequality 44
modulus integrable functions 32
multifunction 169, 286, 497

Subject index

— compact-valued 170, 287
— closed-valued 170, 287
— linearly continuous 170
— linearly lower semicontinuous 170
— linearly upper semicontinuous 170
— measurable 286

n-dimensional space 57
neutral element 57
neighbourhood 139
nonnegatove element 199
— functional 199
nonpositive element 199
nonseparable space 47
norm 65
— of functional 81
— of operator 114
normal sequence of partitions 108
normal sequence of subdivisions 418
normed space 65
norm-preserving extension 91
norms equivalent 69
nowhere dense set 27
null-controllable at time T system 461

observability 231, 240
observable
— functional 232
— operator 245, 254
— system of functionals 239
open set 11, 139
operator
— C-estimable 255
— closed 122
— conjugate 133
— controllable 255
— continuous linear 79
— convex 200
— dual 133
— estimating 255

— input 165
— linear 78
— observable 245, 254
— optimally observable 245
— output 165
— surjective 168
— symmetric 356
optimal
— element 201
— input 201
optimally observable
— — functional 234
— — operator 245
— — system of functionals 241
order relation 93
orthogonal
— basis 335
— complement 66
— element to a subspace 66
— elements 66
— projection 131
— subspaces 66
orthonormal basis 335

performance functional 179
— — quadratic 267
piecewise continuous function 4
point 56
— extreme 159
— of approximative controllability 175
— of bounded controllability 175
— of controllability 168
— of support 191
pointwise degeneracy 457
pointwise degenerated equation 458
polyhedron in general position 275
pre-Hilbert space 67
primal functional 499
projection 245
— orthogonal 131

pseudonorm 69
pseudoseminorm 73

quadratic performance functional 267
quantified functional 307

range of an operator 122
real numbers 20
reflexive space 156
relative controllability 174
representative of an equivalen class 18, 33, 37
Riemann–Stieltjes integral 109

Schauder basis 329
Schwarz inequality 30, 66
second conjugate space 96
selector 289
semigroup 429
— bounded 445
— continuous 429
— of the class C_0 429
seminormed space 73
semi-F-norm 73
semimetric 16
— space 16
semimetrics equivalent 16
seminorm 73
separable space 45, 140
separated sets 96
sequence convergent almost everywhere 34
sequentially
— compact space 49, 143
— continuous function 142
— Γ-compact set 149
— weakly compact set 149
— weakly -∗- compact set 149

set
— bounded 71
— closed 11, 140
— Γ-compact 149
— in general position 276
— nowhere dense 27
— of σ-finite measure 39
— of the first category 27
— of the second category 27
— open 11, 139
— sequentially compact 49
— sequentially Γ-compact 149
— sequentially weakly compact 149
— sequentially weakly -∗- compact 149
— target 197
σ-algebra 285
σ-field 34
σ-finite measure 39
simple function 418
space
— B 66
— B^* 65
— Banach 66
— Bielecki 25
— c 8
— c_0 8
— $C[a, b]$ 6, 15
— $C_n[a, b]$ 6
— $C_{n,\tilde{\varrho}}[a, b]$ 6
— $C'[a, b]$ 7
— $C'_{n,\tilde{\varrho}}[a, b]$ 7
— $C[a, +\infty)$ 7
— $C_{n,\tilde{\varrho}}[a, +\infty)$ 7
— $C'_{n,\tilde{\varrho}}[a, +\infty)$ 7
— compact 52, 143
— conjugate 80
— discrete 3
— $DL^2[0, S]$ 408
— $DM[0, S]$ 408
— dual 80, 133
space
— F- 66

Subject index

— F^*- 64
— Hausdorff 140
— Hilbert 66
— infinite-dimensional 57
— input 165
— l 9
— $L^1[a, b]$ 32
— $L^1_m[0, T]$ 265
— $L^1(\Omega, \Sigma, \mu)$ 37
— $L^2[a, b]$ 32
— $L^2_m[0, T]$ 265
— $L^2(\Omega, \Sigma, \mu)$ 37
— L_{L^2} 362
— l^p 41
— $L^p[a, b]$ 42, 44
— linear 55
— linear complex 56
— linear metric 62
— linear real 56
— linear semimetric 73
— linear topological 146
— locally convex 146
— m 7
— M_{L^2} 362
— $M(\Omega, \Sigma, \mu)$ 44
— $M_m[0, T]$ 265
— metric 1
— metric complete 17
— n-dimensional 57
— nonseparable 47
— of controls 453
— of modulus integrable functions 32
— of square integrable functions 32
— output 165
— pre-Hilbert 67
— reflexive 156
— second conjugate 96
— semimetric 16
— seminormed 73
— separable 45, 140
— sequentially compact 49, 143
— topological 139

— trajectory 165
spaces
— homeomorphic 13, 142
— isometric 15
sphere 1
square integrable function 32
strictly convex set 274
strong
— duality 190
— operator topology 169
strongly
— measurable function 419
— separated sets 96
subspace
— linear 59
— of a metric space 2
— of a topological space 140
— spanned by a set 59
subspaces orthogonal 66
support point 191
supporting hyperplane 83
surjection 168
surjective operator 168
symmetric operator 356
system
— approximatively controllable at the time T 453
— approximatively controllable from 0 at the time T 453
— convex 200
— controllable 168
— controllable from 0 at the time T 453
— linear 165
— long-term controllable 463
  with distributed parameters 168

target set 197
tangent hyperplane 84
Tikhonov product space 146
time 209, 250
topology 139

— invariant 146
— weak 149
— weak -∗- 149
topological space 139
total family of functionals 149
totally ordered set 93
trajectory space 165
triangle inequality 1

uniformly
— bounded evolution operator 482, 493
— exponentially equistable evolution operator 481
— N-equistable evolution operator 482

variation of a function 108
vector measure 282

weak duality 186, 498
— topology 149
— -∗- topology 149
weakly long-term approximatively controllable system 465
weakly long-term approximatively controllable from 0 system 465

zero 57

List of symbols

ϱ 1
$\varrho(x, y)$ 1
(X, ϱ) 1
$C[a, b]$ 6, 15
$C_n[a, b]$ 6
$C_{n,\tilde{\varrho}}[a, b]$ 6
$C'[a, b]$ 7
$C'_{n,\tilde{\varrho}}[a, b]$ 7
$C[a, +\infty)$ 7
$C_{n,\tilde{\varrho}}[a, +\infty)$ 7
$C'_{n,\tilde{\varrho}}[a, +\infty)$ 7
m 7
c 8
c_0 8
l 9
$\lim_{n\to\infty} x_n = x_0$ 9
$x_n \xrightarrow{\varrho} x_0$ 9
$\overline{A}$ 12
Int A 12, 140
Fr A 12
$C(X, Y)$ 15
$C(X)$ 15
$C_c[a, b]$ 15
$L^1[a, b]$ 32
$L^2[a, b]$ 32
$C_f[a, b]$ 32
$\int_\Omega x(t)d\mu$ 35
$\chi_E(t)$ 36

ess sup 44
$M(\Omega, \Sigma, \mu)$ 44
$M[a, b]$ 45
$x + y$ 55
$y - x$ 56
$A \pm B$ 56
tA 56
0_x 57
0 57
$-x$ 57
$\dim X = n$ 57
$\dim X = +\infty$ 57
$\lin A$ 59
$\conv A$ 61
$\| \ \|$ 63
aff (A) 78
n 96
sgn r 101
$\overset{1}{\underset{0}{\text{Var}}} g(t)$ 108
$\{\Delta^m\}$ 108
$\int_0^1 f(t) dg(t)$ 109
$B(X \to Y)$ 114
$B(X)$ 114
D_A 122
R_A 122
G_A 122
$\bigtimes_{\alpha \in A} X_\alpha$ 146

List of symbols

$E(A)$ 159
$m(y)$ 177
$M^{\perp}$ 182
$y \geq 0$ 199
$y \leq 0$ 199
$(C)^{\text{tr}}$ 271
M_{L^2} 362
L_{L^2} 362
$|c|$ 388
$DM[0, S]$ 408

$DL^2[0, S]$ 408
$[t]$ 430
f^* 495
f^{**} 496
f^0 497
$\Gamma^- x$ 498
Γ_A 498
$\overline{f\Gamma}$ 499
$\Gamma_{\exp} A$ 502